Climate Change Research at Universities

Walter Leal Filho
Editor

Climate Change Research at Universities

Addressing the Mitigation and Adaptation Challenges

Springer

Editor
Walter Leal Filho
Faculty of Life Sciences
Hamburg University of Applied Sciences
Hamburg
Germany

ISBN 978-3-319-58213-9 ISBN 978-3-319-58214-6 (eBook)
DOI 10.1007/978-3-319-58214-6

Library of Congress Control Number: 2017938560

Printed on acid-free paper

This Springer imprint is published by Springer Nature
The registered company is Springer International Publishing AG
The registered company address is: Gewerbestrasse 11, 6330 Cham, Switzerland

Preface

Many universities across the world perform state-of-the art research on matters related to climate change, both in respect of mitigation and adaptation. Yet, as shown by the latest 21st Conference of the Parties of the UN Convention on Climate Change (COP 21), held in Paris in December 2015, there is much room for improvements in the role played by universities in the negotiations and in influencing decision-making on a matter of such a global importance.

There are unfortunately relatively few events where a multidisciplinary overview of university-based research efforts and projects on climate change can be showcased, and where researchers from across the spectrum of the natural and social sciences have had the opportunity to come together to discuss research methods, the results of empirical research or exchange ideas about ongoing and future research initiatives focusing on climate change mitigation and adaptation. It is against this background that the Symposium "Universities and Climate Change: the Role of Higher Education Institutions in Addressing the Mitigation and Adaptation Challenges" was organised by Manchester Metropolitan University, UK, and HAW Hamburg, Germany, under the auspices of the International Climate Change Information Programme (ICCIP), and was held in Manchester, UK, in September 2016. The event involved researchers in the field of climate change in the widest sense, not only from traditional climate science, but also from the fields of environment, human geography, business and economics, arts, administration and media studies.

This book is one of the outcomes of the event, and focuses on the role of higher education institutions and research centres in addressing the mitigation and adaptation challenges, hence contributing to the further development of this fast-growing field. The aims of this book are as follows:

I. to showcase the research and projects performed by researchers at universities and research centres from across the world on issues pertaining climate change;
II. to document and promote ideas and experiences acquired in the execution of research projects, especially successful initiatives and good practice;

III. to introduce methodological approaches and projects which aim to offer a better understanding of climate change across society and economic sectors

This book is divided in two main parts:

- Part I Research and Education on Climate Change Mitigation: in this section, a set of papers addressing mitigation aspects have been compiled, focusing on energy issues as well as on means to reduce fossil fuel emissions.
- Part II Research and Education on Climate Change Adaptation: in this section, papers with a strong adaptation component are gathered, illustrating a diversity of means via which adaptation initiatives are implemented.

Whereas the emphasis of the papers in each section is on mitigation and adaptation, respectively, it is noticeable that the borders between the two modalities of response are not very strict and sometimes tend to overlap.

We thank the authors for their willingness to share their knowledge, know-how and experiences, as well as the many peer reviewers, which have helped us to ensure the quality of the manuscripts.

Enjoy your reading!

Hamburg, Germany Walter Leal Filho
Spring 2017

Contents

Part I
Research and Education on Climate Change Mitigation

Integrating Farmer's Traditional Knowledge and Practices into Climate Change Sectoral Development Planning: Case Studies from India

Nidhi Madan

Abstract Change in climate scenario adversely affects water availability; food and energy security; biodiversity etc. which greatly impacts natural resources, health of every living being and their livelihood. Rural communities which are highly dependent on natural resources are significantly affected by climate change. For the rural communities, difficulty in coping with climate change and preparing for climate change risks are further aggravated due to geographic and demographic obstacles; limited economic diversity; higher poverty rates etc. Small farmers based on their traditional knowledge and experiences are implementing various practices in agriculture, water, livestock etc. sectors to adapt the challenges of climate change. These activities are insignificant in combating climate change issues. However, if these small scale activities are integrated with existing sectoral development schemes/programmes, can create a greater impact. This paper presents strategies on how farmer's traditional knowledge and practices are integrated with sectoral development planning by showcasing case studies from various projects/activities in India. This paper provides comprehensive information on the traditional practices for managing natural resources, which would be helpful for policy makers, implementers and researchers to understand the mechanism and opportunities for convergence with the existing schemes/programmes.

Keywords Rural communities · Climate change risk · Traditional knowledge and experience · Natural resource management

N. Madan (✉)
Climate Change Adaptation in Rural Areas-India, Deutsche Gesellschaft für Internationale Zusammenarbeit (GIZ) GmbH, A2/18, Third Floor, Safdurjung Enclave, New Delhi 110029, India
e-mail: nidhi.madan@giz.de

W. Leal Filho (ed.), *Climate Change Research at Universities*,
DOI 10.1007/978-3-319-58214-6_1

1 Introduction

1.1 Climate Change Impacts at Global Level

Human contribution in enhancing the concentration of greenhouse gases (GHG) in the atmosphere is dominant cause of observed warming since 1950. The report projects that global surface temperature are likely to exceed 2 °C whereas sea level rise would range between 2.6 and 8.2 m by the end of the 21st century [Source: Intergovernmental Panel on Climate Change (2014)]. These changes impact the food supply, water resources, infrastructure, forestry, human health etc. Some of these impacts due to the changing climate on key sectors affecting livelihood and economy are as follows:

1.1.1 Water Resources

Scientific studies confirm that climate change results in *intensification and acceleration* of the global hydrological cycle leading to change in rainfall patterns impacting runoff, water storage and water losses due to evaporation and transpiration. These distress mainly the quantity and quality of water resources; extinction/shifting biotic and abiotic species; loss of rural livelihoods and income; coastal flooding and erosion.

1.1.2 Agriculture

Change in climate has both positive and negative influences on *agricultural yield.* However in absence of adaptation efforts, negative influences of climate change on crop yields are more common than the positive influences. Scientific studies have proven that warming and precipitation variability leads to food insecurity which negatively affects wheat, maize, rice and soybean yields.

1.1.3 Human Health

Changing climate *exacerbates health problems* as it enhances the cases of heat-related mortality, water and vector borne illnesses (e.g.: Malaria, Diarrhoea and Malnutrition), and even death due to intensive heat waves and fires, increased likelihood of under-nutrition resulting from diminished food production especially in developing and least developing nations, risks from lost work capacity and reduced labour productivity in vulnerable populations.

1.1.4 Forest

Increased temperature and variable weather events directly and indirectly affect the *growth and productivity of forest*. Major implications include change in distribution of species, extinction of species, shift in the timing of phenological events and invasion of new pests/pathogens.

1.1.5 Infrastructure

Environmental and social systems near infrastructural assets are also impacted by climate change. Vulnerability due to climate change is sector specific and associated with the technology used for construction and operation e.g. rising temperature and low precipitation leads to reduced efficiency of hydro-power plants; infrastructures in coastal areas such as wind turbines are affected by rise in sea level and coastal erosion etc.

1.2 Impact of Climate Change in India

As highlighted above, climate change has unequivocal impact on all the sectors which are not only related to the global economy but also associated with the livelihood of the population. India, a country with special and mixed bio-geographical features such as mountains, coastal area, forest land, wetlands etc., is severely impacted by vagaries of climate change. Map of India is at Fig. 1.

Implications of Climate Change projected for the country indicate rise in annual mean surface air temperature and sea level from 3.5 to 4.3 °C and 1.3 mm/year respectively by the end of the century [Source: Second National Communication submitted to the United Nations Framework Convention on Climate Change (May, 2012)]. It is also projected that possibility of rainfall intensity will also be enhanced with the alteration in the numbers of rainfall days, which will result in less water flow in most of the river basins. Agriculture, an indispensable sector in Indian economy, is also impacted due to the rising temperature; varied precipitation patterns and glacier melting. Climate Change will also impact the overall health sector in India with increase in frequencies of heat wave; increase in the transmission seasons of vector and water-borne diseases like malaria and dengue; increase in migration impacting the regional imbalance etc. Incidence of extreme weather events and change in the composition of the forest are other major consequences of climate change.

Projected impact of Climate Change on key sectors for various regions of the country (Regions are marked in Fig. 1), are tabulated in Table 1 as follows [Source: Climate Change and India: A 4 × 4 Assessment-A Sectoral and Regional Analysis for 2030s (November, 2010)]:

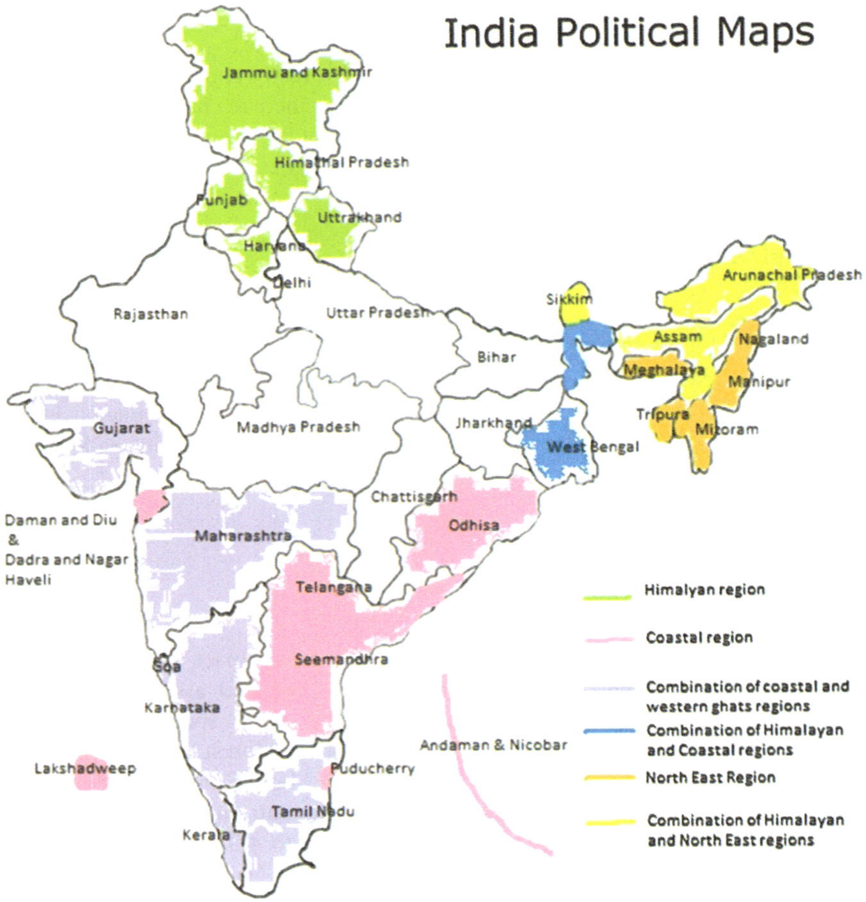

Fig. 1 Map of India highlighting the bio-geographical regions

1.3 Climate Change Impact on Rural Communities

Rural communities are extremely reliant on natural resources such as agriculture, forest, water, fisheries etc. for their livelihood and daily requirements. At present, they are under tremendous stress due to limited carrying capacity of earth and depletion of environmental resources such as water availability and supply, food security, shifts in production areas of food and non-food crops etc.

Climate-related hazards exacerbate these stresses directly impacting rural communities' livelihoods, health, infrastructure etc. [Source: National Action Plan on Climate Change (June 2008)]. In turn these impact disproportionately affect the welfare of the poor in rural areas, such as women because of their lead

Table 1 Projected impact of climate change on key sectors by 2030

S. No.	Bio-geographical region	Impact on key sectors			
		Agriculture	Water	Natural ecosystems and biodiversity	Human health
1	Western Ghats	Productivity of major crops: *Decrease* 1. Rice (by 4%) 2. Maize and sorghum (by 50%) *Increase* Coconut (by 30%)	Variable trend	– 18% of the forest area is likely to undergo change – Net Primary Productivity (NPP) is projected to increase by 20%	No change
2	Coastal area	Productivity of major crops: *Decrease* 1. Rice (by 10–20%) 2. Maize and sorghum (by 15–50%) *Increase* Coconut (by 30%)	Variable trend	– 30% of the forest area is likely to undergo change – NPP is projected to increase by 31%	Reduction in transmission of disease by 34%
3	Himalayan region	Productivity of Apple will decrease	Increase in water yield (by 5–20%)	– 56% of the forest area is likely to undergo change – NPP is projected to increase by 57%	Increase in transmission of disease
4	North-East region	Productivity of major crops: *Decrease* 1. Rice (by 5%) 2. Maize (by 40%)	Variable trend	– 8% of the Forest area is likely to undergo change – NPP is projected to increase by 23%	Increase in transmission of disease

responsibility of acquiring water from long distances and helping their husbands in agricultural activities; people with limited access to land, modern agricultural inputs and infrastructure; illiterate rural families etc.

Rural communities suffer from various geographic and demographic obstacles for mitigating and adapting to the challenges of climate change risks. Remoteness, limited access, limited economic diversity, higher poverty rates, aging population etc. are some of the many hindrances, which tend to increase the vulnerability of rural communities. As projected by the scientific and social communities, the impact on the rural communities will increase progressively with the challenges of climate change, which forces them to migrate where they can thrive and opt for alternative livelihood practices [Source: Shukla et al. (2003)].

Traditional practices for managing natural resources are being implemented in the fields by farmers. However, a few of these practices have been replicated and integrated with state development policies. Therefore, this paper showcases strategy on how pilot activities on climate change adaptation projects implemented in the country integrates with development scheme/plan for replication and upscaling.

2 Approach

The idea behind developing this paper originated based on GIZ-India's technical support under Indo-German project on 'Climate Change Adaptation in Rural India', which aims to integrate climate change adaptation into development planning and strengthen the capacities of key stakeholders for financing, planning, implementing and monitoring of climate change adaptation measures. This paper is backed by comprehensive research conducted via various secondary literature sources and interviews conducted with key stakeholders, while preparing proposals for accessing climate finance. Successful interventions, which are now part of the country's existing schemes/programmes, have been presented in the paper. It highlights the evolution of policy, methodology, interventions and constraints faced while integrating climate change adaptation activities into development practices. Therefore, impactful case studies have been identified, selected and showcased, which would be helpful for other developing nations as well to undertake adaptation planning.

Traditional know-how is found in abundance at the agricultural fields. However, not all such practices have been integrated into the existing plans/programmes. Finding the right case study for the paper was perhaps the biggest challenge while developing the paper. This constraint was overcome by thorough research from secondary sources of literature and cooperation from various key stakeholders.

3 Farmer's Traditional Knowledge and Practices on Key Sectors

Since decades, farmers are implementing and improvising on their practices and technologies in their agricultural fields for their livelihood needs. Following are some of the traditional practices that have been improvised and implemented over the years based on the farmer's experiences [Sources: India's 12th Five Year Plan document (2012); India's Economic Survey (2012-13); Sharma et al. (June 2015); Livestock and Climate Change (2009); Punjab State Action Plan on Climate

Change (February, 2014); Karnataka State Action Plan on Climate Change (December, 2013); Telangana State Action Plan on Climate Change (June, 2016); Haryana State Action Plan on Climate Change (December, 2011)]:

3.1 Agriculture

- Promotion of *Mulching* for maintaining the micro climate of soil, adding nutrients to the soil, reducing evaporation, and preventing soil erosion.
- Promotion of *Green manuring*[1] for enhancing the nitrogen content in the soil, improving soil fertility, controls pest, diseases and weeds.
- Growing of *Millets* e.g.: Marua, Kodo, Grams, Lentils etc., which are high on nutrition content and demand less water.
- Promotion of *organic material* e.g.: Neem, for agricultural pest control.
- Adoption of *Zero tillage* practices for retaining soil carbon, increasing fertilizer use efficiency, reducing soil erosion and saving labour and energy.
- *Intercropping/mixed cropping* to reduce the risk of crop failure during poor monsoon and effective utilisation of land.
- Development of *Seed banks* to maintain the gene pool and conservation of endangered species.

3.2 Livestock

- Propagation of *indigenous varieties of livestock* for enhancing the pool of pure blooded indigenous livestock which are more productive; demands less fodder and resistant to diseases and heat stress.
- Establishing *Bio-gas plants* near cattle sheds for the collection of dung which can be used for generating methane having high calorific value. This also helps in managing cattle waste.
- Adoption of *drought resistant feed and fodder varieties* especially for the drought prone areas to ensure cattle's productivity during less water.

---

[1]Green manuring is the cultivation of a forage/leguminous crop having high nitrogen content during the monsoon which will later be used for wheat or other cash crops.

3.3 Water

- Adoption of a system for *storing rain water* in the low lying areas besides the paddy field for meeting the irrigation demand of the crop.
- Promotion of *drip and micro irrigation practices* for vegetable cultivation.
- Construction of *water storage tanks* such as cascade tanks, check dams, gully plugging, contour and vegetative bunds in the undulating areas to store water, improve soil moisture regime, recharge ground water and reduce soil erosion.

4 Mainstreaming and Integration of Traditional Practices into Development Plans

Based on the research conducted by scientific community and farmer's ground level strong knowledge base gained over generations from practical experiences, farmers are implementing these practices successfully pertaining to agriculture, water, livestock etc. sectors. Experiences from the implementation of practices have gained a huge momentum in farming and some of the practices are an integral part of the district plans, for which adequate technical and finance support are provided by the Government.

A major step in this regard is the development of Action Plan for Climate Change at the Sub-National/State level to address the State specific issues on climate change. As India is a country of 36 States/Union Territories, each with its unique feature, development of Action Plan on Climate Change at the sub-national level has been a significant step in this regard. Each State Action Plan on Climate Change (SAPCC) includes the current and future state's climate change scenarios and likely impacts of climate change of the concerned region on sectors such as water, agriculture, energy, tourism, infrastructure, disaster management, coastal area protection etc. Accordingly, strategies were identified and prioritized to address the issues of the vulnerable areas, sectors and communities and their associated risks.

Development of SAPCCs was a comprehensive exercise prepared through consultations with various experts across sectors; Non-Government Organizations; research institutions etc. It is a guiding document for the activities to be undertaken at the State level for each sector in the next 15–20 years. However, in order to operationalize activities envisaged under SAPCCs, the activities are being implemented/planned for implementation in a project/programme/scheme mode based on the scientific and farmer's ground level knowledge.

5 Results and Analysis

This section showcases case studies which depict some of the examples for operationalization of climate change adaptation activities and their integration with the state development plans/schemes. Case studies are as follows [Sources: Towards Climate Resilient Livestock Production System in Punjab (August, 2015); Resilient Agricultural Households through Adaptation to Climate Change in Mahbubnagar district, Telangana (February, 2016); Pandey Rita (August, 2014); Singh Inderjeet et al. (September, 2013); Climate-Smart Villages in Haryana, India (August, 2014) and Rashtriya Gokul Mission (2012)]:

5.1 Sub-surface Water Preservation in Agricultural Based State (Punjab)

Introduction to the area: Punjab, a northern state of the country, is also called as agriculture flourished state. Major crops grown in the state are Barley, Wheat, Rice, Maize and Sugarcane.

Problems:

- Fall in water table level due to extensive rice cultivation and its early transplantation before mid-June in the rice belt of the state.
- Large areas of the state cultivate rice as it is more profitable, stable and less risky crop with assured agriculture market.

Solution: Delay in sowing of paddy leads to delayed transplanting. This is to ensure that the water requirement is fulfilled from rainfall. Picture of an irrigated land of the State during June, 2015 is shown at Fig. 2.

Strategy:

- Punjab State Farmers Commission (PSFC 2007) took legislative steps by drafting a legislation titled 'Preservation of sub-soil water in the State of Punjab', based on the research done by the State Agricultural University.
- Main purpose of the legislation was to prohibit the sowing of Paddy nursery and transplanting the Paddy before the notified dates.
- Large awareness programmes were conducted for farmers through workshops, media channels etc. highlighting the importance of saving ground water just by shifting the dates of sowing and transplantation.
- Even though many farmer programmes were conducted, making farmers adopt these practices was a challenge. In order to successfully implement the activities, strict action was taken like suspension of officers who did not take action or showed laxity towards violating farmers etc.

Fig. 2 Irrigated land in Punjab during June, 2015

- Based on the successful piloting of the legislation for two years, it was promulgated to wider area in 2008 and a dedicated act was prepared by Govt. of Punjab incorporating all procedures.
- The act was retitled to 'The Punjab Preservation of Sub Soil Water Act, 2009'.

Impact:

- Increase in water table level.
- Water demand for Paddy was met by rainfall unless there was a delay in monsoon.
- Reduced electricity consumption.
- Increase in relative humidity.
- Less amount of fine dust.
- Less pollution due to control of burning of rice straw due to less availability of time for sowing of wheat. Therefore, they use new machinery such as happy seeders etc., which helped in sowing while simultaneously cutting and burying the rice stubbles.
- Saving soil micro-organisms and cost.

5.2 Development of Climate Smart Villages (Haryana)

Introduction to the area: Haryana, a northern state of the country and adjacent to Punjab, is also an agriculture rich state.

Problems:

- Weather variability i.e. increasing temperature and variable rainfall, which affects crop yield.
- Over-exploitation of ground water resources.
- Poor management of crop residues leading to pollution.

Solution: Adoption of sustainable agriculture practices resilient to climate change such as zero tillage; residue management; optimization/diversification of cropping system; water management through laser leveller and micro-irrigation; stress resilient cultivars; seed and fodder banks etc. Figures 3 and 4 show some of best practices (sowing and seeding) for a Climate Smart Village (CSV) in Haryana.

Strategy:

- Project activities were first proposed to be piloted in 27 villages of the State.
- Villages were selected based on the climate vulnerability of the targeted villages through a baseline survey.
- Young farmers/entrepreneurs willing to replicate/upscale the project activities in the future, were selected and their skills were developed on effective implementation of agricultural practices through various training programmes and exposure visits.
- A network of knowledge partners was established for providing agro-advisories to young farmers, so that they can help other farmers of the villages on planning the project activities.
- An innovative excellence model in each village was created with provision of latest equipment such as Happy seeder, Multi-crop planter, Maize thresher etc. These equipments were used by villagers on a rotational basis for implementing the activities in their fields.
- Climate resilient practices were successfully implemented in 27 villages. However, there were still some challenges like financial resources for implementing the activities and farmers capacity building for the application of the innovative technology, which were creating hindrance in replicating the activities in other areas.
- However, based on the successful implementation of the project, Government of India provided financial support for replicating the activities in another 250 villages of the state, which also includes support for the capacity building of farmers.

Impact

- Higher agricultural yield (e.g.: wheat and maize yield increased by 6 and 15% respectively).

Fig. 3 Direct seeding by a farmer as part of CSV by farmers on lands of Haryana

Fig. 4 Picture of Happy Seeder for sowing wheat in the fields

- Adequate management of agricultural residue.
- Effective application of nitrogen and fertilizers in the fields.

5.3 Provision of Shelters to Non-productive Stray Cattle

Problem: Orphanage of non-productive cattle.

Solution: Collecting cattle at a single place to prevent it from roaming, accidents, agricultural loss etc. Figure 5 depicts a model shed for stray cattle in National Dairy Research Institute (NDRI), Karnal.

Strategy:

- To prevent unproductive cows from being sent to the abattoir, a project for collecting cattle in a single place was implemented on a pilot scale. This was done with the existing facilities from research institute for dairy development.
- It was observed through the research conducted within the project that when stray cattle are put inside a cooler environment, they show better physiological status.
- Cooling systems essentially alleviate heat load by using the principle of evaporation, leading to improved feed intake and milk production in areas with high environmental temperatures.
- Cattle dung was also collected to generate bio-gas, which also has co-benefits for mitigating the challenges of climate change by reducing methane emissions.

Fig. 5 Model Shed for stray cattle in NDRI, Karnal

- Based on the successful implementation of the project activities, Government of India launched the Rashtriya Gokul Mission in mid-2014, with an objective to conserve and develop the indigenous breeds in a focused and scientific manner. The mission also supports the development of "Gokul Grams" centres, which provide shelters for stray cattle.

Impact:

- Better environment to stray cattle.
- Mild increase in milk yield.
- Bio-gas generation and reduction of GHG.

6 Conclusion

Case studies presented in the paper signify that successful transformation of ground activities into policies/acts is an essential factor to increase the effectiveness of adaptation planning process. Besides taking into consideration farmer's knowledge and experience while planning adaptation processes for vulnerable communities, it is also essential to consider the following:

- Provide incentives to farmers such as subsidy for new technology, awards for successful implementation of project activities etc.
- Conduct awareness programmes for responsible district officers regarding the new scientific studies on climate change, new technologies to adapt/mitigate the challenges of climate change and their cost benefit analysis etc. This will also create competition amongst officers and hence ease in implementation.
- Penalize violating farmers in the fields, actions against the responsible extension officers etc.
- Encourage farmers to provide financial contributions in adopting new technologies, which helps in ensuring farmer's ownership for sustainable implementation of the activities in the fields.

Acknowledgements I am very much grateful to my GIZ Colleagues (Mr. Kirtiman Awasthi, Senior Policy Advisor, Ms. Angel Jacob, Communication Expert); officials working in Punjab State Council on Science and Technology (Dr. Satnam Singh Ladhar, Additional Director and Dr. Kamalpreet Kaur, Principal Scientist), Department of Environment, Government of Haryana (Dr. Suresh Gahlawat, Director), for their valuable support in concluding and refining this paper.

References

Intergovernmental Panel on Climate Change (2014). Climate Change 2014: Impacts, Adaptation and Vulnerability. Working Group II Contribution to the Fifth Assessment Report of the Intergovernmental Panel On Climate Change. World Meteorological Organization Press. Geneva, Switzerland. (http://ipcc-wg2.gov/AR5/images/uploads/WG2AR5_SPM_FINAL.pdf)

Second National Communication submitted to the United Nations Framework Convention on Climate Change (May, 2012). Ministry of Environment and Forests, Government of India. Environmental Information System Press. New Delhi, India. Pg-8 to 28. (http://envfor.nic.in/downloads/public-information/India%20Second%20National%20Communication%20to%20UNFCCC.pdf)

Climate Change and India: A 4X4 Assessment- A Sectoral and Regional Analysis for 2030s (November, 2010). Ministry of Environment and Forests, Government of India. Environmental Information System Press. New Delhi, India. Pg-73 to 75. (www.moef.nic.in/downloads/public-information/fin-rpt-incca.pdf)

Shukla P.R, Sharma Subodh, Bhattacharya Sumana (2003). Climate Change and India: Vulnerability Assessment and Adaptation. Universities Press. New Delhi, India. Pg-1–27 and Pg-128 to 156

National Action Plan on Climate Change (June, 2008). Prime Minister's Council on Climate Change. Government of India. Environmental Information System Press. New Delhi, India. (http://www.cseindia.org/userfiles/National%20Action%20Plan%20on%20Climate%20Change.pdf)

India's 12th Five Year Plan document (2012). Environment, Forestry and Wildlife. Chapter-7-Faster, More Inclusive and Sustainable Growth. Planning Commission, Government of India. Volume-I. Environmental Information System Press. New Delhi, India. Pg- 202–235, (Link: http://planningcommission.nic.in/plans/planrel/12thplan/pdf/vol_1.pdf)

India's Economic Survey (2012–13). Chapter-12-Sustainable Development and Climate Change, Ministry of Finance, Government of India. Environmental Information System Press. New Delhi, India. Pg-256, (Link: http://indiabudget.nic.in/es2011-12/echap-12.pdf)

Sharma A, Tyagi S, Prasad R, Kumar S, Mandal D.K. (June, 2015).Vision 2050. Project Directorate Cattle. Indian Institute of Farming Systems Research. Indian Council of Agriculture Research Press. Meerut, India. (Link: http://pdfsr.ernet.in/sites/default/files/pdf/vision2050.pdf)

Livestock and Climate Change (2009). Livestock Thematic Papers, International Fund for Agricultural Development. Rome, Italy. (Link: http://www.docstoc.com/docs/22883886/Thematic-paper-on-Livestock-and-Climate-Change-final)

Punjab State Action Plan on Climate Change (February, 2014). Punjab State Council on Science and Technology, Government of Punjab. Chandigarh, India. Pg 53–122. (Link: http://www.moef.nic.in/sites/default/files/sapcc/Punjab.pdf)

Karnataka State Action Plan on Climate Change (December, 2013). Environmental Management & Policy Research Institute, Government of Karnataka. Bengaluru, India. Pg 33–73. (Link: http://www.moef.nic.in/sites/default/files/sapcc/Karnataka.pdf)

Telangana State Action Plan on Climate Change (June, 2016). Environmental Protection Training and Research Institute, Government of Telangana. Hyderabad, India. Pg-66–74 (Link: http://envfor.nic.in/sites/default/files/Telangana.pdf)

Haryana State Action Plan on Climate Change (December, 2011). Department of Environment, Government of Haryana. Chandigarh, India. Pg-56–150. (Link: http://www.moef.nic.in/sites/default/files/sapcc/Haryana.pdf)

Towards Climate Resilient Livestock Production System in Punjab (August, 2015). Punjab State Council for Science & Technology, Government of Punjab. Chandigarh, India. (Link: http://envfor.nic.in/sites/default/files/Punjab.pdf)

Resilient Agricultural Households through Adaptation to Climate Change in Mahbubnagar district, Telangana. (February, 2016). Environment Protection Training and Research Institute

Government of Telangana. Hyderabad, India. (Link: http://envfor.nic.in/sites/default/files/Telangana_0.pdf)

Pandey Rita (August, 2014). Groundwater Irrigation in Punjab: Some Issues and Way Forward. National Institute of Public Finance and Policy, Government of India. Working Paper No. 2014–140. New Delhi, India. (Link: http://www.nipfp.org.in/media/medialibrary/2014/09/WP_2014_140.pdf)

Singh Inderjeet, Bhangoo Kesar (September, 2013). Irrigation System in Indian Punjab. Center for Research in Economic Change, Punjab University. Punjab, India. (Link: https://mpra.ub.uni-muenchen.de/50270/1/MPRA_paper_50270.pdf)

Climate-Smart Villages in Haryana, India (August, 2014). CGIAR Research Program on Climate Change, Agriculture and Food Security, International Maize and Wheat Improvement Center. New Delhi, India (Link: https://cgspace.cgiar.org/rest/bitstreams/34314/retrieve)

Rashtriya Gokul Mission (2012). Department of Animal Husbandry, Dairying & Fisheries, Government of India. Environmental Information System Press. New Delhi, India. (Link: http://dahd.nic.in/sites/default/files/Concept%20Note%20Rashtriya%20Gokul%20Mission.pdf 08_0.pdf)

Roof Top Farming a Solution to Food Security and Climate Change Adaptation for Cities

Gunjan Gupta and Pradeep Mehta

Abstract Two distinct phenomena shape our planet: more than half of the world's human population is urbanised (World Watch Institute 2007); and global warming induced climate change is a grave threat. Modern cities, in ecological terms, have become parasitic energy and resource 'sinks,' consuming 75% of the world's resources on only 2% of the global land area (TFPC 1999). In this way cities 'short-circuit' the natural ecological cycle, harming both the nutrient source and sink. As the rate of urbanisation increases over time, food production sites should be increasingly located near main consumption centres. Roof top farming is one such solution to meet growing demand of safe and healthy food, improving air quality, heat influx, reduced corridors for local flora and fauna. Urban agriculture is gaining relevance all over the world due to its ability to provide direct benefits (food) but also some indirect ecosystem services at a macro level (conservation of biodiversity). This paper presents a survey of roof top farming in South Delhi, India and their importance from ecosystem services, food security and climate change perspective. The paper shows the way forward to popularise roof top farming in cities through outreach programmes, capacity development and policy interventions.

Keywords Climate change · Food security · Urban agriculture · Ecosystem · Urban biodiversity

G. Gupta (✉)
Department of Sustainable Development, The Energy Resource Institute (TERI), New Delhi, India
e-mail: ggunjan055@outlook.com

P. Mehta
Earthwatch Institute, Gurgaon, India
e-mail: pmehta@earthwatch.org.in; pmehtanainital@gmail.com

W. Leal Filho (ed.), *Climate Change Research at Universities*,
DOI 10.1007/978-3-319-58214-6_2

1 Introduction

Two distinct phenomena shape our planet: more than half of the world's human population is urbanised (World Watch Institute 2007); and global warming induced climate change is a grave threat. The former's relation with environment has been the focus of academic discourse from an ecological footprint perspective and an enhanced understanding of global warming has increased linkages between the two phenomena (De Sherbinin et al. 2007). Increased contemporary focus on carbon footprints differs from earlier conceptions of ecological footprints. In attempts to lower their rating as carbon emitters, many cities in the world have tried to shift carbon emitting activities outside their municipal boundaries, thus reinforcing a long term tendency of the expulsion of environmental risks from the core to the periphery. Ecological footprints, on the other hand, suggest a different orientation in which cities, as dynamic spaces of production as well as consumption, have to bear responsibility for the consequences of the sum total of their activities (Franchetti 2013; Cartwright et al. 2012; Da Schio and Brekke 2013).

The growing city-based or dependent production and consumption of crucial resources and also the sheer numbers of poor people dwelling in urban centres in developing countries highlight how important it is for cities to prepare for climatic impacts and lower their anthropogenic contribution towards greenhouse gas emissions (Satterthwaite 2008, 2011; Bicknell et al. 2009). India's urban centres are already confronted by environmental concerns, such as increasing energy consumption, large scale pollution, a scaled up built environment at the cost of green spaces, unmanaged waste generation, unsustainable use of natural resources like water, pressure of increased population density (Mukhopadhyay and Revi 2009).

India has submitted the Second National Communication (NATCOM) to the UNFCCC in 2012. The first National Communication was submitted in 2004. As per the Second national Communication submitted by India to the UNFCCC, these climate change projections are likely to impact health, agriculture, water resources, natural ecosystems and biodiversity. India's strategy for addressing Climate Change is reflected in many of its social and economic development programmes. National Environment Policy, 2006 outlines essential elements of India's response to Climate Change. The National Action Plan on climate change (NAPCC) is coordinated by the Ministry of Environment and Forests. It is being implemented through the nodal Ministries in specific sectors/areas. Under NAPCC India has several other missions covering other sectors that can help mitigate the effects of climate change. These missions will be institutionalised by respective ministries ad will be organised through inter-sectoral groups which include in addition to related ministries, ministry of finance and planning commission, expert from industry, academia and civil societies.

In order to respond to the challenges of climate change effectively, the government has created advisory council on climate change. The council has a broad base of key stakeholder representative including the government, Industry and civil societies. This council also provides guidance on the matters of domestic agenda and review of the implementation of National Action Plan on Climate Change including Research and development Agenda. NAPCC will continue to evolve, based on new scientific and technical knowledge as they emerge in response to evolution of multilateral climate change regime including arrangements for international cooperation's (NAPCC, Government of India).

The agriculture sector in India is already threatened by existing factors such as land use changes, scarcity of water resources, increasing air pollution and loss of biodiversity. In a tropical country such as India, even minimal warming will lead to loss in crop yields (Parry et al. 2007). Further studies conducted by the Indian Agricultural Research Institute (IARI) indicate the possibility of loss of 4–5 million tons in wheat production with every rise of 1 °C temperature throughout the growing period even after considering carbon fertilisation. Losses for other crops are still uncertain but are expected to be smaller, especially for kharif crops (Aggarwal 2008). Research also suggests that erratic monsoons will have serious effects on rain-fed agriculture with projected decreases in the productivity of crops including rice, maize and sorghum (especially in the Western Ghats, Coastal region and North eastern regions), apples (in the Himalayan region) (Kumar et al. 2011). Studies indicate that increased droughts and floods are likely to increase production variability and lead to considerable effects on microbes, pathogens, and insects needed for the upkeep of healthy agricultural systems. The UNFCCC (2007) have indicated that increasing sea and river water temperatures are likely to affect fish breeding, migration, and harvests. Increasing glacier melt in Himalayas could affect availability of irrigation especially in the Indo-Gangetic plains, which, in turn, would have consequences on food production (Darshini, Rajiv et al.).

Rooftop agriculture is one way in which urban areas could attempt to be more balanced and sustainable in their resource consumption. It is possible to produce a variety of fruit, grain, and vegetable crops on rooftops, either in containers or as field crops (TFPC 1999).

Considering the above, a need to study rooftop gardens of Delhi was realised. The present work focuses on the theme of roof top kitchen garden as a measure of sustainable smart city and adaptation to climate change. A study was carried out for assessment and quantification of the potential of rooftop vegetable production in southern part of Delhi. Besides the contribution to food security of the city, the study discusses upon the potential benefits to urban biodiversity, creation of green corridors, and ecosystem services provided by the roof top kitchen garden.

2 Methodology

The study involves Library research which Involves identifying and locating sources that provide factual information or personal/expert opinion on a research question; necessary component of every other research method at some point. A standard outcome of research is a literature review.

Through literature reviews historical records were analysed mainly to back up the ideas and arguments, presented in the research. After doing the library research for the collection of secondary data, Field research was done to collect primary data. It involved personal interviews with key informants and surveys which are also known as PRA (Participatory Rural Appraisal) tools citizen science approach was used to collect data as well as create awareness among the respondents.

Type	Method	Technique
Library research	(1) Analysis of historical records (2) Analysis of documents	Content analysis, articles, journals, Magazines
Field research	(1) Personal interview (2) Survey	Semi structured interview Semi structured interview Questionnaire

2.1 *Rapid Rural Appraisal (RRA)/Participatory Rural Appraisal (PRA) Tools: RRA/PRA Tools Were Used as Part of the Study*

PRA is an approach (and family of methodologies) for shared learning between local people and outsiders to enable development practitioners, government officials, and local people to plan together appropriate interventions.

Participatory Rural Appraisal—a misnomer:

- Participatory—more or less
- Rural—but also urban uses
- Appraisal—but also used in identification, implementation, evaluation, and ESW

2.2 *PRA: Key Principles*

- Participation: local people serve partners in data collection and analysis
- Flexibility: not a standardized methodology, depends on purpose, resources, skills, time
- Teamwork: outsiders and insiders, men and women, mix of disciplines, optimal cost and time efficient, but ample opportunity for Ignorance: analysis and planning
- Systematic: for validity and reliability, partly stratified sampling, cross-checking

2.3 *PRA: Key Techniques*

- Interviews/Discussions: individual's households focus groups, community meetings
- Mapping: community maps personal maps institutional maps
- Ranking: problem ranking preference ranking wealth ranking
- Trend Analysis: historical diagramming seasonal calendars daily activity charts

2.4 *Semi-structured Interviewing/Conversational Interviewing*

The central technique on which any PRA is based is Semi-Structured Interviewing (SSI), or Conversational Interviewing as it is sometimes called. SSI does not involve a formal questionnaire, but instead makes use of a flexible interview guide to help ensure that the interviews stay focused on the relevant issues, while remaining conversational enough to allow participants to introduce and discuss issues that they deem relevant.

Citizen Science Approach: Inclusion of citizens in scientific research is called *citizen science*. Citizen Science promises to bring a fresh perspective to strengthen the environment conservation efforts. Citizen science has the potential to bring youth and science together at the field level and empower them with knowledge, understanding and conviction to build conservation movements at the local level. Data and other information generated through citizen science projects have been shown to be reliable and accurate. There is evidence that data from citizen science research projects are increasingly accepted in the academic literature (UNEP 2014).

This approach was utilized in this study by involving the RTG's owners in collection of the data i.e., production, biodiversity assessment, etc.

2.5 *Limitations of the Study*

There is a lack of literature on Roof Top Gardens nationally as well as internationally. During the study 4 key informants were identified. These key informants belonged to the organisations that help people set up Roof Top gardens in Delhi. They were hesitant to give information about their clients due to which the sample size remained small of 12.

2.6 The Study Area

To conduct the study South part of Delhi was selected. Delhi is located in northern India. Delhi has an area of 1483 km^2. Its maximum length is 51.90 km and greatest width this 48.48 km Delhi shares bordering with the States of Uttar Pradesh and Haryana. Delhi is situated on the right bank of the river Yamuna at the periphery of the Gangetic plains. The ridges of the Aravelli range extend right into Delhi, towards the western side of the city, and this has given an undulating character to some parts of Delhi.

Twelve roof top gardens were selected as part of the study and four other key stakeholders (organisations working on RTG's) were contacted to generate the desired data.

3 Result and Discussion

3.1 Role of RTG's in Biodiversity Conservation and Food Security

A survey was conducted covering twelve rooftop gardens in South, Delhi to document the diversity of pollinators like bees, honey bees, butterflies. The diversity of birds was also documented that have been observed on roof top farms. The results are summarised as under:

Identification of bees and butterflies were done through mobile applications developed by Eathwatch Institute. The results illustrated in show that twelve butterflies were spotted on RTG's. Small branded swift is the most common butterfly as all the twelve respondents confirmed its presence on their RTG. Common grass yellow butterfly has been spotted among eight RTG's. Stripped Tiger has been spotted on nine RTG's. Where as Small orange tip was observed on eight RTG's. Common Jezebel being the least common have been spotted on three RTG's. Other butterflies like lime butterfly, common Mormon, Daniad Egg fly, dark grass blue, painted lady have also been spotted. This clearly shows that RTG's are good habitat for butterflies (Fig. 1).

The results also show that vegetated roof attracts more birds than a barren roof because birds find shade and food on RTG's. Blue rock pigeon (*Columba livia*) and common Myna (*Acridotheres tristis*) are the most common. Whereas birds like blue sunbird (*Cyanomitra alinae*), green sunbird (*Nectariniidae*), white dove (*Zenaida asiatica*), Bulbul (*Pycnonotidae*), Yellow footed green pigeon (*Treron phoenicoptera*), yellow-green.

Vireo (*Vireo flavoviridis*) are also spotted. It is interesting to know that the endangered House sparrows have started showing up. Bird baths kept on the roof are also the reason why birds come and sit on the roof. Birds also find easy perch in roof tops which is missing in big farms. Some other birds observed on the roof are

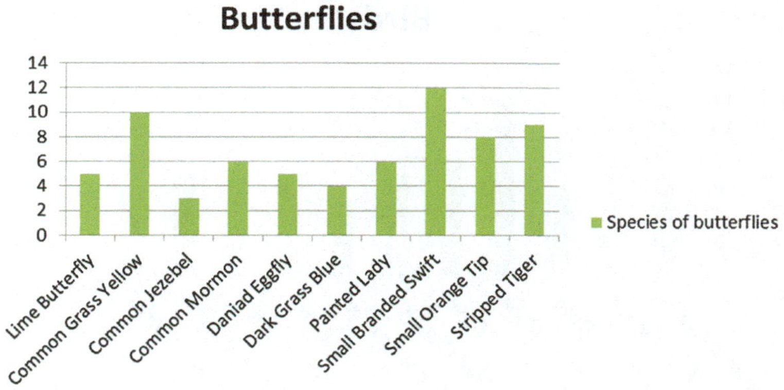

Fig. 1 Butterfly diversity in different RTG'S

brahminy-myna (*Sturnia pagoda rum*), Alexandrine Parakeet (*Psittacula eupatria*), Asian Koel (*Eudynamys scolopaceus*), Asian Pied Starling (*Gracupica contra*), Black drongo (*Dicrurus macrocercus*), Black kite (*Milvus migrans*), and Brown headed barbet (*Megalaima zeylanica*) (Fig. 2).

Two types of bees i.e., *Apis dorsata* and *Apis florea*, were observed on the roof tops. Addition to this, it was observed that bee boxes of *Apis mellifera* was kept for better production of food vegetables and fruits in one of the rooftop. All owners agreed on the presence of a variety of bugs. Lady bugs being the most common. Other bugs include the Adult brown sting bud, black orange bug, brown bug.

In order assess fruit and vegetable diversity in the RTGs, the participants were asked about the variety of fruits and vegetables they grow on their roofs. It was observed that Lady Finger, Beans, Tomatoes, Brinjal were the most common vegetables grown. Reason being that they give a good yield in terms of quantity and continuously for longer duration. It was observed that people experiment less by growing only common vegetables for example vegetables like broccoli, cabbage and sugarcane are not grown by many. Coriander and mint are also grown seasonally by most of the households.

3.2 *Assessment of Ecosystem Services Provided by RTG's*

As RTG's are mainly maintained for vegetable cultivation, but they provide provisioning ecosystem services. Out of total 12 RTG's owners 4 RTG owners briefed that RTG's regulates the temperature not only on roof top but also the room temperature below RTG's. The respondents also briefed that RTG's also regulates carbon by helping in carbon sequestration and providing fresh air. The respondents briefed that the air near RTG's is much fresh than the surrounding air in the locality or on ground. Diversity in RTG's and support to bird population helps in providing

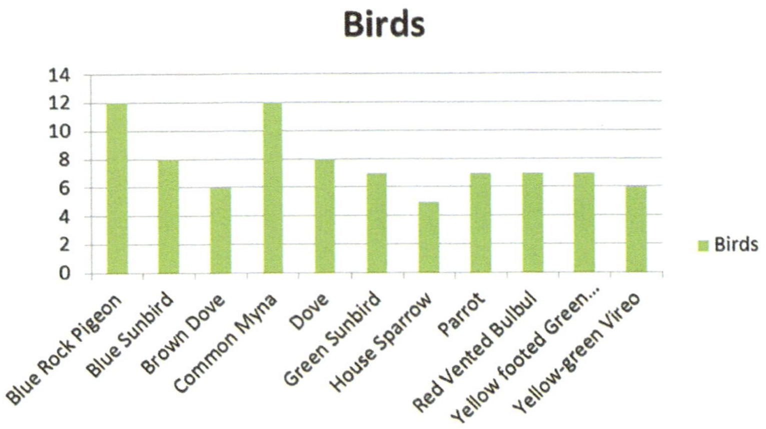

Fig. 2 Diversity of birds on RTG'S

another regulating service that is disease and pest control. Due to the diversity of plants in the RTG's, some plants acts as repellents to insects thus protecting the crop. At the same time, bird population helps in minimising pest control by feeding on larvae of insects and adult insects. The RTG's supports pollination service which was observed during monitoring of RTG's and also reported by the RTG's owners. Different pollinators like *A. dorsata, A. florea* and species of butterflies were observed in the RTG's monitored.

All the RTG's owners expressed that the RTG's provide aesthetic and cultural services. They briefed that the growing of vegetables and plants in the garden gives them a different happiness. Working in the garden keeps them busy and entertained. Most of the owner's also grow sacred plants like basil which provides spiritual benefits to the family members. RTG owners also expressed that they have observed reduction of noise after vegetating their roof.

The study shows that on an average size of roof with one to three plants of each vegetable can give 6 kg of vegetable. Generally, if two tomatoes plant can give 700 g of tomatoes then four to six plants can give us more than 1.5–2 kg. For a family of four people 2 kg of Tomatoes will be sufficient for a week. Similarly vegetables like brinjal, bitter gourd, capsicum, corn, zucchini can give ample quantity of yield. Lady finger must be grown only when there is space for ten to fifteen plants, even though lady finger grows rapidly, but the total amount of yield from a plant will not be suitable for consumption. At a time, one lady finger plant will on an average give four lady fingers. And hence it is advised to plant lady finger in larger number. Companion planting can save space and provide better quality of produce. For example Basil and tomatoes planted together will help make the tomatoes taste better. The basic rule of companion planting is to never grow plants that belong to the same family, together. Different plants occupy different space around them in a diverse way. Some rise vertically and then spread out laterally, some reach out for a little support, some grow deep into the ground, while others sprawl themselves across lazily over the ground. An area of 200 m^2 is sufficient for growing all the seasonal vegetables (Figs. 3, 4 and Table 1).

4 Discussion

This study shows that urban green spaces like RTG's support biodiversity and play a crucial role in biodiversity conservation. They not only support insect and bird diversity but also crop diversity. They are much safer for bees and butterflies as there is minimal spray of insecticide and pesticide when people grow vegetables for their own consumption. Climate change may be a further threat to pollination services (Memmott et al. 2007; Schweiger et al. 2010; Hegland et al. 2009). In order to adapt to climate change RTG's can be one of the solutions for conservation of pollinators.

RTG's also support the population of birds which get easy food, perch and water on RTG's. They not only supports bird population but also acts as pest control as the harmful insects and caterpillars are eaten by the birds. Because most extensive green roofs are inaccessible to the public, they can provide undisturbed habitat for microorganisms, insects, and birds. Similar results have been found in Similar studies have been reported in Germany, Switzerland, and England (Brenneisen 2006). This shows that RTG's have more diversity compared to commercial farms as they are mainly grown for sustenance while the commercial farms are mainly mono cropping or have least diversity. Therefore, RTG's play an important role in maintenance of crop diversity which the rural commercial farms don't. It was also revealed from the study that cultivation in RTG's is more organic and thus supports more biodiversity and is more pest resistant. The greater the biodiversity, the more immune plants are to pests and disease (Gomiero et al. 2011).

Research at Trent University has found on a typical day with a temperature of 18.4 °C a normal roof surface temperature was 32 °C while that of a green roof was 15 °C. Roof gardens keep roofs cool in summer and also help insulate in winter. They require little maintenance and reduce energy bills significantly (Fig. 5).

RTG's reduces the Intensity of solar flux. The Tokyo government estimates that if half their roofs were green it would save a million dollars every day in air conditioning energy use. The majority of the roofs in the world are dark-coloured and as a result have a low albedo and absorb excessive amounts of heat. Earlier there used to be infrastructure that used to maintain the temperature inside the house. But today our architecture is not sustainable. Green roofs can aid reducing urban heat islands effect and help in climate change adaptation. RTG's provide added insulation in the colder months and prevent excessive heat absorption in the summer. By regulating temperature variability; green roofs also reduce energy consumption. The aim is to keep buildings cool and curb carbon footprint.

Fig. 3 Baby corn plant (*Zea mays* L.)

More than 30% of all carbon dioxide emissions in many developed nations is heating or cooling buildings. A study was performed by the University of Cardiff which shows that plant surfaces by transpiration process, do not rise more than 4–5 °C above the ambient. This then translates into a cooling of the environment between 3.6 and 11.3 °C (6.5 and 20.3 °F), depending on the area on earth (gob-eshona). Similar results were reported by the respondents during this study. (Global Warming Videos by Futurist Dr Patrick Dixon).

Hard surfaces in urban areas are more likely to reflect sound, whereas green roofs absorb sound waves because of the nature of the substrate and vegetation. At the airport in Frankfurt, Germany, a 10-cm-deep green roof reduced noise levels by

Fig. 4 Marigold (Tagetes) and red spinach *Amaranthus cruentus* are potted in one *pot* to promote companion planting. Other *pots* has been planted with Mint (Mentha)

Table 1 Quantification of the produce

Vegetable	Plants	Area/pots	Produce (g)
Lady finger	10	5	250
Bitter gourd	3	1	225
Brinjal	2	1	450
Capsicum	4	2	225
Chilli	3	1	60
French beans	3	1	125
Garlic	1	1	40
Onion	3	1	1000
Potatoes	1	1	500
Corn	2	1	1500
Tomatoes	2	1	700
Apple gourd	2	1	250
Zucchini	1	1	600
Cucumber	1	1	150
Black eyed peas	1	1	50
Flat green beans	1	1	50
Cluster beans	1	1	50
Total		22	6225

Fig. 5 A garden bed on the roof

5 dB (Oberndorfer et al. 2007). Other research shows that 12 cm of green roof substrate alone can diminish noise by 40 dB (Peck and Monica 2001).

With the presence of more green area the carbon cycle can function more appropriately. One of the primary benefits is reduction of carbon footprint. More green space means less air pollution. A city with green roofs will have a clean and healthy environment. This will help the city in terms of sustainability and climate change adaptation. Green roofs are, in fact, the best technology to improve sustainability. Increasing urbanisation is leading to large-scale deforestation. One might think it better to check urbanisation than invest in setting up green roofs. But checking urbanisation is a massive step. One can focus on doing little bit to make cities healthier. Cities are already devoid of vegetation. Increasing vegetation in these urban spaces is a good idea which is not in conflict with stopping deforestation or other pro-environment activities in the larger sense. RTG's can regulate the micro as well as macro climate if encouraged on larger scale. "Cities are where change is happening the fastest and we must seize the opportunities we have been presented with to make that change significant and permanent." Miller (2007). Climate change poses serious threats to urban infrastructure, quality of life, and entire urban systems. Not only poor countries, but also rich ones will increasingly be affected by anomalous climate events and trends (World Bank 2010). In 2003, more than 70,000 people died in Europe from a severe heat wave (World Bank 2010). These kinds of extreme events will increase in coming years. The deaths were also considered as an indication as the victims were disproportionately elderly. This acute vulnerability of the elderly, children, and infirm is even more pronounced in the cities of developing countries. The effects of climate change are especially unfair as those most unable to adapt, and those who contributed least to the problem, will be harmed the most (Cities and Climate Change: An Urgent

Agenda, December 2010, vol 10, part III). The International Energy Agency (IEA) estimates that urban areas currently account for over 67% of energy-related global greenhouse gases, which is expected to rise to 74% by 2030. It is estimated that 89% of the increase in CO_2 from energy use will be from developing countries (IEA 2008).

Agricultural production, trade, income, food quality, clean water, sanitation, governance and political stability are all factors influencing one's food security status. Jacques Diouf, FAO Director-General [FAO: The State of Food Insecurity 2006] says that a major question for India concerns the agriculture sector's contribution to national food security. India is the world's second most populous country, and it has the largest number of farmers and rural population. About one-quarter of the world's total food insecure people live in India (India Policy Brief, OECD, and November 2014). At a fast pace and it will not be long before a greater part of developing Country's populations will be living in large cities. Therefore, urban food security and its related problems should also be placed high on the agenda in the years to come.

Urban agriculture contributes to food security, nutrition and livelihoods by providing food for self-consumption, giving us healthy food and allowing for saving on food expenditures. Rooftop agriculture is one way in which urban areas could attempt to be more balanced and sustainable in their resource consumption. It is possible to produce a variety of fruit, grain, and vegetable crops on rooftops, either in containers or as field crops (TFPC 1999).

Many cities worldwide continue to source a large percentage of their food from within city boundaries, though as the global food system is growing, that percentage is declining. Urban agriculture is strongest in developing countries, but even in the Netherlands, 33% of total agricultural production is within urban lands (Brown 2002). Also from an environmental perspective, it is important to point out that food grown in urban areas is more likely to be organic than that produced by the conventional industrial food system. This is due to a number of factors, including the fairly small scale of urban agriculture, the large labour pool, and cost reductions due to lowered transport costs. In addition, urban farms, since they are worked more intensely, can produce up to 15 times more per acre than their rural counterparts (Ableman 2000). Literatures indicate climate components like temperature, precipitation, CO2 concentration and extreme climate events have an effect on food security components. The most direct effect and well researched component of climate change on food security is food availability by reducing net crop production. It is also found that climate change has an impact on food accessibility and utilization but not well studied due to its complexity. Projections indicate that this problem will be more severe in the future than today unless climate change mitigation and adaptation strategies are done (J Earth Sci Clim Change).

5 Conclusion

Roof top farms help in achieving six important sustainable development goals (SDG's). They aid in provisioning of ecosystem services, enrich urban biodiversity and reduce food insecurity. Roof Top gardening is an alternative agricultural approach to farming system that is more responsive to natural cycles and biological interaction. The reason for focusing on urban agriculture is that that RTG's have potential for application in cities and must be seen as a permanent element of urban system because ecological imbalances created due to abuse and overuse of environmental services in the city has left a very thin line between natural and man-made disasters.

The present study shows that there is immense scope of RTG's in urban context as they enhances food security and availability of healthy food. The food produced in RTG's is much safer than the food we get in the market as there is less or no pesticide, insecticide and fertilisers are added. Today the pollinator species are facing extinction worldwide. Farmers are forced to keep bee boxes on their farms for increasing their production. This decline in pollinators is due to overuse of chemicals on agriculture fields. Therefor urban areas provide more conducive environment for pollinators. RTG's help in supporting and conservation of urban biodiversity by providing safer environment and habitat to insects, bees, butterflies and birds. Therefore, they have a key role in biodiversity conservation of urban species. The study also proves that RTG's also help in maintaining room temperature and in saving energy. They have a very important role in climate change mitigation and adaptation. The study showed that there are minimal families who are adopting RTG's which shows that there is huge potential for the promotion of RTG's in urban areas in general and Delhi in particular. The study also showed that there is lack of skilled labour which can develop RTG's, therefore, there is scope of green skilling as part of which employed youths can be trained in developing and maintaining RTG's. This will not only give employment to the youths but will also help in bringing roof tops under green cover.

There is only a small proportion of people who are practicing roof top gardening for growing fruits and vegetables, others only have decorative plants. This is because they are unaware of the fact that one can grow vegetables in containers on roofs. As they are introduced to the potential benefits of RTG's they feel motivated to develop their roofs into a vegetable garden. Rooftop owners have stated that they do feel the difference in quality of home grown food compared to the food available in the market. They have also seen improvement in the available biodiversity on the roof.

RTG's provide different ecosystem services like provisioning, regulating, supporting and aesthetic/cultural due to which they should be adopted by the urban population for their future sustainability. Green roofs come with many perks, such as less air pollution, cooler homes and easy access to green space. We have on our hands vast tracts of rooftop space which mostly lie unused. We must dedicate more serious thought to utilising this wasted space more productively. Buildings absorb a lot of heat, and growing vegetation on the roof can keep it cool. This, in turn, will reduce the use of air-conditioners and fans and can minimize energy uses. Thus green roofs are catching on in Indian cities like Delhi and Bengaluru. While many might adopt a green roof for beautification, it serves a larger purpose. Rooftop cultivation has immense scope for and their services should be made available throughout the country through the offices of the state agriculture department, Horticulture Mission and Vegetable and Fruit Procuring Council. Policies should be made to make RTG's mandatory in urban cities for food security, increase in green cover and biodiversity conservation. If this happens, urban rooftop farming will revolutionize the food deficit state. Transforming our roofs is a practical and comparatively easy way to mitigate climate change. It is also the most effective if not the only way to adapt to a warming climate. Due to their long-term viability and a host of other positive externalities, green roofs are a better option. By restoring vegetation to our cities, we not only reduce carbon and adapt to climate change but we can also reconnect with nature, increase food security and provide habitat for wildlife especially insects.

In cities not everyone has access to roof or spacious balcony, therefore, the limitation of the concept of Roof top gardens. There is a lack of research on roof top gardens in urban areas and hence the research faced certain shortcomings.

Acknowledgements This research was supported by Earthwatch Institute India. I would like to thank my external supervisor Dr Pradeep Mehta (Earthwatch Institute India) who provided insight and expertise that greatly assisted the research.

I thank Bella Gupta from All India Kitchen Garden Association and Kapil Mandewal the CEO of Edible Routes for assistance with providing details of roof top owners in Delhi. They also provided information on improved techniques of setting up vegetated roofs their comments greatly improved the manuscript.

I would also like to show gratitude to All the Roof top garden owners for giving their valuable time to conduct interviews and provide access to their roofs. I am immensely grateful to my internal supervisor Dr Gopal Sarangi (Teri University) for his comments on the report, although any errors are our own and should not tarnish the reputations of these esteemed persons.

References

Aggarwal, V. (2008) "In India, How Do Rooftop Gardens Grow?" THE WALL STREET JOURNAL

Bicknell, J., Dodman, D. and Satterthwaite, D. (2009) Adapting Cities to Climate Change: understanding and addressing the development challenges', London: Earthscan Available at: https://books.google.co.in/books. Accessed 30 October, 2016

Brenneisen, S. (2006) "Space for Urban Wildlife: Designing Green Roofs as Habitats in Switzerland." Green Roofs and Biodiversity: Special Feature

Brown, K. H. (2002) Urban Agriculture and Community Food Security in the United States: Farming from the City Center to the Urban Fringe. Urban Agriculture Committee of the CFSC. http://alivebynature.com/urbanag.html

Cartwright, A., Parnell, S., Oelofse, G. and Ward, S. (2012) Climate Change at the City Scale. Impacts, Mitigation and Adaptation in Cape Town

Da Schio, N. and Brekke, K. F. (2013) The relative carbon footprint of cities, Working Papers du Programme, Villes and Territoires. Available at: http://blogs.sciences-po.fr/recherche-villes/files/2013/03/WP-carbonfootprint.pdf. Accessed 30 October, 2016

De Sherbinin, A., Schiller, A., Pulsipher, A. (2007) 'The Vulnerabilities of Global Cities to Climate Hazards', Environment and Urbanization 19.39. Available at: http://www.ciesin.org/documents/vulofglob_contactshtml.pdf. Accessed 30 October, 2016

Franchetti, M. J., Defne A. (2013) Carbon footprint Analysis: Concepts, methods, implementation, and case studies

Germination Temperature. Available at: http://tomclothier.hort.net/page11.html. Accessed 30 October, 2016

Gomiero, T.; Pimentel, D.; Paoletti, M. G. Environmental Impact of Different Agricultural Management Practices: Conventional Vs. Organic Agriculture. *Critical Reviews in Plant Sciences* [Online] (2011), Volume 30, Issue 1–2: 95–124. Available at: http://www.tandfonline.com/doi/full/10.1080/07352689.2011.554355#tabModule. Accessed 30 October, 2016

Hegland, S. J., Nielsen, A., Lázaro, A., Bjerknes, A. L., and Totland, Ø. (2009) "How does climate warming affect plant-pollinator interactions?" Ecology Letters

Kumar, V. and Mahalle, A. M. (2011) "Investigation of the Thermal Performance of Green Roof on a Mild Warm Climate." INTERNATIONAL JOURNAL of RENEWABLE ENERGY RESEARCH

Memmott, J., Craze, P.G., Waser, N. M. and Price, M.V (2007) "Global warming and the disruption of plant–pollinator interactions." Global warming and ecological interactions: Ecology Letter

Miller, D. (2007) Climate change and cities. Fisrt assessment report of the urban climate research network. Cambridge University Press

Mukhopadhyay, P. and Revi, A. (2009) 'Keeping India's Economic Engine Going: Climate Change and the Urbanisation Question', Economic and Political Weekly 44:31 Available at: http://www.epw.in/journal/2009/31/climate-change-negotiations-special-issues-specials/keeping-indias-economic-engine. Accessed 30 October, 2016

Oberndorfer, E. et al. (2007) "Green Roofs as Urban Ecosystems: Ecological Structures, Functions, and Services." BioScience (American Institute of Biological Sciences)

Parry, M., Canziani., O. F., Palutikof, J. P., van der Linden, P. J. and Hanson, C. E. (2007) Contribution of Working Group II to the Fourth Assessment Report of the Intergovernmental Panel on Climate Change, 2007. Cambridge University Press, Cambridge, United Kingdom and New York, NY, USA

Peck, S. and Monica, K. (2001) "Designed Guidelines for green roof." Compiled by Ontario Association of Architects

Satterthwaite, D. (2008) 'Cities' contribution to global warming: notes on the allocation of greenhouse gas emissions', Environment and Urbanization 20: 539 Available at: http://eau.sagepub.com/content/20/2/539.full.pdf. Accessed 30 October, 2016

Satterthwaite, D. (2011) 'How urban societies can adapt to resource shortage and climate change', Philosophical Transactions of the Royal Society A 369: 1762–83 Available at: http://rsta.royalsocietypublishing.org/content/369/1942/1762. Accessed 30 October, 2016

Schweiger, O. (2010) "Global pollinator declines: trends, impacts and drivers."

TFPC (Toronto Food Policy Council), (1999) Available at: TFPC (Toronto Food Policy Council), (1999) Feeding the city from the back forty: a commercial food production plan for the city of Toronto, (Toronto, Toronto Board of Health). Accessed 30 October, 2016

UNEP, (2014). Realizing the potential of Citizen Science, pg. 38–39. Available at: http://www.unep.org/yearbook/2014/PDF/chapt6.pdf. Accessed 30 October, 2016

World watch Institute, (2007) State of the World: Our Urban Future, New York NY: W Norton and Company, Available at: http://www.worldwatch.org/files/pdf/State%20of%20the%20World%202007.pdf. Accessed 30 October, 2016

World Bank. (2010) "The Impact of Climate Change on Cities: PART II." WORLD BANK.

Modeling and Monitoring of Air Quality in Greater Cairo Region, Egypt Using Landsat-8 Images, HYSPLIT and GIS Based Analysis

Moawad Badawy Moawad, Abdel Aziz Youssief and Khaled Madkour

Abstract The main objectives of this study are to model, monitor and observe air quality in Greater Cairo Region (GCR) based on Landsat-8 OLI and TIRS images, HYSPLIT dispersion model, and GIS techniques to overcome the great lack of data observation and to improve data coverage and comparability. Major land covers were mapped from Landsat-8 images. Air quality map and AOT (Aerosol Optical Thickness) were derived by calculating aerosol path radiance based on exoatmospheric solar constant and atmospheric transmittance. HYSPLIT dispersion model was run to simulate the extraordinary pollution episodes (black cloud) and to determine air mass trajectories. Eventually, GIS spatial analysis was carried out based on in situ data to adjust satellite derived PM_{10_sat}. The study deduced that concentrations of PM_{10} are high in GCR and exceed six times of the WHO air quality standards. It is extremely intensive in spring due to the local dusty wind "called Khamasin" and in autumn due to burning of agriculture residues. Mean annual concentrations of PM_{10} in GCR from 16 April 2014 to 29 January 2015 reveals that ~70% of the region is considered as unhealthy for sensitive groups (200–250 μg/m^3 year).

Keywords Greater Cairo Region (GCR) · Air pollution · Air quality index · HYSPLIT

M.B. Moawad (✉) · A.A. Youssief · K. Madkour
Department of Geography, Faculty of Arts, Ain Shams University, Cairo, Egypt
e-mail: moawad@outlook.com

A.A. Youssief
e-mail: abdelaziz.youssef@art.asu.edu.eg

K. Madkour
e-mail: khaled.madkour@art.asu.edu.eg

W. Leal Filho (ed.), *Climate Change Research at Universities*,
DOI 10.1007/978-3-319-58214-6_3

1 Introduction

Air pollution is a major environmental problem because it harms human health and their environment. Major pollutants are gaseous pollutants, odours, and particulate matters such as dust, mist, and smoke. Pollutant's concentration over urban agglomerations causes severe pollution to the settlers, animals, and plants. The World Health Organization (WHO) estimates that seven million people die in the world every year because of poor air quality. Risks of air pollution include strokes, heart disease, lung cancer, and respiratory diseases like asthma. The most health-damaging pollutants are particulate matter of a diameter 10 μm (PM_{10}) or less that are largely associated with damaging effects on the human body as they can penetrate and lodge deep inside the lungs (WHO 2016).

In Cairo, air pollution is a matter of a serious concern for many authors since it exceeds the world standards in some districts (e.g. Cramer 2002; Abdel-Halim et al. 2003; El-Askary and Kafatos 2007, 2008; El-Askary et al. 2011; Safar et al. 2009; Safar and Labib 2010; Marey et al. 2010, 2011; Shakour et al. 2011; Abou el Fetouh et al. 2013; Robaa 2013; Hassan et al. 2013). On the other hand, the existing environmental measured data for the Greater Cairo Region (GCR) is not comprehensive. Therefore, the main objectives of this paper are to model and monitor PM_{10} in GCR by using integrated environmental datasets derived from Landsat-8 images, HYSPLIT (Hybrid Single Particle Lagrangian Integrated Trajectory) and GIS technique to overcome the great lack of data observations and to improve data coverage and comparability.

2 Study Area and Sources of Air Pollution

GCR is formulated from conurbation of three neighboring governorates, namely Cairo, Giza and Al Qalyuobia. The region is demarcated by desert landscape and agricultural activities. The dominant land covers are residential and industrial areas (Fig. 1). GCR is one of the largest metropolitan areas in the world covering $\sim$1622 km^2 with a total population of $\sim$21 Ma and a population density of $\sim$42,000 persons/km^2 (CAMPAS 2016). Projections of population growth based on 2015 situation reveals that the human population will keep growing, reaching an estimation up to 25 million settlers in 2030 (UN 2014).

GCR lies in the hyper-arid region with aridity index 0.005 (FAO 1985). Daily annual mean temperature of GCR is 21.5°; the annual rain of 4 mm; and the annual mean of relative humidity is 56%. The region occupies a gorge-like area of the Nile Valley with steep sides of two high plateaus ($\sim$250 m asl) from the east and west (El Moqatam and Al Ahram plateaus). This characteristic makes air pollution even worse since it holds the poisoned air like water in a bowl (Marey et al. 2010). However, topography determines the dominant north wind within this gorge-like topography. Northern wind transfers pollutants from the northern pollution sources

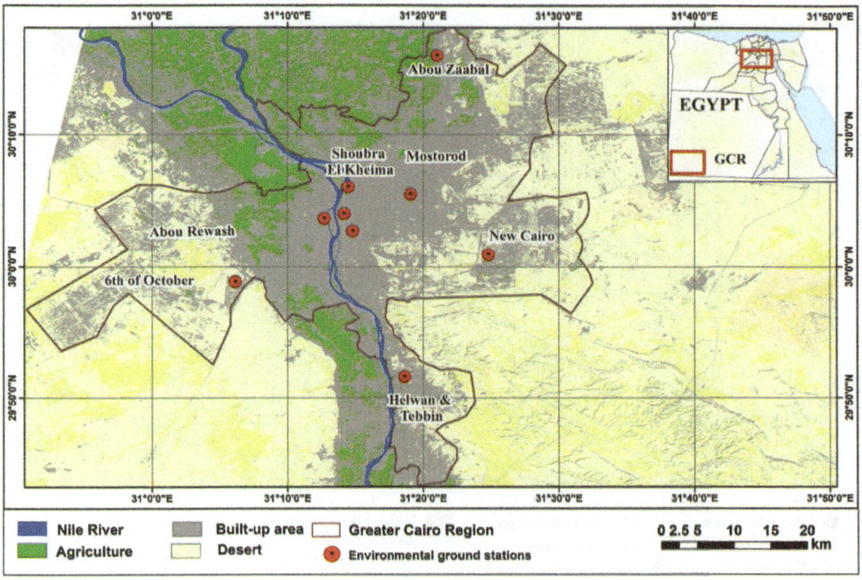

Fig. 1 Location and major land covers of GCR

to the other districts of the region. In spring, blowing of the southern wind leads to transfer pollutants from the southern industrial areas to the north direction. Therefore, it can be said that location of the GCR forms a trap-like of pollutants.

According to the (WHO 2016), CGR is one of the world's worst air pollution, with an annual average concentration of $\sim 179\ \mu g/m^3$. The major sources of air pollution are attributed to transport, energy, waste burning, and industries, which can be categorized as follows (Fig. 2):

- Northern of the GCR includes important industrial areas; namely Shoubra El Kheima, Mostorod and Abou Zaabal (Fig. 3). The most powerful pollution sources in these areas are Shoubra El Kheima power plant, the fertilizer plant of Abou Zaabal, and a multitude of small and medium enterprises as well.
- Southern of the GCR, there are another two industrial areas; namely Helwan and Tebbin (Fig. 3). These two districts include the Tebbin power plant, the Helwan fertilizer company, the heavy industries of iron and steel, coke, and four cement plants.
- West and east of the GCR there are two other industrial agglomerations (i.e. 10th of Ramadan City and El Robeki on the east; 6th of October City and Abou Rewash on the west). Although 10th of Ramadan City and El Robeki do not belong to the GCR, they have negative influences on the air quality of the region
- GCR contains the four largest dump sites in Egypt, namely El Wafaa wel Amal, El Salam, Shubramant, and Abou Zaabal. These damp sites contain $\sim$9.5 million

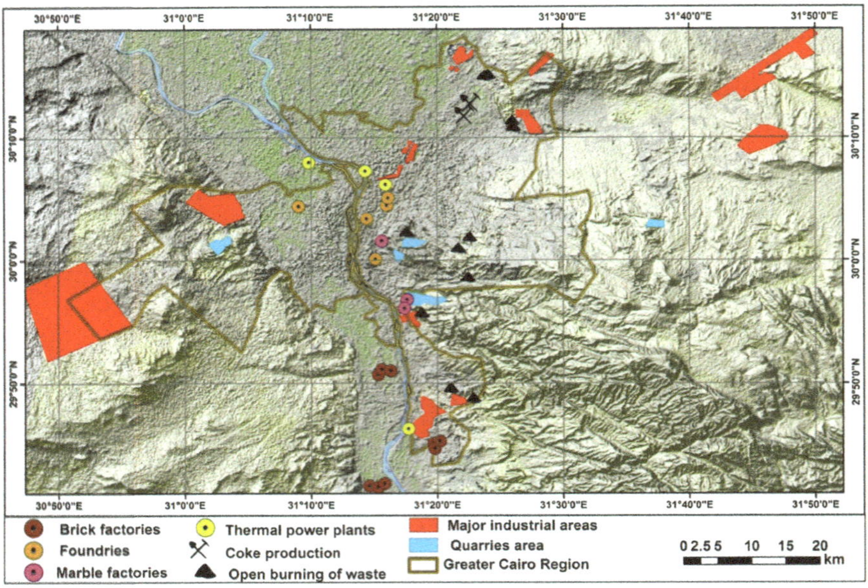

Fig. 2 Topographic setting and sources of air pollution in GCR

tons a year of municipal solid waste (MSW), which represents 47% of the total municipal waste generated in Egypt (GIZ 2010).

- GCR has the largest road network ~10,100 km and ~64 km of underground metro lines.

The region includes as well 750 smelters, 530 brick factories, 1200 metallurgical factories, and 12,500 industrial facilities of medium to small scale (Abou El Seoud 2010). Some 240 facilities still emit small amounts of lead as well. Lead in dust remains a concern, especially during the windy season, when the correlation between the ambient lead measures and lead concentrations in street dust is high (USAID 2004).

GCR includes seven thermal power stations with a capacity of 4585 MW that generate 25,360 GW/h. Four stations are fuelled by natural gas and heavy fuel oil (i.e. Shoubra El Kheima, Cairo East, Cairo West extension, and South Cairo). The other two stations are powered by natural gas and light fuel oil (i.e. Cairo North and Wadi Hof). Only one station is powered by natural gas (i.e. Cairo South II).

It is worth mentioning that the Egyptian crude has a sulfur content of 1.6%, which is a high content relative to other crudes. GCR has one refinery, namely the Cairo oil refinery in Mostorod area, with a capacity of 2 million tons/year. The refinery is planning to add a second-stage oil refinery that will produce over 4 million tons/year, including 2 million tons of Euro 5 diesels, which is the cleanest fuel type in the world.

Fig. 3 Sources of air pollution in GCR. **a** Shoubra El Kheima power plant, **b** dense traffic jam, **c** dust storms, **d** burning of rice straw, **e** burning wastes, and **f** plumes of cement factories

3 Data and Methods

Data and methods used in this study could be summarized as follows (Fig. 4):

3.1 Landsat-8 Images

Landsat-8 images of OLI and TIRS sensors were used in this study to calculate the seasonal variations of pollutant's concentrations in the GCR (Table 1). Major land covers were mapped from Landsat-8 images OLI sensor, using supervised classification (Fig. 1). Air quality map and Aerosol Optical Thickness (AOT) were derived by calculating aerosol path radiance based on exoatmospheric solar constant and atmospheric transmittance.

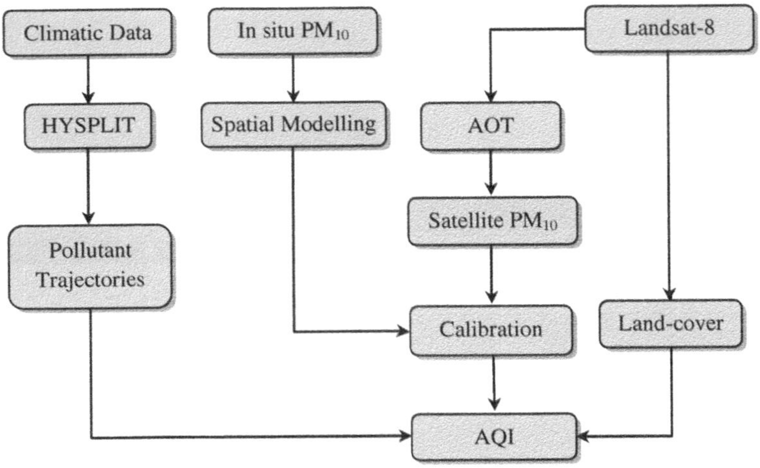

Fig. 4 Study scheme

Table 1 Specifications of Landsat-8 images

Image number		Date of acquisition	Time of acquisition (AM)	Cloudiness
Row	Path			
39	176	29-Jan-15	10:22	5
39	176	16-Apr-14	10:23	0.16
39	176	5-Jul-14	10:23	6.04
39	176	9-Oct-14	10:24	0.74

Pixel-based analysis of PM_{10_sat} was derived from Landsat-8 images (TIRS sensor) according to the following procedures:

First, we estimated aerosol path radiance by the help of dark object subtraction method. It is applicable for shortwave length of bands 2 and 3, in which water vapor absorption is negligible. Aerosol path radiance was estimated after Mishra et al. (2012) by Eq. (1) that reads:

$$L_p = G * DN_{dark} + B - 0.01[E_o * \mu_s * T_z + E_{down}]Tv/\pi \quad (1)$$

where, L_p is aerosol path radiance (W m^{-2} sr^{-1} μm^{-1}); G is band specific gain (m^2 sr µm W^{-1}); DN_{dark} is the darkest DN values in spectral band with at least one thousand pixels; B is band specification bias in DN; E_0 is exoatmospheric solar constant (W m^{-2} μm^{-1}); μ_s is cosine of solar zenith angle; $T_z = e^{-\pi/\mu v}$ atmospheric transmittance in the illumination direction; E_{down} is down welling diffuse irradiation (W m^{-2} μm^{-1}); $T_v = e_s^{-\pi/\mu}$ atmospheric transmittance from the target towards the sensor.

Then, optical thickness for Rayleigh scattering (τ_r) was determined by Eq. (2) after Sturm (1981) as:

$$\tau_r = 0.00879 * (\lambda_c)^{-4.09} \tag{2}$$

where, λ_c is band center wavelength in μm; After that, we calculated AOT by substitution of L_p and τ_r in Eq. (3) proposed by Yil-Yi (1982) as:

$$L_P = \omega_a \left\{ \frac{E_o * \mu_s * P_a}{4\pi(\mu_s + \mu_v)} \right\} \{1 - \exp[-\tau_a(1/\mu_s + 1/\mu_v)]\} * t_{h2o} * t_{o3} * \exp[-\tau_r(1/\mu_s + 1/\mu_v)] \tag{3}$$

where, τ_a is Aerosol Optical Thickness(AOT), ω_a is aerosol single scattering albedo 0.80; P_a is aerosol scattering phase function; μ_v is cosine of sensor viewing angle; t_{H2O} and t_{O3} are the transmittance factors due to water and ozone = 1.

Finally, PM_{10} mass concentration was calculated based on Eq. (4) after Retails and Sifakis (2010) as:

$$PM_{10} = 195.7\tau_a + 14.5 \tag{4}$$

where, τ_a is Aerosol Optical Thickness (AOT). Calculations of PM_{10} concentration in GCR are shown in Table 2 and Fig. 5. PM_{10} concentration map was then averaged to calculate the mean annual air quality map shown in Fig. 6.

3.2 *HYSPLIT Model*

HYSPLIT (Hybrid Single Particle Lagrangian Integrated Trajectory Model) is atmospheric transport and dispersion modeling system developed by NOAA. It simulates how pollutants in the ambient atmosphere disperse and how they react in the atmosphere. In this study, HYSPLIT Trajectory Model was used to depict role of the meteorological factors in determining air quality in GCR during the same times of acquisition of Landsat-8 images. The requirement parameters for the model are gridded fields of meteorological variables at regular temporal intervals (e.g. pressure, wind, and rainfall). HYSPLIT can be downloaded for free with registration from the web site: http://ready.arl.noaa.gov/hyreg/HYSPLIT_pchysplit.php.

3.3 *Adjustment of Satellite Derived PM_{10_Sat}*

Uncertainty satellite derived PM_{10_sat} can be highly variable for many reasons such as retrieval technique, meteorological stability, wind speed, etc. Therefore, the bias of the satellite data needs to be adjusted. We used in situ observed hourly PM_{10_obs} data obtained from the environmental ground stations, Ministry of Health, as a reference data (Fig. 1). We assumed that these data sets are much more reliable as

Table 2 Statistical analysis of the estimated PM_{10} and model efficiency (ME)

Stations	Date											
	PM_{10_obs}	PM_{10_sat}	*Diff.*	*Std.*	$PM_{10\_est(i)}$	*ME*	PM_{10_obs}	PM_{10_sat}	*Diff.*	*Std.*	$PM_{10\_est(i)}$	*ME*
	29-Jan-15						16-Apr-14					
Al Nozha	118	38.9	79.1	55.9	110	0.99	140	123.3	16.7	11.8	120	0.92
El Ataba	223	905.6	−682.6	482.7	214	0.76	432	3316.6	−2884.6	2039.7	378	0.94
Al Sahel	241	897.25	−656.25	464.0	213	0.4	84	470	−386	272.9	94	0.99
Al Amiria	293	1138.9	−845.9	598.1	242	0.67	194	716.6	−522.6	369.5	183	0.65
Abou Zaabal	149	263.9	−114.9	81.2	137	0.95	213	1116.6	−903.6	638.9	213	1
Avg.	204.8	648.9	−444.1	336.4	183.2	0.75	212.6	1148.6	−936.0	666.6	197.6	0.9
	5-Jul-14						9-Oct-14					
Al Nozha	299	542.4	−243.4	172.1	211	0.31	108	71.6	36.4	25.7	110	1
El Ataba	137.9	206	−68.1	48.2	137	1	214	830	−616	435.6	201	0.31
Al Sahel	192	451.5	−259.5	183.5	191	0.47	232	830	−598	422.8	201	0.15
Al Amiria	244	506	−262	185.3	203	0.34	287	964.6	−677.6	479.1	215	0.34
Abou Zaabal	94	19.6	74.4	52.6	96	1	151	296.6	−145.6	103.0	137	0.91
Avg.	193.38	345.1	−151.7	128.3	167.6	0.62	198.4	598.6	−400.2	293.3	172.8	0.54

PM_{10_obs} = the in situ observed PM_{10} (μg/m^3)
PM_{10_sat} = pixel-based estimated PM_{10} from landsat-8 images (μg/m^3)
$PM_{10\_est(i)}$ = the corrected pixel-based estimated PM_{10} (μg/m^3)
Diff. = differences
Std. = standard deviation
ME = model efficiency

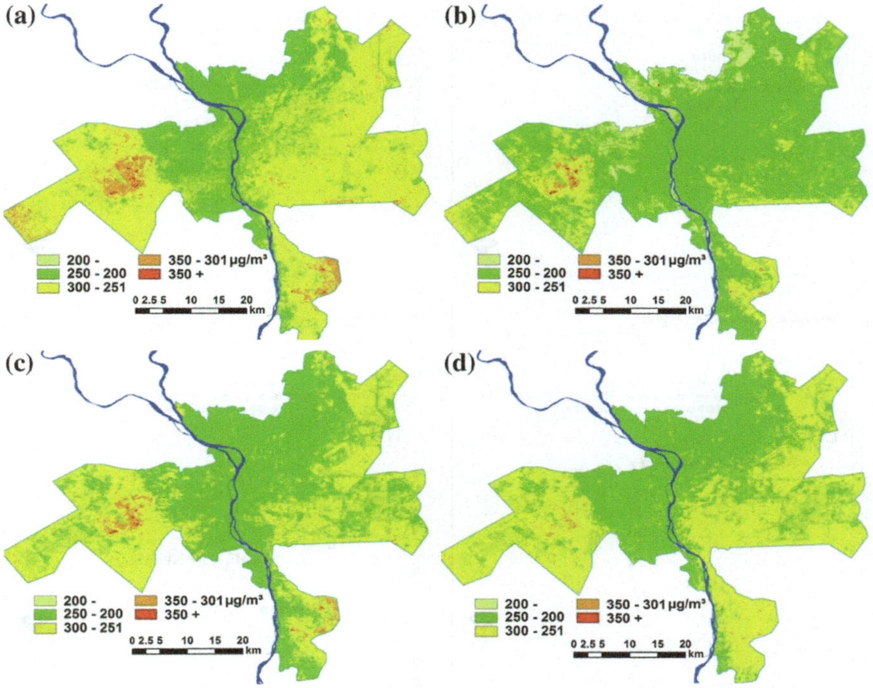

Fig. 5 Concentrations of PM_{10} in GCR derived from Landsat-8 images. **a** 16 April 2014, **b** 5 July 2014, **c** 9 October 2014, and **d** 29 January 2015

they are directly measured at the surface. This in situ observed PM_{10_obs} data differs from the satellite PM_{10_sat} in terms of scale and resolution since they are sparsely located and provide only point-scale measurements, while the PM_{10_sat} is a pixel-based estimation. However, satellite PM_{10_sat} were extracted using explicit x and y coordinates of the observed PM_{10_obs} data and some statistical variables were inspected as shown in Table 2.

It is assumed that some variable of interest (y) can be driven by some other variable (x). Plotting the PM_{10_sat} versus PM_{10_obs} on a scatter diagram revealed linear function with significant correlation ($R^2 = 0.61$). Then, pixel-based PM_{10_sat} was corrected based on the PM_{10_obs} by implementing the regression analysis based on the following equation:

$$PM_{10\_est(i)} = a \times PM^2_{10\_sat(i)} + b \quad (5)$$

where $PM_{10\_est(i)}$ is the adjusted pixel-based PM_{10_sat} at a pixel *i*; and $PM_{10\_sat(i)}$ is the satellite derived PM_{10_sat} at the same pixel *i*; and a and b are coefficients obtained from the linear fitting. Accuracy of the estimation was then performed using model efficiency equation:

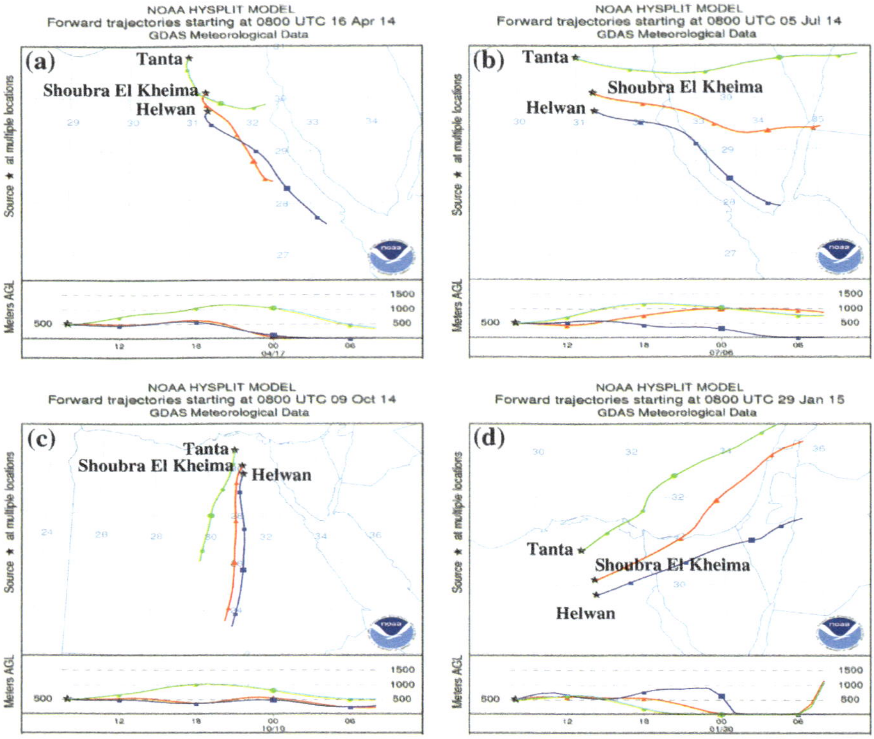

Fig. 6 Airmass trajectories and their role in air pollution transport. **a** 16 April 2014, **b** 5 July 2014, **c** 9 October 2014, **d** 29 January 2015 (from HYSPLIT model)

$$ME = 1 - \left(\sum \left(\mathrm{PM}_{10\_\mathrm{obs}} - \mathrm{PM}_{10\_\mathrm{est(i)}}\right)^2 / \sum \left(\mathrm{PM}_{10\_\mathrm{obs}} - \mathrm{PM}_{10\_\mathrm{avg(n)}}\right)^2\right) \quad (6)$$

where ME is the model efficiency, $\mathrm{PM}_{10\_\mathrm{obs}}$ is the in situ observed PM_{10}, $\mathrm{PM}_{10\_\mathrm{est(i)}}$ is the adjusted pixel-based $\mathrm{PM}_{10\_\mathrm{sat}}$ from landsat-8 images, $\mathrm{PM}_{10\_\mathrm{avg(n)}}$ is averaged of the in situ observed PM_{10} and n is number of observations. Values closer to 1 indicate the superior model performance (Mayer and Butler 1993). Results of the model are shown in Table 2.

4 Results and Discussion

4.1 Seasonal and Mean Annual PM_{10} Concentrations

Seasonal and mean annual concentrations of PM_{10} were analyzed based on the adjusted $\mathrm{PM}_{10\_\mathrm{est(i)}}$ (see Sect. 3.3). Therefore, each scene was dealt as a representative to its season.

Figure 5 clearly shows that concentrations of PM_{10} are high in GCR, generally. PM_{10} concentrations exceed the WHO air quality standards (50 μg/m^3 as maximum 24-h mean and 20 μg/m^3 as average annual concentrations) and the Egyptian standards as well (70 μg/m^3 as maximum 24-h mean and 120 μg/m^3 as average annual concentrations). Spring and autumn are extremely higher where 72.18% (1170.70 km^2) and 58.24% (944.60 km^2) of the total region ranges from unhealthy for sensitive groups (200 μg/m^3) to very unhealthy air quality (600 μg/m^3), respectively. The most suffered districts are that in the vicinity of the industrial areas in the north and north-east Cairo (i.e. Shoubra El Kheima and Abou Zaabal), west (6th of October and Abou Rewash) and south (Helwan, and El Tebbin). Patches of high concentrations also exist in Cairo downtown owing to traffic jam.

On the other hand, in winter and summer concentrations of PM_{10} are relatively lower. Hence, 47.83% (775.80 km^2) and 22.7% (368.3 km^2) of the total region ranges from unhealthy for sensitive groups (170 μg/m^3) to unhealthy (350 μg/m^3), respectively. Concentrations of particulate matter had an obviously negative correlation with air temperature. As air temperature rose, concentrations of particulate matter significantly decreased. Therefore, probability of atmospheric pollution decreases with the increase of air temperature in the summer, while surface temperature is low; the situation is contrary in the winter (Li et al. 2015).

In spring, high concentrations of PM_{10} are due to the dominant dusty local wind blows from the south or the southwest direction which known as "Khamasin". On 16 April 2014, meteorological condition was defined as neutral (D) (Pasquill 1961) and depth of the boundary layer was ~800 m. Scale of the vertical mixing coefficient (turbulent mixing) was estimated as 200 m^2/s during time acquisition of the Landsat image at 10:23 am. Neutral meteorological condition accompanied by low wind speed (1–3 m s^{-1}) led to dignify pollutant concentrations near the earth's surface. Pollutant's trajectories were corresponding to the north wind (Fig. 6a) that pushed pollutants from the middle Nile Delta (Tanta and El Mahala El Koubra cities) and the northern pollutant sources of the GCR inside the GCR. This situation increased concentrations of pollutants near the ground surface.

On 5 July 2014, atmospheric stability was moderately unstable conditions (B) due to low atmospheric pressure in the summer as a result of high temperature and low air density. In unstable conditions, pollutants tend to intensively concentrate near the ground surface because the elevated sources of PM_{10} can lead to intensify ground concentrations than in the case of stable condition (Macdonald 2003). However, depth of the boundary layer was ~1000 m, and scale of the vertical mixing coefficient was estimated as 1000 m^2/s at the same acquisition time of Landsat image at 10:23 am. Moderately unstable condition accompanied with medium wind speed (4–6 ms^{-1}) pushed pollutants to disperse from the middle Nile Delta and northern GCR into the southeast direction inside the center of the GCR (Fig. 6b). Vertical dispersion of the pollutants did not exceed 500 m height due to low wind velocity (1–3 ms^{-1}) within the GCR.

In autumn, most of pollutants are due to pollution episode, which called "Black Cloud". This episode increases concentration of pollutants owing to burning of rice straw in the middle Nile Delta. On 9 October 2014, atmospheric stability fluctuated between slightly and moderately unstable conditions (C, B). Depth of the boundary layer was ~800 m, and scale of the vertical mixing coefficient was ~600 m^2/s at the same time of Landsat image acquisition at 10:23 am. Slightly and moderately unstable conditions were accompanied by low wind speed 1–3 ms^{-1}. Pollutants trajectory (Fig. 6c) moved under the influence of the northeastern wind, which dispersed pollutants from the northern sources in the middle Nile Delta and northern GCR (Shoubra El Kheima) into the center of GCR.

On 29 January 2015, atmospheric stability was neutral (D) during the time acquisition of Landsat image at 10:23 am. This condition was correlated to the high atmospheric pressure as a result of low temperature and high air density. Depth of the boundary layer was about 600 m, and scale of the vertical mixing coefficient was ~250 m^2/s. Effects of neutral stability and air trajectory directions (Fig. 6d) is evident on concentrations of persistent pollutants since the southwestern wind disperses pollutants from the middle Nile Delta and the northern sources (e.g. Shoubra El Kheima) into the northeast direction outside of the GCR. Vertical concentration was traced up to 2100 m above the ground.

In general, mean annual concentration of PM_{10} in GCR from 16 April 2014 to 29 January 2015 (Fig. 7) reveals that ~70% of the region was unhealthy for sensitive groups (200–250 $\mu g/m^3$ year). Rest of the region ranges from unhealthy for sensitive groups (251–300 $\mu g/m^3$ year) to unhealthy categories (+300 $\mu g/m^3$ year). Unhealthy condition distinguished districts of denes traffic jam (downtown), industrial areas (Shoubra El Kheima, Abou Zaabal, 6th of October, Abou Rewash, Helwan, El Tebbin), and oil refinery in Mostorod. According to Safar and Labib (2010), sources of PM_{10} in GCR can be numerated as: (i) ~45% are due to burning of the agriculture and solid wastes, (ii) ~30% are due to sand and dust storms, (iii) ~25% are due to industrial and vehicle emissions. Tanta and El Mahala El Koubra cities in the middle Nile Delta are the most contributors of air pollutions in GCR.

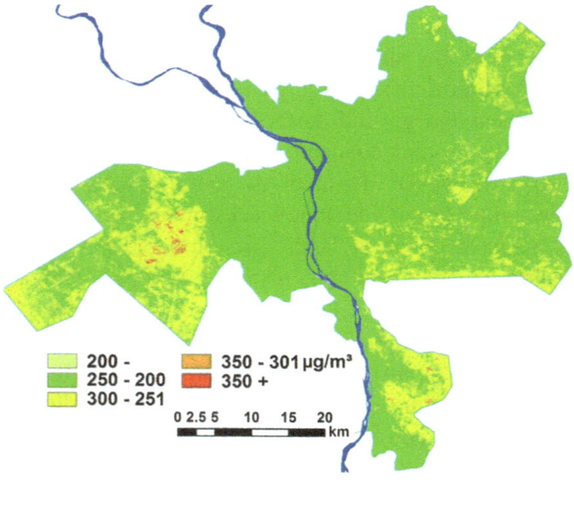

Fig. 7 Mean annual concentrations of PM_{10} in GCR

4.2 Air Quality Index Map of GCR

Air Quality Index (AQI) is a uniform system developed by U.S. Environmental Protection Agency (EPA) to determine whether the air is clean or polluted. The index is normally calculated for five major air pollutants (i.e. ground-level ozone, particulate matter, carbon monoxide, sulfur dioxide, and nitrogen dioxide). It converts the measured pollutant concentrations in the air to a number on a scale of 0 to 500, ranges from good to very unhealthy condition. The higher the AQI value the greater the level of air pollution and the greater health concern (EPA 2016).

In this study, AQI of GCR was calculated only for the adjusted particulate matter (PM_{10}) as stated in Sect. 3.3. The study deduced that the AQI of GCR ranges from unhealthy for sensitive groups to very unhealthy (Figs. 8 and 9), generally. The obvious variation in AQI of GCR is that expansion or downswing of areas represented by a definite AQI category (Table 3). The most notable matters are that: (i) the absence of good and moderate AQI categories from the whole region, (ii) center of GCR belongs to unhealthy to sensitive groups because of traffic congestion, (iii) eastern and western margins of GCR are attributed to unhealthy and very unhealthy classes because of the vicinity deserts and the industrial areas

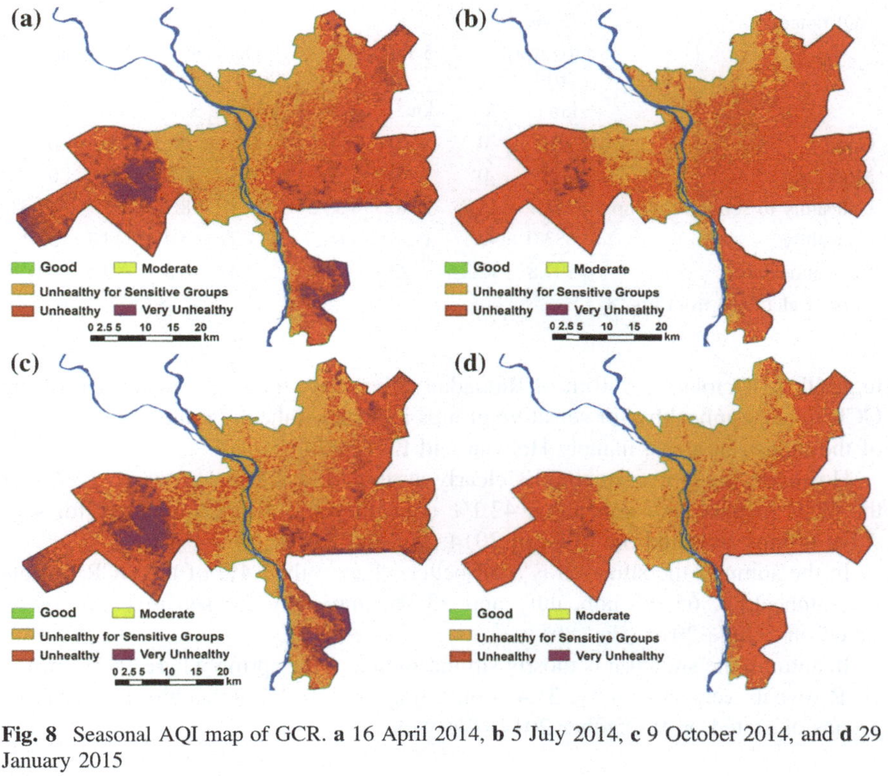

Fig. 8 Seasonal AQI map of GCR. **a** 16 April 2014, **b** 5 July 2014, **c** 9 October 2014, and **d** 29 January 2015

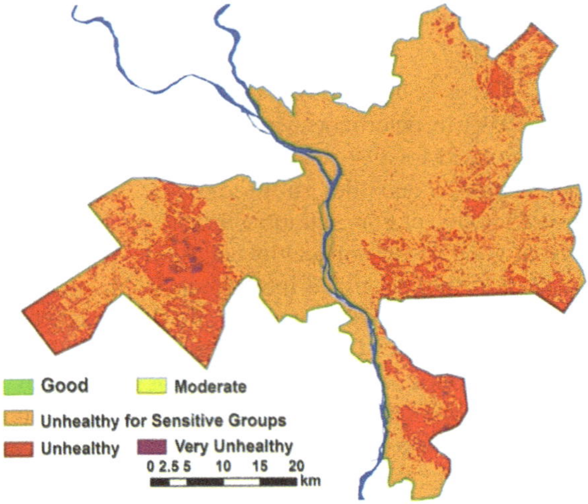

Fig. 9 Mean AQI in GCR from 16 April 2014 to 29 January 2015

Table 3 Air quality index of GCR

AQI range	Dates							
	16 April 2014		5 July 2014		9 October 2014		29 January 2015	
	km^2	%	km^2	%	km^2	%	km^2	%
Good	0	0	0	0	0	0	0	0
Moderate	0	0	0	0	0	0	0	0
Unhealthy to sensitive groups	741.1	45.7	540.7	33.3	779.9	48.1	767.7	47.3
Unhealthy	683.0	42.1	1058.6	65.3	574.7	35.4	844.7	52.1
Very unhealthy	197.8	12.2	22.7	1.4	267.3	16.5	9.5	0.6

Source Calculated from Fig. 8

(e.g. 6th of October and 10th of Ramadan cities), (iv) the southern margin of the GCR reveals unhealthy for sensitive groups and very unhealthy due to the existing of the industrial areas, namely Helwan and El Tebbin.

However, the worst situation is clearly attributed to the spring where 12.2% of the GCR reveals very unhealthy, 42.1% unhealthy and 45.7% unhealthy for sensitive groups as stated on 16 April 2014 (Fig. 8a).

In the summer the situation is a bit better where only 1.4% of the GCR reveals very unhealthy, 65.3% unhealthy and 33.3% unhealthy for sensitive groups as stated on 5 June 2014 (Fig. 8b).

In autumn the situation is mostly similar to that in the spring where 16.5% of the GCR reveals very unhealthy, 35.4% unhealthy and 48.1% unhealthy for sensitive groups as stated on 9 October 2014 (Fig. 8c).

In the winter, 0.6% of the GCR reveals very unhealthy, 52.1% unhealthy and 47.3% unhealthy for sensitive groups as stated on 29 January 2015 (Fig. 8d).

Mean AQI of GCR (Fig. 9) reveals that about 76.3% of the study area has unhealthy air quality conditions to sensitive people (AQI: 101–150). GCR's downtown reveals air quality conditions of unhealthy (AQI: 151–200) due to fuel combustion of heavy traffics, industrial area of Shoubra El Kheima, the oil refinery in Mostorod, the Abou Zaabal and El Obour in northern GCR, that represents 22.6% of region. Rest of the GCR reach the very unhealthy air quality conditions level (AQI: 200–300), mainly in the industrial areas (i.e. Helwan and El Tebbin in southern GCR, 6th of October and Abou Rewash in western GCR). Constructions and expansion of the new suburban (e.g. New Cairo City, 6th of October City) that represents 6% of region have a substantial role in determining air quality conditions in the region as well.

4.3 Adaptation and Mitigation of Air Pollution in GCR

Egypt occupies the eleventh in the world in terms of number of premature mortality linked to outdoor air pollution in 2010 of 35,000 persons/year (Lelieveld et al. 2015). Rapid rate of industrialization and the physical settings of the GCR have increased the current pollution problems. Air pollution in GCR is then characterized by high levels of major air pollutants which exceeds the WHO guidelines, especially in the center of the GCR (Nasralla 2001). PM_{10} in specific is the major pollutant of air quality not only in the GCR but also in the whole Egypt. The annual average is mostly double the Egyptian permissible annual limits (120 $\mu g/m^3$ year) and about six times of the WHO standards.

Most sources of air pollution in GCR are of anthropic origin. These include industrial facilities, thermal power plants, commercial and residential activities, dump and waste burning, transportation and traffic jam. The region is also affected by burning agricultural residues in the neighboring central Nile Delta.

Other natural source of pollution is the dust storms. Such storms are much more active owing to spring depressions leading to a local dusty wind called "Khamasin". This local wind transfers pollutant particles from the neighboring desert to fill up the lower part of the boundary layer over the GCR and increasing levels of air pollution. This would have negative effects on the public health and economic sectors such as airports, agriculture, communications and transportation. It is believed that such storms are responsible for worsened symptoms in asthmatics, and heightened cases of respiratory issues (UNEP 2016).

The most method of adaptation would first base on an early warning system and weather forecasting on a national and regional scale by means of remote sensing. Then, launch and adapting a firm policy for greenbelt plantation around the GCR. In the context, the Egyptian Environmental Affairs Agency (EEAA) launched in 2005 the greenbelt project around GCR for $\sim$100 km; unfortunately the project was not completed (EEAA 2016). Last, protecting cultivation and planting perennials trees such as eucalyptus and olive.

On the other hand, mitigation of air pollution is a crucial issue in GCR since it is related to many sectors. The major governmental projects were successful in eliminating air pollution in GCR through: (i) eliminating lead by producing of unleaded gasoline and relocation of lead smelters, (ii) decreasing SO_2 concentration levels by the fuel shift from heavy fuel oil to natural gas, (iii) dispelling industrial hot spots from the residential areas. These governmental endeavors are appropriate for short and long-term mitigation of air pollution in CGR, but lacking the activation of the laws and promulgated legislations strictly.

5 Conclusions

The major objectives of this study are to model, monitor and observe air quality in Greater Cairo Region (GCR) based on the integration of different datasets (i.e. Landsat-8 images, HYSPLIT model, GIS technique) to overcome the great lack of data observation and to improve data coverage and comparability.

The study deduced that concentrations of PM_{10} are high in GCR and exceeds six times of the WHO air quality standards. It is extremely intensive in the spring and autumn due to local dusty wind in spring and burning agriculture residues in autumn. AQI reveals that 76.3% of the GCR has unhealthy air quality conditions to sensitive people (AQI: 101–150).

References

Abdel-Halim A, Metwally E, El-Dessouky M (2003) Environmental pollution study around a large industrial area near Cairo. Egypt Journal of Radio analytical and Nuclear Chemistry, 257(1): 123–124

Abou el Fetouh Y, El Askary H, El Raey M, Allali M, Sprigg W, Kafatos M (2013) Annual patterns of atmospheric pollutions and episodes over Cairo Egypt. Hindawi Publishing Corporation, Advances in Meteorology, Vol. 2013:1–11

Abou El Seoud A (2010) Egyptian air quality improvement policy. Egyptian Environmental Affairs Agency EEAA, Cairo

CAPMAS (Central Agency for Public Mobilization of Statistics), 2016. Population clock, Egypt now, http://www.capmas.gov.eg/Pages/populationClock.aspx (Last Accessed 25 September 2016)

Cramer J (2002) Population Growth and Local Air Pollution: Methods, Models, and Results, Population and Development Review, Supplement: Population and Environment: Methods of Analysis, 28: 22–52

Egyptian Environmental Affairs Agency (2016) Future projects to enhance air pollution, http://www.eeaa.gov.eg/en-us/mediacenter/reports/projectstudies.aspx (Last Accessed 27 September 2016)

El-Askary H, Kafatos M (2007) Investigation of thermal inversions as a major contributor to the Black Cloud Episodes over Cairo, IEEE International Geoscience and Remote Sensing Symposium IGARSS2007, Barcelona, Spain, July 23- July 27, 2007

EL-Askary H, Kafatos M (2008) Dust storm and black cloud influence on aerosol optical properties over Cairo and the Greater Delta region, Egypt, International Journal of Remote Sensing, 29(24):7199–7211

El-Askary H, Prasad A, Kallos G, El-Raey M, Kafatos M (2011) Analyzing black cloud dynamics over Cairo, Nile delta region and Alexandria using aerosols and water vapor data, air quality-models and applications. Available at: http://www.intechopen.com/books/air-quality-models-and-applications/analyzing-black-cloud-dynamics-over-cairo-nile-delta-region-and-alexandria-using-aerosols-and-water (Last Accessed 23 August 2016)

EPA (2016) Air Quality Index (AQI) Basics. Available at: https://airnow.gov/index.cfm?action=aqibasics.aqi (Last Accessed 30 October 2016)

FAO (1985) The state of food and agriculture 1985, mid-decade review of food and agriculture, FAO, Rome

GIZ (2010) Country report on the solid waste management: Egypt, Sweep net, Cairo

Hassan S, El-Abssawy A, Abd El-Maksoud A, Abdou M H, Khoder M (2013) Seasonal Behaviors and Weekdays/Weekends Differences in Elemental Composition of Atmospheric Aerosols in Cairo, Egypt Aerosol and Air Quality Research, 13: 1552–1562

Lelieveld J, Evans J S, Fnais M, Giannadaki D, Pozzer A (2015) The contribution of outdoor air pollution sources to premature mortality on a global scale. Nature 525: 367–384

Li H, Guo B, Mengfei H, Tian M, Zhang J (2015) Particulate Matters Pollution Characteristic and the Correlation between PM (PM2.5, PM10) and Meteorological Factors during the summer in Shijiazhuang. Journal of Environmental Protection, 6: 457–463

Macdonald R (2003) Theory and objectives of air dispersion modeling, Modeling Air Emissions for Compliance, MME 474A Wind Engineering

Marey H, Gille J, El-Askary H, Shalaby E, El-Raey M (2010) Study of the formation of the "black cloud" and its dynamics over Cairo, Egypt, using MODIS and MISR sensors, Journal of Geophysical Research 115, D21206

Marey H, Gille J, El-Askary H, Shalaby E, El-Raey M (2011) Aerosol climatology over Nile Delta based on MODIS, MISR and OMI satellite data, Atmos.Chem.Phys.,11: 10637–10648

Mayer D, Butler D (1993) Statistical validations. Ecol Modell, 68: 21–32

Mishra R, Pandey J, Chaudhary S, Khalkho A, Singh V (2012) Estimation of air pollution concentration over Jharia coalfield based on satellite imagery of atmospheric aerosol, International Journal of Geomatics and Geosciences, 2: 723–729

Nasralla M M (2001) Geater Cairo air quality profile. A report for EEPP. Cairo

Pasquill F (1961) The estimation of the dispersion of wind borne materials, Meteorological Magazine, 90: 33–49

Retalis A, Sifakis N (2010) "Urban aerosol mapping over Athens using the differential textural analysis (DTA) algorithm on MERIS - ENVISAT data", ISPRS Journal of Photogrammetry and Remote Sensing, 65: 17–25

Robaa S (2013) Some aspects of the urban climates of Greater Cairo Region, Egypt, Int. J. Climatological, 33:3206–3216

Safar Z, Labib W, Gertler A (2009) Evaluation of CO_2 and other pollutant emissions in the Greater Cairo (GC) area from diesel and CNG Buses. paper # 2009-A- 138-AWMA, proceedings of the 102nd Air and Waste Management Association Annual Conference, Detroit, MI, June 16–19, 2009

Safar Z, Labib W (2010) Assessment of particulate matter and lead levels in the Greater Cairo area for the period 1998–2007, Journal of Advanced Research, University of Cairo, 1: 53–63

Shakour A, El-Shahat M F, El-Taieb N M, Hassanein M A, Mohamed AMF (2011) Health impacts of particulate matter in greater Cairo, Egypt. Journal of American science; 7(9): 840–848

Sturm B (1981) The atmospheric correction of remotely sensed data and the quantitative determination of suspended matter in marine water surface layers, Remote Sensing in Meteorology, Oceanography and Hydrology, 147: 163–197

United Nation Environment Program (2016) More action needed on sand and dust storms, Available at: http://www.unep.org/stories/Climate/More-action-needed-on-sand-and-dust-storms.asp#sthash.f6cFSzRX.dpuf (Last Accessed 22 Aug 2016)

United Nations, Department of Economic and Social Affairs, Population Division (2014). World Urbanization Prospects: The 2014 Revision, Highlights (ST/ESA/SER.A/352)

United States Agency for International Development (2004) Cairo air improvement project: Egyptian Environmental Policy Program, Planning for Integrated Air Quality Management: Final Report, Chamonix, Cairo

World Health Organization (2016) WHO Ambient air pollution database. Available at: http://www.who.int/phe/health_topics/outdoorair/databases/AAP_database_summary_results_2016_v02.pdf (Last Accessed 22 October 2016)

Yil-Yi S (1982) A method of correcting the atmospheric effects in satellite measurement of ocean colour, Proceedings of the Annual Technical Conference held in Liverpool: Remote sensing and the atmosphere, Remote Sensing Society, pp 110–116

Going Fossil Free: A Lesson in Climate Activism and Collective Responsibility

Eric S. Godoy

Abstract Colleges and universities already contribute significantly to the fight against climate change, but the UN has recently called upon them to do even more. The purpose of this article is to demonstrate that institutions of higher education play a unique role in combatting climate change and other structural injustices, not only by conducting research and disseminating knowledge, but also by fostering a form of collective political responsibility. A philosophical analysis of different forms of collective responsibility, with specific attention to the Fossil Free divestment movement, reveals how social position facilitates this contribution more so in colleges than in other institutions.

Keywords Climate change · Collective responsibility · Fossil fuel divestment · Student activism

1 Introduction

Colleges and universities[1] already contribute significantly to the fight against climate change. Researchers from a wide range of disciplines identify causes, effects, and possible solutions to the challenges posed by a warming planet. Scholars share the knowledge gained through such work with each other, their students, and the public. Yet the United Nations (UN) has called upon colleges to do even more.

The 21st Conference of the Parties (COP 21) in Paris reached a monumental global commitment of 195 nations to tackle climate change. Article 12 of this Paris agreement explicitly states:

[1]Throughout this article, these two words will be used interchangeably.

E.S. Godoy (✉)
Social Science & Cultural Studies, Pratt Institute, 200 Willoughby Avenue, Brooklyn, NY 11205, USA
e-mail: egodoy2@pratt.edu

W. Leal Filho (ed.), *Climate Change Research at Universities*,
DOI 10.1007/978-3-319-58214-6_4

> Parties shall cooperate in taking measures, as appropriate, to enhance climate change education, training, public awareness, public participation and public access to information, recognizing the importance of these steps with respect to enhancing actions under this Agreement (UNFCC 2015).

This charge is not new. The UN created the Higher Education Sustainability Initiative (HESI) in anticipation of COP 20. HESI also called for improvements in teaching, research, engagement, local efforts at universities, and information sharing (Sustainable Development.org). And in fact, an appeal to educational institutions can be found as far back as "Our Common Future," the report that led to the establishment of the first Earth Summit in Rio 1992 (World Commission on Environment and Development 1987). Universities are the world's best centers for research and knowledge production. Why do so many external to academia deem it necessary to encourage doing more to fight one of the world's gravest challenges? Shouldn't internal motivation drive such research?

Perhaps structural limitations within academia itself discourage this work. Scholars lack incentive to conduct serious research on the deeper social roots of climate change—for instance, research on the political implication of climate science or corporate influence on public opinion—since this type of work is more difficult to publish than research represented with quantitative modeling (Basken 2016).[2] The call for greater action may aim to deconstruct these limitations, and to open up new venues and incentives for such research. However, the possibilities for how colleges can engage with these issues is not exhausted by research, publishing, information sharing, teaching, and community outreach. Colleges offer fertile grounds for activist movements.

Recently over 800 independently operated chapters of the Fossil Free divestment movement have sprung up across the world, many on college campuses (FossilFree.org). Students, faculty, staff, and community members have launched campaigns to rid endowments of fossil fuel holdings. Some demonstrations, involving sit-ins, protests, arrests, and more have received a great deal of attention in the press. At COP 21 the Fossil Free campaign announced over \$3.4 trillion had been divested (350.org), and this number has continued to grow.

Institutes of higher learning have long been a center for many movements; for instance, civil rights, anti-capitalism, anti-war, apartheid, and anti-sweatshop. Recently, they have turned their attention to issues such as police violence and fossil fuel holdings. These issues all have something in common: the individual college or institution cannot improve much by acting on its own. In other words, these issues invoke collective responsibility—a philosophically contentious subject —because they arise from collective action. Such problems require a collective

[2]Perhaps Basken's focus on the hard sciences is too limited. Other fields, such as philosophy, may encounter fewer limitations to pursuing such research. For one instance, see Kovel 2007. However, it is true that IPCC's more recent 5th Assessment report is the first to address the moral implications of climate change by including two philosophers among its authors. So one problem may be that the IPCC does not pay enough attention to work that falls under the social sciences and the humanities.

rather than a private response since they do not arise from individual behavior, but from the social structures that influence such behavior. The response to collective action problems must be collective since no one agent has the power to transform structures in isolation. But why are college campuses so often at the epicenter of movements involving structural injustice?

The purpose of this article is to demonstrate that institutions of higher education play a unique role in combatting climate and other structural injustices, not only by conducting research and disseminating knowledge, but also by fostering a form of collective political responsibility. A philosophical analysis of different forms of collective responsibility, with specific attention to the Fossil Free divestment movement, reveals how social position facilitates this contribution more so in colleges than in other institutions. Section II examines three mutually reinforcing methods for understanding collective responsibility. Section III assesses these methods alongside climate change, a collective action problem that requires a political form of responsibility. Section IV argues that the fossil fuel divestment movement illustrates how colleges can leverage their unique social position to meet this responsibility. Section V concludes that college communities can do more to combat climate change by helping to foster this kind of activity on their campuses.

2 Three Approaches to Collective Responsibility

Most recognize that climate change demands a form of collective responsibility, but such a demand presents many philosophical difficulties.[3] First, collective responsibility challenges principles of justice. *That* climate change is harmful is no longer so controversial as *who* is responsible for its harms. While most agree obligations should fall upon state agents, identifying these state agents among others has been a contentious issue (Brown 1998; Caney 2010; Gardiner 2011; Maniates 2001; Shue 1993; Traxler 2002). Candidates include, for example, those with the most historical emissions (Brown 1998; Caney 2010; Schüssler 2011), those who can reduce emissions in the most cost-effective manner (Posner and Weisbach 2010), or those who exceed equal per capita emission allowances (Singer 2002). Second, according to many climate experts, states are not acting quickly enough.[4] Therefore, many question whether responsibilities fall upon other agents, such as local governments, corporations, other institutions, or even individual citizens. Such responsibilities may be direct, meaning agents should take actions to reduce their own emissions; or they may be indirect, meaning agents should pressure

[3]A number of problems arise with the concept of collective responsibility itself; however, they will not be discussed here. This paper assumes some form of collective responsibility is necessary to combat climate change; the present task is to determine the nature of this responsibility.

[4]James Hansen, former NASA scientist turned activist, has even called COP 21 in Paris a "hoax" (Milman 2015).

governments, which have more power to effect meaningful reductions through regulations and transnational agreements.

The scope of this paper includes neither principles of justly distributed cleanup burdens, nor the question of direct or indirect duties, but a more primary set of concerns: what is the nature of shared collective responsibility? There are at least two reasons to examine this question. First, the form of responsibility limits the shape its content might take; for instance, whether duties can even be direct or indirect, since form determines what kinds of collective agents are involved (i.e., whether responsibility is shared among collective agents). Since state agents are not acting quickly enough, this question becomes particularly important. Second, understanding the nature of collective responsibility will help identify the qualities colleges possess as collective agents that enable such responsibility, qualities that other institutions lack. Identifying this nature will help reveal how colleges are in a unique position to address climate change.

Philosophers have adopted at least three different general methods for describing the structure of collective responsibility and how it arises. Each of these methods is outlined below. The next section will assess each method in light of climate change.[5]

2.1 Group Agency and Individual Agency

The first approach examines the possibility and constitution of group agency. Presuming only agents can be morally responsible, collective responsibility must belong to collective agents. Many philosophers have outlined the criteria for collective agency. For the most part, these criteria mirror those for individual agency. Individual agents constitute a group agent when they share some combination of intentions, desires, and deliberative decision-making procedures (Gilbert 1992; List and Pettit 2011; Tuomela 2013). Collectives organized enough to possess collective responsibility are often referred to as "groups" to distinguish them from a "mere collective" of agents.

The concept of group agency is useful for several reasons. First, certain kinds of actions can only be taken by groups; for instance, a crowd at a sporting event performing the wave, or a team winning a football game. The wave cannot be performed alone just as a single player cannot win a game on her own (Isaacs 2011, p. 34). Second, group agency helps describe how people speak and feel about collectives such as political parties or companies as extra-individual entities (Held 1970, pp. 471–472). People say they disapprove of a political party's stance on an important social issue, or they are angry with *BP* for failing to implement an effective plan to stop a leaking oil rig. Finally, group agency helps avoid a "deficit

[5]These three approaches are discussed in greater detail, alongside the difficulties of sharing responsibility for collective action problems, in Godoy (forthcoming).

of responsibility," which may arise when no individual agent is responsible (List and Pettit 2011, p. 165). Companies can be penalized legally or morally (e.g., incur fines or blame) even when no single person can be found guilty of breaking a law.

2.2 *Practical Reasons for Holding Responsible*

The second approach questions the practical reasons for holding collectives responsible even when they fail to exhibit features of agency. This approach captures the idea that individual agents are able to, often do, and most importantly *should* band together to solve collective action problems. For instance, onlookers should work together to save a drowning swimmer in turbulent waters, perhaps by forming a human chain. A mere collective, even of random agents, has a responsibility to organize action when the result is favorable to doing nothing (e.g., letting the swimmer drown), and when the required action is "obvious to the reasonable person" (Held 1970, p. 476).

The same responsibility applies to established organizations that lack the prerequisites for group agency discussed above—for instance, decision-making procedures. As with small children, holding collectives responsible when they do not display the full signs of moral agency can be instructive and can even encourage them to develop such capacities (List and Pettit 2011, pp. 157, 193). Moral disapproval can encourage a company to develop proper offices and procedures for dealing with harms they may cause. Thus not just agency, but the *potential for* agency gives rise to responsibility.

2.3 *Social Connection and Structural Injustice*

The third approach grounds collective responsibility in neither collective agency nor its possibility alone, but in social position, or connection within a network of agency and collective projects that presume the interaction of many different agents (Young 2011). This form of responsibility is ideal for addressing what Young calls "structural injustice":

> Structural injustice … exists when social processes put large groups of persons under systematic threat of domination or deprivation of the means to develop and exercise their capacities, at the same time that these processes enable others to dominate or to have a wide range of opportunities for developing and exercising capacities available to them (p. 52).

The uncoordinated routine activities of many agents reproduce structural injustices; they are "a consequence of many individuals and institutions acting to pursue their particular goals and interests, for the most part within the limits of accepted rules and norms" (Young 2011, p. 52).

Climate change is a form of structural injustice. First, it results from routine, seemingly harmless activities in the domestic and economic spheres: e.g., travel, production, and investment. Second, climate change certainly puts large groups of people "under systematic threat of domination or deprivation" while also benefiting others. Those who have contributed least to climate change—the world's poor and those yet to be born—stand to suffer its worst consequences. The world's affluent also stands to benefit most from the conveniences of fossil fuel use—industrial production, and cheap transportation and energy. The world's poor are most vulnerable, facing droughts, famine, water shortages, rising sea levels, floods, and storms all with fewer resources to mitigate their effects. Finally, the world's affluent have become so in part because of their unrestricted ability to pollute the atmosphere with greenhouse gases, a byproduct of economic activity and growth.[6]

Structural harm differs qualitatively from harm that arises from individual actors (Young 2011, p. 44); even individual group agents. For instance, investigative journalists revealed that Exxon knew about the link between fossil fuels and climate change since at least 1977 (Banerjee et al. 2015). In this case, each individual member of Exxon may be guilty if she actively misled the public about how fossil fuels cause climate change. However, sometimes focus on individual offending agents alone is unhelpful. If one fails to question the structural factors that encourage "bad" behavior, then "bad" individual actions appear as anomalies rather than predictable byproducts. Indeed, a number of other energy companies shared information about the connection between fossil fuels and climate change (Banerjee 2015). Rather, preventing structural harm requires examining the various social structures that encourage or enable individual agents to cause it. How did so many multinational corporations actively mislead the public about their destructive practices in the pursuit of greater profits? Fossil fuel companies have a strong incentive to deny a link between their product and harm. The same was true of tobacco companies when smoking was linked to cancer. In fact, fossil fuel and tobacco companies confused the public about these links by using similar methods (Oreskes and Conway 2010).

Social connection responsibility is different from other forms of collective and individual responsibility in the following senses. First, responsibility for structural injustice corresponds to "social position," or place of privilege or disadvantage the agent occupies relative to other agents, institutions, and social processes (Young 2011, p. 56). Second, this responsibility is neither collective nor individual, but shared;[7] social position determines how it is shared. Finally, this responsibility aims

[6]Economic growth is almost always paired with increasing carbon dioxide levels. The exceptions are 2014 and 2015. The only other periods of declining carbon dioxide are associated with a slowing economy (International Energy Agency 2016).

[7]For further discussion of the distinction between collective and shared responsibility, see May 1992 (pp. 106–107) and Young 2011 (pp. 110–111).

to transform unjust structures, not hold individuals accountable for "bad" action, so it is met through collective action (Young 2011, p. 111). Agents are morally responsible because they participate in processes that lead to unjust outcomes. However, this responsibility is more than complicity in harm (Kutz 2007), since one does not meet it by withdrawing participation. Rather, meeting social connection responsibility requires examining one's social position and acting with others to transform unjust structures. In other words, this is a forward-looking, political form of responsibility as opposed to a retrospective, individualized form (Young 2011, pp. 105–112).

3 Assessing Methods

This section examines each of the above three models alongside climate change as a form of structural injustice. Each model offers some insight into why universities have responsibilities to address climate change; however, the social connection model explains the role that the social position of colleges plays in giving a unique shape to these responsibilities relative to other institutions.

3.1 Isolated Agents?

Most universities qualify as group agents. As such, they are appropriate subjects of collective responsibility according to the group agent approach. Failure to address responsibilities may warrant condemnation, blame, shame, or other responses from the public. However, this approach tends to individualize responsibility.

For instance, philosophers disagree over whether or how collective responsibility distributes among group members (Cooper 1968; Feinberg 1968; List and Pettit 2011; May 1998; Young 2011), though most agree "judgments assessed on members of the collectivity do not follow necessarily from judgments of collective blame" (French 1998, p. 25). In other words, an employee of ExxonMobil cannot apologize on behalf of the company for lying about its decades-long awareness of the connection between fossil fuels and global warming unless she is authorized to do so (List and Pettit 2011, pp. 35–36). Likewise a citizen cannot apologize on behalf of his state for failing to ratify an international climate treaty. Group agency itself says nothing of how responsibility disseminates among agents: among group members, between group agents, and between members and non-members. Group agency concentrates responsibility within atomized agents (groups or individuals); by doing so, it isolates agency and fails to explain how sharing responsibility for climate change is possible.

Yet since climate change is a form of structural injustice, individuals, group agents, and states all need to cooperate to avert further harmful climate change.

One college divesting of fossil fuels may make headlines, but will not affect much on its own. Furthermore, if only the actions or inactions of individual agents are considered, then the significance of such actions is lost in the vast sea of many other acting agents. Perhaps such actions have *communicative power*—may encourage others to act (Hourdequin 2011)—but not necessarily. Responsibility for structural injustice must be shared among agents.

3.2 *Obvious and Reasonable?*

The second approach addresses some of the issues discussed above. First, it accords with moral intuitions about the obligation to solve problems by working together, constructing new group agents, even spur of the moment, when doing so produces morally preferable outcomes. This approach also encourages agents to further develop capacities to address new moral situations. The first step many schools take when considering divestment is to assemble a working group or committee to assess investments. While committees on socially responsible investment and sustainability have been common features of contemporary universities, new issues such as climate change and divestment strengthen interest in them.

However this approach becomes difficult when we consider problems on a global scale. The "obvious and reasonable" criteria may be clear in the case of a drowning swimmer, but as problems become more complex so do solutions. Agents should act with others to form new groups, but with whom they should be acting is unclear. Social connection adds to the practical reasons approach by giving agents parameters of reasoning about their action corresponding to social position.

3.3 *Connected How?*

The social connection approach holds that agents share a responsibility relative to their social position within the structures that lead to a warming climate. Like the practical reasons approach, this responsibility may entail developing new groups and structures. It may also ask group agents to draw from their position. But both the subject and content of responsibility—who must do what?—is still quite vague. To combat this, Young (2011) suggests four parameters by which to locate and distribute responsibility based on an agent's position within the network: privilege, collective ability, interest, and power (pp. 142–151). The next section examines how these parameters describe the social positions of universities, and how the fossil fuel divestment movement illustrates a form of this political responsibility by leveraging social position to influence change.

4 Social Connection Responsibility on College Campuses

Because universities occupy similar social positions relative to one another, and because they are able to draw upon similar resources afforded by their social position, universities are candidates for sharing responsibility. Below each parameter is examined with reference to the divestment movement.

4.1 Privilege

Endowment-holding institutions reap profits from harmful investments, yet as institutions they avoid the terrible consequences, such as suffering and death, which befall actual people. This places them in a relative position of privilege. By forgoing this privilege, colleges prove such holdings unnecessary for financial health. Structural injustice arises from the routine (in this case, stockholding). Many think it immoral to benefit from the harms of others. Divesting colleges uphold this principle when they surrender their privilege to draw benefits from fossil fuel investments. Social connection responsibility requires forfeiting this privilege in order to challenge the presumption that any agent should benefit from harmful practices. Sharing responsibility entails encouraging others to divest as well. Colleges send a public message to other colleges and stockholders when they join the Fossil Free movement: that they believe divestment is both feasible and good.

4.2 Collective Ability

The divestment movement targets all endowed institutions, which explains why many non-profit and government institutions are also involved. However, colleges are unique in their collective ability. Their decision-making mechanisms, more so than other institutions, often encourage the entire community to participate democratically in issues of governance. Student and faculty governing bodies are often represented at various levels of the institution, even at the board level. These bodies hold decision-makers accountable. They create new directives and committees, such as socially responsible investment committees. It has not been uncommon for board members, students, faculty, and staff to work together on divesting campuses. Compare this structure with that found in the corporate or political world. Employees often cannot criticize their superiors without fear of reprisal or loss of employment. Colleges also care a great deal about community development, ensuring that all students have both a place and a voice in the affairs of the community.[8]

4.3 Interest

"Sometimes agents' interests coincide with the responsibility for justice" (Young 2011, p. 145). Colleges declare their purpose or interest through mission statements. These statements often make reference to increasing public welfare, benefiting society on a national and a global scale, improving the world for future generations, etc.; and more recently, colleges have pledged an interest in sustainability. Many Fossil Free chapters have leveraged their university's mission statements in their arguments to divest. It is difficult to commit to future welfare while engaging in practices that work against it. Finally, colleges have an interest in the truth. The links between burning fossil fuels and climate change, and between profitability and shareholder support, are undeniable.

4.4 Power

Both individual agents within colleges and colleges as collective agents themselves possess unique powers to affect unjust structures merely by provoking debates about them. An organized student group has a great deal of power to influence its university's decision-making processes. First, unlike employees at other institutions, students cannot be fired. Second, students are stakeholders in the reputations of the college; their diploma will bear its mark for their lifetime. Third, it reflects poorly on schools to expel students engaged in peaceful political demonstration. Tenured faculty may have similar power, especially en masse.[9]

While students and faculty possess power within universities, universities themselves hold external power to influence public discourse. Their reputation as centers of knowledge and learning carries intellectual weight both within academia and in public culture at large. Each additional divesting university amplifies the influence of the movement. All endowment-holding universities have the power to divest. By exerting it, they hope to influence the political and economic structures that enable climate change.

[8]This is at least true in principle. The recent Black Lives Matter movement raises questions about the place and voice of non-white students at American universities. Yet the movement strategically leverages this supposed principle of inclusion in order to help make its point.

[9]However, academic freedom is a value that is not equally admired everywhere.

5 Conclusion

This paper began with the claim that universities should be doing more to meet their collective responsibility to combat climate. Three different methods for describing collective responsibility were examined. The social connection model builds upon the group agent and practical reasons methods to highlight how responsibility arises, not in a vacuum, but in a complex network of social relations. Since climate change is a structural injustice, it too arises within a network of routine interactions. Meeting social connection responsibility entails leveraging one's position and working with other agents to transform the structures that encourage climate change. It follows that colleges could foster social connection responsibility in the form of campus discussion, activism, and movement building.

Note this paper makes no claim that divestment is best method for addressing climate change. This would require further studies. Rather, since climate change requires collective political action, colleges can help develop the form of responsibility required to spur positive social change by fostering these types of discussions and engaging with climate movements such as Fossil Free.

Campuses have historically housed a number of movements exercising political responsibility. Take for instance the South African divestment movement. The movement against apartheid was not born of strategy, but dissatisfaction with the governing structures that failed to adequately address injustice (Seidman 2015, p. 1017). Divestment began as an act of moral hand-washing, a communicative act of disapproval. College campuses played a significant role by provoking debates about the moral status of maintaining economic ties to apartheid (Seidman 2015, p. 1023). The seat of the movement's power was not the actual financial effect of divesting colleges, but the power to question—to politicize and moralize—the economic structures that allowed institutions to profit from apartheid. In other words, students and universities drew from their position to leverage social change, change which was only possible by sharing responsibility with other universities and institutions.

Laboratories and classrooms are therefore not the university's only resources for confronting climate change. Challenging structural injustice requires questioning the routines that sustain harmful practices. This can be controversial and depending on the political climate, even dangerous. Colleges can encourage members to engage in the necessary contentious debates, critical self-scrutiny, and experimental reimagining of harmful commonplace habits—and they can pledge to protect those members who do so. They can allow student and faculty representation in decision-making processes and provide information about their investments and practices necessary for informed decisions. The university community itself can become a lab and classroom, and colleges can lead the way instigating the cultural-political transformations necessary for confronting climate change.

References

350.org (n.d.) Divestment commitments pass the $3.4 trillion mark at COP21: 500 + institutions have now joined the movement, campaigners urge governments to follow suit. Available at: http://350.org/cop21-divestment/. Accessed 17 April 2016.
Banerjee N (2015) Exxon's oil industry peers knew about climate dangers in the 1970s, too. Inside Climate News. 22 December. Available at: http://insideclimatenews.org/news/22122015/exxon-mobil-oil-industry-peers-knew-about-climate-change-dangers-1970s-american-petroleum-institute-api-shell-chevron-texaco. Accessed 3 February 2016.
Banerjee N, Song L, Hasemyer D (2015) Exxon's own research confirmed fossil fuels' role in global warming decades ago. Inside Climate News. 16 September. Available at: http://insideclimatenews.org/news/15092015/Exxons-own-research-confirmed-fossil-fuels-role-in-global-warming. Accessed 3 February 2016.
Basken P (2016) On climate change, are university researchers making a difference? Chronicle of Higher Education. 27 January. Available at: http://chronicle.com/article/On-Climate-Change-Are/235035. Access 27 January 2016.
Brown, D (1998) The need to face conflicts between rich and poor nations to solve global environmental problems. In Westra L, Werhane P (ed) The Business of Consumption: Environmental Ethics and the Global Economy. Lanham, MD: Rowman & Littlefield, pp 31–46.
Caney, S (2010) Climate change and the duties of the advantaged. *Critical Review of International Social and Political Philosophy* 13(1):203–28.
Cooper, D (1968) Collective responsibility. *Philosophy* 43(165):258–68, doi:10.1017/S0031819100009220.
Feinberg J (1968) Collective responsibility. *Journal of Philosophy* 65(21):674–88.
Fossil Free (n.d.) Website. Available at: http://gofossilfree.org. Accessed 11 March 2015.
French P, ed (1998) Individual and Collective Responsibility. Rochester, VT: Schenkman Books.
Gardiner SM (2011) A Perfect Moral Storm: The Ethical Tragedy of Climate Change. New York, NY: Oxford University Press.
Gilbert, M (1992) On Social Facts. Princeton, NJ: Princeton University Press.
Godoy, ES (forthcoming) Sharing responsibility to divest from fossil fuels. Environmental Values.
Held V (1970) Can a random collection of individuals be morally responsible? *Journal of Philosophy* 67(14):471–481.
Hourdequin M (2011) Climate change and individual responsibility: a reply to Johnson. Environmental Values 20(2):157–162. doi:10.3197/096327111X12997574391643.
International Energy Agency (2016) Decoupling of global emissions and economic growth confirmed. (Press Release) 16 March. Available at:http://www.iea.org/newsroomandevents/pressreleases/2016/March/decoupling-of-global-emissions-and-economic-growth-confirmed.html. Accessed 16 March 2016.
Isaacs T (2011) Moral Responsibility in Collective Contexts. Oxford: Oxford University Press.
Kovel J (2007) The Enemy of Nature: The End of Capitalism or the End of the World? New York, NY: Zed Books.
Kutz C (2007) Complicity: Ethics and Law for a Collective Age. Cambridge, UK: Cambridge University Press.
List C, Pettit P (2011) Group Agency: The Possibility, Design, and Status of Corporate Agents. Oxford: Oxford University Press.
Maniates M (2001) Individualization: plant a tree, buy a bike, save the world? *Global Environmental Politics* 1(3):31–52. doi:10.1162/152638001316881395.
May L (1992) Sharing Responsibility. Chicago: University of Chicago Press.
Milman O (2015) James Hansen, father of climate change awareness, calls Paris talks 'a fraud.' The Guardian. 12 December. Available at: http://www.theguardian.com/environment/2015/dec/12/james-hansen-climate-change-paris-talks-fraud. Accessed 14 January 2016.

Oreskes N, Conway E (2010) Merchants of Doubt: How a Handful of Scientists Obscured the Truth on Issues from Tobacco Smoke to Global Warming. New York: Bloomsbury.
Posner EA, Weisbach D (2010) Climate Change Justice. Princeton, NJ: Princeton University Press.
Schüssler R (2011) Climate justice: a question of historic responsibility? *Journal of Global Ethics* 7(3):261–278. doi:10.1080/17449626.2011.635682.
Seidman GW (2015) Divestment dynamics: mobilizing, shaming, and changing the rules. *Social Research* 82(4):1015–1037.
Singer P (2002) One World: The Ethics of Globalization. New Haven, CT: Yale University Press.
Shue H (1993) Subsistence emissions and luxury emissions. *Law & Policy* 15(1):39–60. doi:10.1111/j.1467-9930.1993.tb00093.x.
Sustainable Development.org (n.d.) (Website) Available at: https://sustainabledevelopment.un.org/sdinaction/hesi/climatechangeaction. Access April 5, 2016.
Traxler M (2002) Fair chore division for climate change. *Social Theory and Practice* 28(1):101–134.
Tuomela R (2013) Social Ontology: Collective Intentionality and Group Agents. New York, NY: Oxford University Press.
UNFCC (2015) Adoption of the Paris Agreement. 12 December. Paris. Available at: https://unfccc.int/resource/docs/2015/cop21/eng/l09r01.pdf. Accessed April 4, 2016.
World Commission on Environment and Development (1987) Our Common Future. Geneva: Oxford University Press.
Young, IM (2011) Responsibility for Justice. New York, NY: Oxford University Press.

Philippine Higher Education Institutions' Responses to Climate Change

Jocelyn C. Cuaresma

Abstract The paper presents an analysis of higher education institutions (HEIs) as an important sector in addressing climate change issues through carrying out their intrinsic functions of instruction, research and capacity building. The huge amount of resources, expertise and knowledge in the hands of HEIs, particularly state universities and colleges (SUCs) strongly suggest the able fulfillment of their mandate under the Climate Change Act of 2009. A desk analysis of information gathered from websites, government budgets and related resources on 115 SUCs and selected private HEIs show an encouraging amount of responses in terms of integrating climate change issues into curricula, researches, capacity building, community engagement and external linkages, with some HEIs being more accomplished than others. However, efforts remain uncoordinated, and urgency in achieving commitments is absent. Much remains to be accomplished in asserting the role of HEIs in addressing climate change issues and for HEIs themselves to monitor their own pledges to Mother Earth.

Keywords Philippine state universities and colleges · Higher education institutions · Climate change

1 Introduction

1.1 Rationale

The global race to development has brought about a much improved living conditions for peoples and nations at the expense of the natural environment (Brundtland Report 1987, p. 61). The Philippines, still categorized as a developing

J.C. Cuaresma (✉)
National College of Public Administration and Governance, University of the Philippines, Diliman Campus, Quezon City, National Capital Region 1101, Philippines
e-mail: joycepcc@yahoo.com

W. Leal Filho (ed.), *Climate Change Research at Universities*,
DOI 10.1007/978-3-319-58214-6_5

country, has its own contributions to climate degradation at the same time that it suffers from climate change (Crepin 2013: 14, p. 1). Higher education institutions (HEIs) including state universities and colleges (SUCs) have a role to play (Boyd and Hume 2015: 2; Everett and Copeland November 2007; ACUPCC 2012) in educating and capacitating people on disaster risk reduction management (DRRM) and climate change mitigation and adaptation (CCMA), and in developing strategies to adapt to climate risks and mitigate greenhouse gas emissions. The National Climate Change Action Plan (NCCAP) 2011–2028, the government's climate reform agenda, considers capacity building and managing change as one of three critical pillars for scaling up climate actions. Clearly, the higher education sector can take the lead (ACCO 2011: 4) in what the latter intrinsically performs: capacity building through building skills and knowledge and rising awareness on climate change.

This paper examines the responses of higher education institutions (HEIs), with focus on state universities and colleges (SUCs), to fulfill their mandate in mainstreaming climate change (CC) concerns through the delivery of academic, research, capacity building and extension services. The data gathering and analysis was guided by the following questions:

1. What is the policy mandate of SUCs/HEIs in climate change?
2. How much budget subsidy is allotted to SUCs?
3. What are the responses and initiatives of SUCs/HEIs in terms of the following?
    a. Establishment of CCMA facilities
    b. Curriculum and Degree program offerings
    c. Non-degree and/or training program offerings
    d. Researches
    e. Linkages and other CCMA-related engagements.
4. How may HEIs be supported in their efforts to contribute in promoting the sustainability of the climate and our community?

The paper seeks to contribute towards efforts at giving recognition to the intrinsic role of SUCs/HEIs (ACUPCC 2012: 4) as educating and learning institutions. Considering the huge amount of expertise and resources in the hands of SUCs/HEIs, this paper asserts the leadership position that the higher education sector (Molthan-Hill 2015: 6) should take in climate change mitigation and adaptation through ways that they do best.

The paper has five parts: this overview, an analysis of the size of SUCs in terms of the subsidy they receive from the national government, an articulation of the policy mandate of HEIs, a profile and analysis of HEI responses, and the findings, recommendations and conclusions.

1.2 Climate Adaptation and Mitigation: Definitions

The identification of HEI responses is guided by two issuances: the Department of Budget and Management (DBM) and the Climate Change Commission (CCC) Joint Memo Circular No. 2013-01,[1] dated December 27, 2013, and the Commission on Higher Education (CHED) guidelines on climate change expenditure tagging (www.ched.gov.ph). The two issuances defined adaptation and mitigation and provided some examples, to wit:

- Climate change adaptation—activities that tend to reduce the exposure and vulnerability of human or natural systems to the impacts of climate change and climate-related risks. Such activities should increase adaptive capacity and resilience to climate variability (CV). Examples of adaptation measures are:
  - Adoption of CC and CV resilient design standards.
  - Conduct of risk assessment studies and mapping.
  - Social protection for agriculture and fisheries.
  - Programs to reduce land degradation.
  - Reforestation programs.
  - Development and dissemination of climate resilient crop varieties and farming techniques.
  - Investment in early warning systems.
  - Rehabilitation of watersheds.
  - Training and education on CC and CV issues.
- Climate change mitigation—refers to activities that aim to reduce greenhouse gas emissions (GHG), directly or indirectly. Examples are:
  - Renewable energy projects.
  - Projects to improve energy efficiency and raise GHG sequestration.
  - Improved forest management.
  - Improved transport systems.
  - Protection of carbon sinks.
  - Manure management and methane capture in livestock.
  - Capacity building to inventory GHG emissions and carbon footprint.

1.3 Methodology and Limitations

Data are collected through a desk survey of responses from websites of 115 SUCs and some private HEIs related to the mandate of mainstreaming CCMA concerns in

[1]This is amended by JMC No. 2015-01, dated March 24, 2015, Guidelines in Tagging/Tracking Government Expenditures for Climate change in the Budget Process.

the delivery of academic, research, and extension services. It is recognized that some HEI websites are not fully developed and may not contain the needed information. Website information is supplemented with data from government documents, national budgets, and internet-based resources. Indeed, the desk survey yielded a significant amount of information on SUC/HEI initiatives in climate change. Initial talk with a staff from an established climate change center (the University of the Philippines-Marine Science Institute) and brief inquiry through email to the Isabela State University about ISU's Climate Change Center helped generate detailed information and validate findings.

The paper adds to the growing literature on HEIs' accomplishments and contributions to toward climate change. Taking off from the suggestion that SUCs/HEIs are generally untapped and underutilized (ACUPCC 2012) amidst the huge amount of technical expertise, knowhow and resources at their disposal, the paper put forwards possible actions that may cause changes in the current state of affairs. The findings are useful to academics, policy makers as well as the local community, who are all concerned with sustainability and linking resources and expertise to address collective concerns.

Given constraints in time and resources, the volume of responses gathered and the huge number of HEIs prevented a categorization and weighing of their significance, but the amount of information gathered suffice in meeting the main purpose of this paper. The results of the current study shall be used to inform a follow-up study on the impact of responses of selected SUCs to climate change. The relatively large SUCs, based on the size of their budget subsidy from the national government, have uploaded leading information on their respective climate change activities.

1.4 Number of SUCs/HEIs Nationwide

An HEI is an institution of higher learning that primarily offers degree programs as well as engages in research and extension services in support of its teaching function (www.ched.gov.ph). HEIs have been in existence in the Philippines around the early 1600s (e.g., San Jose Seminary in 1601 [www.sanjoseseminary.org]; the University of Santo Tomas in 1611 [www.ust.edu.ph]) and created primarily to promote a highly educated society. There are at least 5 types of HEIs.

1. State University/College (SUC)
2. Local University and College (LUC)
3. CHED Supervised Institution (CSI)
4. Special Public HEIs (SP)
5. Private HEIs.

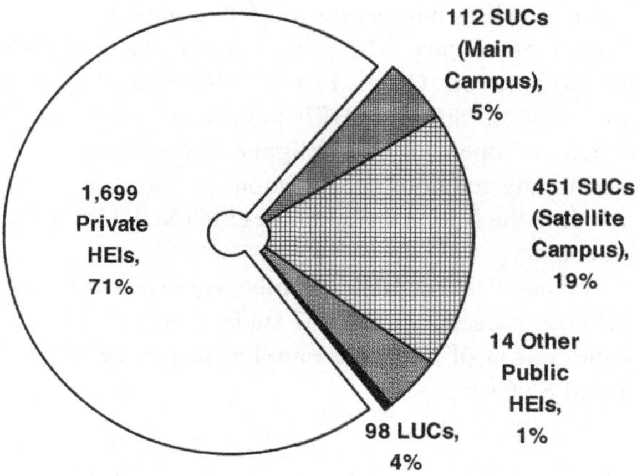

Fig. 1 Types of HEIs, Philippines as of September 2014

The Philippines has a total of 2374 HEIs[2] (see Fig. 1), of which 675 are public or government owned. If equally distributed in the country's 81 provinces, each province will have an average of 29 HEIs, suggesting the abundance of centers of knowledge and excellence in various academic fields such as the natural sciences, biotechnology, engineering, earth science, agriculture, forestry, fisheries, architecture, and land use planning, among others.

State universities and colleges (SUCs) refer to "public higher education institutions established by law and administered and subsidized by government". As of September 2014, SUCs total 112 (or 554 to include both the main and satellite campuses). Every SUC has its own charter, a policy making body called Board of Regents (BOR) in a state university, or Board of Trustees (BOT) in a state college. The Chair of the CHED chairs all the boards of SUCs. (www.ched.gov.ph accessed on April 16, 2016).

Local government units (LGUs) may establish educational institutions through a council resolution or ordinance. As of 2015, the country has 95 local universities and colleges (LUCs). A CHED-supervised institution (CSI) is a non-chartered, public, post-secondary education institution, similarly established by law, and administered, supervised and financially supported by the government. Five special public HEIs have been established to offer higher education and specialized programs to meet the training needs of the government service, namely the Philippine Military Academy, Philippine National Police Academy, the Philippine Public Safety College, the National Defense College, and the Development Academy of the Philippines (DAP).

[2]Source of HEI data: http://www.ched.gov.ph/wp-content/uploads/2014/temp/10-03/home/Higher%20Education%20Data%202014%20-%20Public%20and%20Private%20HEIs.pdf. The 2016 GAA lists 115 SUCs Main Campus (www.dbm.gov.ph).

Private HEIs numbering to 1699 comprise 71.6% of tertiary education institutions in the country. The private sector may establish HEIs subject to the rules of the Corporation Code. Private HEIs are largely classified into sectarian or non-sectarian. Sectarian HEIs are usually non-stock, non-profit, duly incorporated, owned and operated by a religious organization. Non-sectarian HEIs are not affiliated to any religious organization. (www.ched.gov.ph). By major regions, Luzon including the National Capital Region (NCR) is the location of 59.1% of all HEIs in the country.

The huge size of HEIs is further represented by the student population. In 2012, enrollment reached 3,033,967 students in all HEIs nationwide (www.ched.gov.ph). Some 57.7% of them are enrolled in private HEIs; the rest entered government owned-SUCs.

2 Budget Subsidy to SUCs

For 2016, the national government budget subsidy to SUCs amounted to Php47.415 billion, equivalent to a 2.2% share in the national budget or the General Appropriations Act (GAA) (see Fig. 2).

Although a doubling of the total SUC subsidy over the last six years is observed, the significance of the amount changes at the level of individual SUCs, where some SUCs come out very large, and the others, small. The University of the Philippines (UP) System, the country's national university, receives the biggest subsidy of Php11.81 billion in 2016 (Fig. 3). Far second is the Mindanao State University (MSU), with Php2.176 billion, and third, the Polytechnic University of the Philippines (PUP), with a Php1.64 billion subsidy (2016 GAA, www.dbm.gov.ph).

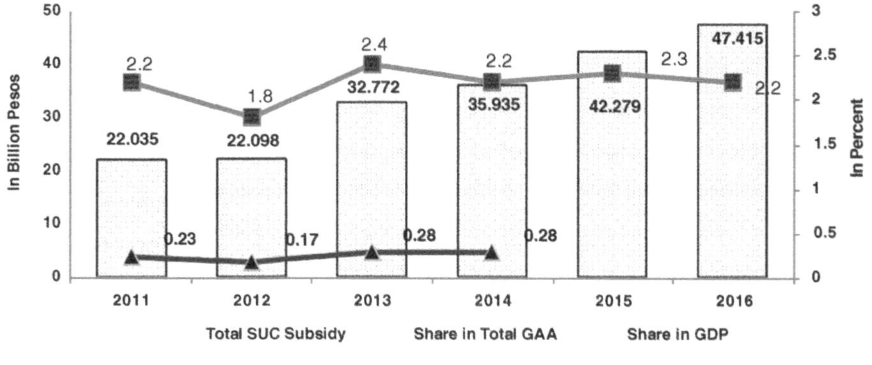

Fig. 2 SUC budget subsidy from the national government, share to total GAA, and share to GDP

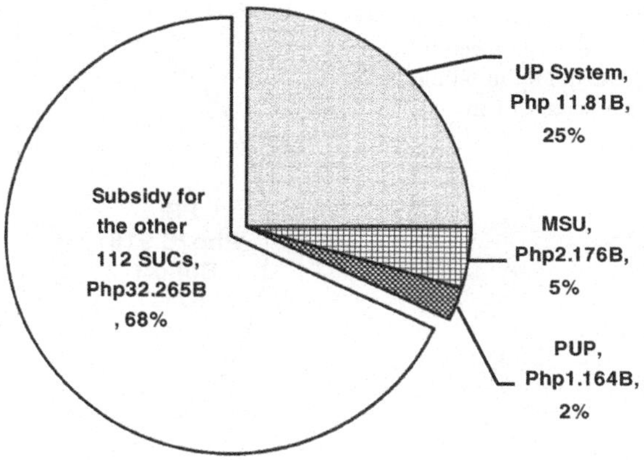

Fig. 3 Three SUCs with the biggest subsidy in 2016, in billion pesos and percent share

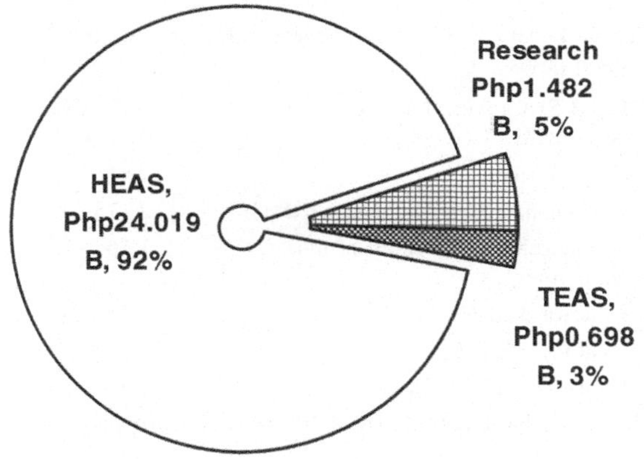

Fig. 4 Distribution of SUC subsidy, by program area, in billion pesos and percent share

By program area, the bulk of the 2016 SUC subsidy goes to Higher and Advanced Education Services (HAES), particularly to Personal Services (PS) in the form of salaries and allowances for teachers (Figs. 4 and 5). The SUC budget subsidy for Research Services (RS) equals only to 3.16% of the total SUC budget, while the share of the SUC budget for Technical, Extension and Advisory Services (TEAS) is smaller at 1.47%. This suggests the limited amount of resources at the disposal of majority of SUCs for research, extension and community services.

By object of expenditure (Fig. 5), 50.7% of the subsidy is allotted for Personal Services (PS), and 24% for maintenance and other operating expenses (MOOE). The Capital Outlay (CO) allocation is smaller at 18.9%. SUCs are authorized to raise their own income and generate additional resources to support their operations. Tracking the amount dedicated for disaster management and CCMA programs and activities is difficult since this is not yet tagged in the individual SUC budget. The CHED has issued CC Expenditures Tagging Guidelines (CCETG) in compliance with the DBM-CCC-Joint Memo Circular of December 27, 2013. Tracking CC expenditures of SUCs is highly difficult at this time.

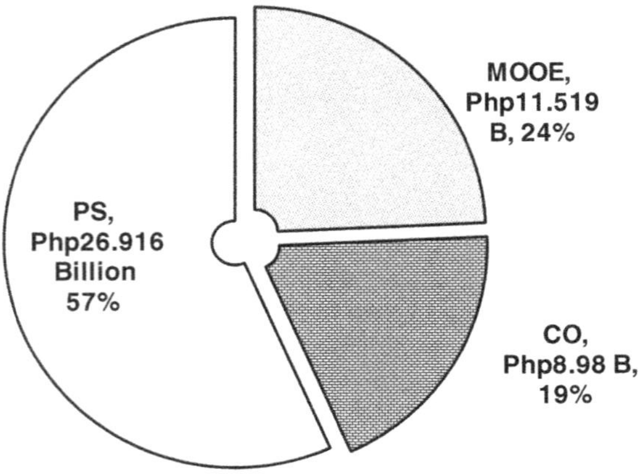

Fig. 5 Distribution of SUC subsidy, by object of expenditure, in billion pesos and percent share

Table 1 Top 3 SUCs with biggest budget subsidy from the national government, 2016 GAA, in billion pesos

Top 3 SUCs with the biggest subsidy (out of 115 SUCs)	Total subsidy, 2016 GAA	HAES	RS	TEAS
1. UP System	11.810	4.863	0.557	0.230
2. Mindanao State University (MSU)	2.176	1.496	0.058	0.022
3. Polytechnic University of the Philippines	1.164	0.585	0.014	0.009
Total for 3 SUCs	15.15	6.944	0.629	0.261
Total for 25 SUCs	28.084	13.723	1.079	0.483
Total—all 115 SUCS	47.415	24.019	1.482	0.698

Source of basic data: 2016 GAA, www.dbm.gov.ph

Except for the UP System, the budget subsidy for RS and TEAS becomes smaller for the remaining 114 SUCs. MSU, for instance, only has Php58 million subsidy for research and Php22 million for technical services. All other SUCs will have to deliver services with smaller budgets (see Table 1).

The 2016 national budget gives the total number of plantilla items in SUCs at 63,302 requiring a salary bill of Php25.251 billion (Figs. 6 and 7). Of the total number, 550 are items for SUC top executives. A total of 57,141 items (or 90.3%) are occupied. One implication is that savings of Php6.529 billion are expected from vacant plantilla items.

3 HEI/SUC Mandate in Climate Change

The Philippine government performed well in putting in place the policy framework on DRRM/CCMA in accordance with its commitments under the United Nations Framework Convention on Climate Change (UNFCCC), the Kyoto Protocol, and

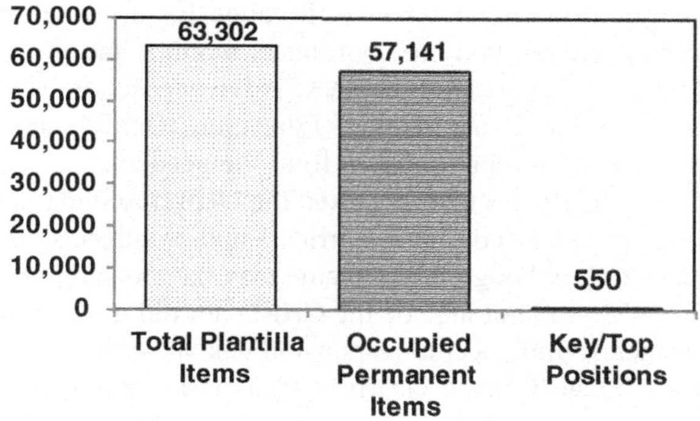

Fig. 6 SUC staffing summary, 2016 GAA

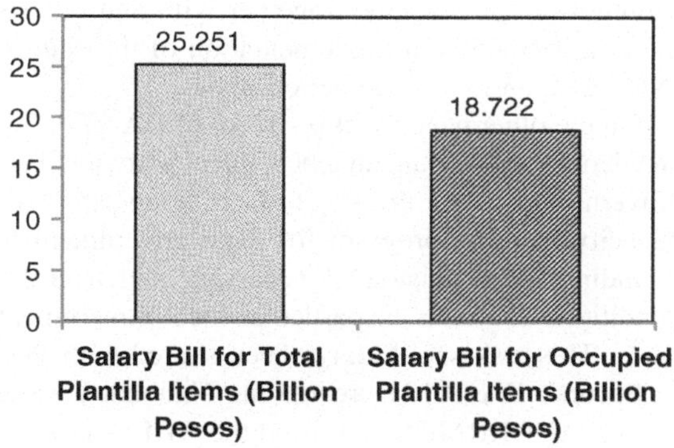

Fig. 7 SUC salary bill, 2016 GAA, in billion pesos

the 2015 United Nations Climate Change Conference (COP21) (Nachmany et al. 2014: 35, 450; Nachmany et al. 2015a, b). The government committed itself to continuously develop and strengthen capacity and capability in addressing the adverse impacts of climate change through financial support, technology development and transfer and capacity building (Philippines, INDC, October 2015).

The paper asserts the inherent role of SUCs/HEIs in the conduct of tertiary education, research and extension services and all aspects of governance (ACUPCC 2012). Their presence in the region and in every province of the country makes them indispensable partners in various aspects of climate change and environmental governance. This part of the paper examines the policy directives on SUCs/HEIs in consideration of CCMA matters in the administration of academic, research and extension programs.

From the 1970s to 2016, the government has issued at least 21 laws and plans on or related to CCMA. Eight of these laws are summarized in Table 2, with emphasis on the role of higher education sector. A review of three most important legislations on climate change shows a hesitant reference to the tertiary education sector. First, RA 9729 (Climate Change Act of 2009) mandated agencies to mainstream climate change into government policy formulations, establish the framework, strategy and program on climate change. It called for the creation of the CCC to be the sole policy-making body to coordinate, monitor and evaluate climate change programs and action plans, and the CCC Advisory Board. The membership of the CCCAB includes the Department of Education (DepEd, representing the basic education sector) and a representative from the academe.

RA 9729 specifically tasked the DepEd to integrate climate change into primary and secondary education curricula and/or subjects. The CHED is not a member of the CCC, although the academe may have one representative. It is unexpected that RA 9729 did not include the CHED nor did it mention SUCs/HEIs to be among the agencies with specific responsibilities since the Framework Strategy and Program on Climate Change clearly assigns tasks within the expertise of the SUCs/HEIs such as Research and Development, Academic Programs, Capability Building and Mainstreaming. This notwithstanding, RA 9729 (Section 16) stated that the contribution of the academe, together with and other stakeholder groups is needed in the development and implementation of the National Climate Change Action Plan (NCCAP), and the local action plans.

On the other hand, Section 15(b) of RA 9729, assigned to the Local Government Academy (LGA), an attached agency to the Department of Interior and Local Government, the tasks of facilitating the development and provision of capacity-building program for local government units (LGUs) in climate change, including the provision of necessary and appropriate technology, with focus on conditions and risks of particular local governments, women and children in rural areas. This task should have been shared with SUCs/HEIs.

Second, RA 10121 created the National Disaster Risk Reduction Management Council (NDRRMC), and listed the CHED as well as DepEd, Technical Education and Skills Development Authority (TESDA), among others, as members, and tasked them "to integrate Disaster Risk Reduction Education into school curricula of secondary and tertiary level of education, whether private or public, formal and non-formal, technical vocational, into the National Service Training Program, in indigenous learning and out-of-school youth courses and programs, in Sangguniang Kabataan (SK) Programs and in mandatory training for government employees" (RA 10121).

Again, RA 10121 made no mention of SUCs/HEIs, but issued under Section 9(i) a direct mandate to the Department of National Defense-Office of Civil Defense (DND-OCD) to establish DRRM Training Institutes in suitable locations to train public and private individuals on DRRM. SUCs/HEIs are not mentioned as cooperating partners to share the task. SUCs/HEIs are also not listed among the responsible agencies/members of the Regional and Local DRRM councils as identified in Section 10 and 11 of RA 10121. The implication is that SUC/HEI efforts related to DRRM/CC may not be called upon to systematically share their knowledge and resources with the local community where they operate.

Third, it is in the crafting of the National Climate Change Action Plan (NCCAP) 2011–2028 in November 2011 where the higher education sector has the opportunity to provide much input. Three SUCs/HEIs were members in the cluster (there

Table 2 Selected laws and plans issued related to climate change

Date issued	Law, plan	Mandate of SUCs/HEIs
June 23, 1999	1. Republic Act (RA) 8749 Philippine Clean Air Act	The academe shall participate in the formulation and implementation of the Integrated Air Quality Improvement Framework and Air Quality Control Action Plan
June 25, 2004	2. Executive Order No. 320, designated the DENR as the National Authority for Clean Development Mechanism (CDM)	HEI/SUC presence is possible through CDM Technical Evaluation Committees (Merillo and GerardaAsuncion 2003)
October 23, 2009	3. RA 9729 Climate Change Act, created the Climate Change Commission (CCC) and the CCC Advisory Board	DepEd, as member of CCC, is tasked to integrate climate change in school curricula. CHED is not mentioned. The academe shall have one representative in the CCC
May 27, 2010	4. RA 10121 Disaster Risk Reduction and Management Act	CHED, DepEd, TESDA are members; tasked to integrate DRR education in school curricula
June 7, 2010	5. EO No. 888 on the adoption of the Strategic National Action Plan (SNAP) on DRR for 2009–2019	CHED and DepEd are assigned the primary responsibility in "education and research"
2011	6. National Climate Change Action Plan (NCCAP) 2011–2028 (required under RA 9729)	No mention of CHED; with presence of some HEIs
August 16, 2012	7. RA 10174 The People Survival Fund Act, creating the People's Survival Fund and the People's Survival Fund Board (PSFB)	The academe/scientific community is represented in the PSFB; the academe representative are prohibited from accessing the PSFB
December 27, 2013	8. DBM/CCC Joint Memo Circular No. 2013-01, on CC budget tracking	For general compliance of all government agencies

were 9 clusters[3]) on "Knowledge and Capacity Development Cluster" particularly the University of the Philippines-National College of Public Administration and Governance, the Development Academy of the Philippines, and the Ateneo School of Government (ASoG). In five other clusters, SUC/HEI presence is possible through the Department of Science and Technology (DOST)-Research and Development Councils. The CHED is not a member of any cluster.

In sum, SUCs/HEIs are given direct mandate on DRRM/CCAM, but their direct participation in policy making bodies is limited, with possible participation through the CHED and/or the DOST-R&D Councils. The implication is that SUCs/HEIs

[3]The 9 Clusters are: Food Security; Water Sufficiency; Human Security; Ecological and Environmental Stability; Climate-Smart Industries and Services; Sustainable Energy; Knowledge and Capacity Development; Finance; and Gender Mainstreaming.

may not be automatically called upon to participate in policymaking bodies, task forces or councils that make decisions and/or implement government DRRM/CCAM programs and projects. This limits the venue where SUCs/HEIs may share scientific knowledge with implementing agencies and the larger community.

4 Analysis of SUC/HEI Responses

Notwithstanding the thin policy reference to SUCs/HEIs and the limited budget support, many SUCs have initiated climate mitigation and climate adaptation measure in their respective jurisdictions.

4.1 Establishment of SUC/HEI Climate Change Centers and Related Facilities

More specifically, at least 21 SUCs and at least 3 private HEIs have established a Climate Change Center or similar facility in 10 regions nationwide (see Table 3). The climate change centers serve as academic centers offering degree programs as well as centers for research and extension services. The various centers are based in the colleges of science and engineering as well as in agriculture, environment and natural resources, and management. The centers offer laboratory courses for students, serve as hubs for research on climate change, but also are part of the SUC/HEI's facilities for capacity building and extension programs for the local and regional participants.

The rationale for creation and objectives of the SUC/HEI-based Climate Change Centers cover wide areas of concern, from local to national and beyond.

- The RCCRDC of *Visayas State University* aim "to undertake and promote ecological education, research and extension in the Visayas" (www.vsu.edu.ph).
- The *CLSU-Environmental Management Institute* seeks to lead and provide quality education, research and extension services on the causes and consequences of climate change (iccem.com.ph).
- The *UPLB-CCDRRSC* aims "to develop national and local climate change adaptation and mitigation strategies for a climate-resilient agriculture, environment, and natural resources sectors" (idsc.uplb.edu.ph, accessed on April 25, 2016).
- The *Benguet State University-CSAC* envisions itself to be a center for building capabilities for climate change mitigation and adaptation in the Cordillera Administrative Region. It seeks to spearhead the development and dissemination of climate-smart agriculture systems and technologies that highland farming

Table 3 SUCs that have established a climate change center and related facilities, and/or program

Region	Name of SUC	Name of climate change center	Year established and website
NCR	1. UP Diliman—Institute of Environmental Science and Meteorology	Environmental Pollution Studies Laboratory	March 27, 2003 http://www.iesm.upd.edu.ph
	2. UP Diliman	UP Solar Photovoltaics Laboratory, College of Engineering-Department of Electrical and Electronics Engineering	http://www.eee.upd.edu.ph/
	3. UP Los Baños (UPLB)	UPLB Interdisiciplinary Program on Climate change in 2007; which became the Climate Change and Disaster Risk Reduction Studies Center	2007; Became a Center in 2013 cidsc.uplb.edu.ph
	4. UPLB	Integrated Natural Resources and Environmental Management Center	cidsc.uplb.edu.ph
	5. UPLB-School of Environmental Science and Management	Climate Change and Disaster Risk Reduction Studies Center (CCDRRSC)	2007, cidsc.uplb.edu.ph
	6. UPLB	Biofuels Research Center	2006, cidsc.uplb.edu.ph
	7. UP Baguio	Knowledge and Training Resource Center for Climate Change and Disaster Risk Reduction	February 2013 http://cordillerastudies.upb.edu.ph/ktrc
NCR	8. Polytechnic University of the Philippines	Center for environmental Studies-Institute for Social Sciences and Development	2009, www.pup.edu.ph
CAR	9. Benguet State University	Climate Smart Agriculture Center (CSAC); DOE-Affiliated Nonconventional Energy Center	2013, www.bsu.edu.ph
I	10. Mariano Marcos State University, la Union	Climate Resilient Agriculture Center	2012, www.mmsu.edu.ph
II	11. Isabela State University	Climate Change Center-Education Research and Development (CCC-ERD)	2015, www.isu.edu.ph
III	12. Central Luzon State University (CLSU), Nueva Ecija	Environmental Management Institute, became the Institute for Climate Change and Environmental Management (ICCEM)	1997; became ICCEM in 2011
IV-B	13. Palawan State University	DOE-Affiliated Renewable Energy Center	2014, psu.palawan.edu.ph

(continued)

Table 3 (continued)

Region	Name of SUC	Name of climate change center	Year established and website
V	14. Bicol University, Albay	Bicol Extension Disaster Education Network (Bicol EDEN); Climate Change Academy	Launched on July 29, 2010, www.bicol-u.edu.ph
V	15. Catanduanes State University	Research Center for Weather and Climate Change	2015, www.catanduanesstateu.edu.ph
V	16. Central Bicol State U of Agriculture (CBSUA), Camarines Sur	Center for Disaster Risk Reduction Management and Climate change Adaptation	2010, cbsua.edu.ph
VII	17. Bohol Island State University	Regional Climate Change Program	2011, www.bisu.edu.ph
VIII	18. University of Eastern Philippines, Northern Samar	Center for Climate Change Adaptation and Disaster Risk Reduction in Samar Island; Center for Environmental Studies and Advocacy	2011, uep.edu.ph
VIII	19. Visayas State University, Baybay, Leyte	Regional Climate Change Research and Development Center (RCCRDC); Institute of Tropical Ecology	1998, www.vsu.edu.ph
XII	20. Sultan Kudarat State University	Innovation and Technology Center for Climate Change	2015, Based on a call for bidding announcement till June 28, 2016 (www.tendersontime.com)
Special SUC			
NCR	21. National Defense College of the Philippines (NDCP), formerly the National Defense College of the Armed Forces of the Philippines (NDCAFP)	Crisis Management Institute (formerly the Emergency Management Institute of the Philippines)	NDCP was established on May 11, 1973; the NDCAFP was created on August 12, 1963; www.ndcp.edu.ph
PRIVATE HEIs			
NCR	22. Ateneo de Manila University (AdMU)	Manila Observatory-Climate Change Assistance (Klima) helps developing countries in capacity building in order to adapt to climate change and fulfill their commitments to the UNFCC	1865, http://www.observatory.ph/climate-change-assistance-klima/
NCR	23. De La Salle University	Environmental Resources Management Center (ERMAC)	2001, www.dlsud.edu.ph/ermac.htm
NCR	24. Miriam College	Environmental Studies Institute offers degree programs (BS, MA, PHD) in environmental management/science	1926, www.mc.edu.ph

communities may use to adapt to climate change (bsu.edu.ph, accessed on April 25, 2016).

- The *Bicol University's (BU)-Bicol EDEN* utilizes education to build resilient communities. It now involves 40 institutions and agencies across the Bicol Region to improve the delivery of services to citizens affected by disasters. Its partnership with US EDEN is important in the information education communication campaign and in accessing technical assistance in disaster mitigation (bueden.url.ph/; www.bicol-u.edu.ph/index.php?p=news-1594).
- Likewise, *Bicol University* hosts the *Climate Change Academy*. The Academy is BU's important contribution to the national effort to fast track the development of climate adaptation capacities. It offers short term courses on DRRM/CCAM specifically tailored for the LGUs according to their needs (bueden.url.ph; www.bicol-u.edu.ph, accessed on April 25, 2016).
- The *National Defence College of the Philippines (NDCP)* established the *Crisis Management Institute*, primarily to offer training workshops on community level disaster risk management, media training on disaster reporting, training of military reservists and student-cadets (www.ndcp.edu.ph, accessed on April 26, 2016).
- The *Climate Change Center* delivers the *Climate Change Assistance (Klima) Program* of the Manila Observatory, a research institution hosted by the Ateneo de Manila University. The Klima Center is one of the many programs of the Manila Observatory, a facility that has been engaged in weather observation since 1865. Klima's mission is to help developing countries in capacity building in order to adapt to climate change and fulfill their commitments to the UNFCC (http://www.observatory.ph/climate-change-assistance-klima/, accessed on April 26, 2016).
- *Miriam College-Environmental Studies Institute* undertakes a number of public education and awareness campaign for the environment, such as the Integration of Solid Waste Management Tools into Specific Settings of European and Asian Communities; Solid Waste Reduction Master Plan, Radio Kalikasan (on air since 1991), Biak-na-Bato Conservation Program, Southern Sierra Madre Wildlife Conservation Center, and the Doña Remedios Trinidad Reforestation Project (www.mc.edu.ph/AdvocacyCenters/EnvironmentalStudiesInstitute.aspx).

4.2 Department of Energy's (DOE) Partnership with HEIs

The DOE partnered with HEIs since the 1990s to promote the utilization and commercialization of non-conventional or renewable energy technologies (www.doe.gov.ph). As of October 31, 2012, at least 18 HEIs/SUCs (Table 4) have been designated as DOE-Affiliated Non-conventional/Renewable Energy Centers (ANEC/AREC) (DOST-ITDI 2012) and became recipients of the Renewable Energy Trust Fund intended to fund R&D in renewable energy. Through the partnership, the HEIs carried out electrification projects for remote barangays where

Table 4 ANEC/AREC in SUCs/HEIs

Region	SUCs-ANEC in mini-hydro technology for rural electrification
CAR	1. Kalinga-Apayao State College
CAR	2. Kalinga State University, since 1996
CAR	3. Benguet State University
I	4. DMMMSU
II	5. Isabela State Univ
III	6. Central Luzon State University-Open University
IV-A	7. Cavite State University—AREC on Biogas technology
IV-B	8. Palawan State University
IV-B	9. Western Philippines University
VI	10. Central Philippine University, since 1995
VIII	11. Visayas State College of Agriculture
X	12. Central Mindanao University, Bukidnon
PRIVATE HEIs	
VII	13. Silliman University
	14. Tiburcio Tancinco Memorial Institute of Science of Technology (TTMIST)
CAR	15. Saint Louis University-College of Engineering, Baguio City
VII	16. University of San Carlos-Mechanical and Industrial Engineering Department, Cebu
X	17. Xavier University, Cagayan de Oro City—College of Agriculture; AREC on Cassava Ethanol Research
	18. Ateneo de Davao University—Center for Renewable Energy and Alternative Technologies (CREATE); Clean Air Youth Alliance (CAYA)

Source DOST-Industrial Technology Development Institute (ITDI). October 31, 2012. Renewable energy development desktop study for the Philippines. http://www.wpu.edu.ph; http://clsu.edu.ph

they introduced the use of renewable energy such as solar home system, micro-hydro generators, biogas system and dual fire green dryer. In the example of Palawan State University-AREC, it introduced the adoption of micro-hydroelectric system, solar energy, wind mill, and bio-gas in far-flung communities of Palawan Province (psu.palawan.edu.ph). The Kalinga State University-ANEC made access to electricity possible for 1942 households in the provinces of Kalinga and Apayao (www.kasc.edu.ph/index.php/affiliated-renewable-energy-center).

4.3 Degree and Non-degree Programs on DRRM/CCAM

The desk survey yielded at least 50 degree programs (bachelor, masters and PhD levels) related to DRRM/CCAM offered in 20 colleges in 10 SUCs/HEIs. Courses on or related to DRRM/CCAM are made part of courses/programs in colleges of science and engineering such as geology, marine biology, meteorology, environmental science, and marine and fishery science, among others. Increasingly, other academic fields have incorporated DRRM/CCAM in their degree programs such as those in land use, agriculture, forestry, air/water quality management as well as in development studies and development management programs. Some HEIs have offered master's programs on Health Emergency Disaster Management (Bicol University), Renewable energy Systems (CLSU). The AdMU offers a Master of Management Program on Climate Change.

In terms of non-degree programs on CC, the course offerings of the UP Open University- Faculty of Management and Development Studies (UPOU-FMDS), the NDCP and the AdMU-School of Government (ASoG) are notable.

- UPOU-FMDS offers the 10-week non-formal course (fmds.upou.edu.ph) on "Integrating Climate Change Adaptation and Disaster Risk Management in Policies, Plans and Investments toward Inclusive and Sustainable Agricultural and Rural Development" (ICCADRM4ISARD). The online course is supported by the Southeast Asian Regional Center for Graduate Study and Research in Agriculture (SEARCA) and UPOU Foundation, Inc. UPOU also offers a 5-module non-formal Online Course on Responding to Climate Risks in Agriculture and Natural Resource Management, which is now on its third offering.
- NDCP-Crisis Management Institute (http://www.ndcp.edu.ph/emip.php) offers training workshops on local/community level disaster risk management; media disaster reporting; training of military reservists and students cadets; managing technological hazards; and enhancing capacities in early warning for disasters. Other course offerings include:
  - 6th basic Disaster Management Training Course;
  - Pacific Disaster Internet Viewing Training;
  - 4th Advanced Disaster Management Training Course.
- ASoG offers the JVO Executive Education Program (http://www.ateneo.edu/asg/executive). This is a 3 module course on Environment Management designed for political leaders, civil servants and corporate managers and meant to foster cooperation between public and private stakeholders.

In addition to offering DRRM/CCAM-related courses, the UPLB stepped up in 2014 and launch 18 interdisciplinary studies centers/programs, which lean heavily on the generation of more science and technology-based solutions to address complex development problems including that of climate change. The UPLB adopted interdisciplinary and transdisciplinary approaches in research, development and extension programs in the fields of agriculture, engineering, environment and natural resources towards climate change as the development challenge/goal (idsc.uplb.edu.ph).

Some SUCs/HEIs in Mindanao have announced their decision to make DRRM/CCAM subjects compulsory in all courses as well as in the National Service Training Program, the Civic Welfare Training Service, and the Reserved Officers

Training Corps at the undergraduate level. (www.minda.gov.ph). The SUCs and a private HEI that have incorporated DRRM and CCAM in their bachelor's degree programs include:

- Davao Oriental State College of Science and Technology
- Southern Philippines Agri-Business and Marine and Aquatic School of Technology
- Notre Dame of Dadiangas University (private HEI).

Eight (8) more SUCs/HEIs have announced on May 30, 2014 to incorporate and make compulsory DRRM and CCAM in all degree programs. (http://www.mindanews.com/mailbox/2014/05/30/mindanao-state-universities-and-colleges-to-offer-drmm-and-climate-change-courses/ accessed on April 26, 2016):

- University of Southeastern Philippines,
- Compostela Valley State College,
- Davao del Norte State College,
- Sultan Kudarat State University,
- Mindanao State University-General Santos City Campus,
- University of Southern Mindanao,
- Cotabato Foundation College of Science and Technology, and the
- Cotabato City State Polytechnic College.

4.4 Research Programs on Climate Change

A random selection of HEI researches on or related to CCMA shows that SUCs/HEIs have developed an extensive wealth of relevant resources (Table 5) (Commission on Higher Education 2014). Some research are specific on a particular ecosystem, marine protected areas (i.e., the Island Garden City of Samal, Ligawasan Marsh), flora and fauna, mining in Compostela Valley, and to particular communities. The research activities are basically undertaken with external funding from the national government (CHED; Department of Agriculture), private organizations (OML) and foreign research and funding institutions (FAO, JICA, SEARCA).

Partnership among SUCs/HEIs, the private sector and the local community in the region and nationwide in the conduct of DRR/CC research programs. SUCs oriented towards the sciences, agriculture, natural resources, environment, land use and related fields have established research centers and institutes designed to conduct and develop new technologies to improve the crops quality and resiliency, food production and promote resiliency of crops and use of organic technologies.

Table 5 Examples of SUC/HEI research projects

SUC/HEI	Title of selected research projects
1. UP Los Baños	• Strengthening the Philippine institutional capacity to adapt to climate change (MDG-F 1656 project) cAssessing the impact of CCA in selected areas in Bicol region (UPLBFI-EEPSEA project) • Design and development of a vulnerability assessment tool: combining scientific and indigenous knowledge for the agricultural sector of Benguet and Ifugao
2. ISU-Higher Education Regional Research Center	• Bicol climate change resiliency of farming systems in Cagayan valley • Environmental quality assessment: biophysical, socio-economic and cultural interrelationships in a protected national park
3. PUP	Risk communication of the national and local DRRM for earthquake preparedness on west valley fault in Marikina City [published in the *European Academic Research*, volume III, issue V, April 2015, pp. 345–367. ISSN 2286-4822]
4. MSU-IIT	Adaptation of a gasoline-fueled conventional engine into a purified biogas-fueled engine for power generation [published in The Mindanao Forum, vol. XXIII, No. 2, 2010, pp. 98–126]
5. MSU-Marawi	Biomass production efficiency and carbon sequestration potentials of Rambutan, Mangium, Falcata, Large-Leaf Mahogany, Narra, Lansones, Durian and Jackfruit planted in Lake Ranao Watershed
MSU-General Santos'	Household vulnerabilities and capacities survey in 11 Ligawasan Marsh communities (Maguindanao and Cotabato provinces)
7. Davao del Norte State College	Save our shores: conservation of endangered giant clams Tridacna spp.: a strategy for environmental protection
8. Central Mindanao University	Performance evaluation of sweet Sorghum lines for bio-ethanol and grains under different ecological zones
9. UP-Mindanao	Evaluation of the ecological and economic impacts of two marine protected areas in the island Garden City of Samal
10. University of Southeastern Philippines	Preliminary geohazard assessment and investigation of Tibungco and Bunawan areas in Davao City
11. Ateneo de Davao	Rural household' social risk management strategies against adverse effects of the changing rainfall patterns and extreme weather events in Agusan Marsh Communities
12. Father Saturnino Urios University	Establishment and implementation of a forest land use plan towards the rehabilitation and protection of Mount Diwata range (Hilong-Hilong) within the Lanuza Bay development alliance
13. Notre Dame of Marbel University	Monthly variations in the physico-chemical characteristics of water in Bulok Creek
14. Notre Dame University	Water analysis of Cotabato City rivers and its implication to human and aquatic life

4.5 Technical, Extension Programs and Community Linkages

The extension services of SUCs/HEIs have taken the form of training programs, seminar workshops, student projects, website development, advocacy, and community engagement such as the examples below.

- *Training Program*. The Ifugao State University-Ifugao Satoyama Meister Training Program (ISMTP) aims to empower local residents through a harmonized heritage conservation, environmental preservation and economic activities within the "Ifugao Rice Terraces" heritage systems (www.ifsu.edu.ph). The ISMTP runs from 2014 to 2016, in collaboration with UPOU, JICA, and the Ifugao Provincial and Municipal Governments.
- *Student Project*. MSU-IIT Engineering students designed two prototype vehicles that operate on high energy efficiency—"Layalang" (Gasoline) and "Tarintik II" (Battery Electric), which were the students' official entry in the Shell Eco-Marathon Competition sponsored by Shell Eco-Marathon Asia (www.msuiit.edu.ph/news/news-detail.php?id=734).
- *Seminar-Workshop*. UPLB-College of Public Affairs and Development's Seminar-Workshop on Climate Change Phenomena and Climate Change Adaptation Action Planning held on July 25–26 2012 (cpaf.uplb.edu.ph) aimed to increase awareness and capability of the local government personnel of Tarlac and Sta. Ignacia LGUs in Tarlac; and municipalities of San Fabian and Sison in Pangasinan. Pangasinan State University and Tarlac State University officials/students have participated.
- *Advocacy*. Romblon State University's Environment and Disaster Risk Management Advocacy for Climate Change Adaptation Program, August 2010 to 2012 (www.rsu.edu.ph), involved 17 towns of Romblon Province.
- *Eco-Park Development*. Isabela State University's establishment of the Cauayan City Agri-Ecotourism Park, in cooperation with the Cauayan City Government DA, DENR, DOST, CHED and CORDEV (www.isu.edu.ph).
- *Website Development*. The UP System launched in May 2016 the resilience.up.edu.ph website (www.up.edu.ph/up-launches-resilience-website/, accessed on May 26, 2016). The REsources for Science-Informed LIteracy and ENgagement towards Building Community ResiliencE is an online knowledge hub to offer information on climate- and disaster-related services and expertise in various fields. It is supported by the CCC and the German Agency for International Cooperation (resilience.up.edu.ph).

4.6 Partnerships with the CCC, Private Sector and Other National Government Agencies

Selected HEIs were found to engage with the CCC, DOST and private organizations in undertaking projects of significant scope (Table 6). A number of these projects are multi-year and receive funding and technical assistance from various government and foreign agencies.

In the project ReBUILD, the tasks of the academe are the capacity building of key officials and personnel in local governments, i.e., surrounding the Cagayan and Jalaur River Basins in Region 2, and in 6 pilot municipalities in the Province of Iloilo in Region VI. In Project NOAH, the partner SUC is tasked to develop a more accurate, integrated and responsive disaster prevention and mitigation system, especially in high risk areas. Under the OML project, the HEI partners are expected to develop a mobile application on weather and hazard alert and tracking system, a flood detector system and guidebook on impact of CCAM measures on agriculture, and an evidence and space-based decision support for DRRM/CCA.

Table 6 Examples of SUC/HEI partnership with the climate change commission and other national government agencies with community engagement

SUC/HEI	Partner agency	Projects
UP Diliman	CCC	ReBUILD (resilience capacity building for cities and municipalities to reduce disaster risks from climate change and natural hazards)
		Project climate twin phoenix (PCTP), supported by UNDP and AusAID
	DOST	NOAH/DREAM Project (nationwide operational assessment of hazards)
Various SUCs/HEIs, LGUs, private sector, NGOs	CCC and GTZ	Support to the Philippines in shaping and implementing the international climate regime (SupportCCC II)
MSU-IIT, Xavier University, LGUs	CCC	Resilience and preparedness toward inclusive development (RAPID) program (RAPID is an expansion of the project Phoenix)
UP Diliman, ADMU and DLSU	OML	Lopez group grants on climate change mitigation (www.energy.com.ph. Accessed on May 13, 2016)

Note GTZ German Technical Cooperation Agency, *OML* Oscar M. Lopez Center for Climate Change

5 Findings and Recommendations

The analysis of HEI responses yields the following observations and brings to light the extensive accomplishments that HEIs have, but at the same time imply that much more information and details are yet to be directly accessed.

5.1 *SUCs Are Complying to the CC Mandate in Spite of Limited Policy and Budget Support*

HEIs have responded in varying capacities to climate change issues, depending on available financial, technical, and faculty and staff resources, among others. SUCs in particular have generally complied with the directive to incorporate climate change in curricular offerings and other activities. At least 24 HEIs have put up climate change centers to perform multiple academic functions.

The policy on engaging SUCs/HEIs is optional, i.e., their (HEI) scientific and technical advise and assistance may be sought in certain areas of concern, (Implementing Rules and Regulations of RA 9729, Rule VII, Section 1). Similarly, investment in public tertiary education is limited. The budget share of all 115 SUCs nationwide accounts for an average of 2.2% of the total new appropriations from 2011 to 2016. The budget for research and extension services is very limited, which makes these services dependent on external funding.

The situation can be improved by amending the law to systematically include SUCs/HEIs in policy making structures at the national and regional levels. SUCs/HEIS should also be systematically included in national government programs on CCMA undertaken at the local community. Identifying SUCs/HEIs as responsible partners in CC efforts will enable the academe to access technical and financial support and to more systematically integrate climate issues into curricula, research and extension services.

To boost the research budget of SUCs, the budget for scholarship coursed through the CHED, which remain unutilized[4] (COA 2014a), should be (re)appropriated directly to SUCs to fund student and faculty research activities. Raising the SUC budget for research should help raise their level of performance and their university ranking in Asia and in the world.[5]

[4]Based on the Commission on Audit Report on CHED (2014a), some Php1.23 billion of CHED's budget for Student Financial Assistance Programs and another Php5.23 billion equivalent to the scholarship of 391,817 students were unutilized.

[5]The 2015 QS University Rankings for Asia ranked the University of the Philippines number 70 among 300 top universities in Asia (www.topuniversities.com, accessed on May 18, 2016).

5.2 SUC/HEI Climate Change Reforms Are Substantial

The 24 HEI-based climate change centers are evidence to this effect. The centers boosted the delivery of degree courses, research and extension services on CCMA, the dissemination of research outputs and the conduct of capacity building programs for various groups. Some SUCs/HEIs have project linkages with the CCC and are partners of agencies in decision making and implementation of CCMA programs and projects. Although many SUCs have less opportunity to be systematically included into nationally sponsored research and extension projects on CCMA, capacity building activities of SUCs/HEIs are in abundance designed for LGUs, CSOs and various audiences where they operate. Examples of capacity building programs and engagement with the local community show the presence of external support, technical and financial, from national, private and foreign partners.

The growing number of HEI-based climate change facilities suggests the need for a systematic assessment of current academic technical capacity to educate and build awareness on climate issues and concerns, amount and level of resources, program coverage, impact and sustainability. One purpose of the assessment may be to determine the level of resources and capacity at the regional and possibly provincial levels. This should pave the way for a more rational distribution and access to external support to benefit all HEIs nationwide.

SUCs/HEIs should actively collaborate with LGUs in capacity building on climate change, particularly for newly elected officials and their staff. The academe can take the opportunity to directly engage with LGUs and offer technical assistance as well as public service forums to provincial, city, municipal and barangay governments officials to build awareness and generate support on climate change adaptation, preparation and updating of climate change action plans of LGUs.[6] A portion of the People's Survival Fund (2016 GAA) can be allocated for capacity building of local government personnel and the members of the local community. An executive/legislative issuance is needed to this effect.

5.3 Conclusion and Next Steps

HEIs should to take a lead role in things they do best: educating society on solutions and strategies, creating solutions through research, and capacitating society to do a more proactive role. The government should recognize the mandate of HEIs by mandating their inclusion in policy making bodies at the national and local levels, systematically engage their expertise in informing government itself and the public on climate change issues and solutions, and provide adequate funds therefor.

[6]Findings of the Commission on Audit (2014b) on Typhoon Yolanda, a.k.a. Haiyan, suggest the need to build awareness among people on the extent of risks and threats of disasters such as a storm surge. SUCs/HEIs offer the best venue to this end.

This research provided encouragement to investigate more deeply into how SUCs/HEI may better participate and with certain amount of urgency in addressing climate change issues. The data presented offer some trends in climate-related activities of HEIs, but is limited in details. Further data collection and in-depth study of HEIs through case studies is needed to proceed to a detailed determination of the impact and sustainability of HEI climate actions.

Likewise, the gathering of data from websites, though crude, inspired the need to gather more data on climate change action plans of HEIs, and on HEI accounting of its own carbon foot prints. These two points can be the subject of future research.

Finally, on October 1, 2015, the Philippines submitted its intended nationally determined contribution (INDC) including a conditional greenhouse gas reduction target of 70% below business as usual (BAU) levels by 2030. The accomplishment of the target is set on the condition that the Philippines financial resources, technology and capability building are available and made available to the Philippines. The support of SUCs/HEIs and all sectors of government, the economy and society are necessary for government to effectively pursue and achieve its goals.

References

ACUPCC/Second Nature Education for Sustainability. 2012. Higher Education's Role in Adapting to a Changing Climate. American College and University Presidents' Climate Commitment.

Association of Climate Change Officers. 2011. Climate Change Leadership in Higher Education Institutions. Complimentary Summary. Washington, D.C.

Boyd, Mary Clarke and Therese Hume. 2015. Addressing the Challenges of Climate Change: The Potential Role of Development Education in the Tertiary Sector. In Stephen McCloskey (ed), *Policy and Practice: A Development Education Review*,Center for Global Education, pp.63–86.

bueden.url.ph. Bicol EDEN. Reducing the Impact of Disasters through Education. Accessed on April 25, 2016. Link: http://bueden.url.ph/.

Climate Change Commission. National Framework Strategy on Climate Change 2010–2022.

Commission on Audit. 2014a. 2014 COA Audit Report on the Commission on Higher Education.

Commission on Audit. September 5, 2014b. Disaster Management Practices in the Philippines: An Assessment. COA Audit Report.

Commission on Higher Education. 2014. Higher Ed Alerts. Snapshots of Higher Education in the Philippines. Series No. 1, 2014. Link: http://www.ched.gov.ph/wp-content/uploads/2014/temp/09-22/CHED/Higher%20Ed%20Alerts%20Final.pdf.

Crepin, Christophe. 2013. *Getting a grip on climate change in the Philippines: extended technical report*. Public Expenditure Review (PER). Washington, D.C., World Bank. http://documents.worldbank.org/curated/en/2013/06/17917468/getting-grip-climate-change-philippines-extended-technical-report.

Department of Science and Technology-Industrial Technology Development Institute (DOST-ITDI). October 31, 2012. Renewable Energy Development Desktop Study for the Philippines. 64 pages.

Email survey, Isabela State University, through Dr. Joan Carinugan, Assistant Professor, College of Business, Accountancy and Public Administration, ISU.

Everett, Brian and Rob Copeland. November 2007. Climate Change a Trade Union Responsibility in Higher education. University and College Union, UK. 7 pages.

General Appropriations Acts from 2011 to 2016

Interview with Ms. Bianca Tecson, Staff, Marine Science Institute, College of Science, University of the Philippines, Diliman Campus, Quezon City.

Merillo, Ma. Gerarda Asuncion D., Inter-Agency Committee on Climate Change Environmental Management Bureau-DENR, Philippines. September 2–5, 2003. "Addressing Climate Change and the CDM: The Philippine Initiatives." Presented in the 13th Asia-pacific Seminar on Climate Change held at Miyazaki, Japan.

Molthan-Hill, Petra, Team Leader NTU Green Academy Project 'Food for Thought'. 2015. Preparing for the climate change summit—Insights from the Higher Education Sustainability Initiative. Nottingham, Trent University. 10 pages.

Nachmany, Michal, Sam Fankhauser, Jana Davidová, Nick Kingsmill, Tucker Landesman, Hitomi Roppongi, Philip Schleifer, Joana Setzer, Amelia Sharman, C. Stolle Singleton, Jayaraj Sundaresan and Terry Townshend. 2015a. Climate Change Legislation in the Philippines. An Excerpt from *The 2015 Global Climate Legislation Study. A Review of Climate Change Legislation in 99 Countries*. Grantham Research Institute on Climate Change and the environment, GLOBE-The Global Legislators Organization, and Inter-Parliamentary Union for Democracy, For Everyone. 13 pages.

Nachmany, Michal, Sam Fankhauser, Jana Davidová, Nick Kingsmill, Tucker Landesman, Hitomi Roppongi, Philip Schleifer, Joana Setzer, Amelia Sharman, C. Stolle Singleton, Jayaraj Sundaresan and Terry Townshend. 2015b. The 2015 Global Climate Legislation Study. A Review of Climate Change Legislation in 99 Countries. Summary for Policy-makers. Grantham Research Institute on Climate Change and the environment, GLOBE-The Global Legislators Organization, and Inter-parliamentary Union for Democracy, For Everyone. 45 pages. Link: http://www.lse.ac.uk/GranthamInstitute/wp-content/uploads/2015/05/Global_climate_legislation_study_20151.pdf.

Nachmany, Michal, Sam Fankhauser, Terry Townshend, Murray Collins, Tucker Landesman, Adam Matthews, Carolina Pavese, Katharina Rietig, Philip Schleifer and Joana Setzer. 2014. The GLOBE Climate Legislation Study: A Review of Climate Change Legislation in 66 Countries. Fourth Edition. London: GLOBE International and the Grantham Research Institute, London School of Economics. Link: http://www.lse.ac.uk/GranthamInstitute/wp-content/uploads/2014/03/Globe2014.pdf.

Republic of the Philippines. Intended Nationally Determined Contributions (INDC). Communicated to the UNFCCC on October 2015. 6 pages. Link: http://www4.unfccc.int/submissions/INDC/Published%20Documents/Philippines/1/Philippines%20-%20Final%20INDC%20submission.pdf.

UN Documents Gathering a Body of Global Agreements. Our Common Future, From One Earth to One World. From *A/42/427. Our Common Future: Report of the World Commission on Environment and Development*. Complied by the Committee on Education of the Conference of NGOs from United Nations web sites. Link: http://www.un-documents.net/our-common-future.pdf.

Various republic acts, executive orders and issuances.

Websites of 115 state universities and colleges and selected private HEIs.

www.energy.com.ph. May 31, 2013. *Lopez Group Grants P 13 Million to 7 Research Projects on Climate Change Mitigation*. Accessed on May 13, 2016. Link: http://www.energy.com.ph/lopez-group-grants-p13-million-to-7-research-projects-on-climate-change-mitigation/.

www.mindanews.com. May 30, 2014. Mindanao state universities and colleges to offer DRRM and climate change courses. Link: http://www.mindanews.com/mailbox/2014/05/30/mindanao-state-universities-and-colleges-to-offer-drmm-and-climate-change-courses/ accessed on April 26, 2016.

www.observatory.ph. Climate Change Assistance (Klima). Link: http://www.observatory.ph/climate-change-assistance-klima/, accessed on April 26, 2016.

Adapting Sri Lanka to Climate Change: Approaches to Water Modelling in the Upper Mahaweli Catchment Area

M.P. Gunawardena and M.M.M. Najim

Abstract Water flow and water quality changes in surface water bodies due to climate change is a major problem worldwide, typically in river eco-systems. To address these issues, proper monitoring of river flow variations and water quality is necessary, but daily monitoring and data collection is greatly time consuming and costly. Therefore, a study was done with the objectives to calibrate and validate the HEC-HMS 3.5 and HEC-RAS 4.1.0 models to simulate flow variation and water quality variations respectively in Nillambe Oya, Sri Lanka. The development of such models are a significant requirement in the present scientific world to predict river flow and water quality variations as an adaptation for the climate change. The HEC-HMS 3.5 model was used to generate flow which is needed for HEC-RAS 4.1.0. Daily flow data for five years (from October 1991 to September 1996) were used to calibrate the HEC-HMS 3.5 model and another set of daily flow data for five years (from October 1994 to September 1999) were used to validate the model. HEC-RAS 4.1.0 model was calibrated using water quality data collected and analyzed in five days. The validity of the calibrated model was evaluated using water quality data collected in the other four days. The calibrated and validated HEC-HMS 3.5 model reliably predicted the flow in Nillambe Oya. The calibrated and validated HEC-RAS 4.1.0 model simulated Dissolved Oxygen (DO), Biological Oxygen Demand (BOD) and Nitrate Nitrogen (NO_3-N) splendidly. Orthophosphates was also predicted reliably by the model, but the simulation of organic Phosphorus was not that sensible.

Keywords Catchment · Flow · HEC-HMS · HEC-RAS · Modelling · Water quality

---

M.P. Gunawardena (✉)
Department of Biotechnology, Faculty of Science, Horizon Campus, Malabe, Sri Lanka
e-mail: medhisha@gmail.com

M.M.M. Najim
Department of Zoology and Environmental Management, Faculty of Science, University of Kelaniya, Kelaniya, Sri Lanka
e-mail: vcseu@seu.ac.lk

W. Leal Filho (ed.), *Climate Change Research at Universities*,
DOI 10.1007/978-3-319-58214-6_6

1 Introduction

1.1 Impacts of Climate Change on Surface Water Resources in Sri Lanka

There is mounting evidence to conclude that Sri Lanka's climate has already changed and there is a significant impact due to such changes. If depicted through an example, Sri Lanka's mean air temperature has increased by 0.016 °C per year during the period of 1961–1990, while the global average temperature rise within the period of 1956–2005 is only 0.013 °C per year (Chandrapala 1996b). When considering Sri Lanka's hundred year warming trend (from 1896 to 1996), it is only about 0.003 °C per year (IPCC 2001). However, the warming trend during the last ten years within that period (1987–1996) is 0.025 °C per year (Fernando 1997). This is most conclusive evidence to prove that there is an increased rate of warming trend in the resent, which will definitely continue into the near future as well. On the other hand the mean annual precipitation of Sri Lanka has decreased by 144 mm during 1961–1990 when compared it with 1931–1960 period, which is about a 7% decrease in mean annual rainfall (Chandrapala 1996a; Jayathilake et al. 2005). When considering the rainfall changes in the central highlands, there is a significant decrease in mean annual precipitation (Herath and Rathnayake 2004). In addition, the Mahaweli headwaters in central highlands indicate a 39% decrease in mean annual precipitation from 1880 to 1974 (Shantha and Jayasundara 2005). Furthermore, Shanthha and Jayasundara (2005) have predicted that there will be a 17% reduction in rainfall in Upper Mahaweli catchment in the central highlands by 2025. Therefore, it is evident that the climate change is ongoing process in Sri Lanka and it will be continued in future as well. But the problem lies in the reliability of predicting the climate for the next fifty years and how well the country is adapts to face such changes. There are some ongoing studies that aim to predict the effects of future climate change in Sri Lanka, but the identification of how such changes might impact on major resources such as surface water bodies are still unknown. Eriyagama et al. (2010) have clearly stated that, since the central highlands contribute the largest volume of water for hydropower generation and also for the irrigation purposes in the country through multipurpose reservoirs in the Mahaweli catchment, the climate change in this area will result in severe adverse impact on the economy. Not only the economy, but these changes will also most likely to have negative impacts on the well-being of the society and the ecology. The major reservoirs build in the Upper Mahaweli catchment provides 29% of the country's energy generation and 23% of the irrigation water requirement of the country (Shantha and Jayasundara 2005). Due to the reduction in water flow in this region, water allocation between two sectors will be challenging (Eriyagama et al. 2010). Therefore, it is a vital requirement to predict climate change using reliable methods and also develop models to predict the river water flow variations in Upper Mahaweli catchment. This will provide a better understanding of the extent to which we need alternative power generation, in addition to our dependence on hydropower in the future, to provide sufficient water flow allocations for irrigation and agriculture which our country is largely dependent on. Therefore, Sri Lanka needs more detailed studies to assess the vulnerability of the country's water resources to the climate change. Eriyagama et al. (2010) also have stated that Sri Lanka has to formulate reliable outline of future climate change patterns and establish a national water resources audit. For the purpose of this, flow and water quality modelling are major requirements. In addition, some of the major questions that Sri Lanka is yet to address is, weather the country will be able to sustainably satisfy its national water requirements for agriculture, industry and domestic consumption and how the nation's surface water quality will diverge during the twenty first century due to climate change.

1.2 Flow Modelling

When the flow pattern of a river changes, flooding frequency, duration and intensity of floods, water extraction from rivers and water flow required for hydropower generation etc. are also changed. Understanding flow variability and predicting river flows are therefore vital. In order to predict flow patterns, sound sets of data are necessary. Collection of accurate data is greatly time consuming, resource over riding and costly. Therefore, the most appropriate and economical approach is flow modelling using computer software, even though it is much difficult in the initial stages to collect the required data and calibrate the models for the local conditions.

Flow modeling is a widely used tool in developed countries but in developing countries it is marginally practiced giving an insignificant attention. There are many modelling software available to perform flow modelling, but only few good models are available for free usage. HEC-HMS is one of the widely used open source models, which is developed by the United States Army Corps of Engineers (Yilma and Moges 2007; Kalita 2008; Arekhi 2012; Sardoii et al. 2012; Yan 2013). As in other developing countries, in Sri Lanka also, flow simulation and prediction using models are hardly tested. However, only few have tested the applicability of HEC-HMS model to different local conditions in Sri Lanka (Hunukumbura et al. 2008; Ratnayake et al. 2010; Halwatura and Najim 2013). Therefore, calibration and validation of HEC-HMS model to local conditions will be a significant boost to less concerned hydrological modelling systems. Further, it is important to apply these models to the upcountry and the mid country of the wet zone of Sri Lanka which feeds most of Sri Lankan rivers.

Water flow from the Nillambe Oya catchment is used to satisfy major purposes of the area, such as hydropower generation at Doluwa hydropower station, drinking water source for Udunuware and Yatinuwara water supply schemes and for agricultural purposes. Therefore, prediction and simulation of the flow is critically important to balance the water requirements between these extractions while satisfying the environmental needs. Simulated flows of Nillambe Oya through calibration and validation of HEC-HMS model can be used to comment on hydropower generation capability or predict water share that can be allocated for hydropower generation, agriculture and other requirements while giving priority for drinking water supply.

1.3 Water Quality Modelling

The quality of natural water in rivers, lakes and reservoirs and below the ground surface depends on a number of interrelated factors. These factors include geology, climate, topography, biological processes and land use. The impurities determine the characteristics of a water body. Therefore, in order to determine the water quality of a river all these factors and their interactions have to be considered and understood. As the environment is dynamic, and as different environmental factors are changing from time to time, intense monitoring which is more time consuming and costly is necessary. Therefore, once the relevant data are collected, models can be developed which will reduce the constraint of more intense monitoring. Therefore, many water quality modelling systems are being developed such as AGNPS (Young et al. 1987), ANSWERS (Beasley et al. 1980), CREAMS (Knisel 1980), EMSS (Vertessey et al. 2001; Watson et al. 2001), HSPF (Johanson et al. 1980), IHACRES-WQ (Jakeman et al. 1990; Dietrich et al. 1999), IQQM (Simons et al. 1996), LASCAM (Sivapalan and Viney 1999), SWRRBWQ (USEPA 1994) and HEC-RAS. Among these models HEC-RAS includes more recent modelling approaches and much complex modelling system (Zhang and Johnson 2012). Therefore, water quality modelling using HEC-RAS is implemented in some countries (Chihhao et al. 2009; Suliman 2010; Chihhao et al. 2012; Halaj et al. 2013; Zhang and Wu 2013). Due to the complexity of HEC-RAS, it is not used for water quality modelling in Sri Lanka. Similarly, there are not many studies done using other models as well. Therefore, calibration of this model in order to simulate water quality will motivate water quality modelling requirement in Sri Lanka as well as determine the mode applicability to the country as an adaptation for water quality changes due to climate change.

2 Methodology

2.1 Study Area

Nillambe Oya sub-catchment located within the Upper Mahaweli catchment with typical land uses of the central highlands of mid country in Sri Lanka, was selected as the study area. The Nillambe Oya is a main tributary of Mahaweli river located at latitudes between 80-37′ and 80-42′E and longitudes between 7-07′ and 7-14′N (Fig. 1). The study area covers 6171.2 ha of land. The topography of the area could be explained as steep and moderate slopes and 40% of the area is above 30% slope.

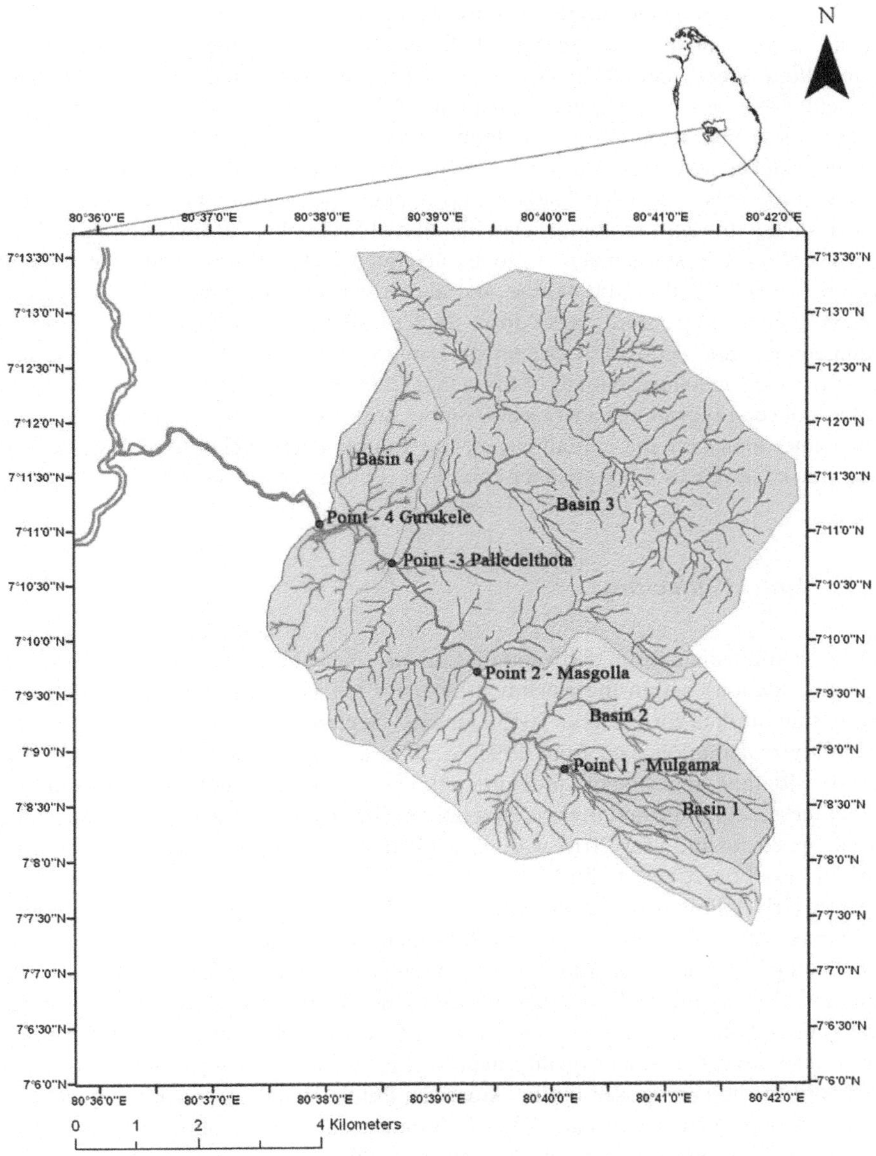

Fig. 1 Study catchment and sampling points

Red Yellow Podsolic and Reddish Brown Latasolic are the major soil types found in the area. Land cover condition of the catchment can generally be divided as agricultural areas, natural forest patches, scrublands and home gardens. The main agricultural crop is tea, which covers about 58.9% of the land area in the catchment. Vegetables, other export crops and home gardens occupy nearly 10.9% of the area. Forest patches that consist of unclassified forests, plantation forests with pine and eucalyptus trees and scrub lands occupy about 21.2% from the catchment area According to the agro ecological classification of Sri Lanka, the area belongs to three agro ecological zones namely wet zone mid country (WM2, WM3) and wet zone up country (WU2) within 300–900 m and above 900 m elevation, respectively. The mean annual temperature varies from 28.4 to 30.7 °C. Minimum and maximum temperatures are recorded in December–January and April–May, respectively. The area receives an annual rainfall of more than 2200 mm. In summer monsoon and in inter monsoon season, the area receives more or less equal amount of rainfall, but in the winter monsoon, the rainfall is very low from mid-December to mid-March and the area experiences a dry weather condition (Punyawardena 2008).

2.2 *Data Collection*

Daily rainfall data from two rainfall stations (New Forest Estate and Gurukele gauging station) within the catchment area for ten years from 1991 to 2000 and daily rainfall data for the year 2013 (January–September) were used in this study. Monthly evaporation data from 1991 to 2000 and 2013 for Kothmale were also used. Monthly atmospheric pressure, minimum and maximum air temperature, cloudiness, minimum and maximum relative humidity and minimum and maximum wind speed for the year 2013 were also obtained. Short wave radiation was estimated based on Allen et al. (1998).

Daily river flow data for ten years (1991–2000) at Gurukele gauging station of Nillambe Oya was obtained from the Environment and Forest Conservation Division of the Mahaweli Authority of Sri Lanka, Kothmale. Minimum flows in the monthly flow hydrograph were considered as the monthly base flow at the Gurukele gauging station. Flow data for the study period from May 2013 to October 2013 were also collected. Water quality data for the study period were measured using standard methods proposed by the American Public Health Association, American Water Works Association and Water Pollution Control Federation (Franson et al. 2005), on the nine sampling days from May 2013 to October 2013. The water quality was measured at four points (Mulgama, Masgolla, Palledelthota and Gurukele) along the river representing different reaches of the Nillambe Oya. Three samples were taken at each location at each sampling event. The water quality parameters measured were Dissolved Oxygen, Biological Oxygen Demand, Ammonium Nitrogen, Dissolved nitrate Nitrogen, Dissolved nitrite Nitrogen, Dissolved inorganic Orthophosphate, Total Phosphorus and dissolved organic Phosphorous.

2.3 River Reaches

River was divided into four reaches according to the morphology of the river and sample collection point was identified at the downstream end of each reach. Sampling locations were selected considering different factors such as accessibility, distance among sampling locations, flow pattern and other obstacles.

2.4 Data Map Preparation

Data maps that are necessary for preparation of the digitized maps (land use and contour) in digital format were collected from the Department of Survey (issued year 2013). Spatial data preparation was done using Arc GIS 10.0 software package. The topographical maps with 1:10,000 scale were first registered according to the SL_grid_99 projection system (Datum-Kandawala). Data maps that were needed as input data for the flow estimation (catchment area, sub catchments and stream network) were digitized in different layers. Landuse maps for the catchment area were prepared. Area covered by each landuse type, the total area of the catchment and stream length were measured based on the digitized maps. Within the main catchment, four sub catchments were digitized according to the sample collection points (Gurukele, Palledelthota, Masgolla, Mulgama) for HEC-HMS 3.5 and HEC- RAS 4.1.0 model calibration and validation processes.

2.5 Model Conceptualization and Calibration—HEC-HMS 3.5

The daily stream flows were computed using the HEC-HMS 3.5 model. Data maps prepared using Arc GIS 10.0 software package were used in the model. Watershed and meteorology information were combined to simulate the hydrologic responses. Entire catchment was used to accomplish the calibration goal. Daily rainfall data for five years (October 1991 to September 1996), monthly base flows of the river, monthly evaporation data of the catchment and the catchment area were inserted to the model as input data Fig. 2.

The model was calibrated using Initial and Constant Loss and Snyder Unit Hydrograph Method, because Halwatura and Najim (2013) have shown that Snyder Unit Hydrograph method is the best suited method for flow simulations in wet zone streams of Sri Lanka. Model parameters imperviousness, peaking coefficient and standard lag were changed based on the data gathered through digital maps, data collected from different sources and measured in the field, and the model calibration

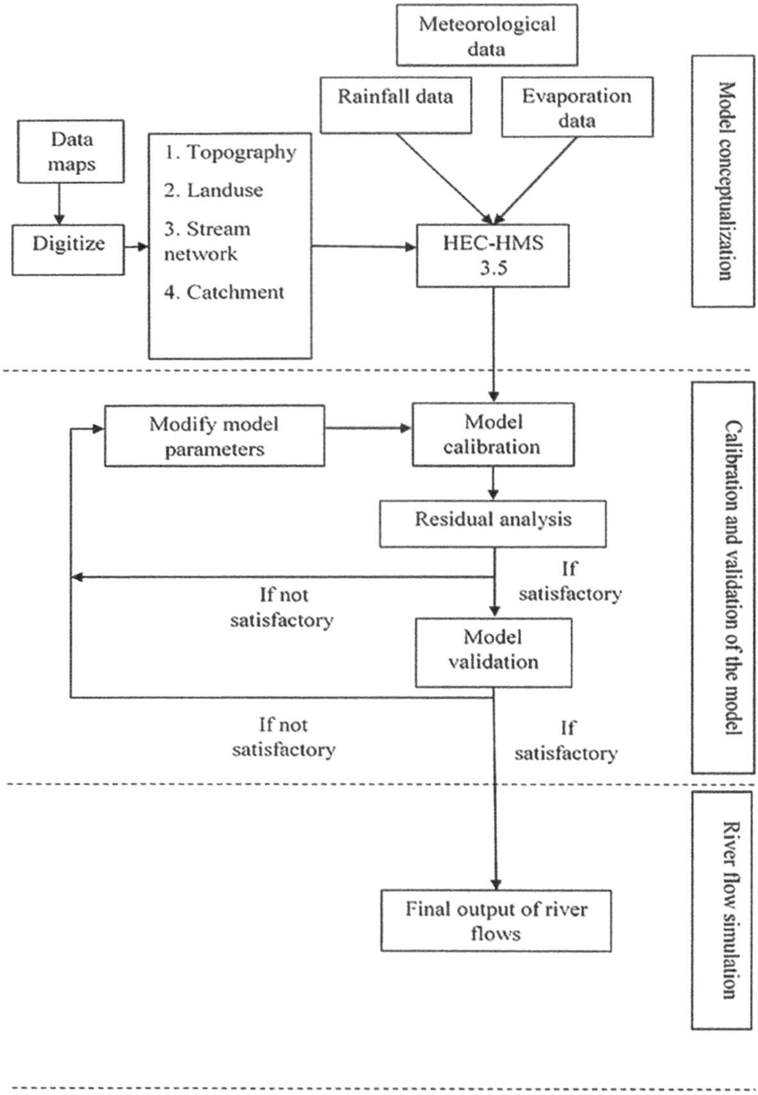

Fig. 2 HEC-HMS 3.5 model conceptualization, calibration and validation process

which was performed separately by giving eight different combinations of possible parameters. The flows simulated for each combination of parameters were tested statistically.

2.6 Model Validation—HEC-HMS 3.5

Calibrated model was used in the validation process with another set of rainfall data for the next five years (October 1994 to September 1999). Calibrated model parameters were not changed during the validation process. Statistical evaluation was also performed to the flows observed and simulated by the calibrated model. Eight different simulation processes that were used in the calibration processes were also applied in the validation process and the results were compared statistically. The best results obtained by HEC-HMS model were considered in the water quality modelling using HEC-RAS 4.1.0 model.

2.7 Statistical Evaluation—HEC-HMS 3.5

The model was evaluated by residual method (Pauly 1980; Bennett et al. 2013). Bennett et al. (2013) specified that one of the most prevalent method for model evaluation are residual methods, which computes the difference between observed and modelled data points. The residual plot is a simple graphical plot used to analyze model residuals. Of the many possible numerical calculations on model residuals, by far the most common are bias and Mean Square Error. Bias is simply the mean of the residuals, indicating whether the model tends to under- estimate or over-estimate the measured data, with an ideal value zero.

The model was initially run with the calculated model parameters. R-squared values for each calibration and percentage residuals within ±1SD and ±2SD ranges were calculated. To find the most suitable values for the best performance of the model that gives the most reliable R-squared value and percentage residual values within the range ±1SD and ±2SD, the model parameters were adjusted until R-squared value and the results of the statistical evaluation gave highest percentage within ±1SD and ±2SD range. The parameters that were changed in each of the simulation processes are Initial loss, Constant rate, Imperviousness, Standard lag and Peaking coefficient.

2.8 Model Conceptualization and Calibration—HEC-RAS 4.1.0

HEC-RAS 4.1.0 model which was developed by United States Army Corps of Engineers was used in modelling water quality. Geometry data, steady flow data, water quality data and meteorology data were combined to simulate the water quality. The main river reaches were used to accomplish the calibration goal. Water quality data of five sampling days during year 2013, daily flow data for year 2013, information on the meteorology station and monthly average meteorology data for year 2013 were inserted to the model.

2.9 Model Validation—HEC-RAS 4.1.0

Calibrated model was used in the validation process with another set of water quality data of four different sampling days. Statistical evaluation was performed to the water quality parameters observed within this four days and simulated using the calibrated model. For the model calibration and validation, eighty nine different simulations were performed in order to determine the most suitable method for the study catchment. Most suitable simulation processes were applied for the calibration and validation process and the results were compared and the best simulation processes were selected Fig. 3.

2.10 Statistical Evaluation—HEC-RAS 4.1.0

The model was evaluated by residual method as explained previously in HEC-HMS 3.5 model evaluation. The model was initially run with the calculated model parameters and R-squared values and percentage residuals within ±1SD and ±2SD ranges were calculated. To find the most suitable values for the best performance of the model that gives the highest R-squared value and the most reliable percentage residual values within the range ±1SD and ±2SD, the variables were adjusted until the results of the statistical evaluation gave the highest percentage within ±1SD and ±2SD range.

2.11 Limitation

The HEC-RAS model is not capable of modelling the water quality changes due to landuse changes in the catchment. The model is only capable of predicting water quality changes within the river system.

3 Results and Discussion

3.1 Flow simulation using HEC-HMS 3.5

Calculated residuals (%) for the five years of calibration (October 1991 to September 1996) and validation (October 1994 to September 1999) corresponding to different combination of variables are given in Table 1. All the first five calibrations produced good results with high R squared values more than 0.62 and high percentage residuals within ±1SD and ±2SD. Calibration processes 5 and 1 have shown the best performance.

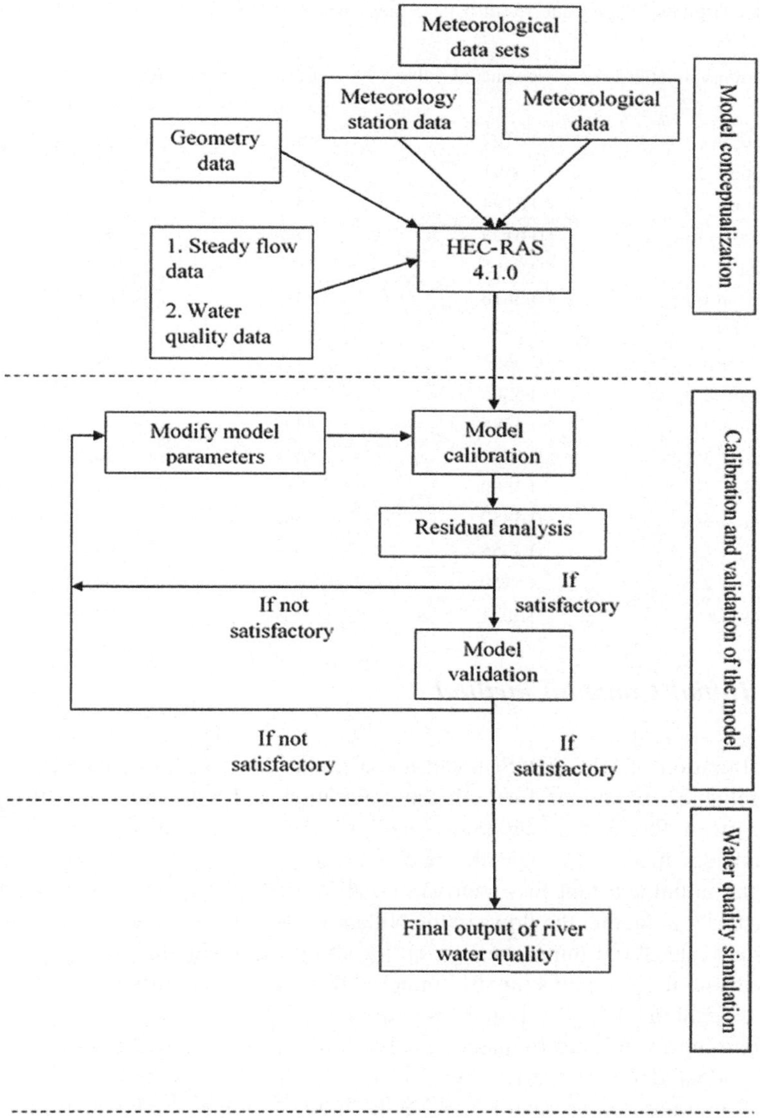

Fig. 3 HEC-RAS 4.1.0 model conceptualization, calibration and validation process for water quality modelling

All the results for the validation process also produced good results with high percentage residuals within ±1SD and ±2SD. The percentage residuals within ±1SD and ±2SD for validation 5 is much better than the values for the other validation processes. The results from the validations show that all the eight validation events have given statistically acceptable results for the model performance.

Table 1 R-squared values and residual percentage within ±1SD and ±2SD for the calibrations and validations

Calibration/validation	R-squared value	Residual percentage	
		within ±1SD	within ±2SD
Calibration 1	0.6797	88.8	96.7
Calibration 2	0.6491	85.4	96.1
Calibration 3	0.6194	85.1	96.0
Calibration 4	0.6193	85.1	96.0
Calibration 5	0.6465	88.0	97.2
Calibration 6	0.6818	84.1	94.9
Calibration 7	0.6873	84.4	94.3
Calibration 8	0.6854	84.4	94.4
Validation 1	0.5865	86.9	95.8
Validation 2	0.5588	84.6	95.1
Validation 3	0.5692	84.3	94.9
Validation 4	0.5690	84.3	95.0
Validation 5	0.5857	86.6	96.4
Validation 6	0.5655	85.3	94.4
Validation 7	0.5469	86.0	94.0
Validation 8	0.5521	85.4	94.1

3.2 *Initial Constant method*

The calibration of HEC-HMS model using Initial constant loss method has given more reliable simulations for both the calibration and validation. The R-squared values given for all the calibrations and validations are higher with the initial constant loss method along with the Snyder unit hydrograph method. Therefore, using the initial constant loss method, the HEC-HMS 3.5 model can be calibrated satisfactorily to predict the flows in the Nillambe Oya. An initial loss of 1.2 mm and a constant rate of 0.9 mm/h were given for all the eight simulations and for all the four sub-basins. By adjusting the imperviousness of each sub-basins the model could be calibrated for the Initial loss method.

According to residual frequency analysis, in the first five calibration processes and all the validation processes, the model has over predicted the flow very slightly and in calibration processes 6, 7 and 8 the model has under predicted the flow in very minor quantities. However overall flow simulation was satisfactory giving more than 84% residuals within ±1SD. The first five calibrations and validations gave more than 95% of residuals within ±2SD range. In general all the simulations gave more than 94% within ±2SD range. The remaining 5–6% residuals outside ±2SD range is due to sudden increases in the peaks in the observed flows which the model have not quantitatively represented well. Sudden peaks are under predicted by the model in most of the times, except in simulations 6 and 8. But the model is capable in predicting the increased flow pattern in all the sudden peaks in all the simulations. Yan (2013) also applied the HEC-HMS to Uganda Manafwa River and found that neither the Curve Number nor Initial and Constant loss method is perfectly suitable to model both the short-term and long term simulations (only capable of perfectly modelling long term simulations). The Curve Number method is capable to better model the precipitation-runoff processes in the short term simulations. The Initial and Constant loss method tends to underestimate water volume runoff in the short term simulations from what is observed (Yan 2013). Therefore, HEC-HMS model is capable of modelling overall long term flow simulations rather than particular events.

Arekhi (2012) applied the HEC-HMS Model to Kan watershed in Iran to model the runoff by using three different loss methods and showed that the initial and constant loss method is the optimum method that simulated the surface runoff, while Green and Ampt and deficit constant loss methods were the second preferences. Sardoii et al. (2012) attempted to calibrate the loss estimation methods in the HEC-HMS for the simulation of surface runoff in Amirkabir dam watershed in Iran and found that Green and Ampt loss method had given better results than the Soil Conservation Services curve number method (SCS) and initial and constant loss method. Accordingly, the selection of a suitable loss method for model calibration depends on physical factors of the study area and also depends on the predicting fitness of different model parameters according to available data. For example to use deficit constant loss method, initial deficit (amount of water needed to saturate soil layer to maximum storage) and maximum storage (calculated as depth of active soil layer multiplied by soil porosity) must be known, but at the same time initial constant loss method requires only initial loss, while both the methods required constant rate and imperviousness to calculate the loss.

Initial constant loss method has not been used in Sri Lanka for calibration of the HEC-HMS model previously. This method requires only the initial loss as an additional data requirement when compared to the other methods available in the HEC-HMS. Therefore, initial loss method was selected to calibrate the model and to evaluate its performance to Nillambe Oya, along with mostly accepted and widely used Snyder unit hydrograph method in Sri Lanka (Hunukumbura et al. 2008; Ratnayake et al. 2010; Halwatura and Najim 2013).

3.3 Snyder Unit Hydrograph method

In Snyder unit hydrograph method, changing of the peaking co-efficient have not resulted in significant differences in R-squared values but adjusting of the standard lag has caused difference in R-squared value. Standard lag has played a major role in simulating flow using Snyder unit hydrograph method for Nillambe Oya catchment. The values given for standard lag demonstrate that the standard lag for Nillambe Oya catchment is having fairly higher values (sub-basin 1—35 and 39.66, sub-basin 2—36 and 39.66, sub-basin 3—44 and 44.01, sub-basin 4—27 and 36.28 for better simulations). Halwatura and Najim (2013) applied the HEC-HMS model to the Attanagalu Oya in wet zone of Sri Lanka using Deficit constant loss method together with Clark unit hydrograph transform method and Snyder unit hydrograph method, respectively, and they have reported that Snyder unit hydrograph method performed better than Clark unit hydrograph method. Snyder unit hydrograph method has been used as the transformation method by Hunukumbura et al. (2008), to estimate the runoff from Upper Kotmale basin, and yet the model has predicted reasonable flows. But for various landuse types, the model has not predicted flows accurately. Ratnayake et al. (2010) applied the HEC-HMS for flood prediction in the Nilwala basin and the model was calibrated using Clark's, Snyder's and SCS transformation methods along with Green-Ampt loss method and found that Snyder's method performed better than other methods in calibration and verification for Nilwala basin. Therefore, Snyder unit hydrograph method is well suitable for predicting flows in wet zone of Sri Lanka, as well as to the upper Mahaweli catchment. HEC-HMS has been used by Kalita (2008) to estimate flooding in South Brahmaputra using Snyder unit hydrograph method and reliable results were obtained. HEC-HMS was applied to Ethiopian Nile River with Snyder unit hydrograph method by Yilma and Moges (2007), and they have stated that it is the best model for short period flood forecasting in the Ethiopian Nile river basin. Therefore, Snyder unit hydrograph method is capable of modelling both the short term and long term flow simulations along with a suitable loss method.

3.4 Water quality modelling using HEC-RAS 4.1.0

Eighty nine simulations were performed to select the best simulation process that predicts water quality with reasonable accuracy. Table 2 shows ±1SD and ±2SD residual percentages and R-squared values for each water quality parameter for the best simulation process. Except for water temperature and Biological Oxygen Demand, all the other water quality parameters gave more than 80% residuals within ±1SD range. All the water quality parameters have more than 90% of residuals within ±2SD range including water temperature and Biological Oxygen Demand. Obtained residual plots and calculated residuals (%) for the 9 days of

Table 2 R-squared values and residual percentage within ±1SD and ±2SD for the simulation

Water quality parameter	R-squared value	Residual percentage	
		Within ±1SD	Within ±2SD
Water temperature	0.91	61.1	94.4
Dissolved oxygen	0.93	80.6	97.7
Biological oxygen demand	0.95	61.1	94.4
Nitrate nitrogen	0.94	80.6	91.7
Orthophosphate	0.63	80.6	91.7
Organic phosphorus	0.31	86.1	91.7

calibration and validation corresponding to the best simulation process is graphically shown in Figs. 4 and 5.

The HEC-RAS model has modelled water temperature well. Residuals within ±1SD is low (Table 2) because the model has over predicted some temperature events (only two events) slightly, causing 5.6% residuals outside the ±2SD range. When the atmospheric temperature variations are provided to the model, the model has predicted the variation of water temperature appropriately. At Nillambe Oya, the water temperature variation is only 6.6 °C (maximum temperature value is 27.7 °C and minimum temperature value is 21.1 °C) during the study period. It can be said that the model is capable of modelling water temperature perfectly even when the water temperature variation is low. Therefore, HEC-RAS model can be used as a perfect tool for water temperature variation forecasts in tropical conditions. However, the water temperature predicting capability using HEC-RAS in temperate regions need to be evaluated separately. Jensen and Lowney (2004) also have stated that HEC-RAS can be used to model water temperature.

Water quality module of HEC-RAS includes a set of nutrient simulation modules (NSM) (Zhang and Johnson 2012). NSM 1 computes riverine organic and inorganic nitrogen and phosphorus species, BOD and DO; NSM 2 computes nitrogen, phosphorus and DO along with many other complex elements and NSM 3 integrates a dynamic bed sediment analysis component, which simulates the chemical and biological processes undergone at the sediment water interface after sediments are settled. In HEC-RAS model, by means of quickest-ultimate transport algorithm, the conservation of mass equation is solved (Zhang and Wu 2013).

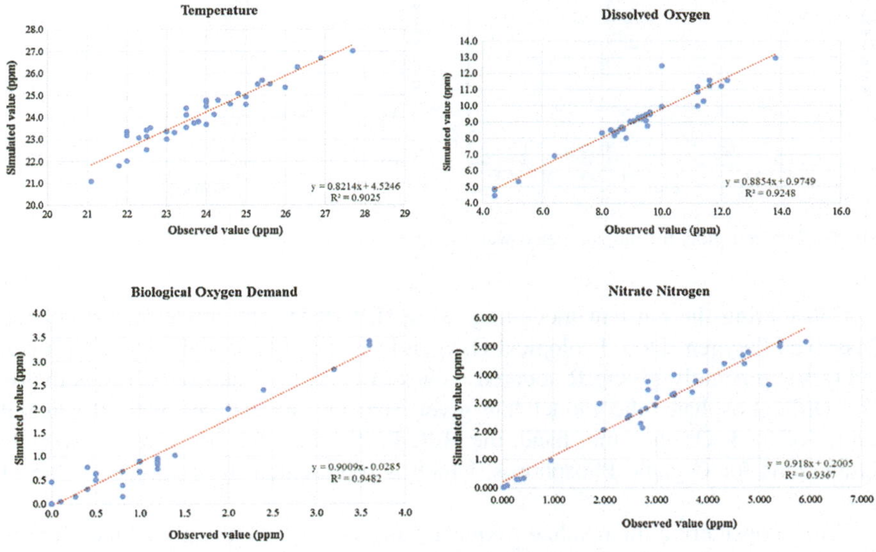

Fig. 4 Observed and Simulated water quality at Nillambe Oya

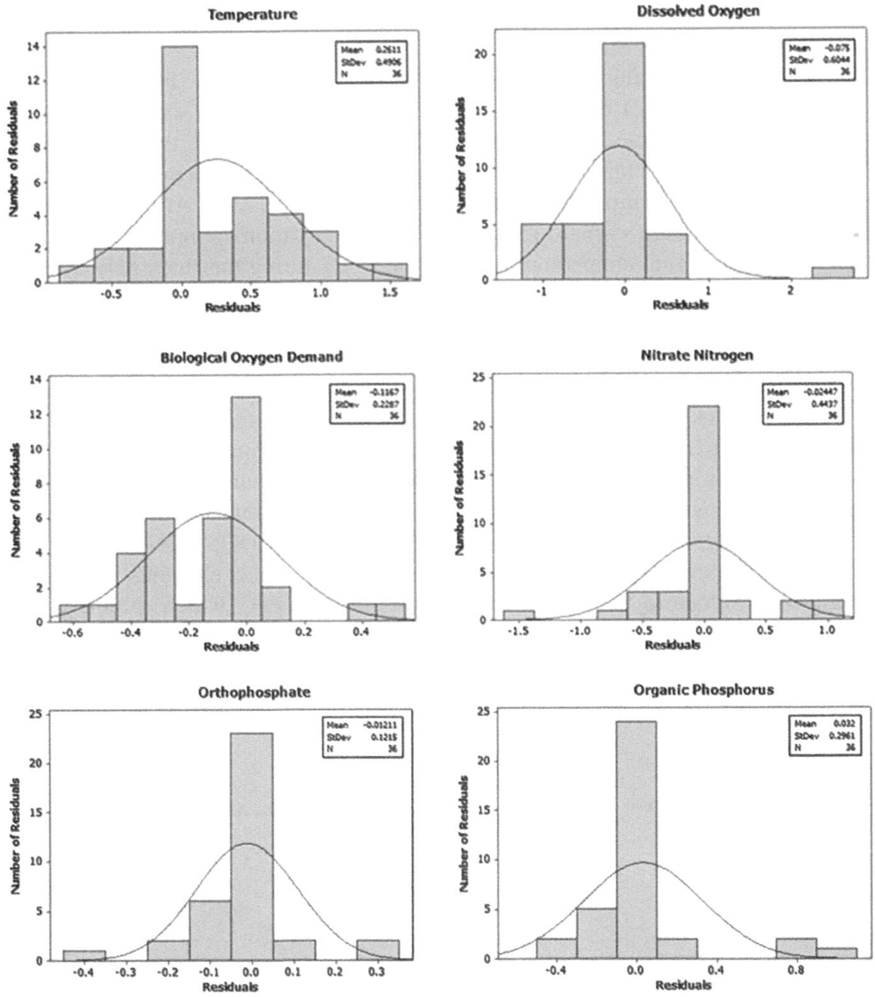

Fig. 5 Residual plots for the simulated water quality parameters

Considering the nutrient modelling using HEC-RAS, the model has simulated Dissolved Oxygen (DO), Biological Oxygen Demand (BOD) and Nitrate Nitrogen (NO_3-N) splendidly giving R-squared values of 0.92, 0.95 and 0.94, respectively. For Orthophosphate, the model has given sensible simulations with R-squared values of 0.63. On the other hand, the HEC-RAS model has not given reasonable simulations for Organic Phosphorus, which is represented by R-squared value of 0.31.

When considering the residual frequency analysis of DO, the model has slightly under predicted some values giving 80.6% residuals within ±1SD. The model has performed well in DO simulation compared to other nutrient simulations by giving 97.2% of residuals within ±2SD range. Only one event has significantly over predicted a DO concentration which could be a result of low DO level due to some abnormal condition. Therefore, HEC-RAS model can reliably predict DO levels in Nillambe Oya under normal conditions.

The residual frequency analysis of BOD shows that the model has slightly under estimated some events and slightly over estimated few events (61.1% residuals within ±1SD range). The model has significantly under estimated two events causing 5.6% residuals outside ±2SD (94.4% residuals are inside the ±2SD range). Therefore, HEC-RAS can be used to predict BOD levels in Nillambe Oya, but slight variations could be reported between the realistic values and modelled values. When the residual frequency distribution of Nitrate Nitrogen is considered, there are some slight under predictions and very few slight over predictions (resulting in 80.6% residuals within ±1SD range). Under ±2SD range there is 91.7% residuals, as the model has only significantly over predicted two events and significantly under predicted one event. Even though the model can be used to predict variations in Nitrate Nitrogen in Nillambe Oya, slight differences could be observed between the simulated values and observed values but only for few events due to some sudden changes in the environment.

Residual frequency distribution of Orthophosphate gave 80.6% residuals within ±1SD range, which is due to some slight under predictions by the model. There were 91.7% residuals within ±2SD range because the model has significantly over predicted two events and under predicted one event. The HEC-RAS model can simulate variations in Orthophosphate levels in Nillambe Oya to a certain extent (R^2 value is 0.63) with reasonable accuracy. The significant difference in simulated and observed values of Orthophosphate in some occasions can be due to the influence of the catchment processes. For example, the significant under prediction by the model may be due to increased Orthophosphate level in the river at that point, resulting from Phosphorus containing fertilizer usage in the corresponding sub-catchment in particular situation.

As per residual frequency distribution of Organic Phosphorus, there are 86.1% residuals within ±1SD, which is because the model has slightly under predicted Organic Phosphorus in few events, nevertheless the model has significantly over predicted three events resulting 8.3% residuals outside ±2SD (91.7% residuals are within ±2SD range). Though the residual frequency distribution is acceptable, it is difficult to conclude that the HEC-RAS model is capable of modelling Organic Phosphorus variations in Nillambe Oya due to the low R-squared value.

In general, HEC-RAS model has simulated water quality of most of the events satisfactorily. The HEC-RAS model is capable in simulating the transportation and fate of nutrients and their dynamic changes in riverine nutrient concentrations (riverine modelling system). It cannot simulate changes in water quality concentrations due to changes in the watershed characteristics. Therefore, these events with significant difference may be due to the influence of changes in watershed characteristics that may directly result in water quality in the Nillambe Oya. Therefore, HEC-RAS model needs to be integrated with other models that have the capability of modelling water quality changes due to alterations in catchment conditions, and the results of such models could be used in HEC-RAS for better water quality simulation of all the events. Zhang and Wu (2013) have applied HEC-RAS model integrated with SWAT model (watershed modelling system) to Upper Mississippi River basin and they were able to simulate water quality dynamics thoroughly. Therefore, it is obvious that HEC-RAS can perform well when it is integrated with a watershed modelling system.

Chihhao et al. (2012) have applied a combination of modified Steeter-Phelps equation and HEC-RAS to simulate the water quality in a tidal river when available data is limited to assess the impact of tidal variations. HEC-RAS in combination with Qual2 K model has simulated water quality indices that are agreeable with water quality monitoring data in a tidal river in Northern Taiwan (Chihhao et al. 2009). Halaj et al. (2013) have used HEC-RAS to model the process of pollutant transport in streams and determine the dispersion characteristics represented by longitudinal dispersion coefficient. HEC-RAS model with QUAL2E model has been used to simulate water quality and assess the performance of vegetation buffer strips in Johor River (Suliman 2010).

The HEC-RAS model has been used to fill the gaps in water quality data, to generate water quality indices, to estimate pollutant spreading in rivers, to evaluate performance of water quality improving methods and also to simulate water quality, (Chihhao et al. 2009, 2012; Suliman 2010; Halaj et al. 2013). However, there is no published evidence on the use of the HEC-RAS model in Sri Lanka for water quality assessment.

4 Conclusion

The HEC-HMS 3.5 computer model can be reliably used to simulate Nillambe Oya flows with proper calibration and validation. As the transformation method, Snyder unit hydrograph method simulates flows reliably in the study catchment, along with initial loss method. Therefore, as a climate change adaptation strategy in Sri Lanka, mainly in the upper Mahaweli catchment, HEC-HMS 3.5 model can be reliably used to estimate the flow volumes that are available for hydropower generation, drinking water supply and irrigation purposes while maintaining the ecological harmony with the riverine ecosystem.

The HEC-RAS model can more reliably be used to simulate water temperature variations of the Nillambe Oya with proper calibration and validation. The water temperature variation of Nillambe Oya varies between 21 and 28 °C and model simulated this 7 °C variation more consistently. The HEC-RAS model can also be used to simulate water quality of the Nillambe Oya once appropriate calibration and validation is performed. Calibrated and validated HEC-RAS model can be used to simulate Dissolved Oxygen (DO), Biological Oxygen Demand (BOD) and Nitrate Nitrogen (NO_3-N) splendidly. HEC-RAS can also be used to simulate Orthophosphates reasonably. On the other hand, the HEC-RAS model has not given reasonable simulations for Organic Phosphorus. Accordingly, the HEC-RAS model can simulate long term water quality variations rather than particular events due to its inherent ability of modelling water quality properties, only within the riverine system. Therefore, in order to simulate particular events more accurately, HEC-RAS model has to be incorporated with another watershed based water quality model. The development of such models will be a significant requirement in the present scientific world to predict river flow and water quality variations due to climate change. Hence, it will be the primary step towards water resources management, as an adaptation policy for climate change.

Acknowledgements Our heartfelt thanks go to the Director General, Department of Meteorology and Director General of the Mahaweli Authority for providing us meteorological data and flow data, respectively.

References

Allen, R. G., Pereira, L. S., Raes, D. and M. Smith, 1998. Crop evapotranspiration–Guidelines for computing crop water requirements—FAO Irrigation and drainage paper 56. FAO—Food and Agriculture Organization of the United Nations, Rome, Italy.

Arekhi, S. 2012. Runoff modeling by HEC-HMS model (Case Study: Kan watershed, Iran). International Journal of Agriculture and Crop Sciences 4(23): 1807–1811.

Beasley, D. B., Huggins, L. F. and E. J. Monke, 1980. ANSWERS—a model for watershed planning. Transactions of the American Society of Agricultural Engineering 23: 938–944.

Bennett, N. D., Crokea, B. F. W., Guariso, G., Guillaume, J. H. A., Hamilton, S. H., Jakeman, A. J., Marsili-Libellic, S., Newhama, L. T. H., Nortona, J. P., Piercee, S. A., Robsonf, B., Seppeltg, R., Voinovh, A. A., Fathi, B. D. and V. Andreassiand, 2013. Characterising performance of environmental models. Environmental Modelling & Software 40: 1–20.

Chandrapala, L. 1996a. Calculation of areal precipitation of Sri Lanka on district basis using Voronoi Tessalation Method. Proceedings of National Symposium on Climate Change, Central Environmental Authority, Colombo, Sri Lanka. March 7–8, 1996.

Chandrapala, L. 1996b. Long term trends of rainfall and temperature in Sri Lanka. In: Climate Variability and Agriculture, Abrol, Y. P.; Gadgil, S.; Pant, G. B. (eds.). New Delhi, India: Narosa Publishing House. Pp 153–162.

Chihhao, F., Chun-Han, K. and W. Wei-Shen, 2009. An innovative modeling approach using Qual2 K and HEC-RAS integration to assess the impact of tidal effect on River Water quality simulation. Journal of environmental management 90(5):1824–1832.

Chihhao, F., Wang, W., Liu, K. F. and T. Yang, 2012. Sensitivity analysis and water quality modeling of a tidal river using a modified Streeter–Phelps equation with HEC-RAS-Calculated hydraulic characteristics; Environmental Modeling and Assessment 17(6): 639–651.

Dietrich, C., Green, T. R. and A. J. Jakeman, 1999. An analytical model for stream sediment transport: application to Murray and Murrumbidgee reaches, Australia. Hydrological Processes 13(5): 763–776.

Eriyagama, N.; Smakhtin, V.; Chandrapala, L.; Fernando, K. 2010. Impacts of climate change on water resources and agriculture in Sri Lanka: a review and preliminary vulnerability mapping. Colombo, Sri Lanka: International Water Management Institute. 51p. (IWMI Research Report 135). doi:10.5337/2010.211

Fernando, T. K. 1997. Recent variations of rainfall and air temperature in Sri Lanka. Proceedings of the 53rd Annual Session of Sri Lanka Association for the Advancement of Science (SLASS). Section E1. Pp. 318.

Franson. M. A. H., Clesceri, L. S., Greenberg, A. E. and R. R. Trussell, 2005. Standard methods for the examination of water and wastewater. Seventh edition, American Public Health Association, American Water Works Association and Water Pollution Control Federation, American Public Health Association, 1015 Fifteenth street NW, Washington, DC 2005.

Halaj, P., Bárek, V., Báreková, A. and D. Halajová, 2013. Application of HEC-RAS water quality model to estimate contaminant spreading in small stream. Journal of International Scientific Publications: Materials, Methods & Technology 7(1): 482–491.

Halwatura, D. and M. M. M. Najim, 2013. Application of the HEC-HMS model for runoff simulation in a tropical catchment, Environmental Modelling & Software Vol 46: 155–162.

Herath, S.; Ratnayake, U. R. 2004. Monitoring rainfall trends to predict adverse impacts—a case study from Sri Lanka. Global Environmental Change 14: 71–79.

Hunukumbura, P. B., Weerakoon, S. B. and S. Herath, 2008. Runoff Modeling in the Upper Kotmale Basin. In: Traversing no man's land, Interdisciplinary Essays in Honor of Professor MaddumaBandara (Hennayake, N., Rekha, N., Nawfhal, M.,Alagan, R. and C. Daskon, eds), University of Peradeniya, Sri Lanka, 169–184 pp.

IPCC. 2001. Climate Change 2001: Impacts, Adaptation, and Vulnerability. Contribution of Working Group II to the Third Assessment Report of the Intergovernmental Panel on Climate Change, McCarthy, J.; Canziani, O.; Leary, N.; Dokken, D.; White, K. (eds). Cambridge, UK: Cambridge University Press.

Jakeman, A., Littlewood, I. and P. Whitehead, 1990. Computation of the instantaneous unit hydrograph and identifiable component flows with application to two small upland catchments. Journal of Hydrology 117: 275–300.

Jayatillake, H. M;, Chandrapala, L.; Basnayake, B. R. S. B.; Dharmaratne, G. H. P. 2005. Water resources and climate change. Proceedings of Workshop on Sri Lanka National Water Development Report. Wijesekera, N. T. S.; Imbulana, K. A. U. S.; Neupane, B. eds. Paris, France: World Water Assessment Programme (WWAP).

Jensen, M. R. and C. L. Lowney, 2004. Temperature modeling with HEC-RAS. In Critical Transitions in Water and Environmental Resources Management, American Society of Civil Engineers, 1–10 pp.

Johanson, R. C., Davis, H. H., and J. C. Imhoff, 1980. User's Manual for the Hydrologic Simulation Program—Fortran (HSPF) version No. 5.0, EPA-600/9–80-105. US EPA Environmental Research Laboratory, Athens, GA.

Kalita, D. N. 2008. A Study of Basin Response Using HEC-HMS and Subzone Reports of CWC. In: Proceedings of the 13th National Symposium on Hydrology. National Institute of Hydrology, Roorkee, New Delhi.

Knisel, W. G. 1980. A Field Scale Model for Chemicals, Runoff and Erosion from Agricultural Management Systems. United States Department of Agriculture Research Report 1980, USDA, USA.

Pauly, D. 1980. The interrelationship between natural mortality, growth parameters, and mean environmental temperature in 175 fish stocks. Journal of International Exploration 39(2): 175–192.

Punyawardena, B.V.R. 2008. Agro Ecological Regions of Sri Lanka. Department of Agriculture, Peradeniya, Sri Lanka.

Ratnayake, U., Sachindra, D. A. and K. D. W. Nandalal, 2010.Rainfall forecasting for flood prediction in the Nilwala basin, International Conference on Sustainable Built Environment (ICSBE-2010), Kandy.

Sardoii, E. R., Rostami, N., Sigaroudi, S. K. and S. Taheri, 2012. Calibration of loss estimation methods in HEC-HMS for simulation of surface runoff (Case Study: Amirkabir Dam Watershed, Iran). Advances in Environmental Biology 6(1): 343–348.

Shantha, W. W. A.; Jayasundara J. M. S. B. 2005. Study on changes of rainfall in the Mahaweli upper watershed in Sri Lanka due to climatic changes and develop a correction model for global warming. Paper presented at the International Symposium on the Stabilisation of Greenhouse Gas Concentrations. Hadley Centre, Met Office, Exeter, UK.

Simons, M., Podger, G. and R. Cooke, 1996. IQQM—a hydrologic modelling tool for water resource and salinity management. Environmental Software 11(1): 185–192.

Sivapalan, M. and N. R. Viney, 1999. A conceptual model of sediment transport: application to the Avon River Basin in Western Australia. Hydrological Processes 13: 727–743.

Suliman, A. H. 2010. Using HEC-RAS and qual2e to assess Johor River water quality. Doctoral dissertation, Faculty of Civil Engineering, University Teknologi Malaysia, Bungi, Malaysia.

US Environmental Protection Agency 1994. SWRRBWQ Window's interface users guide. Washington, D.C. 20460.

Vertessey, R. A., Watson, F. G. R., Rahman, J. M., Cuddy, S. D., Seaton, S. P., Chiew, F. H., Scanlon, P. J., Marston, F. M., Lymbuner, L., Jeanelle, S. and M. Verbunt, 2001. New software to aid water quality management in the catchments and waterways of the south-east Queensland region. In: Proceedings of the Third Australian Stream Management Conference, 611–616 pp.

Watson, F., Seaton, S. and J. Rahman, 2001. Deploying environmental software using the Tarsier modelling framework. In: Proceedings of the Third Australian Stream Management Conference, 631–638 pp.

Yan, M. 2013. Uganda Manafwa River early flood warning system development hydrologic watershed modeling using HEC-HMS, HEC-RAS, ArcGIS. Doctoral dissertation, Massachusetts Institute of Technology, USA.

Yilma, H. and S. A. Moges, 2007. Application of Semi-Distributed Conceptual Hydrological Model for Flow Forecasting on Upland Catchments of Blue Nile River Basin, A Case Study of GilgelAbbay Catchment. Catchment and Lake Research, 200 pp.

Young, R. A., Anderson, W. P., Bosch, D. D. and C. A. Onstad, 1987. AGNPS, agricultural nonpoint source pollution. A watershed analysis tool. In: Conservation Research Report 35. US Department of Agriculture, Washington, DC, USA.

Zhang, Z. and B. Johnson, 2012. Aquatic nutrient simulation modules (NSM). ERDC/EL-12-X, Draft, U.S. Army Engineer Research and Development Centre, Vicksburg, MS.

Zhang, Z. and M. Wu, 2013. Evaluating the transport and fate of nutrients in large scale river basins using an integrated modelling system. In: Landscape Ecology for Sustainable Environment and Culture (Fu, B. and K. Bruce Jones, eds), Springer, 10:187–204.

The Challenges and Opportunities for Higher Education Institutions at the Science–Policy Interface

Edward A. Morgan

Abstract As one of the major 'producers' of climate change research, higher education institutions have had a key role to play in getting the climate change on the policy agenda. As the call for action on climate change intensifies, researchers within institutions will face new demands, with new challenges and opportunities, to inform policy to support both mitigation and adaptation. However, the uncertain, complex and interdisciplinary nature of climate change as an issue, combined with the inevitably highly politicised context of climate change action, means that involvement with policy will have a number of potential pitfalls. This paper draws lessons from recent research into the role of science within water resource management, to consider the challenges and opportunities higher institutions might face at the science–policy interface for climate change research. It notes that simply providing knowledge might be limited by uncertainty and politicisation, and suggests that collaborative, co-learning approaches might be more useful. Such approaches will require resource intensive interactions between scientists and stakeholders, as well as suitable governance arrangements. However, the paper argues that this focus on collaboration provides opportunities and could have multiple benefits.

Keywords Climate change · Science–policy interface · Co-learning · Governance · Research · Universities

E.A. Morgan (✉)
Cities Research Institute, Griffith University, Nathan, Brisbane, QLD 4111, Australia
e-mail: ed.morgan@griffith.edu.au

E.A. Morgan
Griffith Climate Change Response Program, Griffith University, Gold Coast Campus, Southport, QLD 4222, Australia

W. Leal Filho (ed.), *Climate Change Research at Universities*,
DOI 10.1007/978-3-319-58214-6_7

1 Introduction

The scientific community has played and important in getting anthropogenic climate change on to the policy agenda. As the 'producers' of research, higher education institutions have been at the forefront of climate change research, and consequently have become involved in the policy (and political) processes that have come with it. The five IPCC reports draw upon the work done by thousands of scientists, based in a range of higher education institutions (as well as other research institutions), into the science behind climate change adaptation and mitigation. Notably, however, climate change has taken a long time to get on to the policy agenda and getting concerted action has proven very difficult. Despite an ongong 'supply' of knowledge from research about climate change, the political nature of the issue has made responding difficult.

At the same time, university researchers and research centres are increasingly being asked to demonstrate their public service and measure their impact (Ball 2012; Grant 2012; Javey 2012). Their public outreach and their influence on policy are commonly cited as the 'impact' that university research can have. The rise of neo-liberalism in most developed countries has reached universities and they are increasingly being expected and allowed to raise their own funds and operate as businesses in a competitive market. As a result, measures of success and impact in both education and research are becoming increasingly important, placing increased demand on researchers for publications, securing research funding and being involved in 'public service' (Ball 2012; Dowling 2008; Fuller 2000).

This paper discusses how these two factors—the political and complex nature of climate change and the changing role and expectations of universities—provide both challenges and opportunities for higher education institutions and those working within them in the context of the science–policy interface. The paper starts by highlighting the politicised and complex nature of climate change and how this affects the use of knowledge. It then briefly discusses the interface between science and policy, and the role universities can have. The discussion then turns to the challenges and opportunities these factors create. The discussion draws upon research into the different roles science can play in policy, as highlighted by recent research into the use of science in water management. It argues that thinking about the role of science and scientists as stakeholders in the governance of climate change opens opportunities for researchers to impact on policy. This would mean a changing role for higher education institutions, which are already facing a range of pressures to change.

2 Climate Change and Politicisation

Climate change is a political issue, as evidenced by ongoing debate about the science, its potential impacts and arguments over what action (if any) to take (Giddens 2009). This can make the involvement of researchers and the wider scientific community more challenging (Grundmann 2007; Pielke 2004, 2007; Sarewitz 2004).

The implications of anthropogenic global warming and climate change impact upon everyone and an effective response likely requires substantial behaviour change and possibly transformative change in economic and social systems (IPCC 2014; Pelling et al. 2015). The impacts of such a shift (which could be both positive and negative) will be felt by almost everyone, although of course not equally (IPCC 2014). Although evidence and consensus has steadily built over the decades, there have been continued efforts to cast doubt on the science and the need for action (Oreskes and Conway 2011). Scientists have become caught up in the politics and drawn criticism for engaging in the politics and the way in which they have done so. In fact, it has been argued that scientists have been too political in urging action (Kellow 2007; Pielke 2004), and also that they have not been strong enough in their advocacy for action (Grundmann 2002).

The attempts of the scientific community to remain knowledge providers that 'speak truth to power' have been challenged by the very fact that the issue is so politically fraught. Any slight disagreement or uncertainty has been used to show a lack of consensus and therefore implied doubt; in response scientists (and those calling for action) have worked hard to emphasise the consensus (Oreskes 2004; Doran 2009; Cook et al. 2013). The IPCC, which explicitly aims to be a purely advisory body and not to advocate for any particular solution, has been challenged repeatedly for being a political body. In fact, it has always been more than simply experts providing knowledge, but a hybrid or boundary organisation negotiating the interface between science and policy (Hoppe et al. 2013). However, the involvement of scientists has, in many ways, exacerbated the problem, as has been noted for a range of environmental issues (Sarewitz 2004). The issue, which is fundamentally about our behaviour and our values, has often been reduced to a battle over science and uncertainty. The politics of climate change have become a series of 'knowledge fights' (van Buuren and Edelenbos 2004). Different factions with different views on the issue have tried to present themselves as the rational, evidence-based actor, presenting experts and knowledge that back their argument.

Although the international community responded to the possibility of climate change by establishing the UNFCCC, international agreement on what to do in response has been slow to develop, with action even slower to materialise. The recent Paris Agreement has been seen by some as an important step forward, at the very least in getting all nations to agree (in principle) to take action but whether it will result in the sort of significant and ongoing action that is needed remains to be seen (Clemencon 2016). However, that in a number of countries there is still significant disagreement over the need for action and what action is sufficient.

Despite the efforts of scientists in providing more and more knowledge the response to climate change is determined by the politics, and hence getting research to have an impact means being involved in this politicisation.

3 Universities and the Science–Policy Interface

The term science–policy interface is used to describe the interactions between scientists and policy-makers (Hoppe et al. 2013; van den Hove 2007). These interactions have attracted attention from a variety of perspectives, including policy sciences, philosophy, science and technology studies etc.

The push for evidence-based policy-making, continuing the tradition of rational policy-making, puts scientific knowledge and research at the forefront of policy-making (Head 2008). However, a number of studies have noted challenges in using science in policy-making. The expectation that scientists or other experts can simply provide facts, or speak truth to power, has been widely questioned and criticised (Landry et al. 2001; Lovbrand and Oberg 2005; Slobet al. 2007). The idea that knowledge can be transferred from scientists/experts to policy-makers as a simple, linear supply–demand transaction has been criticised as too simplistic (Hoppe 2010). Alternative models that consider science and policy as overlapping communities or see the interface as a web of interactions (Kasperson 2011; Slob et al. 2007). Similarly, a range of different roles for scientists within policy processes has been identified (Michaels 2009; Pielke 2007; Spruijt et al. 2014; Turnhout et al. 2013). These discussions note that scientists can be far more than simply providers of knowledge that aim to speck truth to power, but can be involved as arbiters, brokers, participatory knowledge producers or even advocates.

These different roles become increasingly importance if policy-making is considered through a network governance perspective (Rhodes 2007). Responses to climate change are increasingly focused on a network governance approach that considers the wide variety of actors that are involved in making and implementing policy (Bauer et al. 2011; May and Plummer 2011). It recognises the role of industry, experts, communities, markets and other institutions on policy development and implementation. Scientists and other experts will interact with any or all of these actors within the governance network, or indeed may become an actor within it (Hoppe 2010; Miller 2001).

This is important for universities, and not simply because they are institutions that carry out a wide range of research, much of which will be relevant to the governance of climate change. There is also growing pressure on universities and individual researchers to justify their research through its impact beyond the scientific community (Grant 2012; Javey 2012). The neo-liberal turn in Western democracies demands that universities show their (usually economic) value to society, and their contribution to the market (Ball 2012; Fuller 2000). Universities and other higher education institutions are being expected to find other sources of funding, as increasingly stretched government funding is cut from the higher education sector. Universities are increasingly competing for (often reduced) research funding, building collaborations with government and industry to secure alternative research funding, and looking to produce knowledge that is policy-relevant or that can have some sort of visible or measurable impact.

As a result, universities, or at least some parts of universities, are becoming boundary organisations (Parker and Crona 2012; Wehrens et al. 2014). Boundary organisations carry out boundary work at the science–policy interface. These organisations interpret, translate, broker and mediate between scientists and policy-makers (Guston 2001; Hoppe 2010). Such organisations allow the participation of actors from both arenas, thus creating better interaction between the two, while still being accountable to both. In fact, it has been suggested that these organisations actually help to define and negotiate the boundary between science and policy (Guston 2001; Hoppe 2010). They do the work that is important and necessary for effective interaction with between scientists and policy-makers, but that neither group can effectively carry out themselves. Those working in universities are increasingly expected to perform this sort of boundary work in order to ensure their research has impact.

This role for universities at the boundary between science and policy creates both significant challenges, but also some opportunities for university researchers. Involvement in climate change, being such a global, pressing and political topic, highlights many of the challenges. However, this paper argues that a broader consideration the use of science and recognition of the complexity of the science–policy interface can provide opportunities for universities to have impact in climate change policy. It draws on the literature around the science–policy interface, and recent empirical research into the use of science in policy to highlight both challenges and opportunities for universities and researchers.

4 Methodology

This discussion draws on recent research into the science–policy interface that focused on the use of science in water resource management. The project used a case study approach to investigate how science was used by a range of stakeholders in the governance of water resources in Southeast Queensland, Australia (Morgan 2014). The study used semi-structured interviews with 22 key stakeholders, document analysis and triangulation to theory to study how science was used within water management.[1] The project examined the use of science and the interactions between scientists and a range of stakeholders involved in water management, including policy-makers, planners, decision-makers, scientists and community members. It asked how and why science was used, and then drew conclusions on

[1]Some of the results of this study have already been published where further details on the study can be found (Morgan and Grant-Smith 2015; Morgan and Osborne 2016).

different roles of science within policy, especially where issues are uncertain and complex and governance arrangements are diffuse, as is the case with climate change (and especially climate change adaptation).

The insights into the science–policy interface gained from this empirical study support and expand existing literature. These insights are used here to analyse the challenges and opportunities for higher education institutions to be involved in climate change policy, and particularly climate change adaptation policy. The better understanding of how science is used within policy gained from the empirical work is used to consider the role researchers in universities might be able to play to have greater impact on policy. The research highlighted the complexity of the use of science in policy, which provides new insights into the challenges and opportunities faced by university researchers at the science–policy interface.

5 Discussion

5.1 Challenges

The highly political nature of climate change can make it difficult to get 'evidence' accepted into policy processes, even where governments claim to be focused on evidence-based policy. Even where the existence of climate change is broadly accepted, there are still questions over how to mitigate carbon emissions and how quickly to do it. Adaptation to climate change faces similar problems, in part because the impacts at the local scale are still relatively uncertain and responses involve a complex mix of social, ecological and economic changes. The conflict between pursuing continued economic growth—especially important for those in developing countries—and current reliance on carbon intensive industries for growth is a deep political and policy challenge.

The political nature of climate change and the fact that the response has a broad range of impacts means that research in the area is at risk from politicisation. Any research in the area can be dismissed by those that disagree with it, as ongoing attacks on climate science show. Similarly, the research is used by those wanting action to challenge government policy and the status quo. It is possible for interest groups to tie the issue they are interested into climate change adaptation. The uncertainty around the exact nature, magnitude and impacts of climate change are used to both question the science and demand action. The development of knowledge fights can be seen in a range of areas where there are already strong views or where ideas challenge particular values. Responding to this politicisation by simply trying to fill a knowledge gap is unlikely to be effective. Research into the rejection of recycled water, for example, highlights that attempts to focus on and communicate the science fail because the issue is highly emotive and determined by values (Morgan and Grant-Smith 2015). People's values (in this case around water) were challenged by the 'yuck' factor of drinking water that was seen as deriving from sewage—no science however well communicated could challenge this fear. Similarly, climate change challenges our values about use of resources, economic growth and consumption behaviours. Consequently, simply providing more knowledge to people gives no guarantee of action if that knowledge challenges deeply held values or interests (economic or social).

This politicisation presents a further challenge for universities and those working within them. Universities and researchers rely heavily on government funding. If universities are seen as challenging power, they risk having funding reduced or removed by any government that dislikes their apparent politics. To some extent university research must follow the priorities of the government of the day, and an issue such as climate change is likely to move up and down the agenda of different governments (especially compared to, say, defence or health research). The drive for higher education institutions to demonstrate their worth in a neo-liberal market exacerbates this. It results in increasing pressure to show that the research will contribute to economic growth. If the research is highlighting the problems with fossil fuel-based economic growth or calls for changes in consumption it will be difficult to get that message heard in the corridors of power.

The complexity of the issue is also a challenge. Where issues are less complex, it is more straightforward to use science within policy. If governments are open to advice, there is scope for simply providing knowledge or acting as a knowledge broker to provide a range of policy options. Mitigation and adaptation research has to consider the social, economic and ecological impacts of climate change, as well as responses to them. Such research is complex, interdisciplinary and is both harder to communicate and unlikely to present straightforward answers for policy. Research into the use of science in water management noted that hydrological modelling was highly embedded into water policy and was used to supply knowledge about water supply and demand. In contrast, ecological science around environmental water quality was far less developed and was more complex. There were few ecologists within government, and getting ecological science taken up into policy required substantial boundary work within relatively long-term relationships between scientists and policy-makers (Morgan 2014). Climate change has little history in policy-making and the issues are complex, hence it is likely that translating, brokering and even knowledge co-production will be needed if research is to influence policy.

This complexity and politicisation means that researchers will face the question of whether to become involved in advocacy. The problem with scientists or researchers acting as advocates is that (a) it is perceived to damage the credibility of science in general (Pielke 2007) and (b) 'experts' can often hold considerable sway in decision-making processes, power that does not have democratic legitimacy (Kellow 2007; Lackey 2007). Fears over a technocracy have called into question the position and legitimacy of experts. If researchers call for particular adaptation or mitigation actions, where do they draw their legitimacy from, and what is their responsibility if the actions turn out to be wrong? However, scientific knowledge can be used by others in advocacy, as is often the case in environmental protests (Morgan and Osborne 2016). This can help establish legitimacy and credibility for activists within technocratic policy processes. In some cases scientists will become involved in advocacy. This will require the skills to negotiate the politics, and researchers that become involved risk political attacks (potentially with serious consequences). However, given the right context, there may be 'safer' ways to be involved in advocacy. For example, if there is an argument over a legally protected species, scientists can advocate 'on behalf of' the species by focusing on the risks to that species, which can be established by science (Morgan and Osborne 2016). Similarly, if action on climate change was legislated for, or at least widely accepted within governance arrangements, then scientists might be able to advocate more strongly for action. Note, however, that any advocacy of a particular action risks the perception of a technocratic approach, which might be seen as lacking legitimacy.

5.2 *Opportunities*

Despite the challenges that climate change research might face, consideration of the role of science and scientists in policy open up opportunities for researchers and universities. Firstly, although governments are often the focal points of policy, a governance view of policy emphasises the role of a range of stakeholders: government, industry, communities etc. Higher education institutions are a stakeholder in a range of policy processes, most obviously education. In some respects they are a stakeholder in climate change governance, as 'producers' of research and 'suppliers' of knowledge to the debate, but there is room for this role as stakeholder to expand.

There is a focus on interacting with and influencing national governments around climate change. However, the response to climate change is about more than what governments do. Industry, non-governmental organisations (NGOs), the media, communities and individuals all have a role to play. Climate change adaptation, in particular, will need to take place at the local and regional level, and will require the involvement of a range of community groups, NGOs and government agencies operating collaboratively (May and Plummer 2011).

In order to work more successfully with any of these groups, applying a knowledge provision, supply-demand approach may not be sufficient. Engaging effectively may require researchers to aim to be stakeholders in ongoing, long-standing processes. Rather than aim to supply knowledge either to highlight an issue or to provide a possible solution, researchers can work with other stakeholders to create a shared understanding of the issue and develop a range of potential options (Morgan and Grant-Smith 2015). This extends the idea of researcher as 'honest broker' towards a knowledge co-producer role (Turnhout et al. 2013) to involve them more in the governance process. Importantly, having researchers involved in co-learning also allows space for other knowledges to be included into policy (Morgan and Cole-Hawthorne 2016). It requires researchers (as well as other stakeholders) to approach the process as equal participants, and helps avoid issues of technocratic power, and by creating a shared understanding political conflicts can be resolved as part of the process.

The idea of co-learning is paralleled in the literature as an ideal for good environmental governance (Blackmore 2007; Head 2010; Lockwood et al. 2010). It also reflects the growing interest in and call for social learning for sustainability, resource management and water governance (Collins and Ison 2009; Rijke et al. 2012; Steyaert and Jiggins 2007). However, such collaborative processes are not easy to establish. They are time and resource intensive, require long-standing interactions to allow trust to develop, and are easily disrupted by power imbalances or changes to personnel and external factors (Durose and Rummery 2006; Morrison et al. 2004; Morrison 2007; Wanna 2008).

It is suggested here that higher education institutions can have a role to play in helping develop these processes. They can act as boundary organisations where such interactions can develop and take place. If research and co-learning are part of the same process, universities can be places where collaboration takes place and researchers can be facilitators. They have the scope to develop long-standing partnerships with stakeholders, are often trusted as 'independent' and can potentially balance power differentials. They can extend their boundary organisation role to be places where co-learning can take place. Water quality management provides an example where university researchers worked closely with others to establish co-learning, and allowed them to have significant impact on a complex issue through a collaborative approach (Cottingham et al. 2010; Morgan 2014). Researchers were able to align their outcomes with the needs of policy-makers thanks to a long-standing collaborative approach. This is not to suggest that all research needs to take this approach. Such a participatory approach to research needs to be carefully designed and carried out, and will have limitations (Mackenzie et al. 2012). However, for research focused on responses to climate change, collaborative, participatory co-learning approaches can ensure that research can influence policy and governance.

Note, however, that such a role for universities would require a change in focus. Different sorts of outputs would have to be recognised and grant funding would have to consider the importance of such long-standing co-learning partnerships. Similarly, it is not compatible with a neo-liberal market/competition-focused university. Impact in this context is likely to be long-term and hard to measure with simple metrics. However, for an issue such as climate change, which is complex, requires an adaptive, learning-based approach, and is long term and ongoing, the potential for impact of research in co-learning is significant. It provides an opportunity to design projects that can meet the needs of both researchers (in terms of publications and 'impact') and stakeholders (in terms of giving them access to policy-relevant and cutting-edge research).

6 Conclusions

Higher education institutions have had a key role to play in providing much of the background science that raised the issue of climate change and helped put it on the international policy agenda. However, responding to climate change, and especially adapting to climate change, is both highly complex and highly political. At the same time, higher education institutions are being asked to justify their place in society, with a focus on the impact of research. This article notes that these two pressures provide a particularly challenging for universities as they attempt to influence climate change policy. The disputed and political nature of the issue means uncertainty can be used to limit action or research can be used to justify political decisions. However, it also notes that there are opportunities for these institutions to expand their role as boundary organisations. Drawing on research into the science–policy interface, it argues that focusing on a collaborative, co-learning approach to interacting with a range of stakeholders, rather than simply acting as experts for policymakers, could provide a more effective way of using research to support responses to climate change. Universities could become the boundary organisations that facilitate this adaptive, learning approach. Such an approach, however, would require a shift in focus for universities that may be difficult if measures of success remain volume of publication and simple measures of immediate impact. Although, the focus on impact might be a welcome addition to measuring the 'success' of research, it remains unclear what impact is and how it could be measured. Climate change is a long-term, ongoing issue; with action on mitigation only just beginning, adaptation research may not have impact for decades yet. Nonetheless, responding to climate change requires long-term, adaptive approaches that require the best-available knowledge, and the interactions between universities and society can have a role to play in developing these approaches. Better understanding of these interactions will help universities play a role in climate change. There are a range of different initiatives being used by different universities, and these provide an opportunity to study interactions and draw lessons on what works. Importantly, however, it is likely that there will be no one answer, best approaches will be context dependent and 'success' might look different. We must avoid simplistic measures of impact and success and accept that the influence of universities and research may very well be indirect and long-term

References

Ball, S. J. (2012). Performativity, Commodification and Commitment: An I-Spy Guide to the Neoliberal University. *British Journal of Educational Studies*, *60*(1), 17–28. doi:10.1080/00071005.2011.650940

Bauer, A., Feichtinger, J., & Steurer, R. (2011). *The governance of climate change adaptation in ten OECD countries : Challenges and approaches*. Institute of Forest, Environmental and Natural Resource Policy: Vienna

Blackmore, C. (2007). What kinds of knowledge, knowing and learning are required for addressing resource dilemmas?: a theoretical overview. *Environmental Science & Policy, 10* (6), 512–525. doi:10.1016/j.envsci.2007.02.007
Clemencon, R. (2016). The Two Sides of the Paris Climate Agreement: Dismal Failure or Historic Breakthrough? *The Journal of Environment & Development, 25*(1), 3–24. doi:10.1177/1070496516631362
Collins, K., & Ison, R. (2009). Jumping off Arnstein's ladder: social learning as a new policy paradigm for climate change adaptation. *Environmental Policy and Governance, 19*(6), 358–373. doi:10.1002/eet.523
Cook, J., Nuccitelli, D., Green, S. A., Richardson, M., Winkler, B., Painting, R., Way, R., Jacobs P., Skuce, A. (2013). Quantifying the consensus on anthropogenic global warming in the scientific literature. Environmental Research Letters, 8(2), 24024. https://doi.org/10.1088/1748-9326/8/2/024024
Cottingham, R., Delfau, K. F., & Garde, P. (2010). Managing diffuse water pollution in South East Queensland An analysis of the role of the Healthy Waterways Partnership. Brisbane: International Water Centre
Doran, P. T., & Zimmerman, M. K. (2009). Examining the scientific consensus on climate change. Eos, 90(3), 22–23. https://doi.org/10.1029/2009EO030002
Dowling, R. (2008). Geographies of identity: labouring in the "neoliberal" university. *Progress in Human Geography, 32*(6), 812–820. doi:10.1177/0309132507088032
Durose, C., & Rummery, K. (2006). Governance and Collaboration: Review Article. *Social Policy and Society, 5*(02), 315. doi:10.1017/S147474640500299X
Fuller, S. (2000). *The governance of science*. Buckingham, UK: Open University Press.
Giddens, A. (2009). *Politics of Climate Change*. Cambridge, UK: Polity Press.
Grant, W. J. (2012, May 4). The "impact" of research carries weight (but ripples matter more). *The Conversation*. Retrieved from http://theconversation.com/the-impact-of-research-carries-weight-but-ripples-matter-more-6820
Grundmann, R. (2002). Transnational policy networks and the role of advocacy scientists: from ozone layer protection to climate change. *Proceedings of the 2001 Berlin Conference on the Human Dimensions of Global Environmental Change "Global Environmental Change and the Nation State*, 405–414. Retrieved from http://eprints.aston.ac.uk/1496/
Grundmann, R. (2007). Climate change and knowledge politics. *Environmental Politics, 16*(3), 414–432. doi:10.1080/09644010701251656
Guston, D. H. (2001). Boundary Organizations in Environmental Policy and Science: An Introduction. *Science, Technology & Human Values, 26*(4), 399–408. doi:10.1177/016224390102600401
Head, B. W. (2008). Three Lenses of Evidence-Based Policy. *Australian Journal of Public Administration, 67*(1), 1–11. doi:10.1111/j.1467-8500.2007.00564.x
Head, B. W. (2010). Water policy—Evidence, learning and the governance of uncertainty. *Policy and Society, 29*(2), 171–180. doi:10.1016/j.polsoc.2010.03.007
Hoppe, R. (2010). From "knowledge use" towards "boundary work". Sketch of an emerging new agenda for inquiry into science-policy interaction. In R. J. in't Veld (Ed.), *Knowledge Democracy. Consequences for Science, Politics and Media* (pp. 169–186). Heidleberg: Springer. Retrieved from http://works.bepress.com/robert_hoppe1/15
Hoppe, R., Wesselink, A., & Cairns, R. (2013). Lost in the problem: The role of boundary organisations in the governance of climate change. *Wiley Interdisciplinary Reviews: Climate Change, 4*(4), 283–300. doi:10.1002/wcc.225
IPCC. (2014). *Climate Change 2014: Synthesis Report. Contribution of Working Groups I, II and III to the Fifth Assessment Report of the Intergovernmental Panel on Climate Change*. (R. K. Pachauri & L. A. Meyer, Eds.). Geneva, Switzerland.
Javey, A. (2012). Measuring academic impact. *ACS Nano, 6*(8), 6529. doi:10.1021/nn303711b
Kasperson, R. (2011). Characterizing the Science/Practice Gap. In R. Kasperson & M. Berberian (Eds.), *Integrating Science and Policy* (pp. 3–22). London: Earthscan.
Kellow, A. (2007). *Science and Public Policy*. Cheltenham: Edward Elgar.

Lackey, R. T. (2007). Science, scientists, and policy advocacy. *Conservation Biology: The Journal of the Society for Conservation Biology, 21*(1), 12–17. doi:10.1111/j.1523-1739.2006.00639.x

Landry, R., Amara, N., & Lamari, M. (2001). Utilization of social science research knowledge in Canada. *Research Policy, 30*(2), 333–349. doi:10.1016/S0048-7333(00)00081-0

Lockwood, M., Davidson, J., Curtis, A., Stratford, E., & Griffith, R. (2010). Governance Principles for Natural Resource Management. *Society & Natural Resources, 23*(10), 986–1001. doi:10.1080/08941920802178214

Lovbrand, E., & Oberg, G. (2005). Comment on "How science makes environmental controversies worse" by Daniel Sarewitz, Environmental Science and Policy, 7, 385–403 and "When Scientists politicise science: making sense of the controversy over The Skeptical Environmentalist" by Roger A. Pielke Jr., Environmental Science and Policy, 7, 405–417. *Environmental Science & Policy, 8*(2), 195–197. doi:10.1016/j.envsci.2004.12.007

Mackenzie, J., Tan, P.-L., Hoverman, S., & Baldwin, C. (2012). The value and limitations of Participatory Action Research methodology. *Journal of Hydrology, 474*, 11–21. doi:10.1016/j.jhydrol.2012.09.008

May, B., & Plummer, R. (2011). Accommodating the Challenges of Climate Change Adaptation and Governance in Conventional Risk Management: Adaptive Collaborative Risk Management (ACRM). *Ecology And Society, 16*(1), 47.

Michaels, S. (2009). Matching knowledge brokering strategies to environmental policy problems and settings. *Environmental Science & Policy, 12*(7), 994–1011. doi:10.1016/j.envsci.2009.05.002

Miller, C. (2001). Hybrid Management: Boundary Organizations, Science Policy, and Environmental Governance in the Climate Regime. *Science, Technology & Human Values, 26*(4), 478–500. doi:10.1177/016224390102600405

Morgan, E. A. (2014). *Science & Sustainability: The Use of Science and the Science–Policy Interface in Sustainable Water Resource Management*. Griffith University.

Morgan, E. A., & Cole-Hawthorne, R. (2016). Applying a shared understanding between Aboriginal and Western knowledge to challenge unsustainable neo-liberal planning policy and practice. *Australian Planner, 53*(1), 54–62. doi:10.1080/07293682.2015.1135815

Morgan, E. A., & Grant-Smith, D. C. C. (2015). Tales of science and defiance: the case for co-learning and collaboration in bridging the science/emotion divide in water recycling debates. *Journal of Environmental Planning and Management, 58*(10), 1770–1788. doi:10.1080/09640568.2014.954691

Morgan, E. A., & Osborne, N. (2016). It's the Lungfish, Stupid: Knowledge Fights, Activism, and the Science-Policy Interface. *Geographical Research*, n/a–n/a. doi:10.1111/1745-5871.12181

Morrison, T. H. (2007). Multiscalar Governance and Regional Environmental Management in Australia. *Space and Polity, 11*(3), 227–241. doi:10.1080/13562570701811551

Morrison, T. H., McDonald, G. T., & Lane, M. B. (2004). Integrating Natural Resource Management for Better Environmental Outcomes. *Australian Geographer, 35*(3), 243–258. doi:10.1080/0004918042000311304

Oreskes, N. (2004). Science and public policy: what's proof got to do with it? Environmental Science & Policy, 7(5), 369–383. https://doi.org/10.1016/j.envsci.2004.06.002

Oreskes, N., & Conway, E. M. (2011). *Merchants of Doubt: How a Handful of Scientists Obscured the Truth on Issues from Tobacco Smoke to Global Warming*. New York: Bloomsbury Publishing.

Parker, J., & Crona, B. (2012). All things to all people: Boundary organizations and the contemporary research university. *Social Studies of Science, 42*(2), 262–289. doi:10.1177/0306312711435833

Pelling, M., O'Brien, K., & Matyas, D. (2015). Adaptation and transformation. *Climatic Change, 133*(1), 113–127. doi:10.1007/s10584-014-1303-0

Pielke Jr, R. A. (2004). When scientists politicize science: making sense of controversy over The Skeptical Environmentalist. *Environmental Science & Policy, 7*(5), 405–417. doi:10.1016/j.envsci.2004.06.004

Pielke Jr, R. A. (2007). *The Honest Broker: Making Sense of Science in Policy and Politics*. Cambridge: Cambridge University Press. doi:10.1017/0S0008423908080992

Rhodes, R. A. W. (2007). Understanding Governance: Ten Years On. *Organization Studies, 28*(8), 1243–1264. doi:10.1177/0170840607076586

Rijke, J., Brown, R., Zevenbergen, C., Ashley, R., Farrelly, M., Morison, P., & van Herk, S. (2012). Fit-for-purpose governance: A framework to make adaptive governance operational. *Environmental Science and Policy, 22*, 73–84. doi:10.1016/j.envsci.2012.06.010

Sarewitz, D. R. (2004). How science makes environmental controversies worse. *Environmental Science & Policy, 7*(5), 385–403. doi:10.1016/j.envsci.2004.06.001

Slob, A. F. L., Rijnveld, M., Chapman, A. S., & Strosser, P. (2007). Challenges of linking scientific knowledge to river basin management policy: AquaTerra as a case study. *Environmental Pollution, 148*(3), 867–74. doi:10.1016/j.envpol.2007.01.048

Spruijt, P., Knol, A. B., Vasileiadou, E., Devilee, J., Lebret, E., & Petersen, A. C. (2014). Roles of scientists as policy advisers on complex issues: A literature review. *Environmental Science & Policy, 40*, 16–25. doi:10.1016/j.envsci.2014.03.002

Steyaert, P., & Jiggins, J. (2007). Governance of complex environmental situations through social learning: a synthesis of SLIM's lessons for research, policy and practice. *Environmental Science & Policy, 10*(6), 575–586. doi:10.1016/j.envsci.2007.01.011

Turnhout, E., Stuiver, M., Klostermann, J., Harms, B., & Leeuwis, C. (2013). New roles of science in society: Different repertoires of knowledge brokering. *Science and Public Policy, 40*(3), 354–365. doi:10.1093/scipol/scs114

van Buuren, A., & Edelenbos, J. (2004). Conflicting knowledge: Why is joint knowledge production such a problem? *Science and Public Policy, 31*(4), 289–299. Retrieved from http://repub.eur.nl/pub/1651/

van den Hove, S. (2007). A rationale for science-policy interfaces. *Futures, 39*(7), 807–826. doi:10.1016/j.futures.2006.12.004

Wanna, J. (2008). Collaborative government: meanings, dimensions, drivers and outcomes. In J. O'Flynn & J. Wanna (Eds.), *Collaborative governance: a new era of public policy in Australia?* (pp. 3–13). Acton, A.C.T: ANU E Press.

Wehrens, R., Bekker, M., & Bal, R. (2014). Hybrid Management Configurations in Joint Research. *Science, Technology & Human Values, 39*(1), 6–41. doi:10.1177/0162243913497807

Study of the Vulnerability of Basic Social Infrastructure of the Mexico's North Border for Purposes of Resiliency and Adaptation to the Adverse Effects of Climate Change

Víctor Manuel Lopez-Lopez, Artemisa M. Reyes-Gallegos and Cynthia Lopez-Galvez

Abstract The infrastructure to deliver public services represents the backbone of global society and its development since local climate change official documents consider as strategic the basic social infrastructure for primary networks related to the provision of water, energy, transport and communications. This represents a good opportunity to approach the increasing weather-climate risks to the social infrastructure projects. On the other hand, according to the World Bank it has decreased the amount of life loss due to natural hazards and disasters, but in contrast, each decade the population increases, thus affecting in their belongings and the social infrastructure exposed to the danger of the climate. Related to the above and taking into account the commitments and aspirations in the field of climate change in Mexico, we proposed a research project in our universities related to vulnerability of the basic social infrastructure of Mexico′s north border, for purposes of resilience and adaptation to the adverse effects of climate change. The general project aim is to identify and classify the basic social infrastructure at the north border States of the Country related to the socio-economic and biophysical vulnerability (climate change impacts) to suggest adaptation and resilience measures. These paper aims to expose the results of these initiative and points out some opportunities for university students and the incorporation of the variable climate change in major projects, including the National Infrastructure Program that worth 600 billion dollars.

V.M. Lopez-Lopez (✉)
Instituto Politécnico Nacional (IPN), Unidad Profesional "Adolfo López Mateos" del IPN. Delegación G.A.M, Edificio 12, Mexico City, Mexico
e-mail: victorlopez@ai.org.mx

A.M. Reyes-Gallegos
Universidad Nacional Autónoma de México, Mexico City, Mexico
e-mail: artemis.rg@comunidad.unam.mx

C. Lopez-Galvez
Student at Universidad Jesuita de Guadalajara, Guadalajara, Mexico
e-mail: muronyvamp@hotmail.com

W. Leal Filho (ed.), *Climate Change Research at Universities*,
DOI 10.1007/978-3-319-58214-6_8

Keywords Climate change · Adaptation · Social infrastructure

1 Introduction

By its geographic location, topography and socio-economic aspects, Mexico is especially vulnerable to the impacts of climate change and climate variability. As a proof of this, the hydrometeorological extreme events that regularly beat the country resulting in serious damage and disasters in several productive sectors in recent years.

The capacity for prevention and response to these climatic events are essential to generate safer and more prepared constructions to face the uncertainty of the conditions of the environment; achieved through studies, projects and constructions to identify and try to reduce vulnerability to climate change in regions and cities. Focusing on official (from government) documents prepared for this purpose (National Strategy for Climate Change, for example) and the quality of the current and future physical infrastructure, must to take into account the climate trends that affect the country.

These paper aims to expose such initiatives and suggests some measures for the preparation of the university students and the incorporation of the variable climate change, at least into major projects that are including in the current National Infrastructure Program which is worth 600 billion dollars.

The product of this initiative can offer Mexican inhabitants and public decision-makers of the main Mexican cities a point of reference about the vulnerability and potential risks from climate change, which population face.

The United Nations Framework Convention on Climate Change (UNFCCC), has established two major categories of response to climate change: mitigation and adaptation. Both are mutually complementary actions to confront this climate phenomenon, because while the first tries to avoid the cause (GHG emissions), the adaptation attends the effects occurring already, to overcome the negative impacts of the no longer possible reverse climate change consequences.

To date, international efforts to mitigate the causes of climate change have been unsuccessful and as a result, researchers and academics are paying more attention to the actions of adaptation.

Globally the Fifth Assessment Report (AR5) of the Intergovernmental Panel on Climate Change (IPCC 2014), had analyzed the involvement of climate change on natural and human systems. This derives from a growing concern of the international scientific community by the implications of climate impacts on poverty, economic growth, the ecosystems, the safety of people and their property and in general by the human development.

The Mexican General Law on Climate Change (articles 27 to 30) refers specifically to the adaptation issue, establishing the reduction of the society and ecosystems of climate change vulnerability (SEMARNAT 2012). This deal with the strength of the resilience of human and natural systems by material measures, as well as the minimization of risks and damages taking into account the current and future scenarios of the climate phenomenon.

It is that kind of responses that this paper includes, placing them in the basic community infrastructure since the contemporary society is becoming more dependent on the uninterrupted, safe, and affordable supply of services like water, energy, transport, telecommunications, and information. The infrastructure of these services must now deal with increasingly frequent weather events and extremes (IPCC 2007, 2014), which should not follow to conceiving and building for climates that do not already exist.

2 General Objective

From the COP 16, held in year 2010 in Cancun, agreements were settled so that governments prepare the world population to live with the challenges posed by global climate change, i.e., facing the inevitable and increasing climate change impacts, adaptation measures should be assumed. Hence, the aim is to establish adaptation responses to the physical infrastructure, classifying the impacts of climate change according to the vulnerabilities of specific region in which was break down the country: The north western region of Mexico.

3 Methodology

According to the Intergovernmental Panel on Climate Change, adaptation implicates forms of response to address the positive and negative consequences of climate change, having implicit adjustments that can be related to human systems (community infrastructure in this case) and to natural systems (ecosystems), to deal with the consequences of current and future climate variability, extreme weather events and changes in the climate at medium and long terms.

The National Climate Change Strategy (of Mexico) 2013–2018 (SEMARNAT 2013), sets out the methodological orientation to address adaptation actions, which specifically was applied in the project "Adaptation to the impacts of climate change in the Mexican country by means sustainable project measures", carried out by the authors of this paper.

In another order, the National Plan of Development 2013–2018 (PND 2014) names social infrastructure to the community-based services network, which can be broke down as follows:

- Water infrastructure: water supply and sewerage networks.
- Energy infrastructure: high, medium and low voltages, fuel transport lines, and other sources of energy.
- Transport infrastructure: land, sea, air.

- Telecommunications infrastructure: television networks, fixed-line telephony, fiber optic, mobile cells.

The methodological approach adopted is related to the information of the basic physical infrastructure (water, energy, transport and communication), and the associated risks (vulnerability and exposure) to the climate change impacts posed by documents and official publications on the subject.

Eventually, we incorporated young graduate students from the Higher School of Engineering and Architecture (IPN-National Polytechnic Institute) to research projects whose main aim is to gain awareness and to prepare them to assume future infrastructure projects taking into account the climate change variable.

4 Results and Discussion

Derived from its geographic location, topography and socio-economic aspects, Mexico is especially vulnerable to the impacts of climate change and climate variability (SEMARNAT 2006). As proof of this statement, extreme hydrometeorological events that regularly plague the country, such as droughts, cyclones, and atypical floods have caused serious damage and disasters in several productive sectors, especially in recent years.

Prevention and response to these climatic events require safer and more prepared areas, to deal with the uncertainty of the environment, which can be achieved through studies, projects and constructions to identify and to try to reduce vulnerability to climate change in regions and cities; official documents were prepared for that purpose focusing on people (National Climate Change Strategy, for instance). For a good quality of the current and future physical infrastructure, resilience and resistance will have to take into account trends in the climate that have affected the country in recent decades, and international experience as well.

Contemporary society is becoming increasingly dependent of uninterrupted, secure, and affordable supply of water, energy, transport, telecommunications and information services, which creates conditions of habitability and economic development.

Derived from this demand, and because climate change variable should to be taking in consideration in developing countries, adaptation to the impacts of future climates should be consider in the planning, design, construction, operation and deconstruction of infrastructure projects.

This paper aims to expose the results of carried out initiatives and suggest some measures, as well as the incorporation of the climatic variable in major projects such as those included in the National Infrastructure Program (of Mexico).

5 The Research Project

National Climate Change Strategy 10-20-40 Vision of the Federal Government of Mexico (NCCS 2013), has basic aims in reducing social vulnerability and increasing resilience on basic infrastructure to the effects of climate change, through the characteristics of the impacts it will dictate the way to address them, as well as the type of infrastructure to be prioritized, and the risks to which the infrastructure is exposed.

The large hydrometeorological phenomena that took place on Mexico in year 2013, it helped to ratify the advantage to pay priority attention to basic or vital infrastructure related to water, energy, transport and communications. This basic infrastructure interconnects with the rest of the infrastructures of all productive sectors, ecosystems, croplands, and cultural heritage of the country.

Challenges faced by the adaptation of the above mentioned infrastructure, also offers opportunities for innovative technology development. Those technology have being set at the context of the sustainable use of natural resources and in a dual or multiple use of certain infrastructures.

In the area of physical infrastructure, adaptation to the climate change impacts has at least two ways of response: to tackle climate impacts that occur in long term, and to develop resilience in such infrastructure in the face of weather events, which occur in short term. A combination of both responses was incorporated in this research.

To carry out the study of the characteristics of vulnerability in the Mexican Republic, the total territory was divided into eight regions, because the local vulnerability is subject to geographical conditions and historical hydrometeorological in each region where the communities in study are located. This regionalization is indicated in Table 1.

Vulnerability ratings grouped by geographical regions allowed to observe the places where there are social, and infrastructure higher climate risks. As a result, the infrastructure of the Northeast, Northwest and Peninsula regions are the least

Table 1 Mexican Republic regionalization

Regions name	States name
North Center	Aguascalientes, Guanajuato, Querétaro, San Luis Potosí, Zacatecas
South Center	Distrito Federal, Estado de México, Morelos
North East	Coahuila, Nuevo León, Tamaulipas
North West	Baja California, Baja California Sur, Chihuahua, Durango, Sinaloa, Sonora
Western	Colima, Nayarit, Michoacán, Jalisco
Easter	Hidalgo, Puebla, Tlaxcala, Veracruz
Peninsula	Campeche, Quintana Roo, Tabasco, Yucatán
South	Chiapas, Guerrero, Oaxaca

Source Modified from (IMCO 2012)

vulnerable to climate change, generally speaking. In both regions, South Central and East, are seen as highly vulnerable areas, in comparison to other Mexican regions.

The region with lower-quality infrastructure to tackle climate change impacts reaches only 30 points from 50, the region that follows has 35 points and the others show a performance close to 40 points. The type of infrastructure that has been developed in these regions and urban composition of lower quality comparatively to other regions (for example those at the center of the country), foster the vulnerability of such regions in the face of climate change impacts.

The vulnerability is greater in those places where hydrometeorological extreme events are unlikely, because the population and authorities are not suitably prepared for such events.

At the very end the northeast and northwest regions are outlined because those two regions coincide with the Mexico's north border, which is our study object.

6 Conclusions

- The first step to tackle the impacts of climate change is to set out national adaptation strategies, supported on regional and local strategies of adaptation formulated in different ways according to the specific features of each region. The case study addressed in our project has been supported both on the Mexican National Climate Change Strategy and the General Law of Climate Change, that make mandatory programs and actions for adaptation to regional and municipal authorities. However, there is great ignorance on the climate change issue on local government level and they have little qualified staff to address the adaptation measures required by law.
- Possible solutions to climate change impacts on the basic infrastructure of the country will require the assessment of specific risks (mostly hydrometeorological risks), which will based on the biophysical and socio-economic vulnerability and the population exposure to those impacts.
- Infrastructure projects considered in programmatic documents such as the National Infrastructure Program will inevitably experience in its life cycle the effects of climate change predicted by the IPCC, which follows the challenge to incorporate the climate change risk in their design, implementation and maintenance.

  The absence of the climate change variable into codes, and building regulations, increases the vulnerability of the high risk regions infrastructure.
- The project findings offer to inhabitants and public decision-makers of Mexican regions, a starting point about vulnerability and potential risks caused by climate change to which the populations should to face.

- Mexican Institutions of higher education, universities and national research centers are on the way to contribute into infrastructure projects with the integrated climate change process, which includes strategies of mitigation, adaptation, vulnerability and environmental risks.

References

IMCO (Instituto Mexicano para la Competitividad, A C.) (2012). *Índices de vulnerabilidad climática de las ciudades mexicanas*. México.

IPCC (Intergovernmental Panel on Climate Change) (2007): *IPCC Fourth Assessment Report: Climate Change 2007 (AR4)*. En internet, URL: http://www.ipcc.ch.

IPCC (Intergovernmental Panel on Climate Change) (29 March 2014). *Working Group II contribution to the IPCC Fifth Assessment Report (AR5) Climate Change 2013: Impacts, Adaptation and Vulnerability*. Yokohama, Japan. Available from: http://ipcc-wg2.gov/AR5/report/final-drafts/.

NCCS (2013) National Climate Change Strategy. 10-20-40 Vision. Mexico: Federal Government of Mexico.

PND (Plan Nacional de Desarrollo 2013–2014). Fecha de consulta 13 de abril de 2014). En internet, URL: http://www.dof.gob.mx/nota_detalle.php?codigo=5299465&fecha=20/05/2013.

SEMARNAT (2006). *México Tercera Comunicación Nacional ante la Convención Marco de las Naciones Unidas sobre el Cambio Climático*. México.

SEMARNAT (2013). *Estrategia Nacional de Cambio Climático. Visión 10-20-40*. Gobierno de la República. México.

SEMARNAT (Secretariat of Environment and Natural Resources) (2012). *Mexico's Climate Change Law and Polices*. SEMARNAT, Mexico.

Climate Change Mitigation and Adaptation Studies in Nigeria Universities: Achievements, Challenges and Prospects

John-Felix K. Akinbami and Catherine Abiola O. Akinbami

Abstract Various studies on the impacts of climate change on the different sectors of the national economy have been carried out over some decades and such studies are still ongoing in Nigeria. This paper gives an overview of studies that have been carried out on various aspects of climate change by the higher institutions and research institutes in the country and what these have achieved in national climate governance and policy making. This is with the aim of assessing the achievements of such publications in climate policy and management as well as the likely missing gaps in and further prospects for climate research and studies by the Nigerian universities. The methodological approach is based on a theoretical background using secondary data obtained majorly from the internet and other published and unpublished literature. Amongst other issues, the results reveal some good measure of achievement of these research outputs in shaping the climate policy and management in the country and at the same time expose the challenges and some of the missing gaps which call for the attention of the Nigerian universities researchers.

Keywords Climate change · Mitigation and Adapatation Studies · Climate governance · Nigerian universities · Sustainable socio-economic development

1 Introduction

Climate change refers to a change which is attributed directly or indirectly to human activities that alter the composition of the global atmosphere and which are in addition to natural climate variability observed over comparable time periods

J.-F.K. Akinbami (✉)
Centre for Energy Research and Development,
Obafemi Awolowo University, Ile-Ife, Nigeria
e-mail: akinbamijfk@gmail.com; jakinbami@cerd.gov.ng

C.A.O. Akinbami
Institute for Entrepreneurship and Development Studies,
Obafemi Awolowo University, Ile-Ife, Nigeria
e-mail: abiolaakinbami1@gmail.com

W. Leal Filho (ed.), *Climate Change Research at Universities*,
DOI 10.1007/978-3-319-58214-6_9

(UNFCCC 1992). Climate change is one of the most serious environmental and human threats undermining the achievement of the Millennium Development Goals (MDGs) and the international communities' efforts to reduce extreme poverty. It has, therefore, emerged as a major challenge to development in the 21st century and conveys some urgency which demands scholarly attention.

Climate change poses challenges on a new scale for humanity, particularly for the populations of lower income countries. Although there are some uncertainties of climate change science and the lack of downscaled data makes it hard to predict how the climate will change at a very local level, however, scientific studies are continuing to understand this phenomenon and to deliver more accurate and downscaled projections. Climate mitigation, low carbon pathways and latterly adaptation responses are being researched and piloted by the international community and national governments, as well as civil society and local communities (Nelson 2011). In this regard, various analyses are emerging in the academic, policy and practitioner literature on various aspects of climate change.

Cook et al. (2013) analyzed the evolution of the scientific consensus on anthropogenic global warming (AGW) in the peer-reviewed scientific literature, examining almost 12,000 climate abstracts from 1991 to 2011 matching the topics 'global climate change' or 'global warming'. They found that 97.1% endorsed and only a minority of 0.7% rejected the attribution of global warming to human activities, the consensus position that humans are causing global warming. Their analysis indicates that the number of papers rejecting the consensus on AGW is a vanishingly small proportion of the published research. Furthermore, the study also demonstrated the fact that scientific consensus on anthropogenic global warming (AGW) has been growing in the covered period (1991–2011). Contrarian claims that there is no consensus among (serious) scientists regarding AGW can therefore be clearly rejected as other studies have also affirmed (Oreskes 2004; Reusswig 2013). This is against the 'debunk' climate science scenario that has happened, especially in the US (McCright and Dunlap 2003). Climate science has been a highly politicized science, though not because climate scientists are advocates of a particular political mission but the issue they are dealing with is clearly political in nature. For instance, if global warming was caused by natural factors alone (natural cycles, activity of the sun, volcanoes etc.), adaptation to it would still be a necessary human response, but nothing from the broad range of activities that we call 'mitigation' would be necessary or make even sense. But if AGW is a fact, and if avoiding dangerous climate change is a meaningful or even necessary goal, then the de-carbonization of the global economy has to be the answer (Reusswig 2013).

Climate science has become part of a much wider social discourse on climate change (Reusswig and Lass 2010). While these discussions have been going on globally, the developing countries particularly in the Sub-Saharan Africa have been on the receiving end of the impacts of climate change due especially to anthropogenic activities being carried out. For instance, in Nigeria, according to a recent survey study by the Department of Climate Change, Federal Ministry of Environment, 50–60% of the country's population could be affected by climate change in the nearest feature. According to the report, global warming related events such as floods, strong winds, droughts and tidal waves could cause massive relocation of people, noting that about 15% of the country's population was presently affected by climate variation and sea level changes. The study also noted that the declining agricultural productivity that has been a major trigger for population movement in the country could be worsened, especially in the semi-arid and arid zones of the northern part of the country (Ihejirika 2016).

Other evidences abound of the effects of climate change in the country. For instance, Agwu and Okhimamhe (2009) reported that the evidence of the unpleasant impact of climate change abound in the southeast of Nigeria. These include increased cases of flooding and numerous gully erosion sites which have resulted to loss of arable farmlands, farm stead, economic tree, biodiversity and others. Also, in the southern ecological zone of Nigeria largely known for high rainfall, Bello et al. (2012) observed that the area is currently confronted by irregularity in the rainfall pattern, while the Guinea savannah is experiencing gradually increasing temperature. Higher rainfall observed in some southern parts of the country coupled with rise in sea level also resulted to crop losses due to water logging, loss of arable land and increased pest infestation (FME 2003). Kalejaiye-Matti et al. (2010) observed that signs of desertification and savanalization are now becoming evident in Oyo, Osun, Ondo and some other parts of the south-western states of Nigeria which hitherto fell within rainforest. In the Sahel zone of northern Nigeria, the most pronounced climate change-related forms of land degradation are wind erosion and related sand dune formation, drought and desertification (Farauta et al. 2012).

In this regard, various studies on climate science and the impacts of climate change on the different sectors of the national economy, different geographical regions and people groups have been carried out over some decades and such studies are still ongoing in Nigeria. These studies span through various aspects from vulnerabilities to mitigation and adaptation strategies to climate change. Sociological and anthropological, scientific and technological, theological and economic, as well as gender and capacity building dimensions to climate change and its impacts have been examined.

Various techniques which included analysis of climate variable data collected over a long period of time, technological and economic modelling, space technology applications, questionnaire administrations and personal interviews and observations have been employed in developing adaptation strategies to cope with the effects of climate change as well as mitigation options to either reduce or eliminate the impacts of climate change. These have led to the development of National Communications, National Adaptation Strategies and Plans of Action and various other national policies for the country. However, despite these achievements in climate change studies in the country, there is still a lot to be done to effectively combat the immediate, foreseeable and remote impacts of climate change in Nigeria.

This paper gives an overview of studies that have been carried out on various aspects of climate change by the higher institutions and research institutes in the country and what these have achieved in national climate governance and policy making. Efforts are also made to examine the challenges to and prospects of climate change studies in the country. This is with the aim of enhancing more robust analysis as well as adequate uptake and integration of these research outputs and outcomes for sustainable national socio-economic development and growth.

Consequently, the objectives of this study are to:

- Identify the various research outputs on climate change from the Nigeria universities.
- Classify these outputs into various aspects of climate change to assess their primary focus.
- Assess the uptake and integration of the research outputs into the national socio-economic development.
- Identify the possible missing links in climate change studies and research in the country.
- Develop a policy framework for improved climate change studies and their uptake into the national economy.

2 Rationale for the Study

The paper is an attempt at reviewing some of the literature on various studies and research on climate change by various scholars in Nigerian universities and research institutes and identifying the various aspects of climate change that have been examined. This is with the aim of identifying the likely gaps in climate change studies and research so far and challenges confronting climate change research in the country. It also reviews the achievements and prospects of such studies to socio-economic development in the country. The study aims to contribute to a framework for informing key stakeholders and decision-makers on and making them more aware of the status of climate change research and phenomenon in the country with a view to stimulating more robust discussions around this topical issue in research and policy circles. It is hoped that the results of the study will help to sensitize the stakeholders on the need for developing a framework of integrating the various research outputs into national socio-economic development and the private sector and development partners on possible areas of climate research focus for funding in alleviating the burdens of the consequences of climate change in Nigeria.

3 Study Area—Geography, Climate and Vegetation

Nigeria is situated in the West African region and lies between longitudes 3° and 14° and latitudes 4° and 14°. It has 36 states and a Federal Capital Territory (FCT) (Fig. 1) with a land mass of 923,768 km^2 and measures about 1200 km from east to west and about 1050 km from north to south. Its coastline covers a distance of about 800 km which confers on her a maritime power. It shares borders with the Republic of Benin in the west and with Cameroon in the east right down to the shores of the Atlantic Ocean which forms the southern limits of Nigerian Territory (see Fig. 2). In the north it borders on the Republics of Niger and Chad. Nigeria features 5 geographical regions: a low coastal zone along the Gulf of Guinea, hills and low plateaus north of the coastal zone, the Niger-Benue river valley, a broad stepped plateau stretching to the northern border that has elevations exceeding 1200 m, and a mountainous zone along the eastern border, which includes the country's highest point, Chappal Waddi (2419 m). Much of the country is laced with productive rivers. Nigeria's ecology varies from tropical forest in the south to dry savanna in the far north, yielding a diverse mix of plant and animal life.

Nigeria, like other tropical lands, has only two seasons namely: the Dry season and the Rainy or Wet season. The dry season is accompanied by a dust laden air mass from the Sahara Desert, which is locally known as Harmattan, or by its main name, the Tropical Continental (CT) air mass and is associated with lower temperatures. The rainy or wet season is heavily influenced by an air mass originating

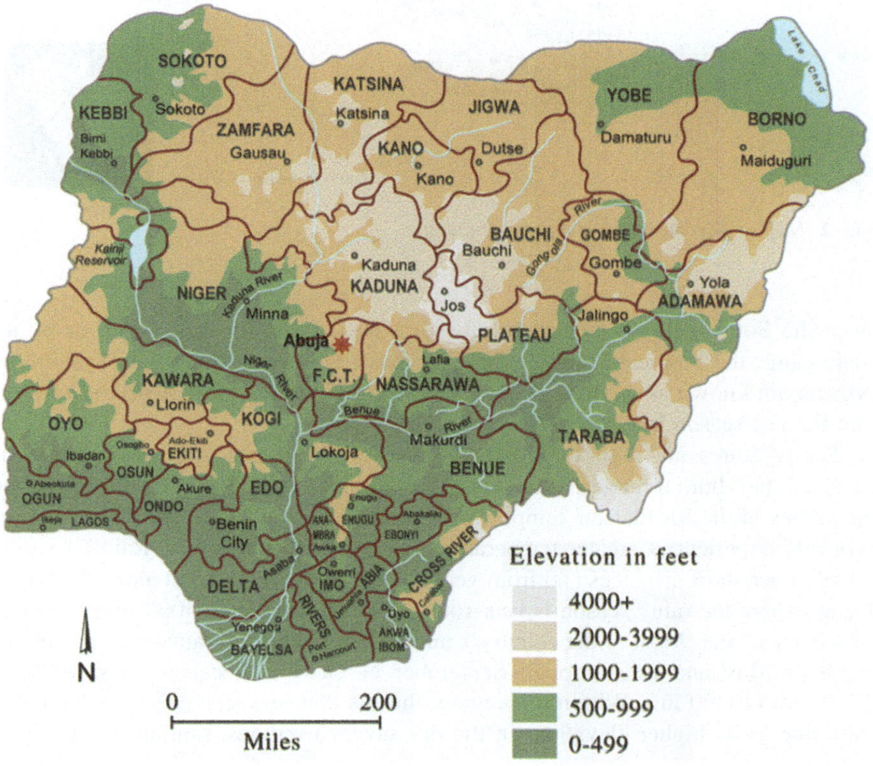

Fig. 1 Map of Nigeria showing Thirty-Six States and a Federal Capital Territory

Fig. 2 Map of Nigeria showing its border countries

from the South Atlantic Ocean, locally known as the south west wind, or by its main name, the Tropical Maritime (MT) air mass. These two major wind systems in Nigeria are known as the trade winds. The wet season lasts from April to October; and the dry season lasts from November till March.

Temperatures are high throughout the year with a very narrow variation in seasonal and diurnal ranges, averaging from 25 to 28 °C (77–82 °F). In the higher elevations of the Jos Plateau, temperatures average 22 °C (72 °F). Northern Nigeria typically experiences greater temperature extremes than the south. Rainfall varies widely over short distances and from year to year. Parts of the coast along the Niger Delta, where the rainy season is year-round, receive more than 4000 mm (160 in.) of rain each year. Most of the country's middle belt, where the rainy season starts in April or May and runs through September or October, receives from 1000 to 1500 mm (40–60 in.). Within this region, the Jos Plateau receives somewhat more rain, due to its higher elevation. In the dry savanna regions, rainfall is especially variable over distance and time. The region along Nigeria's northeastern border receives less than 500 mm (20 in.) of rain per year, and the rainy season lasts barely three months. The second half of the dry season, February–March, is the hottest period of the year when temperatures range from 33 to 38 °C. The extremes of the wet season are felt on the southeastern coast; while the extremes of the dry season, in aridity and high temperatures, are felt in the north east of the country.

Vegetation varies dramatically at both the national and local level in relation to physiographic elements such as climate, soil, elevation, slope and human impact on the environment (Adejuwon 1979; Adesina 2001). There are 3 different types of vegetation in Nigeria which include: forests, savannahs and montane land. Vegetation associated with altitude is described as montane vegetation (Akinbami 2015). The forest and savannahs are divided into 3 zones each. The Forest subtypes include Salt-water swamps (Mangrove), Fresh-water swamps and High forest. The Savannah subtypes are Guinea, Sudan and Sahel. The southern most portions of the forests are mangrove swamp. North of this swamp is a fresh water swamp which contains vegetation from the mangrove swamp. In the northern portion of the forest is the rain forest with its many species of tropical hardwoods.

One of the savannah zones is the Guinean forest-savannah mosaic which is made up of plains of tall grass which are interrupted by trees. Another zone is a Sudan savannah, which is similar to the Guinean forest-savannah mosaic, but it contains shorter grass and shorter trees. The last zone is a Sahel savannah in Nigeria's very dry northeastern zone, which contains patches of grass and sand. Throughout these

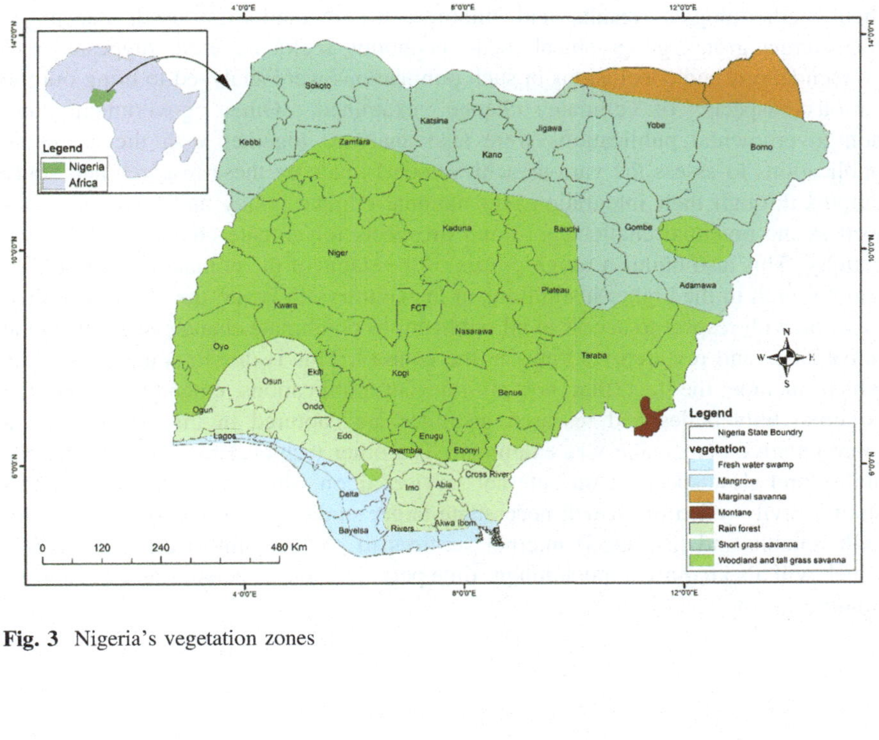

Fig. 3 Nigeria's vegetation zones

drier savannahs, drought and overgrazing have led to desertification—the degradation of vegetation and soil resources. The savannah, especially Guinea and Sudan, are the major grains, grasses, tubers, vegetable and cotton growing regions. The Tropical evergreen rain forest belt bears timber production and forest development, production of cassava; and plantation growing of fruit trees—citrus, oil palm, cocoa, and rubber, among others. Figure 3 shows the different vegetation types in the country.

4 Methodology

This study aims at discussing the achievements of various studies and research work by the Nigerian universities and research institutes on climate change in climate governance and management in the country as well as the challenges to and prospects of climate change studies to socio-economic development in the country. The methodological approach adopted is theoretical and is based on secondary data obtained through the Google search engine to obtain publications published in peer-reviewed journals using various topic searches for climate change studies in Nigeria. Other article types reviewed also included books, conference proceedings papers, study reports, postgraduate (M.Sc. and Ph.D.) thesis reports and other documents that have emanated as research and study outputs from various universities and research institutes in the country. The publications were purposively selected based on their relevance to the focus of this study. Their abstracts as well as their methodologies, results and findings, the methodology used, the major focus/target group, geographical focus, economic sector covered, suggested recommendations and conclusions in such publications were analyzed to bring out the various aspects of climate change examined. Other government and non-governmental publications were also analysed together with the academic publications to assess the various achievements made by these research and study outputs through their integration into national climate policy and governance, as well as the potential challenges to and prospects for climate change study in the country. This also builds a rough profile of the status of the climate change studies and research in the higher institutions in the country However, the study is limited in scope with respect to access to all publications on climate change issues from the universities and research institutes in the country. This is due to some constraints which include the fact that not all relevant publications emanating from the Nigerian universities and research institutes are captured in this study; only a representative percentage was examined. The other technical reason is the unreliability and slow speed of the internet services upon which the data used for this study heavily relied on. It will necessitate larger studies being undertaken together with reliable and fast speed internet services to capture almost all such studies which will also require a much longer time period, resource persons and substantial funding to implement.

5 Results and Findings

The search generated quite a large number of papers. Those that were not published by academic scholars either in the universities or research institutes were eliminated. This brought the number of publications analysed to 68 written by 176 authors, while quite a large percentage were published in various journals, some were published as conference proceedings, others as study reports or postgraduate thesis (M.Sc. and Ph.D.). A close examination of the studies revealed that their focus spans through vulnerabilities to mitigation and adaptation strategies to climate change. Sociological and anthropological, scientific and technological, theological and economic, as well as gender and capacity building dimensions to climate change and its impacts were examined by these studies. They were published within a time period of 30 years between 1986 and 2016 (see Table 1). The articles were classified under various topical themes such as: climate change and capacity building, climate change and educational curriculum, climate change awareness, climate change and gender, emissions inventory, climate change and vulnerability, adaptation and mitigation strategies, adaptation strategies, climate change impacts, climate trends, climate change and conflict resolution and greenhouse gas mitigation (see Table 2). Most of the articles accessed and assessed are skewed towards mitigation options of greenhouse gases in some sectors of the economy especially energy, transport, industrial and agricultural sectors; emissions inventory in various sectors of the economy; climate change impacts; and adaptation strategies to climate change in various sectors of the economy and aggregately. While most of the articles had a national scope, some of them had regional scope such as South West, South East, Niger Delta, Middle Belt, and North. Some covered either a Local Government Area or a State or just a Community in their area of study coverage.

The various researchers employed various methodologies in implementing the set objectives of the research focus. These included Global Climate Model (GCM) (Abiodun et al. 2011), case studies, observations and photographic prints (Adedeji and Ogunsote 2012), focus group discussions and timeline analysis and ranking (Sadiku and Sadiku 2011), multi-stage sampling (Adebayo et al. 2012), spreadsheet models such as COPATH and COMAP (Adesina et al. 1999; Momodu et al. 2011); mixed method (qualitative and quantitative together) (Nzeadibe et al. 2011), simulation and linear optimization models such as MADE-II and MARKAL

Table 1 Frequency of publications per specific period

Publication period	Quantum of publications	% of total publications
1980s	2	2.9
1990s	33	48.5
2000s	9	13.3
2010s	24	35.3
Total	68	100.0

Table 2 Classification of publications into topical themes

	Topical theme	Quantum of publication	% of total publications
1.	Climate trends	7	10.3
2.	GHG mitigation	13	19.1
3.	Adaptation strategies	5	7.4
4.	Climate change scenarios	1	1.5
5.	Climate change and historical perspective	1	1.5
6.	Climate change and conflict resolution	1	1.5
7.	Climate change and technological capability and innovations	1	1.5
8.	Climate change impacts	10	14.7
9.	Adaptation and mitigation strategies	3	4.4
10.	Climate change and vulnerability	2	2.9
11.	Emissions inventory	13	19.1
12.	Climate change and macroeconomics	1	1.5
13.	Climate change and gender	2	2.9
14.	Climate change and awareness	3	4.4
15.	Climate change and educational curriculum	1	1.5
16.	Emissions inventory and mitigation	2	2.9
17.	Climate change and capacity building	2	2.9
Total		68	100.0

(Adegbulugbe 1991; Adegbulugbe et al. 1996; Akinbami 2001; Akinbami and Ibitoye 2003; Akinbami and Lawal 2012).

5.1 Uptake and Integration of the Studies into National Climate Governance

It may be difficult to assess the achievements of these research outputs with respect to their uptake and integration into the national policy and governance for sustainable socio-economic development, however, some of the observed national strides in climate change policy and management in the country which could be attributed to some of these research outputs and the researchers themselves include:

- First National Communication (FME 2003).
- Second National Communication (FME 2014).
- National Adaptation Strategy and Plan of Action on Climate Change for Nigeria (NASPA-CCN) (FME 2011).
- National Environmental, Economic and Development Study (NEEDS) for Climate Change in Nigeria (FME 2010).

- Nigeria's Intended Nationally Determined Contribution (INDC) (FME 2015).
- Lagos State Climate Change Policy (LASGME 2012).
- Increased awareness of climate trends and impacts on various ecosystems has led the law makers to be debating the Nigeria Climate Change Bill sponsored by a member of the House of Representative.
- Nigeria Climate Change Policy.
- Advocacy for the establishment of a National Climate Change Commission.

5.2 *Missing Gaps and Recommendations*

Inspite of the achievements of these research outputs from the various universities and research institutes in the country in shaping climate policy and governance, there are some observed missing gaps in climate change studies in the Nigerian universities. There is the need for comprehensive adaptation studies for each state of the Federation which should also include various communities in both rural and urban settlements. This should ultimately lead to developing evidence based climate change policy for each state as Lagos State has done. There appears to be a gap between the research outputs and the communities of research focus in disseminating the research outputs to them and in taking steps to facilitating the implementation of such outputs through fundable climate project development proposals. Research outputs from these efforts should be disseminated to the studied communities. This should be done through increased interactions between university/research institutions involved in environmental studies and management on the one hand, and the communities under study on the other hand (Momodu et al. 2011).

More studies are needed in the area of climate modeling especially at city levels for dynamic climate policies both at the national and city levels. Mitigation and adaptation pilot projects are needed which would be monitored for scaling up and replication. In this regard, business models have to be developed for sustainability of such climate development projects. Consequently, climate entrepreneurs are essential for investments in climate mitigation and adaptation projects. This is essential in both rural and urban communities and it will also promote both job and wealth creation. However, funding for green projects can be a little complicated, especially from investors that sometimes see the green industry as an unsafe and obscure industry. But with the efforts of continuous advocacy and increased educational awareness by both government and relevant non-governmental organisations (NGOs), investors interest will also continue to rise. Assistance of the development partners will be essential in this regard.

Green buildings and cities are also needed as both mitigation option and adaptation strategy. In this regard, as suggested by Akinbami and Lawal (2012), research and development, demonstration and dissemination in design and materials for energy conscious architecture and energy conscious city layout should be well funded by both government and the private sector.

Another area that demands urgent research focus is conflict resolution especially climate change impact induced conflicts. This is even of utmost urgency especially with the recent bloody clashes between the nomadic herdsmen and crop farmers across the country as a result of the nomadic cross-country grazing culture of these herdsmen which has caused a lot of hardship and trauma to the crop farmers and their communities. As observed by Sayne (2011), for conflict-sensitive adaptation to deliver best value it may have to exhibit some characteristics. First, it will be cross-cutting in vision and delivery. This is because adapting to climate change is often said to mean "adapting development". Yet conflict-sensitive adaptation raises the bar even higher. Thus, traditional development work should be connected to action on security, peace-building, trade, communications, and disaster management. This will involve an integrated kind of study that is all encompassing. Second, successful adaptation will be based on inclusive planning and implementation. Hence, all stakeholders must be involved in the planning, implementation and integration of the uptakes into the socio-economic development of the communities involved, since different stakeholders are endowed with various mix of skills and interests.

For instance, many communities and households have local knowledge crucial to crafting effective responses. Therefore, involving them directly in the planning and implementation of adaptation measures would create opportunities for strengthening resilience and the social fabric and if approached in a conflict-sensitive manner, climate change adaptation initiatives could even help reconcile divided groups by offering a common enemy and a shared set of tasks. Furthermore, successful adaptation will have to confront uncertainty. Science will never provide an exact climate change forecast. However the sophisticated models, predictions may sometimes prove unreliable or even contradictory. Likewise, analyzing conflict after the fact is always simpler than foreseeing it. Vexing knowledge gaps will abide regardless of the number of studies undertaken. Hence, outputs of research work and studies must be carefully implemented with great caution and the public institutions handling the country's adaptation policy will need to plan and permit flexibility and experimentation. They should also adopt "no regrets" policies that yield good results even if the environment does not change in expected ways. Consequently, extensive and vigorous studies are needed to know how best to possibly address climate induced conflicts across the country.

6 Conclusions

Climate research in Nigerian universities and research institutes has gone a long way in bringing to the fore the causal agents and various impacts of climate change in various sectors of the national economy. The studies have been of a national scope as well as regional and community in scope and their focus spans through various aspects of climate change using different methodologies to obtain results in line with the objectives of such studies. The body of research work and studies in the higher institutions has raised the level of awareness on the consequences of climate change both within the government circle and the citizenry. This has also enhanced the uptake and integration of these studies into the national climate governance and policy to some extent although there is still room for improvement. However, inspite of the positive impacts of these studies on the national socio-economic development, some missing gaps have been identified which also provide some direction for future research prospects for climate change studies in Nigerian universities and research institutes for national sustainable socio-economic development.

References

Abiodun, B. J., Salami, A. T., & Tadross, M. (2011). Climate Change Scenarios for Nigeria: Understanding Biophysical Impacts. *Climate Systems Analysis Group, Cape Town, for Building Nigeria's Response to Climate Change Project. Ibadan, Nigeria*. Nigeria: Nigerian Environmental Study/Action Team (NEST).

Adebayo, A. A., Onu, J. I., Adebayo, E. F., & Anyanwu, S. O. (2012). Farmers' Awareness, Vulnerability and Adaptation to Climate Change in Adamawa State, Nigeria. *British Journal of Arts and Social Sciences, 9* (II), 104–115.

Adedeji, Y. M. D., & Ogunsote, O. O. (2012). Anticipated Contributions of Modern Building Materials to Sustainability: Climate Change Scenario in Hot and Humid Regions of Nigeria. *Architecture Research, 2*(4), 55–59. doi:10.5923/j.arch.20120204.04.

Adegbulugbe, A. O. (1991). Energy Demand and CO_2 Emissions Reduction Options in Nigeria. *Energy Policy*, 940–945.

Adegbulugbe, A. O., Ibitoye, F. I., & Akinbami, J. -F. K. (1996). Greenhouse Gases Emission Reduction in Nigeria, Least-cost Reduction Strategies and Macroeconomic Impacts, *submitted to the US Country Studies Program, 1996* (Vol. 3a): Energy Sector, USCP, Washington DC, USA.

Adejuwon, J. O. (1979). *Geography of the Tropics*: Longman Africa, Ibadan, Nigeria.

Adesina, F. A. (2001). Forestry sector response to global Climate Change. *The African Challenge, Conflicts and Development Watch, 2*, 19–29.

Adesina, F. A., Siyanbola, W. O., Oketola, F. A., Pelemo, D. A., Momodu, S. A., Adegbulugbe, A. O., & Ojo, I. O. (1999). Potential of Agroforestry Techniques in Mitigating CO_2 Emissions in Nigeria: Some Preliminary Estimates. *Global Ecology and Biogeography, 8*, 163–173.

Agwu, J., & Okhimamhe, A. (2009). Gender and climate change in Nigeria: A study of four communities in North central and Southeastern Nigeria. (pp. 1–71). Lagos, Nigeria: Heinrich Boll Stiftung Publisher, Lagos, Nigeria.

Akinbami, C. A. O. (2015). Assessment of Social Dimension of Climate Change on Nigeria Rural Women Livelihood Practices: Implications for Entrepreneurship Development. *A CIRCLE Post-doctoral Fellowship Report submitted to African Academy of Science*. Nairobi, Kenya.

Akinbami, J. -F. K. (2001). Renewable Energy Resources and Technologies in Nigeria: Present Situation, Future Prospects and Policy Framework. *Mitigation and Adaptation Strategies for Global Change, 6*(2), 155–181.

Akinbami, J. -F. K., & Ibitoye, F. I. (2003). Greenhouse Gas Mitigation Options Assessments for Nigeria: 2000–2004. *Energy Sector Report submitted to NEST/Global Change Strategies International Inc.* (Vol. 1), Ibadan, Nigeria.

Akinbami, J. -F. K., & Lawal, A. F. (2012). Opportunities and Challenges to Electrical Energy Conservation and CO_2 Emissions Reduction in Nigeria's Building Sector. In M. F. Daniel Hoornweg, Marcus J. Lee, Perinaz Bhada-Tata, and Belinda Yuen (Ed.), *Cities and Climate Change: Responding to an Urgent Agenda* (Vol. 2, pp. 386–406). Washington, D.C.: The International Bank for Reconstruction and Development/The World Bank, USA.

Bello, O. B., Ganiyu, O. T., Wahab, M. K. A., Afolabi, M. S., Oluleye, F., Ig, S. A., . . . Abdulmaliq, S. Y. (2012). Evidence of climate change impact on Agriculture and food security in Nigeria. *International Journal of Agriculture and Forestry, 2*(2), 49–55.

Cook, J., Nuccitelli, D., Green, S. A., Richardson, M., Winkler, B., Painting, R., Skuce, A. (2013). Quantifying the consensus on anthropogenic global warming in the scientific literature. Environ. Res. Lett. Vol. 8, pp. 1–7:.

Farauta, B. K., Egbule, C. L., Agwu, A. E., Idrisa, Y. L., & Onyekuru, N. A. (2012). Farmers adaptation initiatives to the impact of climate change on agriculture in Northern Nigeria. *Agricultural Extension, 16*(132–144).

FME. (2003). *First National Communication—Nigeria. Federal Ministry of Environment (FME), Abuja, Nigeria.* Abuja, Nigeria.

FME. (2010). *National Environmental, Economic and Development Study (NEEDS) for Climate Change in Nigeria.* Abuja, Nigeria.

FME. (2011). *National Adaptation Strategy and Plan of Action on Climate Change for Nigeria (NASPA-CCN), Prepared by the Building Nigeria's Response to Climate Change (BNRCC)*: Federal Ministry of Environment, Abuja, Nigeria.

FME. (2014). *Second National Communication—Nigeria.* Abuja, Nigeria: Federal Ministry of Environment, Abuja, Nigeria.

FME. (2015). *Nigeria's Intended Nationally Determined Contribution (INDC).* Abuja, Nigeria: Federal Ministry of Environment, Abuja, Nigeria.

Ihejirika, P. I. (2016). Climate Change to Affect 60% Of Nigerian Population—Survey, http://leadership.ng/news/504043/climate-change-affect-60-nigerian-population-survey. (Accessed Feb 24, 2016).

Kalejaiye-Matti, R. B., Nassar, S. A., & Audu, H. O. (2010, 18th–22nd). *Climate change: Causes, implications and corrective strategies.* Paper presented at the 44th Annual Conference of Agricultural Society of Nigeria, LAUTECH, Ogbomoso, Nigeria.

LASGME. (2012). *Lagos State Climate Change Policy, Lagos, Nigeria.* Lagos, Nigeria. Lagos State Government Ministry of Environment (LASGME) (2012), Lagos State Climate Change Policy, Logos, Nigeria.

McCright, A. M., & Dunlap, R. E. (2003). Defeating Kyoto: the conservative movement's impact on US climate change policy. Soc. Probl pp. 348–373.

Momodu, A. S., Siyanbola, W. O., Pelemo, D. A., Obioh, I. B., & Adesina, F. A. (2011). Carbon Flow Pattern in the Forest zones of Nigeria as Influenced by Land Use Change. *African Journal of Environmnetal Science and Technology, 5*(9), 700–709.

Nelson, V. (2011). Gender, Generations, Social Protection & Climate Change: A Thematic Review. A Study Report Prepared for the Overseas Development Institute (ODI), London, UK.

Nzeadibe, T. C., Egbule, C. L., Chukwuone, N. A., & Agu, V. C. (2011). Climate Change Awareness and Adaptation in the Niger Delta Region of Nigeria. *The African Technology Policy Studies Network (ATPS) Working Paper Series* (pp. 32 pp). Nairobi, Kenya.

Oreskes, N. (2004). Beyond the ivory tower: the scientific consensus on climate change. *Science, 306 1686.*

Reusswig, F. (2013). History and future of the scientific consensus on anthropogenic global warming. *Environ. Res. Lett., 8*(031003), 3 pp.

Reusswig, F., & Lass, W. (2010). Post-carbon ambivalences—the new climate change discourse and the risks of climate science. *Sci. Technol. Innov. Stud., 6*(2), 155–181.

Sadiku, N. A., & Sadiku, I. B. S. (2011). *Indigenous Efforts by African Farmers in Ensuring Sustainability in Agricultural Productivity in the Face of Changing Climate.* Paper presented at the Environmental Management Conference, Federal University of Agriculture, Abeokuta, Nigeria.

Sayne, A. (2011). Climate Change Adaptation and Conflict in Nigeria *Special Report* (pp. 16 pp): United States Institute of Peace, Washington DC, USA.

UNFCCC. (1992). United Nations Framework Convention on Climate Change (Vol. FCCC/INFORMAL/84, pp. 25 pp): United Nations, New York, USA.

Landslide Loss and Damage in Darbung Village, Gorkha District, Nepal

Bishnu Kumari Sapkota

Abstract The impact of climate change beyond adaptation is known as loss and damages that result from extreme events to slow onset processes. Landslides is a major climatic stressor in Nepal mainly due to the unique combination of active tectonic setting, high rates of weathering and abundant rainfall, aggravated by human interference in form of rapid urbanization and infrastructure development. This study estimates the total loss and damage due to the occurrence of Khanikhola landslides and also establishes ongoing adapting strategies of people towards debris flow in the community. The paper presents the calculation of loss and damage of livelihood assets. Of the total household, 14, 72, 75 and 20% reported a loss of livestock, loss of cultivated crops, loss and damage of land and loss and damage to their houses respectively. The RUSLE model predicted erosion rates with an average value of 36.30 t/ha/year. The total cost of loss and damage was found to be US$4,524,940. It is envisaged that the output of the study will be key inputs in designing of sustainable mitigation measures to minimize the impact of landslides and the associated risks.

Keywords Landslides · Climatic stressor · RUSLE model · Loss and damage

1 Introduction

Natural disasters occur and affect people's lives and livelihoods in almost all parts of the world. Some populations are more vulnerable than others and disparity exists between nations and communities within a country. Furthermore, within communities different households may be affected differently and even within households the vulnerability of individual household members may vary (KC 2013). Among the natural disaster, landslides are a significant component of many major natural

B.K. Sapkota (✉)
Central Department of Environmental Science,
Tribhuvan University, Kirtipur, Kathmandu, Nepal
e-mail: me_bis@hotmail.com

W. Leal Filho (ed.), *Climate Change Research at Universities*,
DOI 10.1007/978-3-319-58214-6_10

disasters and are often responsible for the greater loss of life and livelihoods than is generally recognized. This is large because they are often recorded according to the triggering event such as an earthquake or hurricane even though the losses from landslides may exceed all other losses from the overall disaster (Spiker and Gori 2000).

It is predicted that climate change will increase the severity and frequency of climate-related disasters like floods, landslides, drought, and severe storms. Disaster statistics shows that the frequency and intensity of extreme natural events have been increased in recent years (UNDP 2004). According to the briefing note 'Climate Change and Disaster Risk Reduction' of International Strategy for Disaster Reduction (2008), natural hazards by themselves do not cause disasters but it is the combination of an exposed, vulnerable and ill prepared population or community with a hazard event that results in a disaster. Climate change will, therefore, affect disaster risks in two ways, firstly through the likely increase in weather and climate hazards, and secondly through increases in the vulnerability of communities to natural hazards, particularly through ecosystem degradation, reductions in water and food availability, and changes to livelihoods. Climate change will add yet another stress to those of environmental degradation and rapid unplanned urban growth, further reducing communities' abilities to cope with even the existing levels of weather hazards.

The adverse effects of climate change undermine the economic development, human security, and people's fundamental rights (UNDP 2007). It deteriorates the poverty situation and obstructs the achievement of the Millennium Development Goals (MDGs) of the least developed countries, who are highly vulnerable to the climate-induced disasters (Vashist and Das 2009). Currently the planet is not on track to limit dangerous climate change; in fact, it is rapidly heading towards the world 4–6 °C warmer by the end of this century compare to pre-industrial level (IEA 2011). Yet the global community remains uncertain as to how it will enter and manage this unknown territory of increasing climate impact and related loss and damage.

Loss and damage is an issue of growing importance for the international community, as no country will escape the impacts of climate change (IPCC 2014). Loss and damage results from a spectrum of climate change impacts, from extreme events to slow onset processes (Warner et al. 2012). While extreme events are difficult to attribute to climate change, the risks of some climate-related events, such as heat waves, extreme precipitation and coastal flooding are already moderate and are expected to increase as temperatures rise (IPCC 2014). Loss and damage emanating from climate change impacts can be economic in nature, such as loss of income or damage to property and assets, and non-economic, which include the cultural, social and mental impacts of climate change, as well as the loss of biodiversity and ecosystem services, amongst others (Morissey and Oliver-Smith 2013).

Nepal is a country with varying topography starting at the relatively flat and low (60 m) Tarai in the south and steeply increasing to thousands of meters of elevation in the Himalaya. Each year during the monsoon season, massive rain events, some dropping as much as 550 mm of rainfall in 24 h (Neupane 2008) can send huge flood pulses downstream through the steep and mountainous terrain causing flash floods and landslides in the hilly regions of Nepal. Landslide ranked third both in terms of human deaths and injuries. A total of 60 deaths (18%) and 33 injuries (6.7%) were recorded in 2012 (GoN 2012). Landslides are the most severe types of geo-hazards in Nepal, where more than 80% of the area is mountainous. Deeply weathered and fractured rocks and greatly incised rivers and streams contribute to excessive mass wasting in the mountainous terrain. High precipitation during the monsoon season (June–September) is another detrimental factor, which causes landslips, debris flows and flash floods. Apart from such natural conditions, anthropogenic activities have also greatly contributed in accelerating soil erosion and mass movements in Nepal. These activities include large-scale deforestation, unplanned human settlement, overgrazing, irrigation of steep slopes and infrastructural development (e.g. highways, roads, irrigation canals and dams) without proper engineering investigation (Chauhan et al. 2010). Thus, the appropriate strategies are needed to cope with the impacts of ongoing disaster condition. However, very few researches have been carried out in Nepal in understanding disaster impacts focused on loss and damage.

Nepal is facing unprecedented disaster risks as people are being exposed to more frequent and severe hazards owing primarily to unmanaged population growth, increasing the level of poverty and marginalization, environmental degradation, poor planning and preparedness and the impacts of climate change (GoN 2012).

Due to it's geographical and other climatic conditions, rugged and steep topography, extreme weather events and fragile geological conditions, the country is regarded as a disaster hotspot because of the vulnerability of the population together with the regular and frequent occurrence of different natural hazards. The country's social context characterizes with a low level of development as well as low level of institutional capacity consequent to intensify the impact of disasters (GoN 2010).

Disasters affect men and women differently because of the different roles they occupy and the different responsibilities are given to them in life and because of the differences in their capacities, needs and vulnerabilities. Family size may change at household level due to disasters. For example in Chitwan district, Nepal during the floods, the extended family system collapsed, leaving the women and elderly without support (Ariyabandu and Wickramasighe 2005).

Nepal is a mountainous country situated in the heart of the Himalaya. This young mountain chain is also famous worldwide for its very active tectonics. Furthermore, it lies in an area of strong monsoon precipitation, with the rainfall distribution being such that over 90% occurs within a short three months. Extreme rainfall events are very common which can bring over 400 mm of rain within 24 h. Thus, these geological and climatic factors make the mountain slopes of Nepal highly vulnerable to rainfall triggered landslides (Dahal et al. 2010), causing damage to roads, buildings and other structures, and disrupting the activities of the local people. They also threaten the lives of the people themselves for example, on average over 300 people are killed by landslides each year in Nepal alone

(Hearn et al. 2003) causing severe economic damages exceeding US$10 million (GoN 2010).

In Nepal, in between 1995 and 2002, the totals of 6854 lives were lost by the landslide and flood disasters (ADB/ICIMOD 2006) with billions of dollars economic loss of land and infrastructures.

Landslides have been characterized as one of the most destructive geological processes often resulting in a major loss of life and economic damage (Brabb 1993; Lee and Jones 2004). It has been frequently postulated that one of the underlying causes of the increase in landslides in mountain areas in less developed countries is road building (Petley et al. 2007).

In Nepal, for example, in only the half monsoon period of 2007 (i.e., from June 10 to August 15), 70 people were killed by landslides (NRCS 2007). Studies indicate that loss due to landslides and related problems in the Himalayan region alone constitutes about 30% of the world's total landslide-related damage value (Li 1990).

In Nepal, the total number landslides in the period from 2004 to 2010 are recorded as 211 and that of a number of death from landslide is recorded as 801 (Petley 2012).

In 1968, a huge landslide dammed the Budhi Gandaki River at LabuBesi in central Nepal for 29 h and created a 60-m deep lake. When it breached, the debris flow and flood washed away most of the houses and bridges downstream. Arughat Bazaar was the most affected with the heavy loss of life and property (Sharma 1990).

In 1988 a huge landslide at Darbung dammed the Myagdi Khola for some time and the subsequent flood after breaching claimed many lives and caused extensive damage to property. A similar landslide at the same site had buried Darbang Bazaar some 62 years before, killing about 500 people (Yagi et al. 1990).

In 1993, the continuous heavy monsoonal rainfalls caused tremendous land sliding at several locations in the corridor on Prithivi Highway, severely damaging the highway and closing it for several days (Bhandary et al. 2013).

In 2003, nearly 40% length of the Narayanghat-Mugling Highway was damaged by debris flows (i.e., slope collapses occurred away from the road, but the debris mass damaged the road and bridges) and landslides (i.e. roadside slope collapses) induced by heavily concentrated rainfall exceeding 400 mm a day (Bhandary et al. 2013).

Flash flooding in the Kaski district of northwestern Nepal resulted in the death of at least 31 people, left dozens missed, and caused great loss of property including homes, businesses, crops, and livestock. The flood occurred in a tributary of the Seti River originating from the east of Mount Machhapuchhre. It is believed to be due to the outburst of a landslide-dammed lake. Most of the loss and damage of life was near the confluence with the Sardi Khola (ICIMOD 2012).

In the past 30 years, 4511 people died as a result of landslides. In addition, 18,414 houses were destroyed, affecting 555,000 people (UNEARTH News 2014).

In 2014 falling rock masses caused a devastating landslide in the Sunkoshi River at Jure village of Sindhupalchok district in northern Nepal, destroyed dozens of houses and killed 156 people (UNEARTH News 2014).

In 2014, a series of massive floods hit 17 districts in mid and far western regions of Nepal, at least 113 persons died, 155 were injured, and more than 77 were missed, 12,282 houses were completely damaged, 13,695 partially damaged and more than 40,055 houses inundated leaving 12,276 families displaced (ACT Alliance 2014).

Annually, nearly 24 million tons of soil is washed away from the country (ADB 1998). A study of Practical Action Nepal (2009) suggests that the loss of 1.7 mm of productive soil annually in Nepal, reducing already scarce productive agriculture.

2 Methods

The study basically consists of two methods as; primary data collection, secondary data collection.

2.1 Primary Data Collection

2.1.1 Direct Observation

Direct observation is conducted to learn about the normal behavior of surrounding on their natural environments. Direct observation in this study involves careful watching and records of happening action.

2.2 Key Informant Interview

Key informant interview was introduced to the study site with the list of a questionnaire to explore the subject in depth.

2.3 Household Questionnaire Survey

According to the disaster report of Darbung (2002), it was found that Khanikhola landslides incident of 2002 had affected four wards (4, 6, 7 and 9) where 34 households were found as totally damaged, 12 households partially damaged, 173 households were affected and now 500 households were in vulnerable condition. Total of 719 households were found and the households were selected through the stratified random sampling for the survey and were carried out.

2.4 Focus Group Discussion

Focus Group Discussion was conducted among the people who are directly affected by the happening disaster due to climate-induced landslide.

2.5 Soil Loss Rate

The Revised Universal Soil Loss Equation (RUSLE) given by Renard et al. (1991) is used to predict erosion due to a landslide in landscape/watershed using GIS 9.3.

$$A = R * K * LS * C * P \tag{1}$$

where,

A Average annual soil loss in Mg/ha/year
R Rainfall/runoff erosivity (MJ mm ha^{-1} h^{-1} $year^{-1}$)
K Soil erodibility (Mgh/MJ/mm)
LS Soil Length and Steepness Factor
C Cover-management
P Support practice factor.

2.6 Overall Loss and Damage

Landslide is the major climatic stressor in the study area. Following equation is given by Warner and Geest (2013) is used to calculate the overall loss and damage of study area.

$$\begin{aligned}\text{Loss and damage } (\%) = {} & (\text{ES } (\%) * \text{EI } (\%) * \text{IdM } (\%) * \text{AM } (\%)) \\ & + (1 - \text{IDM } (\%)) * \text{ES } (\%) * \text{EI}(\%)\end{aligned} \tag{2}$$

where,

ES Experienced Stressor
EI Experienced Impact
IdM Impact despite Measures
AM Adopted Measures.

2.7 *Secondary Data Collection*

Secondary data as required to fulfill the objectives of the study were collected and reviewed throughout the whole study period to support the primary data obtained from the locals during the analysis process. Secondary information included demographic profile and political map from CBS, district profile, topographical map, hydro-meteorological data mainly precipitation data of Gorkha stations from DHM, and different reports on the loss and damage scenario from ICIMOD, CARE Nepal, UNDP, World Bank, line agencies of GoN and local governments.

2.8 *Limitations*

- Price may vary for the financial estimation of loss and damage of public and private property due to inflation, growth and other economic activities.
- As the case study is one of the mid-hills of Nepal and not necessarily representative of the entire country.

3 Results

3.1 *Socioeconomic Status*

3.1.1 General Description of Respondents

Table 1 presents general descriptions of respondents. In household questionnaire survey total 85 households were selected based on stratified random sampling. Among the respondents, 45 (52.9%) were female and 40 (47.1%) were male. Ethnically 51.8% were indigenous (Magar, Tamang, Gurung, Rai, Newar), 42.4% were Brahmin/Chettri (Khanal, Tiwari, Duwadi, Aryal, Thapa, Rana), 4.70% were

Table 1 General description of respondents

	No. of respondents	Gender (%)		Ethnicity (%)				Literacy rate (%)	
		M	F	Indigenous	Brahmin/Chettri	Dalit	Muslim	Illiterate	Literate
HH Survey	85	53	47	51.8	42.4	4.7	1.2	33.33	66.67

Dalit (Sarki), and 1.2% was Muslim (Miya). Among the respondents 33.33% were illiterate, 33.33% were literate, 19.05% were SLC, 9.52% were intermediate and 4.76% were Bachelor/Master.

3.1.2 Employment and Income

Households engaged in agricultural activities such as cultivating crops, growing fruit trees, raising livestock were 85.9%. The product of these activities is mainly used for household purposes, however, in terms of income generation, it comprises only about 38.8%. Among the respondents, 11.8% rely on service, 9.41% households were involved as labors and 7.06% were involved in the small family business (Fig. 1).

3.1.3 Land and Farm

Of the total, 88% of household have their land and house they farm and live in whereas 12% respondents did not have not their own land and house (Fig. 2). The average land size per household was 0.29 ha. Crop production mainly consists of cash crop such as Paddy, Maize, Wheat, Millet and cereals such as black lintel soybeans (Maas and Bhatmaas). Of those engaged in agriculture, 57.53% have irrigation facility. Crop cultivation is mainly carried out for household consumption, 24.7% of respondents stated that the main purpose of production was for sale. The total average sale of agricultural produce is $44.74 per year per household.

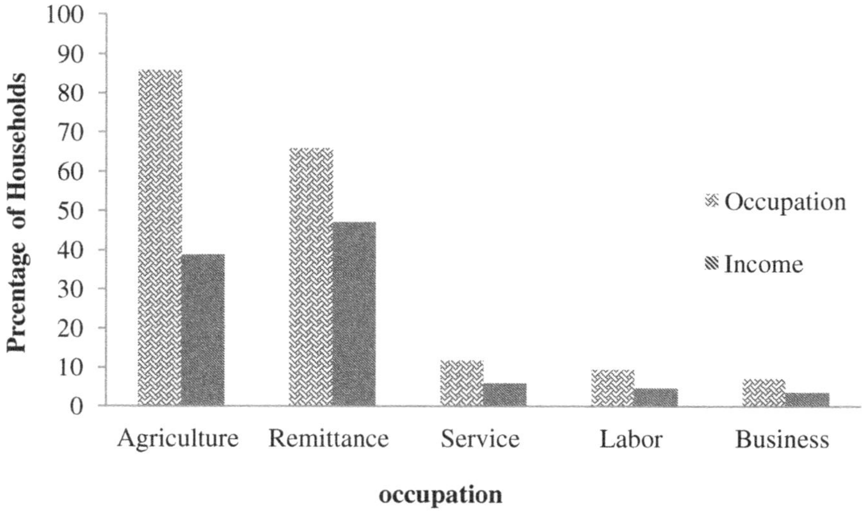

Fig. 1 Occupation and income

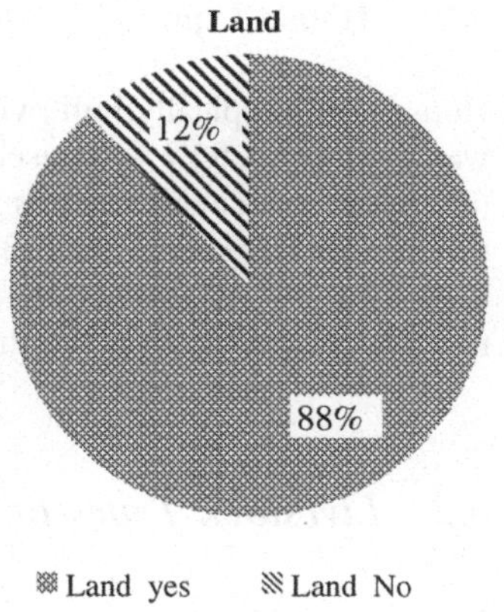

Fig. 2 Household having own land and no land

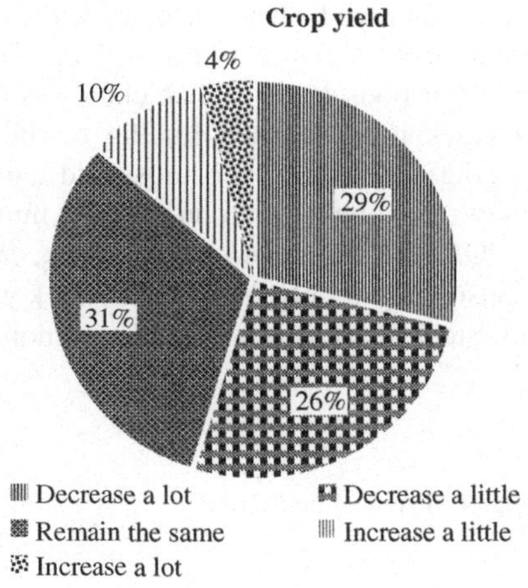

Fig. 3 Crop yield status

3.1.4 Crop Yield Status

In the survey, 28.8% of respondents said that crop yields were decreasing a lot, and 26% said crops were decreasing a little. The main reason given for this decrease was a scarcity of water. 31.5% of the households indicated that crop yields remained the same, whereas 9.6% indicated that crops were increasing a little, and 4.1% indicated that they were increasing a lot (Fig. 3). The main reasons given for the increase was increasing uses of chemical fertilizer.

3.1.5 House Type

Houses were typically built with slate/tin roofs and mud walls. Slate/tin for roofing was used by 58.82% of households, followed by roofs made of the natural product (24.70%) and concrete roofs (16.47%) (Fig. 4). Mud walls were used by 85%, followed by cement walls (13%) only in a few cases (2%) natural materials were used for the walls. Some households use cement for walls even though the roof is not plastered so as to make the house more resistant to landslides (Fig. 5).

3.2 Livestock Types and Number

Livestock raising was carried out by nearly 93% of the households interviewed, with respondents raising buffalos, bull, cow, ox, goat, chicken, pigs. Buffalo rearing was the main livestock activity; 46% of all households own buffalos, with an average number of 2 buffalos per household. Approximately, 45% of the households surveyed owned a goat, with an average of 3 goats per household. Of the total, 40% of household have a cow in their house, with an average number of 1 cow per household. The chicken was raised by 28% of household, with an average 4 chicken per household. Pig rearing was done by 25% of household with an average of 2 goats per household and only 7 households, equivalent to 8%, said they raised bulls, with an average number of 2 bulls per household (Fig. 6).

Of those who owned livestock, 93% said the main purpose was household consumption; only 7% said livestock were intended for sale. The average income of livestock rearing was $65.24 per month.

Fig. 4 Types of house roof

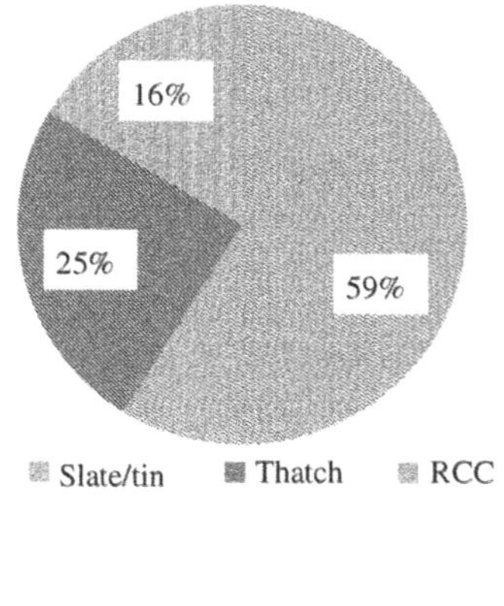

Fig. 5 Types of house wall

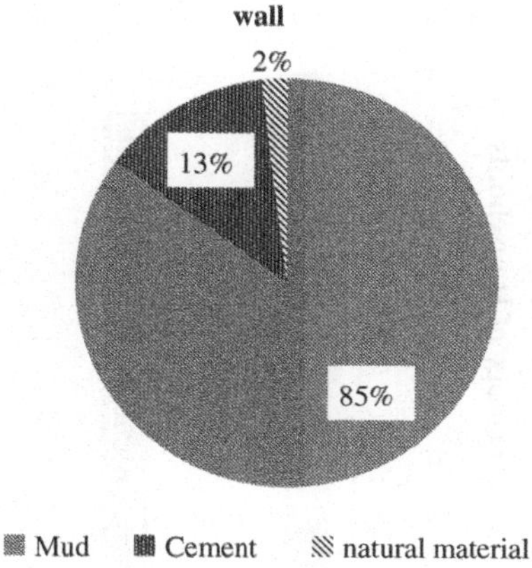

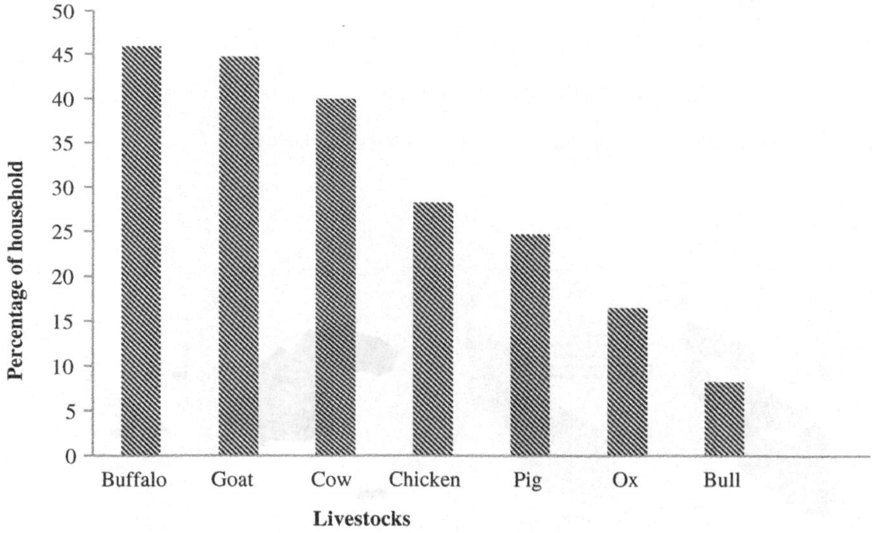

Fig. 6 Percentage of households having different livestocks

3.3 Utilities and Facilities

The majority of respondents (65%) believed their income was average; 23% believed their household income was below average and 12% believed it to be more than average.

As an indication of their relative income to utilities and facilities, most households owned electricity (84%) followed by drinking water facilities (76%) and availability of toilet was found in 69% of household. The main assets owned by households were telephones (91%), radios (79%), and televisions (77%) (Fig. 7).

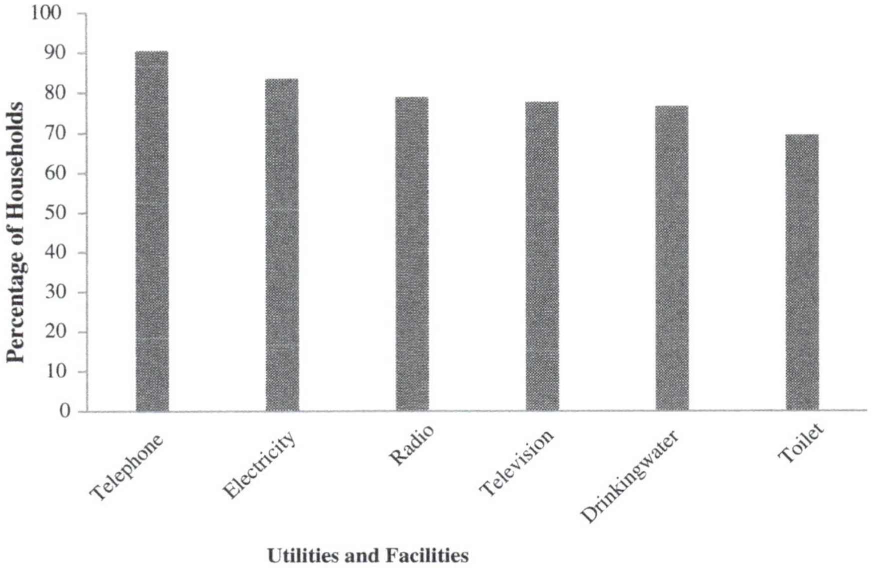

Fig. 7 Percentage of household having different utilities and facilities

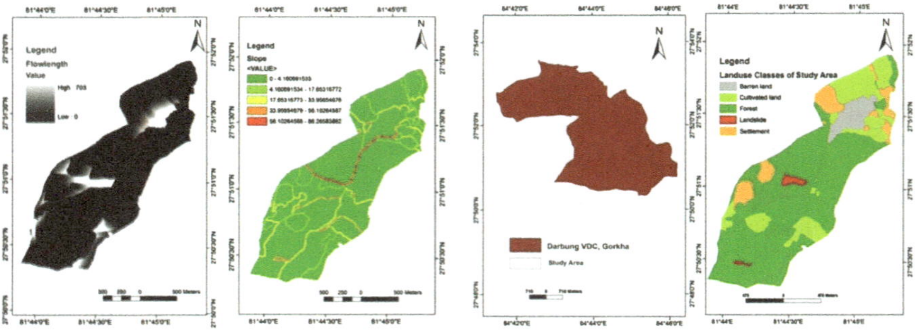

Fig. 8 Flow length, slope, land use of landslide area in Darbung, Gorkha

3.4 Soil Loss Rate

Rainfall erosivity, soil erodibility, slope length and steepness, cover management and support practice factors were calculated. The RUSLE calculated the annual average soil loss for the Khanikhola watershed from Eq. 1 using six factors and it is estimated as A = 36.30 Mg per ha per year. (36.30 mega gram per hectare per year) which is equal to 36.30 ton per ha per year (Fig. 8).

3.5 *Livelihood Assets*

3.5.1 Livestock Loss

Of total household, 14% had lost their livestock due to the effect of Landslide, 2002. Total numbers of chickens, pigs, goats, buffalos lost were found to be 25, 17, 3 and 1 respectively (Table 2).

3.5.2 Cultivated Crop Loss

A severe landslide had affected cultivated crops of 72% of household. Cultivated crops like paddy, maize, wheat were lost because the month of occurrence of landslides was March, which is the period of cultivation of those crops in the study area. Cultivated crop lost was estimated to be 16,000 kg of paddy, 4800 kg of maize and 4500 kg of wheat (Table 2).

3.6 *Land and House Loss and Damage*

Of total household, 70% reported that their land was lost and 80% of household reported that their land was damaged. The total loss of land was estimated to be 10.68 ha and that of land damage was estimated to be 12.21 ha (Figs. 9 and 10). The categories of land and house loss and damage are shown in Figs. 9 and 10. By the landslide, 40% of households were found to be loss while 4% found to be damaged.

Table 2 Livestock and cultivated crop loss

Livestocks	Lost (number)	Cultivated crops	Lost (kg)
Chicken	25	Paddy	16,000
Pig	17	Maize	4800
Goat	3	Wheat	4500
Buffalo	1		

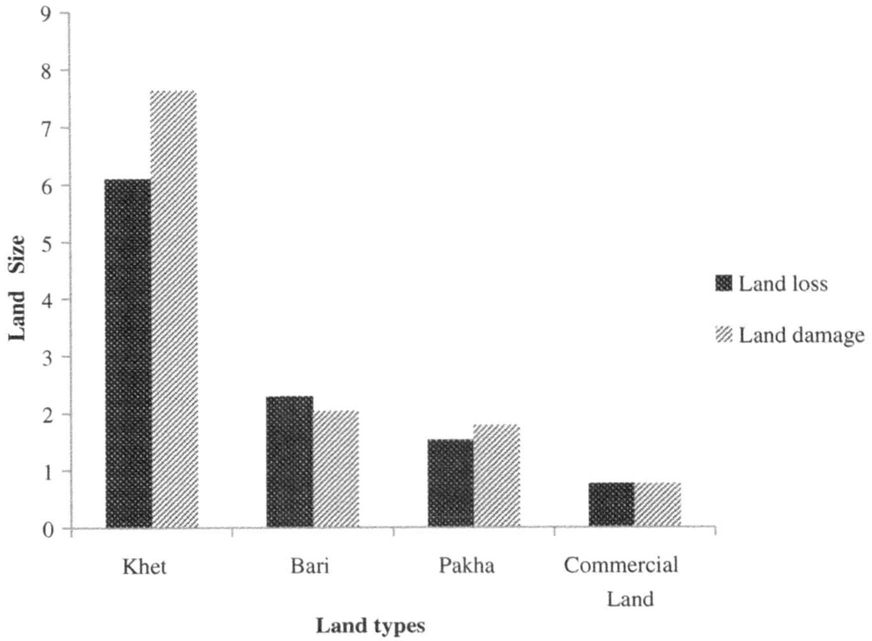

Fig. 9 Loss and damage percentage of land

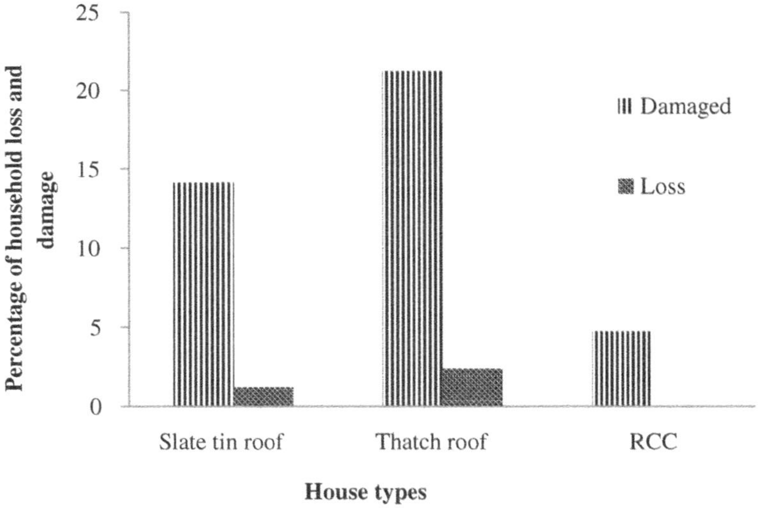

Fig. 10 Loss and damage percentage of house

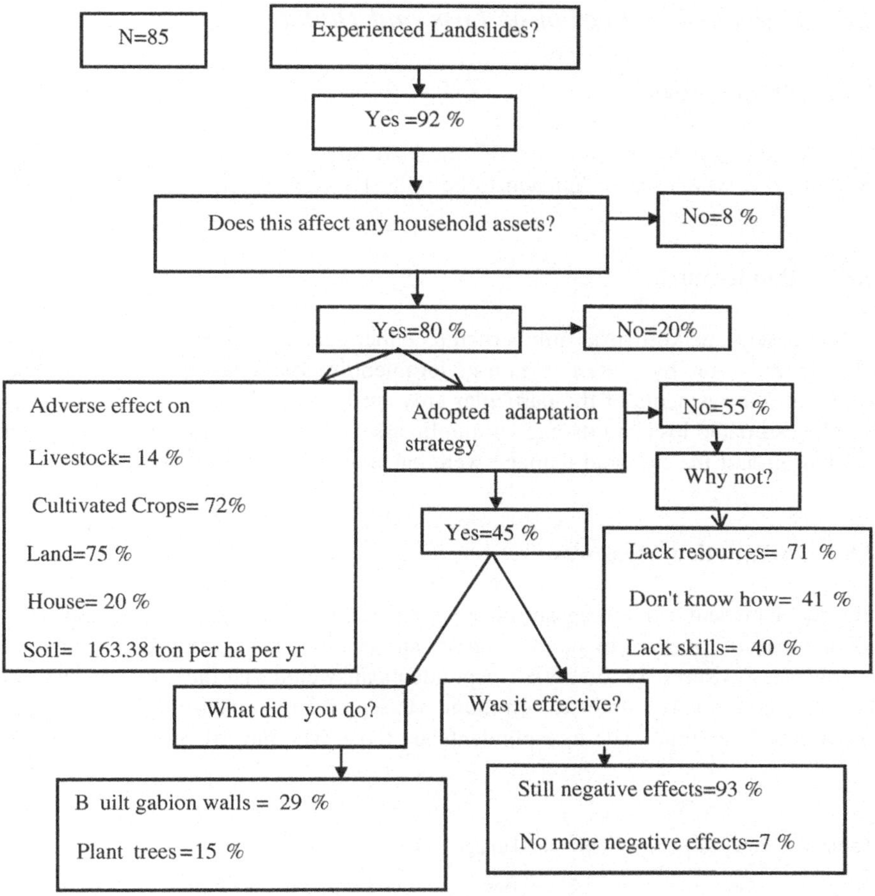

Fig. 11 Tree diagram of experience and impact of landslides

3.7 Overall Loss and Damage

The study area is experiencing landslide as particular climate-related stressors which have been indicated by the erratic rainfall which directly impacts on the household economy. The percentage of experienced stressor was 92%, the experienced impact was 80, 45% households adopted coping or adapting measures, 93% of households expressed that they were still in the impact of landslide despite some adapting measure. Households' incurring residual loss and damage was calculated using Eq. 2 which is found to be 71%. The component of overall loss and damage is represented in Fig. 11.

3.8 *Estimated Valuation of Loss and Damage*

3.8.1 Private Assets

Loss and damage of private assets by the landslides of 2002 are calculated in the current price which was calculated to be $3,551,134.59 (Table 3).

3.8.2 Public Assets

Public assets are known as infrastructure either made by a government agency, communities, or by different non-government/international organization from which all the residents of the particular area are benefitted.

The estimated loss and damage of public assets were obtained from focus group discussion and the loss and damage were calculated as $666,430.77 (Table 4).

3.8.3 Adaptation Measures

Figure 11 present a tree diagram of experience, the impact of landslide and adaptation measure followed by people. Survey showed that 45% of households adopted adaptation measures. In total 20% respondents answered that they are contributing labor or economic resources to community-based or government/NGO-gabion wall construction program. These respondents are those who have their houses adjacent

Table 3 Estimated price for loss and damage of private assets

S. no.	Assets	Categories	Estimated price of loss and damage ($)
1.	Livestocks	Chicken	121.7
		Pig	316.90
		Goat	167.7
		Buffalo	466.04
2.	Cultivated crops	Paddy	596.53
		Maize	1565.88
		Wheat	1593.84
3.	Lands	Khet	2,013,273.36
		Bari	838,863.90
		Pakha	181,753.85
		Commercial Land	335,545.56
4.	Houses	Slate/Tin roofed	69,905.32
		Thatched roofed	64,312.90
		RCC	37,282.84
Total			3,551,134.59

Table 4 Estimated price for loss and damage of public assets

S. no.	Assets	Quantity	Estimated price of loss and damage ($)
1.	Road	10 km	93,207.10
2.	Post office	1 RCC building	9320.71
3.	Health post	1 RCC building with equipment	93,207.10
4.	Mango garden	1.25 ha	466,035.50
5.	Water tank	1	4660.35
Total			666,430.77

to the river, 15% answered they are planting trees in their surrounding and 11% of respondents said that they built their house in an effective way to check the effect of landslide.

It was found that 93% of those who have been following adaptation measures indicated that these measures were insufficient. The majority of households i.e. 85% felt that the most effective solution to landslides would be large-scale gabion walls supported by the government as well as communal action, moving to other areas (migration) by 8% of household and planting trees by 7% of household.

4 Discussion

Nepal falls among least developed countries in the world, with almost one-quarter of its population living below the poverty line. Agriculture is the mainstay of the economy, providing a livelihood for three-quarters of the population and accounting for about one-third of gross domestic product (Khadka and Verma 2012). In Darbung Village Development Committee (VDC), the major occupation was agriculture where 85.9% households were engaged in agricultural activities such as cultivating crops, growing fruit trees, raising livestock. Livestock raising being carried out by nearly 93% of the households interviewed, with respondents raising buffalos, bull, cow, ox, goat, chicken, pigs. Of those who owned livestock, 93% said the main purpose was household consumption; only 5% said livestock were intended for sale; the remaining percentage was bull used for a mate. The average income of household from livestock rearing was $186.41 per year.

Presently, remittances are the largest source of household income with 67% of respondents indicating that their household received remittances with the average amount sent per household per month being $139.81. Remittance is central to Nepal's contemporary economy. While Nepalese have long migrated outside the nation's borders to earn money, the insurgency and the following period since 1996 have accelerated and intensified out-migration by 2010. It is estimated that 4 million Nepalese were working in foreign countries according to Nepal's national census

(GoN 2010). Therefore, the country has made a recent stride in poverty alleviation largely due to the high amount of remittances, now received by 55.8% of all household an astonishing drop by points from 31% under the poverty limit to 13% under the poverty limit (CBS 2012).

In Nepal, nearly 60% of rural households are 'functionally landless' with insufficient land to meet their basic food requirements (Wily et al. 2008). Of total surveyed household, 88% of household own the land and house whereas 12% respondents do not have their own land and house. The average land size per household was 0.29 ha. Houses are typically built with slate/tin and mud walls. slate/tin for roofing was used by 58.82% of households, followed by roofs made of the natural product (24.70%) and concrete roofs (16.47%). Wood walls are used by 85%, followed by cement (13%) only in a few cases is the natural material used for the walls. Some households use cement for walls to make the house more resistant to landside.

The impacts of extreme weather events reinforce the landslides by breaking down protection such as gabions walls, houses are damaged and agricultural land becomes unsuitable. Of total household, 14% of respondents reported the loss of livestock, 72% respondents reported a loss of cultivated crops, 75% reported the loss and damage of land and 20% respondents reported of having loss and damage to their house.

In the present study, the RUSLE model predicted erosion rates with an average value of 36.30 t/ha/year for the study area. These rates were found lower when compared with the erosion rates found in Shrestha (1997) for the similar watershed conditions, which estimated an average soil erosion rate 56 t/ha/year in areas under rain-fed cultivation. It may be due to fragile geography and lower canopy cover of forest and rain-fed agriculture practice in the study area.

In the study area, the total cost of reconstructing houses, land and infrastructure was estimated to be $2,278,953.30 (Disaster Report 2002). From the study, the cost of loss and damage of private property (livestock's, cultivated crops, land and houses) was estimated to be $3,551,134.59 and the loss and damage of public property were estimated to be $666,430.77 and the total cost of loss and damage was found to be $4,217,565.35.

In some cases, families have poor resilience, when a household is living in a subsistence economy. For instance, when loss and damage strike a household, the family's savings or subsistence capital can be wiped out. Such households and communities face barriers that erode livelihoods, food security and asset bases and that prevent them from accessing appropriate, sufficient adaptation options to manage climatic risks. Of those households who were affected by landslides, 45% have carried out adaptation measures.

5 Conclusion

In the study area, socioeconomic status of population was taken as major drivers to tackle with post-disaster period and to prepare well for other. In the study, socio-economic indicators such as gender, ethnicity, education, land, farm, house, crop yield, employment, income, utilities and facilities shows that there were majority of typical middle-class family that they can offer preparedness for any disaster for their own safety, as from each household it seems that at least one family member was in abroad, where the income is balanced by remittances and agriculture.

The study area is mid hills, where there are mostly sloppy areas with the high chances of occurrences of landslides, carrying the fertile soil from top to bottom, therefore, the flat land, near to the river is more fertile than the top of the hills. Due to the landslides, average soil loss rate is found to be high. It is also evident that there are varying underlying causes of people's vulnerability to adapt and then poses a challenge for reducing or minimizing vulnerability. Proximity to the landslide prone area further demonstrated that effects of landslides in one sector can affect other sectors of society, for instance, the soil loss rate of the area was attributed to the impact of landslide which further decreases the crop production rate align with making community highly vulnerable to climatic extremes.

The study found the loss and damage of livelihood assets such as livestock loss, cultivated crop loss, land and house loss and damage which shows the significant loss and damage to livestock in number, cultivated crops in weight, land in area and house in number. Darbung VDC has experienced severe landslides in 2002 with frequent landslides before and after. The landslides had caused severe damage to agricultural land with heavy loss of cultivated crops. Most of the domestic animals have been lost and some houses were swept away while others were completely/partially damaged. The entire area of the VDC has experienced a rapid change, primarily caused by debris flow.

Due to erratic rainfall in the study area, a landslide is characterized as a climatic stressor. Most of the populations were found to be directly impacted and still in risk of reoccurrence of landslides.

Financial loss due to a loss of livelihood assets of the study area was categorized into two groups; private and public assets where the loss of private assets were found to be high. Poor families with insufficient finance, land, or other assets are likely to lose the minimal properties they have and face increasing indebtedness and poverty. Such vulnerable populations are disadvantaged in terms of accessing resources, which exposes them to increased risks during disasters and in the wake of climate-related events. These risks include physical dislocation and psychological trauma, the loss of household resources.

People have tried to adapt to the changing climate by building gabions, house, and planting trees. The adaptation measures are self-governing and implemented at household level, and some community and government-planned gabion walls have been built.

Acknowledgements I would like to express my profound gratitude and sincere appreciation to a supervisor, Dr. Dinesh Raj Bhuju for his continuous support, patient guidance, and mentorship throughout the study period. I would also like to thank my co-supervisor, lecturer Ramesh Prasad Sapkota, for his encouragement and insightful advice and comments throughout the study duration. I would like to extend my sincere gratitude to the Head of Department, Professor Dr. Kedar Rijal, for his regular supervision in the timely commencement of the study. I am much thankful to village development committee secretary of Darbung, principal, Mr. Krishna Prasad Khanal of Ratneshwor Secondary School, staffs of Natural Disaster Relief Committee, Darbung for providing valuable information and all the respondents from Darbung VDC for their patient and relevant responses to the questionnaire. Last but not the least I would like to thank my family for their warm love, generous support, and encouragement to carry out this research and successful completion.

References

ACT Alliance (2014). *Emergency response and early recovery support to flood and landslide affected households in Western Nepal* Retrieved from http://reliefweb.int/sites/reliefweb.int/files/resources/www.actalliance.org_resources_appeals_NPL141Prel_Floods-Landslides_Nepal.pdf.

ADB (HMG/MOFSC) (1998). *Master Plan for Forestry Sector (MPFS), Nepal.* Ministry of Forest and Soil Conservation. Kathmandu.

Ariyabandu, M.M. & Wackramasinghe, W.M. (2005). *Gender Dimension in Disaster Management*: A guide for South Asia: Sri Lanka.

ADB/ ICIMOD (2006). *Environmental Assessment of Nepal: Emerging Issues and Challenges.* Kathmandu, Nepal: Asian Development Bank, ICIMOD.

Bhandary N.P., Yatabe R., Dahal R.K., Hasegawa S., & Inagaki H. (2013). *Area distribution of large-scale landslides along highway corridors in central Nepal, Georisk*: Assessment and Management of Risk for Engineered Systems and Geohazards, DOI:10.1080/17499518.2012.743377, First version.

Brabb, E. E. (1993). Proposal for worldwide landslide hazard maps. *Landslides, Seventh International Conference and Field Workshop.*

CBS (2012). *National Population and Housing Census 2011: National Report Vol. 01.* Central Bureau of Statistics. National Planning Commission Secretariat, Government of Nepal.

Chauhan, S., Sharma, M., & Arora M.K. (2010) *Landslide susceptibility zonation of the Chamoli region, Garhwal Himalayas, using logistic regression model. Landslides* 7:411–423.

Dahal, R.K., Hasegawa, S., Bhandary, N.P., & Yatabe, R. (2010). *Low-cost road for the development of Nepal and its engineering geological consequences*, Retrieved from http://www.hils.org.np/ranjan/papers/Ranjan_Low-cost_road.pdf.

Disaster Report Darbung (2002). Government of Nepal, Ministry of Urban Development, Department of Urban Development and Building Construction, division office, Gorkha.

GoN (2010). Nepal Disaster Report 2009: *The hazardscape and vulnerability.* Kathmandu: Ministry of Home Affairs and Disaster Preparedness Network-Nepal, Government of Nepal.

GoN (2012). Nepal Disaster Report 2013: *The hazardscape and vulnerability.* Kathmandu: Ministry of Home Affairs and Disaster Preparedness Network-Nepal, Government of Nepal.

Hearn, G., Petley, D., Hart, A., Massey, C., & Chant, C. (2003). Landslide Risk Assessment in the Rural Sector Guidelines on best practice.

ICIMOD (2012).*Flash flood in Nepal causes loss of life and property.* Retrieved from http://www.icimod.org/?q=7122.

IEA (2011). World Energy Outlook, International Energy Agency Factsheet, page 1, Retrieved from http://www.iea.org/weo/.

ISDR (2008). *Climate Change and Disaster Risk Reduction.* Retrieved from www.ipcc.ch. International Strategy for Disaster Reduction.

IPCC (2014.). Impacts, Adaptation and Vulnerability. IPCC Working Group II Contribution to AR5. Summary for Policymakers. Intergovernmental Panel for Climate Change: Retrieved from http://www.ipcc.ch/.

KC., S. (2013). Community vulnerability to floods and landslides in Nepal. *Ecology and Society.* Retrieved from. http://dx.doi.org/10.5751/ES-05095-180108.

Khadka, M. & Verma, R. (2012). *Gender and biodiversity management in the greater Himalayas: Towards equitable mountain development*. ICIMOD.

Lee, E. M. & Jones, D. K. C. (2004). *Landslide Risk Assessment*. London: Thomas Telford Publishing.

Li, T. (1990). *Landslide management in the mountain area of China*. ICIMOD, Paper No. 15, 50.

Morissey, J.& Oliver-Smith, A. (2013). *Perspective on Non-economic Loss and Damage: Understanding values at risk from climate change*. International Centre for Climate Change and Development (ICCCAD), Dhaka, Bangladesh. Retrieved from <http://www.lossanddamage.net>.

Neupane, N. (2008). *To study the various factors affecting the summer monsoon rainfall in Nepal.* Research Report, The University of Texas at Austin, Texas, USA. Retrieved from http://www.geo.utexas.edu/courses/387h/Lectures/term_Neupane.pdf.

NRCS (2007). *Map of Disaster situation of floods and landslides 2007* (covering 10 July to 15 August, 2007). Nepal Red Cross Society, Kathmandu.

Petley, D. (2012). *Global patterns of loss of life from landslides*; The Geological Society of America. Retrieved from http://geology.gsapubs.org/content/early/2012/07/31/G33217.1.

Petley, D., Hearn, G. J., Hart, A., Rosser, N. J., Dunning, S. A., Oven, K., &Mitchell, W. A. (2007). *Trends in landslide occurrence in Nepal. Natural hazards*, 43 (1): pp. 23–44.

Practical Action Nepal (2009). *Temporal and Spatial Variability of Climate Change over Nepal (1976–2005)*, Practical Action, Kathmandu.

Renard, K.G., Foster, G.R., Weesies, G.A., & Porter, J.P. (1991). RUSLE: *Revised Universal Soil Loss Equation. Journal of Soil and Water Conservation*, 46(1), 30–33.

Sharma, C.K. (1990). *Geology of Nepal Himalaya and Adjacent Countries*. Kathmandu.

Shrestha, D.P. (1997). *Assessment of soil erosion in the Nepalese Himalaya*: A case study of Likhu Khola Valley, Middle Mountain Region. Land Husbandry, 2(1), 59–80.

Spiker, E. C. &. Gori P. L. (2000). National Landslide Hazards Mitigation Strategy: *A Framework for Loss Reduction*: USGS Open-File Report, 00–450.

UNDP (2004). A Global Report: Reducing Disaster Risk: *A challenge for development*, United Nations Development Programme, Bureau for Crisis Prevention and Recovery, New York.

UNDP (2007). *Climate change and the MDGs*, United Nations Development Programme, Retrieved April 3, 2007 from http://www.undp.org/gef/adaptation/dev/02a.htm.

UNEARTH News (2014). *Recent flooding and landslides in Nepal underline complex relationship between natural disasters and poverty.* Retrieved from http://unearthnews.org/2014/08/21/recent-flooding-and-landslides-in-nepal-emphasize-relationship-between-natural-disasters-and-poverty/.

Vashist, S. & Das, P.K. (2009). *South Asia needs greater cooperation to fight climate change, Clime Asia:* Climate Action Network-South Asia newsletter, BCAS, Dhaka.

Warner, K. & van der Geest, K. (2013). *Loss and damage from climate change: Local -level evidence from nine vulnerable countries*. International Journal of Global Warming Vol. 5, No. 4, pp. 367–386.

Warner, K., van der Geest, K., Kreft, S., Huq, S., Harmeling, S., Koen, K. &de Sherbinin, A. (2012). *Evidence from the Frontlines of Climate Change: Loss and Damage to Communities Despite Coping and Adaptation, Loss and Damage in Vulnerable Countries Initiative*, Policy Report, Report No. 9, United Nations University Institute for Environment and Human Security, Bonn.

Wily, L.A., Chapagain, D., & Sharma, M. (2008). *Land Reform in Nepal: Where is it coming from and where is it going*? DFID Kathmandu.

Yagi, H., Maruo, Y., Saijo, K.S.& Nakamura, S. (1990). *The September 1988 large landslide in the vicinity of MCT, Darbang, Nepal. Journal of the Japan Geological Society* 26, 45–49.

Adaptations to Climate Change in Bangladesh: Development of a National Inventory

Debanjali Saha, Mohammed Abed Hossain, Md. Shahriar Shafayet Hossain, M. Shahjahan Mondal and Rezaur Rahman

Abstract Bangladesh is highly vulnerable to climate change impacts and adaptation is the only viable option to deal with this phenomenon. This paper provides an overview of the climate change adaptations in the country which includes spatial and sectoral distribution, aims and timings, stresses and shocks addressed, etc. An inventory was prepared with documented climate change adaptations in Bangladesh. A first cut analysis was performed based on the inventory. The analyses found that most of the adaptations took place in coastal districts. Disaster risk reduction, water resource management, infrastructure and agriculture are the major sectors of adaptations in terms of number. Government of Bangladesh was the major provider (87%) followed by non-governmental organizations (11%). Chronic stresses like salinity, waterlogging, drought, etc., were the focus of majority of adaptations while sudden shocks like cyclone and large flood were addressed in community level as well as national level. Major trait has been reactive than anticipatory. Information on specific stress/shock factors addressed and the spatial/population coverage of adaptations were often lacking due to inadequate documentation. The inventory is a step towards improved documentation aiding evaluation of practiced adaptations that may inspire policy makers to devise effective adaptation measures to combat climate change in Bangladesh.

D. Saha (✉) · M.A. Hossain · Md.S.S. Hossain · M.S. Mondal · R. Rahman
Institute of Water and Flood Management, Bangladesh University of Engineering and Technology, Dhaka 1000, Bangladesh
e-mail: debanjalisaha@iwfm.buet.ac.bd

M.A. Hossain
e-mail: abed@iwfm.buet.ac.bd

Md.S.S. Hossain
e-mail: shahriar.buet08@gmail.com

M.S. Mondal
e-mail: mshahjahanmondal@gmail.com

R. Rahman
e-mail: rezaur@iwfm.buet.ac.bd

W. Leal Filho (ed.), *Climate Change Research at Universities*,
DOI 10.1007/978-3-319-58214-6_11

Keywords Climate change · Adaptation · Inventory · Community · Chronic stresses · Sudden shocks

1 Introduction

In the present trend of urbanization, globalized economy development and technological revolution, climate change has come out as an undesired yet emphatically declared occurrence. Though the impacts of climate change can be evident worldwide, the extent depends mostly on development factors like poverty, social stratification, political perplexity, etc. (Ayers et al. 2014). In this regard, impact management strategies are driven by two means of defense mechanisms; mitigation and adaptation. As a developing country with limited resources and less contribution to global policy formulation, Bangladesh can consider adaptation as the foremost line of defense. According to IPCC (2014), adaptation is the process of adjustment to actual or expected climate and its effect. It is a continuous stream of actions against climatic variability and climate change that reflects existing social norms and processes (Adger et al. 2005). Observed impacts of climate change which are actually the forerunner of events to come proves the necessity of proactive adaptation measures (Adger et al. 2005; Baigini et al. 2014). Societies and communities have contemplated the necessity of adaptation against altered future climatic condition in reaction to prominent evidences, and in some cases in anticipation of worse possibilities.

As there are evident changes in world's climate and their impacts on physical systems are already observed, different adaptation practices have already taken places and others are in progress (Rosenzweig et al. 2007; Tompkins et al. 2010). Bangladesh prioritizes adaptation and disaster risk reduction measures (MoEF 2009). This country has progressed in adaptation planning and practices in different sectors. Throughout the whole country governmental, non-governmental and community initiatives have brought about diversified adaptation practices against climate change. Some of these measures have been taken way back in response to climatic variability which helped in combating climate change impacts in the present times, while some are pretty recent initiatives in anticipation of increased severity of major hydro-climatic hazards like cyclone, storm surge, flood, river erosion, drainage congestion, drought, etc. (Mondal et al. 2015). Climate change has been a significant concern in different policies of Bangladesh since the last decade of twentieth century onwards which was exhibited in some visionary policy documents like National Environment Management and Action Plan (1995), National Water Policy (1999), etc. But explicit mentioning of climate change became much more frequent at the later part of the last decade (Haq et al. 2015). Different policy documents emphasized on institutional capacity building, research and development activities and good governance to gain adaptation capacity to deal with climate change. Through the policies and frameworks in action different large scale projects and initiatives were undertaken by the government. Bangladesh Climate Change Trust (BCCT) and Bangladesh Climate Change Resilience Fund (BCCRF) are some major Government of Bangladesh (GoB) initiatives of implementing climate change adaptations in Bangladesh which have drawn attention as well as raised awareness among the community. However, there has been no mention of the requirement of adaptation documentation in the policy documents so far. Different policies and plans like National Adaptation Program of Action (2005), The Bangladesh Climate Change Strategy and Action Plan (2009), National Plan for Disaster Management (2010), etc., have mentioned many adaptation options but the need for a thorough documentation of adaptation measures in practice at different parts of the country has not been exclusively mentioned in any policy or framework.

In contrast, most of the studies conducted in Bangladesh focused on potential adaptation options (Ali 1999; Faruque and Ali 2005; Ahmed 2006; Rawlani and Sovacool 2011; The World Bank 2011; Asia Foundation 2012; Dev 2013; Haque et al. 2015) where the need for documentation of the existing adaptation practices and the suitability of such was not of importance. Thus there is no systematic study on adaptation from a holistic perspective where the geographical and sectoral arrangement of the practices, involvement of institutions, timing of adaptations and their drivers can be understood. Also, there is serious lack of information stemming from inadequate documentation on the part of the implementers and providers about the adaptation measures taken in Bangladesh which creates knowledge gap among institutions resulting in repetitive implementation of adaptation options without regard for the suitability and thus leading to potential maladaptation.

A notable study making use of the adaptation inventory was Tompkins et al. (2010) which was based on an inventory of adaptation practices in UK against climate change. The inventory (Tompkins et al. 2009) was prepared by the Tyndall Centre for Climate Change Research, UK. Their study focused on some selected sectors of adaptations and cross-sectoral adaptation measures from the wide array of information in the inventory. No such work has so far been done in Bangladesh which is regarded as one of the pioneers in climate change adaptation among the developing countries. This paper focuses on the practiced adaptation measures in response to climate change in Bangladesh. Based on an inventory of documented adaptation practices analyses were performed to reveal distribution of adaptations in terms of geographical locations, thematic sectors, major providers, major drivers of adaptation and stressors responsible for them, etc. Proper knowledge of adaptation practices in the whole country can ensure efficacy of present actions and improved responsiveness, institutional coordination, effective upscaling of measures, better utilization of fund and also can lessen policy-planning gaps. In this regard, this study can help the researchers, adaptation providers, planners and policy makers to create a society better equipped in combating climate change and its ordeals.

2 Methodology of the Study

This study was based on an inventory of adaptation practices which started as part of an international collaborative project in the Institution of Water and Flood Management (IWFM), BUET, Bangladesh titled "DEltas, Vulnerability and Climate Change: Migration and Adaptation (DECCMA)". Work under DECCMA included only the adaptation practices in the coastal region of Bangladesh which was later extended with the same framework and design in preparation of the national adaptation inventory. Building on the developed protocol by Tompkins et al. (2009), a new protocol was developed to make the inventory of evidences of practiced adaptation measures which have been already documented. The documents included published reports, journal papers, conference proceedings, project documents and institutional reports. Most of the documents were collected from a number of online sources and hard copy reports were collected from different international and national academic and research institutes, international development agencies, national government agencies, international and national non-governmental organizations, etc. Collected information were arranged in a total of 44 columns in a universal spreadsheet template which finally came out as the national adaptation inventory. The columns were arranged in several thematic sections. These sections covered a number of queries focusing on different parameters. The answer to the queries were based on available documentation of each adaptation practice. The parameters under each thematic section considered in this paper are outlined in Table 1 in the following:

Other than these parameters the inventory also provided information about the barriers of adaptations, sustainability perspective, gender focus, capacity of adaptation measures to reduce disaster risk, vulnerability and increase large scale system resilience. Information on different columns were of boolean type (e.g., gender focused? yes/no), list data (e.g., reactive/anticipatory, taking action/building capacity) and open ended. Information for some of the parameters in different thematic sections were often inadequate particularly for areas outside of delta. For example, gender focus of the adaptation measures was not discussed in many of the documents as well as information on the barriers to implementation was missing or inadequate to document. So, the present paper was limited to the thematic sections where the parameters were properly documented.

Table 1 Thematic areas and parameters

Thematic sections	Parameters
Location and sector of adaptation	Geographical location, thematic sector
Adaptor and beneficiary	Types and names of adaptation provider and beneficiary
Forms of adaptation	Summary and output of adaptation, aim of adaptation
Drivers of adaptation	Historical evidence of adaptation, adaptation timing: reactive or anticipatory, adaptation taken in response to chronic stress or sudden shock, what are the stress and shock events

Some key terms like aim of the adaptation (taking action/implementing change, building capacity), the timing of adaptation (reactive or anticipatory), etc., were included in the inventory. These terms were previously defined by DECCMA Project and these definitions were used when associating these terms with the adaptation measures. Most of the questions attributed to a single adaptation practice were answered from the information collected through literature and reports, while in some cases ancillary information was utilized or expert judgement was applied. The overall inventory provided a complete picture of the status of adaptation in Bangladesh after answering all the queries under different sections. It is to be noted that this national inventory is a living document where new entries as well as updating of the previous entries is taking place whenever a new literature is available.

The DECCMA inventory, which is a subset of this inventory, has already been used by Saha et al. (2016) which focused on the agricultural adaptations in the coastal region of Bangladesh where the present situation of coastal agriculture sector in response to climate change was evaluated. The agricultural adaptations associated in this study were taken from the DECCMA inventory, i.e., inventory of adaptations in the coastal delta region which has been proved to be very helpful in this study.

Based on the data recorded in the inventory the analysis of adaptation measures were conducted where some selected attributes and thematic areas were taken for detail analysis. The columns of the national adaptation inventory selected for analysis in this paper include geographical location, thematic sector, provider of adaptations, historical evidence and timing of adaptations, stress and shock in response to which adaptation is taken, etc. Based on the information of these columns this paper provided a synthesis outlining the distribution of adaptation practices, the providers' focus on adaptations, how the adaptations evolved with time, etc. Unlike the study of Tompkins et al. (2010), this paper tried to look at the arrangement of adaptation practices among the delta and non-delta regions of the country, sectoral focus of adaptation providers, reasons behind the emergence of adaptations in recent decades and also pointed at the lack of proper documentation necessary for evaluation and upscaling of adaptation practices. Thus, this study may be a guideline for future similar studies and also a supporting document for the relevant governmental and non-governmental organizations for taking steps forward in improving the climate change adaptation components where this inventory can provide all the necessary ingredients.

3 Key Findings from the Adaptation Inventory

In the inventory of documented adaptation practices in different districts and areas of Bangladesh, about 589 adaptation measures were collated. These adaptations represent both government and NGO participation in structural interventions as well as non-structural ones including awareness raising and capacity building activities from the late nineties to recent years. These measures cover different thematic sectors and are observed to be taken in response to regular hazard events or in anticipation of increased severity of natural disasters for protection of the vulnerable community. Following sections represent the analyses of the adaptation practices in terms of different criteria.

3.1 Spatial Distribution of Adaptations

Almost all the districts of Bangladesh have experienced some adaptation measures since the last few decades where some have witnessed more in number than the others. Spatial distribution of the adaptation practices in Bangladesh represents that at least 30 adaptation measures were taken in each of the districts while in some districts more than 100 adaptations were practiced. The districts experiencing more adaptations are Khulna, Satkhira, Bagerhat, Chittagong, Jessore and Noakhali. All these districts are within the coastal region, i.e., the delta region of Bangladesh. According to Integrated Coastal Zone Management Plan (ICZMP) of Water Resources Planning Organization (WARPO), coastal delta region is comprised of 19 districts which are: Bagerhat, Barguna, Barisal, Bhola, Chandpur, Chittagong, Cox's Bazar, Feni, Gopalganj, Jessore, Jhalokathi, Khulna, Lakshmipur, Narail, Noakhali, Patuakhali, Pirojpur, Satkhira and Shariatpur. These districts are more exposed to the Bay of Bengal and consequently to the devastating effects of the natural calamities like cyclonic storm surge, tidal flood, salinity intrusion and other hazards much more than the districts further inward. This compels the delta region to take more adaptation measures than the non-delta region (61% adaptations in delta region) which can be somewhat observed from Fig. 1.

Interaction with frequent disasters on a daily basis and high vulnerability of people living in the delta region resulted in more adaptation initiatives to increase resilience of the community. This fact also made the coastal community more aware about the necessity of adaptation. This is the reason that delta region of Bangladesh covers about 60% adaptations in the inventory. Among the exposed coastal districts, Satkhira, Khulna and Bagerhat cover more than 10% of the total number of adaptation measures which can be attributed to the attention these districts received in context of climate change adaptation since super cyclone Sidr in 2007 and Aila in 2009. Also, a significant number of adaptations in Chittagong district includes infrastructural interventions like construction of embankment, drainage structures, bank protection works, irrigation facilities, etc. Due to increasing frequency and intensity of hazards, many adaptation measures are being restarted in the delta region for more improved protection and less vulnerability. Though the hazards in the non-delta region are not so diversified as in the delta, the need for protection and resilient socio-economic condition is still the same for the community people. This calls for higher focus of the government and donor agencies to the non-delta region so that a sustainable and well adapting societal condition can be achieved even under the alarming threat of climate change.

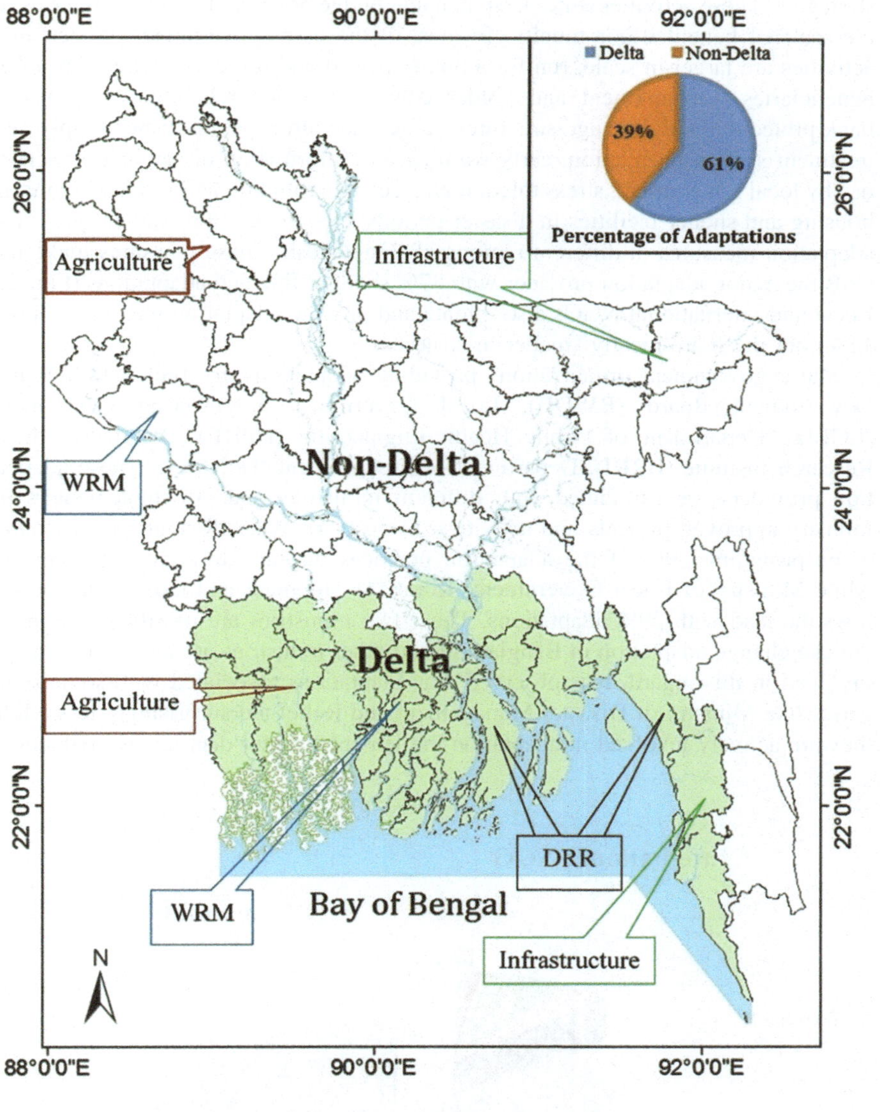

Fig. 1 Spatial focus of adaptations in Bangladesh

3.2 Adaptations in Context of Providers and Thematic Sectors

Adaptation measures in Bangladesh are being supported and implemented by GOs, NGOs and sometimes by the community people since the beginning of adaptation strategies. These activities differ in spatial and budgetary extent and also in terms of the scale of benefited community. In most of the cases government projects and activities are larger in scale, run for a longer period and serve to a bigger group of beneficiaries. Embankment and polder construction and rehabilitation projects, bank protection and drainage structures, large scale afforestation, water supply and treatment system installation, early warning and information dissemination activities by local government, stress tolerant crop seed distribution and farmers' training, housing and shelter facilities in disaster periods, etc., are the government provided adaptation measures in different regions of Bangladesh. These measures make the GoB the major adaptation provider with 87% share of the total adaptations (Fig. 2). Local and international NGOs also initiate and provide adaptation measures (about 11%) but these are mostly smaller in scale.

Major government organizations providing adaptations are Bangladesh Water Development Board (BWDB), Local Government Engineering Department (LGED), Department of Public Health Engineering (DPHE), Bangladesh Rice Research Institute (BRRI), Department of Environment (DoE), etc. These adaptation providers are associated with different ministries and are implementers of ministry approved projects and activities. Ministry of Water Resources was found to be providing 39% of the adaptation practices through different implementers while Ministry of Local Government, Rural Development and Co-operatives follows the lead with 29% adaptations. These two ministries are mostly focusing on climate change adaptation in Bangladesh and other ministries are also getting more involved in this regard. It is observed that adaptations associated with some ministries like Ministry of Disaster Management and Relief are surprisingly less while they are actively involved in adaptation implementation for disaster risk reduction.

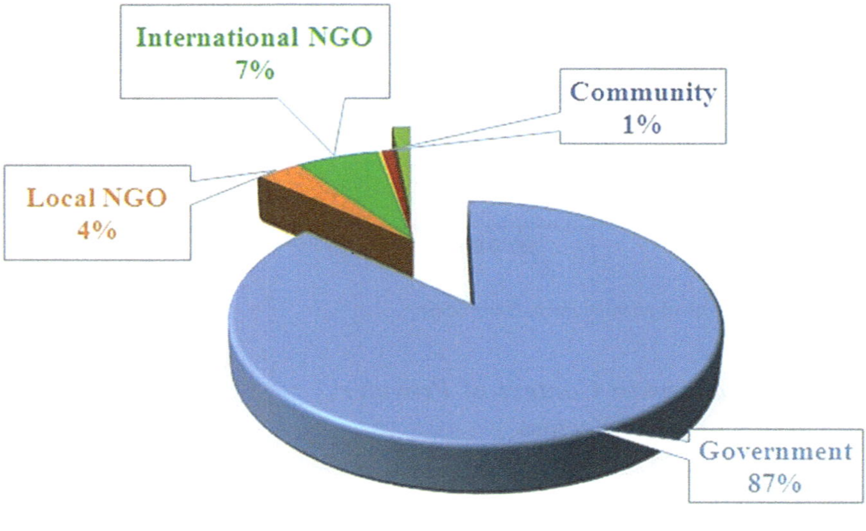

Fig. 2 Distribution of adaptation providers

But due to inadequate documentation of these activities this inventory could not include a large number of adaptation measures by many providers and implementers. This fact emphasizes the need of proper documentation of government activities in climate change adaptation practices for true recognition and evaluation of adaptations for future strategic action plan.

Adaptations recorded in the inventory were classified under different thematic sectors including disaster risk reduction (DRR), water resource management (WRM), infrastructure, agriculture, environment, capacity building, housing, etc. Many adaptation practices overlapped two or more sectors because these measures had multiple purposes and aims which can be classified in different sectors or areas. Also, adaptation measure taken in one thematic sector as a planned adaptation has become an autonomous one for another sector, which indicates the serendipitous effect of these practices. For example, embankment construction and improvement is an adaptation measure aimed at disaster risk reduction predominantly, but it has an additional effect on agricultural enhancement of the protected area. Among the sectors mentioned above, DRR was found to have experienced almost one-fourth of the total adaptations and seems to be the most important thematic area where adaptation strategies are taken and plans are formulated. Adaptation measures aimed at disaster risk reduction comprise of construction of embankment, revetment, groyne, construction and rehabilitation of polder, drainage canal and regulator, cyclone shelter, etc. The second most important thematic sector appears to be WRM which covers 20% of the total adaptations. Drinking water supply system, rainwater harvesting, water purification systems including pond sand filter and desalinization plant, water reservoir construction, etc., are the major adaptations in WRM sector. Infrastructural development and agriculture are the next two sectors which cover 17 and 13% adaptations, respectively.

The foremost reason of DRR and WRM covering about 45% of the adaptation measures all over the country is that most of these activities are concentrated in the delta region. This region is affected by the major natural disasters and this fact has drawn the attention of adaptation providers and funders since long. From Fig. 3 it can be noticed that about 73% of the adaptations for disaster risk reduction are focused on the delta region. Also in the WRM sector about 60% of the adaptation were taken in the delta. Severity of natural disasters and increasing risk of disasters due to climate change have inspired many government interventions in this area most of which were funded by Bangladesh Climate Change Trust (BCCT) and Comprehensive Disaster Management Plan (CDMP) programs. Coastal Embankment Improvement Project (CEIP) is also another attempt from the government to rehabilitate the existing conditions of coastal polders. These activities represent the interest of adaptation providers in funding distribution and planning implementation. However, other sectors show a balance in receiving adaptation activities in both delta and non-delta regions.

If we look at the sectoral adaptations in terms of government and NGO involvement then two distinct pictures can be observed (Fig. 4a, b). Agriculture receives higher percentage of adaptations by NGOs than government which can be justified by the fact that NGOs introduced stress tolerant crop cultivation,

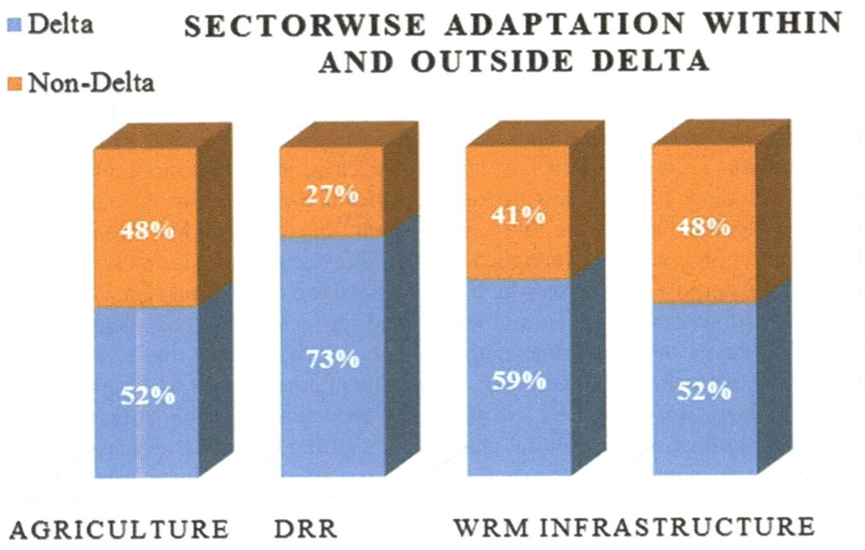

Fig. 3 Distribution of adaptation in delta and non-delta regions in major thematic sectors

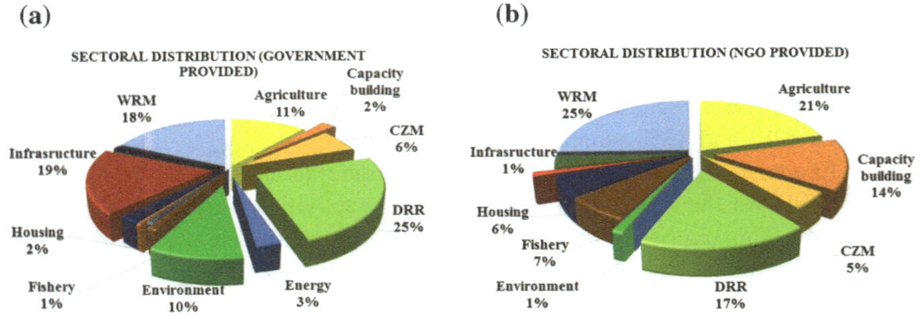

Fig. 4 Sectoral distribution of adaptations by **a** Government and **b** NGOs

micro credit system for marginal farmers for new crop variety experimentation, integrated crop and vegetable cultivation techniques and other similar activities. These projects may be smaller in scale but they have been effectively implemented and have improved adaptive capacity of the local people. Similarly, community level drinking water supply and small scale water treatment and purification facilities are some locally effective measures by NGOs in the WRM sector which associated them with high percentage of adaptations than the government. Also, capacity building and training activities by local and international NGOs are known adaptation techniques which have given them a share of 14% among the adaptation practices whereas government share is only 2%. On the other hand, environment and DRR are two major thematic sectors receiving higher government support than the NGOs. Most of the projects in these two sectors are large scale and therefore need wide institutional involvement and coordination and significant economic support as well. As a result, these practices represent exclusive government involvement. These projects are expected to provide long term service in vulnerability reduction and resilience enhancement which is the major reason for high government focus. However, this distribution is subjected to the documented adaptation measures in Bangladesh and there can be a whole new picture in front of us if various unaccounted activities by the adaptation providers are properly documented and made available for inclusion in this sort of national inventory.

3.3 Nature of Adaptations and Stresses and Shocks Responsible

Adaptation practices have emerged due to long term chronic stresses or short lived sudden shocks. About 75% of the adaptation measures were triggered by long term chronic stresses in Bangladesh including salinity, river bank erosion, waterlogging, regular flood, drought and so on. In comparison, number of adaptation measures to combat environmental shocks like cyclonic storm surge and severe flood is low. However, spatial extent and investment demand of these two may be quite different. Majority of these were reactive in nature while some were anticipatory. In some cases, some adaptation practices were first initiated as a reactive measure against a chronic stress in a location. But with time the need for improvement and the threat of higher intensity and frequency of a regular natural hazard have inspired uptake and upscaling of the former reactive adaptation measure, thus making it anticipatory in nature. For example, coastal polders and embankments were constructed as a reactive measure against regular tidal flooding which is a chronic stress in the delta region. But recently major cyclonic storm surge events, which can be attributed as sudden shocks of the delta area have brought about some projects for coastal embankment improvement which are aimed at providing additional protection from sudden shock events, and also these can be defined as anticipatory measures with recognition of future extreme events to come. However, percentage of anticipatory adaptations measures are still low. Recent activities have tried to alter the balance of this fact but still there is a lot to accomplish for creating a resilient society against climate change induced stresses and shocks.

3.4 Significance of Adaptation Documentation and Necessity of Such Inventory

Climate change impacts are being revealed in recent times and much more attention is being given to this phenomenon in this millennium than previous times which is reflected in temporal distribution of the adaptation measures. That is why climate change adaptation measures mostly took place in the last decade. Bangladesh has been actively addressing climate change issues which can be clearly visible from the recent enactments of different plans and strategies. Bangladesh is one of the first countries to formulate Climate Change Strategy and Action Plan (BCCSAP) in 2009 which is a major landmark document among the developing countries. Establishment of Bangladesh Climate Change Trust Fund (BCCTF) and Bangladesh Climate Change Resilience Fund (BCCRF) to allocate funding for a large number of adaptation measures by GoB and NGOs are noteworthy achievements of Bangladesh. Comprehensive Disaster Management Programs (CDMP) by UNDP have contributed significantly in putting the country on the path of climate change adaptation and disaster resilience. All the efforts from the government, NGOs, international donor agencies and the community people upgraded the status of Bangladesh as a well-adaptive country against the threat of climate change. But it is still not sufficient, for climate change is anticipated to bring global catastrophe which a developing country like Bangladesh may struggle to handle. Though there are a large number of activities going on around the country, but effectiveness and future contribution of the expensive, large scale interventions depend on the documentation and sharing of success stories and barriers as well.

Climate change adaptation activities have reportedly started since last few decades, but the coping mechanisms adopted by community people as well as government, though not being aware of the climate change phenomenon, started since long. Many of the interventions now deemed as addressing climate change effectively through reduction of vulnerability were in fact having serendipitous effect where the earlier target was climate variability induced natural hazards. This makes it difficult to distinguish between climate hazard focused structural interventions and climate change adaptations. Also, lack of proper documentation of activities by adaptation providers ceases to give the true timeline distribution of adaptations in Bangladesh.

Aim and sectoral focus of adaptations provided by government and NGOs changed time to time according to policy-planning and also the need of the affected community. Many of the implemented projects gained sheer success while some created maladaptive features as well. Nevertheless, the regular practices were continued without necessary modification or documentation of the associated issues. There are many recurring problems in our country that need continuation of previous works while new threats and difficulties require upscaling or improvement of previous adaptive measures. But inadequate documentation of the methodologies, post-project success stories or barriers of any adaptation practice/project restricted the opportunity of emulating from the best or learning from the shortcomings. This can be explained with the case of Tidal River Management (TRM) project. TRM was a community derived adaptation initiative in Jessore district of south-west coastal part of Bangladesh to resolve drainage congestion in the area. It was largely successful for the drainage congested areas of Keshabpur, Manirampur and Abhaynagar upazilas (sub-districts) at the beginning. But the documentation and the follow-up of the project was limited to academic circles, while largely neglected by implementing agencies. This resulted in a mismatch between the funding timeframe and the life cycle of the TRM at the site, inadequate addressing of the challenges and slow development of institutions to support it, which resulted in negative representation of the TRM in government planning and policy circle even though it was highly regarded in academic circles and by development partners as a successful adaptation. Such community friendly projects have been implemented and are running effectively in many locations which the relevant organizations are sometimes completely unaware of. This inhibits the execution of effective adaptation measures by replicating or scaling up of best management practices. Maintenance of an elaborated bibliography of adaptation practices as an inventory can help the government and other implementers to uptake significant adaptation initiatives and evaluate existing projects/practices for better accomplishment of adaptation strategies in Bangladesh.

4 Conclusion

From the analysis of adaptation practices documented in the national adaptation inventory it was evident that the adaptations are currently disproportionately arranged between delta and non-delta regions. The adaptations were more abundant in DRR, WRM, infrastructure and agriculture sectors where the major provider was the Government of Bangladesh. Delta region being the most likely vulnerable areas under climate change scenario has been in focus for DRR and WRM adaptations as is evident from the inventory. This signifies the focus of the policy makers on most vulnerable areas and sectors. But the experience of the preparation of the inventory shows some weaknesses in the current adaptation paradigm. Bangladesh has been promoting different adaptation strategies in diverse sectors for a long time and enough proof of that is evident from the resilient community of the present. Challenges lie on prioritizing on the sectors of adaptation for different future climatic scenarios and deciding on the successful adaptations which would be suitable for uptake, sustenance and upscaling for long time development planning. Bangladesh Government is having short term (e.g., 7th five year plan), mid-term (SDG 2030 goals and plans) and long term perspective plans for the country (e.g., Bangladesh Delta Plan 2100) which are explicitly looking into future climate scenarios, possible adaptations and necessary investment planning. The investment for adaptation funding from likely donors e.g., Green Climate Fund (GCF), Climate Resilience Fund, Global Environment Facility (GEF), and from the GoB would demand value for money for the developing country facing many challenges one of which being prioritization of fund. Deciding on adaptations with far reaching positive outcomes would be critical. The inventory is a start in the right direction of building on past experiences, shortcomings and achievements that can help in the adaptation planning or even the better, adaptive planning.

Acknowledgements This work is carried out under the 'Deltas, vulnerability and Climate Change: Migration and Adaptation (DECCMA)' project under the Collaborative Adaptation Research Initiative in Africa and Asia (CARIAA) program with financial support from the UK Government's Department for International Development (DFID) and the International Development Research Centre (IDRC), Canada. The views expressed in this work are those of the authors and do not necessarily represent those of DFID and IDRC or its Boards of Governors.

References

Adger, W. N., Arnell, N. W. and Tompkins, E. L. (2005). Successful adaptation to climate change across scales, Global Environmental Change, 15: 77–86, DOI:10.1016/j.gloenvcha.2004.12.005.

Ahmed, A. U. (2006). Bangladesh Climate Change Impacts and Vulnerability: A Synthesis, July 2006, Climate Change Cell, Department Of Environment, Component 4b, Comprehensive Disaster Management Programme, Dhaka-1207, Bangladesh, ISBN: 984-32-3467-7.

Ali, A. (1999). Climate change impacts and adaptation assessment in Bangladesh, Climate Research, 12: 109–116.

Asia Foundation (2012). A Situation Analysis of Climate Change Adaptation Initiatives in Bangladesh, September 2012, Asia Foundation, Dhaka-1206, Bangladesh.

Ayers, J., Huq, S., Wright, H., Faisal, A. M. and Hussain, S. T. (2014). Mainstreaming climate change adaptation into development in Bangladesh, Climate and Development, 6(4): 293–305, DOI:10.1080/17565529.2014.977761.

Baigini, B., Bierbaum, R., Stults, M., Dobardzic, S. and McNeeley, S. M. (2014). A typology of adaptation actions: A global look at climate adaptation actions through the Global Environment Facility, Global Environmental Change, 25: 97–108, DOI:http://dx.doi.org/10.1016/j.gloenvcha.2014.01.003.

Bangladesh (1995). Bangladesh - National environment management action plan (NEMAP), Dhaka, Bangladesh: NEMAP Secretariat, Ministry of Environment and Forest, Government of Bangladesh, Dhaka, Bangladesh, accessed at: http://documents.worldbank.org/curated/en/329001468741610744/Bangladesh-National-environment-management-action-plan-NEMAP.

Dev, P. K. (2013). Water logging through soil-less agriculture as a climate resilient adaptation option, In: Climate Change and Disaster Risk Management, Filho, W.L. (ed.), pp. 681–692, Springer-Verlag Berlin Heidelberg, eBook ISBN: 978-3-642-31110-9, DOI:10.1007/978-3-642-31110-9.

Faruque, H. S. M. and Ali, M. L. (2005). Climate Change and Water Resources Management in Bangladesh, In: Climate Change and Water Resources in South Asia, Mirza, M. Q. and Ahmad, M. Q. (eds.), pp. 231–254, Taylor and Francis, London, UK, eBook ISBN: 978-0-203-02077-7, DOI:10.1201/9780203020777.ch9.

Haq, I., Omar, M.A.T., Zahra, Q.A. and Jahan, I., 2015. Evaluations of Adaptation Policies in GBM Delta of Bangladesh. DECCMA Working Paper, Deltas, Vulnerability and Climate Change: Migration and Adaptation, IDRC Project Number 107642. Available online at: www.deccma.com, accessed on 25 August, 2016.

Haque, K. N. H., Chowdhury, F. A. and Khatun, K. R. (2015). Participatory environmental governance and climate change adaptation: mainstreaming of tidal river management in south-west Bangladesh, In: Land and Disaster Management Strategies in Asia, Ha, H. (ed.), pp. 189–208, Springer India, eBook ISBN: 978-81-322-1976-7, DOI:10.1007/978-81-322-1976-7.

IPCC (2014). Climate Change 2014: Synthesis Report. Contribution of Working Groups I, II and III to the Fifth Assessment Report of the Intergovernmental Panel on Climate Change [Core Writing Team, Pachauri, R.K. and Meyer L.A. (eds.)]. IPCC, Geneva, Switzerland, 151 pp.

MoEF (2009). Bangladesh Climate Chnage Strategy and Action Plan 2009, Ministry of Environment and Forests, Government of the People's Republic of Bangladesh, Dhaka, Bangladesh. Xviii + 76 pp, accessed at: http://www.climatechangecell.org.bd/Documents/climate_change_strategy2009.pdf.

Mondal, M. S., Rahman, M. R., Mukherjee, N., Huq, H. and Rahman, R. (2015). Hydro-climatic hazards for crop and cropping system in the chars of the Jamuna River and potential adaptation options, *Natural Hazards*, 76(3): 1431–1455, DOI:10.1007/s11069-014-1424-9.

National Adaptation Program of Action (2005). Ministry of Environment and Forests, November 2005, Government of the People's Republic of Bangladesh, Dhaka, Bangladesh, accessed at: http://unfccc.int/resource/docs/napa/ban01.pdf.

National Plan for Disaster Management (2010–2015). Disaster Management Bureau, Disaster Management & Relief Division, April 2010, Government of the People's Republic of Bangladesh, Dhaka, Bangladesh, accessed at: http://www.lcgbangladesh.org/derweb/doc/Final%20Version%20Nataional%20Plan%20for%20Disaster%20(2010-2015).pdf.

National Water Policy (1999). Ministry of Water Resources, Government of the People's Republic of Bangladesh, Dhaka, Bangladesh, accessed at: http://faolex.fao.org/docs/pdf/bgd146075.pdf.

Rawlani, A. K. and Sovacool, B. K. (2011). Building responsiveness to climate change through community based adaptation in Bangladesh, Mitigation and Adaptation Strategies for Global Change, 16(8): 845–863, DOI:10.1007/s11027-011-9298-6.

Rosenzweig, C., G. Casassa, D. J. Karoly, A. Imeson, C. Liu, A. Menzel, S. Rawlins, T. L. Root, B. Seguin, P. Tryjanowski (2007). Assessment of observed changes and responses in natural and managed systems. Climate Change 2007: Impacts, Adaptation and Vulnerability. Contribution of Working Group II to the Fourth Assessment Report of the Intergovernmental Panel on Climate Change, M. L. Parry, O. F. Canziani, J. P. Palutikof, P. J. van der Linden and C. E. Hanson, (eds.), Cambridge University Press, Cambridge, UK, 79–131.

Saha, D., Hossain, M. S. S., Mondal, M. S. and Rahman, R. (2016). Agricultural Adaptation Practices in Coastal Bangladesh: Response to Climate Change Impacts, Journal of Modern Science and Technology, 4(1): 63–74, ISSN: 2201-6686.

Tompkins, E. L., Adger, W. N., Boyd, E., Nicholson-Cole, S., Whitehead, K. and Arnel, N. (2010). Observed adaptation to climate change: UK evidence of transition to a well-adapting society, Global Environmental Change, 20: 627–635, DOI:10.1016/j.gloenvcha.2010.05.001.

Tompkins, E.L., Boyd, E., Nicholson-Cole, S. Weatherhead, K., Arnell, N., Adger, W.N. (2009). An Inventory of Adaptation to climate change in the UK: challenges and findings. Tyndall Working Paper 135, July 2009. Tyndall Centre for Climate Change Research, UK.

The World Bank (2011). The cost of adapting to extreme weather events in a changing climate, Bangladesh Development Series, Paper No. 28, The World Bank, December 2011, Dhaka 1207, Bangladesh.

Implementing Heat-Related Adaptation Measures in the Tri-City Area Bergisches Städtedreieck

Alice Neht, Kathrin Prenger-Berninghoff and Dirk Vallée

Abstract In order to meet the challenges of climate change, the overarching strategies of climate change mitigation and adaptation play an equally important role in urban development (IPCC 2014). This leads to the overall question of how municipalities realize climate change adaption strategies in the context of their spatial development. The project 'BESTKLIMA—Implementation and quality assurance of the climate adaptation concept in the Bergisches Städtedreieck' aims to develop and test transferable spatial development solutions for climate adaptation in the cities of Remscheid, Solingen, and Wuppertal. The project follows the research approach of climate change governance using a moderated bottom-up method, which involves the actors and their knowledge of methods and facts (BMVBS and BBSR 2013). One objective of the project is to promote the integration of measures that tackle thermal stress into administrative practice. The question is now how precicely to implement the heat-related measures that were decided upon in the previously adopted *regional climate change adaption strategy*. For this purpose, constellations of actors and responsibilities in handling heat waves are identified based on expert interviews. Initial results show that the cities share approaches to forestry and health care but differ in their approaches to spatial planning as well as green and open space development when it comes to heat.

Keywords Climate adaptation · Heat · Inter-municipal cooperation · Climate change governance

A. Neht (✉) · K. Prenger-Berninghoff · D. Vallée
Faculty of Civil Engineering, ISB—Institute of Urban and Transport Planning, RWTH Aachen University, Mies-van-der-Rohe-Str. 1, 52074 Aachen, Germany
e-mail: neht@isb.rwth-aachen.de

K. Prenger-Berninghoff
e-mail: prenger-berninghoff@isb.rwth-aachen.de

D. Vallée
e-mail: vallee@isb.rwth-aachen.de

W. Leal Filho (ed.), *Climate Change Research at Universities*,
DOI 10.1007/978-3-319-58214-6_12

1 Heat in Cities

As a result of anthropogenic climate change and the changing temperature and precipitation patterns, the occurrence of extreme weather events such as long periods of heat or draught will increase significantly. Current climate models of the International Panel of Climate Change (IPCC) forecast a rise of the global average temperature from 2.6 to 4.8 °C by the end of the century compared to the level of 2014 (IPCC 2014). In Germany, temperatures will rise by 3.5 °C on average or between 1.5 and 5 °C in summer and between 2 and 4.5 °C in winter compared to the reference period 1961–1990. However, temperatures in Germany will not only rise on average, but also grow in intensity. In summer the number of hot days will increase from 8–12 to 15–25 days a year (DWD 2015; UBA 2013). 'Hot days' is the climatological term for days on which the maximum temperature of the day reaches or exceeds 30 °C (DWD 2015).

The impact of these changes will particularly be felt in urban areas and will exacerbate the urban heat island effect as the high building density and the resulting high degree of sealing increase the likelihood of heat accumulation and draughts (IPCC 2014; Weischet and Endlicher 2008; Schönwiese 2003). In addition, the rising heat affects people's health and livelihood. Higher temperatures lead to heat-related health issues and air pollution in cities (IPCC 2014). The increase of hot days in summer means that there will be fewer cool nights, which are necessary for human regeneration (Baumüller 2012; Schönwiese 2003). According to Dosch (2015), "as long as no damages have occurred [...] the topic is ignored in everyday political and administrative practice" (own translation). But the events in the summer of 2003 have shown that particularly intense heat waves come with long periods of draught and lead to a build-up of heat especially in high-density inner cities (Dosch 2015; Weischet and Endlicher 2008).

One solution to this problem is to integrate urban climate concerns into urban planning early on, creating, protecting, or ideally increasing cool air flow and evaporation chill (Dosch 2015; Baumüller 2012; Chmielewski et al. 2005). This solution can be put into practice using an urban-planning heat management which aims to protect or, where possible, expand fresh- and cold-air corridors, green and water spaces, and adapted building regulations (Dosch 2015; IPCC 2014; Weischet and Endlicher 2008). The focus and the measures must be selected depending on the type of settlement and the possibilities of existing and new buildings. Ideally, measures such as the "evaporation capacity of plants in combination with retaining and storing rainwater and providing it to plants contribute to cooling overheated urban areas, bridging periods of draught, and thus mitigate extreme heat" (Dosch 2015; own translation).

This paper focuses on how municipalities deal with the challenge of addressing thermal stress in the context of climate change adaption in their spatial development. The research approach contains the analyses of the status quo of the case study in the Tri-City Area Bergisches Städtedreieck concerning the regional climate change adaptation strategy and the heat-related measures that have been already realized. Based on the comparison between measures proposed in the strategy paper and the realized measures, this paper investigates which measures from the context of the strategy have not been realized and why, as well as what kind of resources municipalities need in order to reach their goal of a successful climate change adaptation concerning thermal stress. Our research is based on the hypothesis that cooperation between the municipalities is fundamental for successful climate adaption.

2 Municipalities and Thermal Stress

The following chapter explains the rationale behind this hypothesis. Local administrations play a decisive role in successful climate adaptation. They are able to integrate climate adaptation into regulations, investments, and—by political action—adjust it to local conditions (IPCC 2014). Possibilities for adaptation must be assessed specifically for each city since each municipality has its own urban climate conditions (Baumüller 2012). Climate adaptation has been part of the municipalities' tasks since 2011, after a first amendment of §1a of the Baugesetzbuch (German building code) (Baugesetzbuch 2015). This task also requires municipalities to confront current and future heat stress (BMVBS and BBSR 2011). This results in four challenges for the municipalities: The spatial cause-and-effect radius of climate risks such as heat waves, the funding of measures, the design of measures and concepts in the face of uncertainty, and the lack of acceptance in everyday practice.

Heat waves are not confined by administrative borders, which means that dealing with thermal stress requires numerous layers of action and stakeholders (Dosch 2015; Fröhlich et al. 2011). The most promising strategies are therefore inter-municipal ones in which the local administrations agree on preventative and reactive (sets of) measures within a certain context which matches the cause-and-effect radius of specific effects of climate change (Baumüller 2012). Another challenge is posed by the direction of municipal financial and human resources towards developing climate risks in times of strained financial means. This just makes the need for immediate climate adaptation more pressing: While it is a complex task, adaptation is still possible—and in the long term more convenient than doing nothing (IPCC 2014). As future developments are only incorporated into planning processes using probability assessments, any decision has to be made under uncertainty (Bundesregierung 2008). One of the current challenges is therefore to identify suitable measures by the so-called 'no-regrets principle'. This principle requires concepts and measures that act preventatively and offer social benefits regardless of climate change (Birkmann et al. 2013). In practice, preventative measures in particular often meet barriers because they lack acceptance. In order to promote acceptance and thus ensure the implementation of measures, a climate change governance approach is taken which will involve actors and their knowledge. As this is meant to be a voluntary involvement, it is necessary to make decisions by consensus, involve key actors, communicate effectively, and to integrate political decision-makers or agreements to make sure contributions and results are binding. The research project BESTKLIMA also follows the climate governance approach since meeting the challenges of climate change requires a thorough exchange between the relevant actors (BMVBS and BBSR 2013). BESTKLIMA addresses the four mentioned challenges, as the existence of a regional adaptation strategy alone does not guarantee the strategy's implementation into everyday administrative action.

3 Research Project BESTKLIMA

The development and evaluation of action plans for climate adaptation at the national level has already been initiated by the 'Federal Environmental Agency'. The research project 'KlimaExWoSt—Urban Strategies for Adapting to Climate Change' (2009–2013) builds the scientific basis for BESTKLIMA. The projects 'klimAix—Climate-adapted commercial area development in the StädteRegion Aachen', 'Karlsruhe Neighborhood Association—Urban development versus climate comfort', and 'Nuremberg climate adaptation: Summer in the city—Effectively combating climate change' in particular constitute intersections with state-of-the-art research that will lead to a further development of promising climate adaptation in Germany (Benden et al. 2012; UBA 2012).

Within the research project 'BESTKLIMA—Implementation and quality assurance of the climate adaptation concept in the Tri-City Area Bergisches Städtedreieck', transferable solutions will be developed for the organization and control of complex adaptation projects at the urban-regional level from 2014 to 2017. These solutions are developed in the cities of the Tri-City Area—Remscheid, Solingen, and Wuppertal—and will be provided to other regions as a guideline.

3.1 The Predecessor

In addition to developing skills and a quality management system/monitoring for urban and regional adaptation progress, the project aims for the implementation of knowledge gained in the previous project 'Adaptation to climate change for the cities of Solingen and Remscheid'—a climate adaption concept—in collaboration with the regional actors. Nine different measures were described in the former project to the end that the municipalities were supported and advised in dealing with thermal stress. Concerning environmental issues, measures like adaptational silvicultural strategies as well as the assurance and improvement of urban green spaces with climatic relevance are suggested. Regarding the building sector, the indoor climate comfort and cold-air generation and outflow zones should be secured, urban and climate modelling should be established, and preventive measures regarding thermal stress in public transport and road space should be implemented. Furthermore, a heat wave preparation concept and the installation of a local heat warning system as well as the use of heat waves as a scenario in operational planning should be realized.

3.2 Methodology

The investigation is conducted at two levels: The content level covers administrative practice. In this context, the status quo of how administrations deal with thermal stress is analyzed by a desktop analysis. This results in suitable data and information being provided for the municipal administrations, which in turn is the basis for a prioritization of the need for action and the implementation of the measures in cooperation with the actors (BMVBS and BBSR 2011). At the level of the actors, those actors who are relevant in the field of thermal stress and the transformation process are identified in accordance with the climate change governance approach. This leads to the limitations of the investigation at the actor level: The deliberate choice of specific actors is based on current knowledge about the actors' role in climate change adaption processes but still leads to predetermined results. The decision for expert interviews as a socio- scientific method is based on the hypothesis of this research, i.e. that cooperation between the municipalities is crucial for prospective climate adaption, and therefore the inside perspective is explored through the expert's knowledge. Data collection was completed in 47 qualitative expert interviews. A specific interview guideline was used and the data was analyzed by means of an actor analysis (Zimmermann 2006; Hübner-Schmidt et al. 2003).

4 Thermal Stress in the Administrative Context in the Tri-City Area

The institutional conditions shaped by administrative structures are the framework of the implementation of measures. The way in which these structures take shape in town services, departments or offices in the municipal administrations depends on the respective municipality (Pahl-Weber et al. 2015).

The administrations of Remscheid, Solingen, and Wuppertal are very similar. In each of the three cities, the administration is divided into five administrative departments. These are in turn subdivided into town services. However, those departments and town services relevant to climate adaptation are structured differently in the three city administrations. The town services 'Environmental protection' and 'Urban development/urban planning' belong to different departments in Remscheid but are part of one department in Wuppertal and Solingen (Stadt

Remscheid 2015; Stadt Solingen 2015a; Stadt Wuppertal 2015). In the three cities, urban drainage is not part of an administrative department but is managed by operations committees (Stadt Solingen 2015b). This means that the technical services Remscheid, the technical services Solingen, and 'Wuppertal water and sewage' are considered municipal institutions similar to an owner-operated municipal enterprise.

Climate adaptation and measures for dealing with thermal stress are handled with an inter-agency approach since climate adaptation concerns numerous social areas (Beierkuhnlein and Foken 2007). The analysis of the content covers the fields of forestry, nature conservation, green and open spaces, water management and drainage, spatial and land-use planning, architecture and construction, urban development concepts and strategies, health, communications and public relations, and cooperation and networking among the actors. In order to structure the results clearly, these social fields are grouped into the fields 'Environment', 'Building', and 'Social Affairs'.

The following figures represent a comparison of the three cities' measures for dealing with thermal stress. The overview of measures is meant to provide a visualization of each city's level of activity and state of knowledge.

4.1 Action Field 'Environment'

Concerning forestry, the three cities are already restructuring their forests and planting sustainable tree species. Remscheid has also established forest safety points that make locating and fighting wildfires faster and speed up the rescue of visitors or staff who have had an accident. In addition, the forest association 'Wald 2.0', which works to conserve the utility and the protective and recreational function of forests, was accepted into the national joint project 'Forest Climate Fund' through the Joint Forestry project of the 'Ministry of the Environment of North Rhine-Westphalia'. For Solingen, politicians have also suggested the strengthening of the forest conservation policy.

Concerning nature conservation, a measure could only be identified for the city of Remscheid. The federal journal on nature conservation of the city has covered the topic of thermal stress several times. This journal also includes essays relevant for the topic of green and open spaces in public. In addition, it defines and locates recreational areas and green spaces in the city. With the action program 'Green into the city', Remscheid has furthermore dedicated itself to developing an implementation tool addressing the weaknesses found regarding inner-city green spaces and their various functions. Independently of this program, the town service 'Environment' works on the continuous development of urban green spaces. The plan is to expand recreational areas on flat terrain in order to create more accessible recreational areas for citizens with limited mobility (e.g. small children, elderly people). All three cities are planning to create green spaces contributing to urban climate, e.g. 'climate control islands' or 'cool islands'. Moreover, the experts in the

Table 1 Comparative overview of the measures of the action field 'Environment'

	Remscheid	**Solingen**	**Wuppertal**
Forestry	● ● ●	● ⊜	●
Nature conservation	●		
Green and open spaces	● ● ● ◍	● ⊜ ⊜ ⊜	⊜ ⊜ ⊜
Water management and drainage		⊜	● ● ⊜

implemented ● planning ◍ proposition ⊜

Source Own data and depiction 2015

interviews agreed that ensuring cold-air generation areas and cold-air flow was an important task for the future.

Concerning water management and drainage, Wuppertal changed the tariff system, which now entitles people to reduced charges or being exempt from charges if the leachate is harmless. The Wupperverband, a river basin authority, is also looking to adapt the dam management to ensure the drinking water supply and its quality during heat waves. In Solingen and Wuppertal, daylighting streams (Solingen) and creating new bodies of water above ground (Wuppertal) were suggested (see Table 1).

4.2 Action Field 'Building'

Concerning urban development concepts and strategies, there are climate reports on the Ittertal in Solingen and on heat islands and cold-air generation areas in Wuppertal. Remscheid and Solingen share a climate adaptation concept, which is further supplemented by an integrated climate mitigation concept in Remscheid. While Remscheid lacks the expert basis regarding urban climatology, the city still actively contributes at the conceptual level and develops ideas and suggestions for the future, e.g. establishing a connection between projects on heat wherever possible to generate synergies between the measures. One suggestion is to give urban climate concerns a higher priority in future concepts. Based on the climate report and the climate adaption concept, the Bergisch-Rheinischer Wasserverband (the second river basin authority in the area) suggested identifying particular hotspots in Remscheid. For their spatial and land-use planning practices, all three cities have an urban climate analysis, which includes the location and function of spaces relevant to the climate. Within this context, Solingen and Wuppertal focus on conserving fresh- and cold-air corridors, which is also being planned in Remscheid. The cities differ widely regarding which level of spatial development the topic of thermal stress is allocated to in the plans. Wuppertal and Remscheid include the topic in their landscape plans, while dealing with thermal stress must explicitly be taken into account in the creation of development and land-use plans in Remscheid and Solingen. A representative of the administration of Wuppertal suggested adopting this approach at the level of land-use planning in Wuppertal as well. As far as they can, Remscheid and Solingen want to incorporate the recommendations made so far into land-use planning as mandatory ecological principles. This further development of legal means is supposed to be included in development plans or regulations and in additional urban planning contracts. Furthermore, all three cities promote the unsealing of spaces. In order to provide an improved overview of the spaces available, Remscheid characterized the different types of settlement areas based on sealing and vegetation structures for the public interest bodies, resulting in a backup and vacant-lot cadaster.

At the object level, in the fields of architecture and construction, the three cities are already supporting the funding and organization of roof greening. In Solingen there is the additional idea to take heat shielding into account in future insulation measures. In Wuppertal, the plan is to introduce a regulation for the use of brighter façade paint (see Table 2).

Table 2 Comparative overview of the measures of the action field 'Building'

	Remscheid	**Solingen**	**Wuppertal**
Urban development concepts and strategies	● ● ⊜ ⊜	● ● ⊜	● ●
Spatial and land-use planning	● ● ◌ ◌ ● ● ● ●	◌ ● ● ● ●	⊜ ● ● ● ●
Architecture and construction	●	● ⊜	● ⊜

implemented ● planning ◌ proposition ⊜

Source Own data and depiction 2015

4.3 Action Field 'Social Affairs'

Concerning health, both Remscheid and Wuppertal already have measures in place, such as organized warnings to care facilities in the event of heat waves or deploying the Solingen fire department to supply drivers stuck in traffic jams with drinking water during heat waves. It was also suggested that Solingen and Wuppertal avoid building sensitive facilities in particular urban hotpots in the future. In addition, vulnerable groups of people and facilities should be identified and given special protection during heat waves. For Wuppertal, the idea was put forward to focus on preparing for afflictions and diseases that are likely to increase in frequency due to climate change.

Concerning communications and public relations, Solingen has informational material, i.e. health advice flyers, while Remscheid provides the public with health advice via press material released during heat waves. In addition, information is offered on the city's website throughout the year.

Cooperation and networking among the actors regarding thermal stress is handled very differently in the three cities. In Wuppertal, heat-related events such as a planners' conference, a convention, and a workshop were conducted. Solingen and Remscheid additionally offered a related workshop conversation during the project prior to BESTKLIMA. In Remscheid, there is also the 'Climate Alliance', which provides climate advice for citizens' everyday life. All three cities are also planning to set up a skill network 'Climate Adaptation' as well as local measures to raise awareness. Representatives of the group of actors from associations and societies suggested that Remscheid set up a digital network for the green space management of the city to connect all the actors working in that field (see Table 3).

Table 3 Comparative overview of the measures in the action field 'Social Affairs'

	Remscheid	Solingen	Wuppertal
Health	●	● ● ≡ ≡	≡ ● ● ≡ ≡
Comunications and public relations	●	●	●
Cooperation and networking	≡ ● ● ▒ ▒	● ▒ ▒	▒ ● ● ● ▒

implemented ● planning ▒ proposition ≡

Source Own data and depiction 2015

5 Who Branches Out Concerning Thermal Stress?

In order to determine the potential of the actors in this project, interests and knowledge of key actors and central and decentral actors were analyzed. In addition, the existing connections between the actors were identified. Examining the interconnections in particular reveals the best possible constellations and synergies as well as potential difficulties so that these can be intensified or avoided respectively during the course of the project. The interconnections are divided into informal activities (concrete projects and implementation measures, coordinated administrative agreements, information and communication approaches, etc.) and formally controlled activities (mandatory regulations within the administration, council orders, bylaws, etc.) (BMVBS and BBSR 2013). The fundamentals laid down by Hübner-Schmidt et al. (2003) and Zimmermann (2006) were chosen from the numerous approaches to actor analysis as they are particularly suited to regional climate adaptation processes. The results of the analysis very much depend on the subjective interests and knowledge of the people interviewed and can therefore differ widely (Solga and Blickle 2009). For the actor analysis, representatives from the administration, politics, the economy, science, and associations and societies were interviewed.

5.1 Actors in Remscheid

In Remscheid, the Department of Urban Development, Building, and the Economy is only indirectly involved in the topic and deals with thermal stress in cities in the context of framework development planning. The 'Department of Order, Security, and Law' is responsible for the town service 'Environment'. The department primarily focuses on coordination and on raising awareness among the population. For this purpose, it also has a good overview of measures related to heat. The town service 'Environment' is concerned less with thermal stress than with water-related aspects. In general, however, the town service 'Environment' has extensive knowledge on municipal climate adaptation. The topic of heat is somewhat taken into account when it comes to expert reports and internal trainings. The technical service 'Green and Open Spaces' of Remscheid deals with the invasion of forest pests. A climate-adapted restructuring of the forests is already being undertaken and further knowledge on green urban development exists. The town service 'Health' has recorded a higher death rate among the elderly during heat waves, which required taking stronger precautions and an improved warning system planning, which is why there is now knowledge on protecting groups of people particularly vulnerable to heat.

Outside of the administration, the 'Federal Association for Nature Conservation and the Environment' and the 'Rhenish Agricultural Association' deal with the topic of heat in the context of climate adaptation. The former is concerned with the way in which biodiversity and faunal and floral diseases are affected by climate change. The ‘Federal Association for Nature Conservation and the Environment’ has extensive knowledge on numerous environmental protection issues, especially on the importance of green spaces for urban climate. The ‘Rhenish Agricultural Association’ justifies its specific focus with the importance of the climate for agriculture (e.g. heat waves). It is particularly knowledgeable regarding the problems of soil sealing, regarding cultivation methods which reduce the risk of crop failures as a result of draught, and regarding countermeasures for erosion due to the rising frequency of storms.

In terms of enterprise, the Remscheid public works are concerned with heat since periods of draught affect the water level of drinking water reservoirs. The ‘Department of Urban Development, Building, and the Economy’ (abbr. Urban Planning), the town service ‘Environment’ (abbr. Environment) and the ‘Department of Order, Security, and Law’ (abbr. Security) are defined as key actors in dealing with thermal stress. These key actors are distinguished by their significant specialist knowledge and may have essential responsibilities and skills. In addition, they are well-connected with the other actors both formally and informally (Zimmermann 2006). Some of the existing relationships in Remscheid are a given due to the structure of the administration, so that a connection between Security, Environment, and Green and Open Spaces already exists. The relationships between Urban Planning, Environment, and Green and Open Spaces, on the other hand, have only developed in the context of thermal stress. The central actors include the technical services ‘Green and Open Spaces’ (abbr. Green Spaces).

Central actors are defined as being relatively closely connected to the key actors and directly affected by a project. This affectedness may be a benefit, a disadvantage, or immediate influence on measures essential to the project (Zimmermann 2006). The decentral actors are the town service ‘Health’ (abbr. Health) and the ‘Federal Association for Nature Conservation and the Environment’ (abbr. NCE). Decentral actors are only indirectly or temporarily affected by or involved in a project and its related measures. Decentral actors therefore only have a minor influence on the climate adaptation process (Zimmermann 2006). The town service ‘Health’ is related to the town service ‘Environment’, whereas the ‘Federal Association for Nature Conservation and the Environment’ is connected to the technical services ‘Green and Open Spaces’ through informal activities (see Table 4). The public works cannot be classified.

5.2 *Actors in Solingen*

The management of the ‘Department of Planning, Transport, the Environment, and Housing’ consider themselves affected as the department includes the town services ‘Planning, Mobility and Monument Preservation’ (abbr. Urban Planning), ‘Environment and Nature’ (abbr. Environment), and ‘Health’ as well as the fire service. The town service ‘Planning, Mobility and Monument Preservation’ is only

Table 4 Overview of the actors in Remscheid

		Urban planning	Environment	Green and open spaces
Key actor	Urban planning	-	Informal activities	Informal activities
	Environment	Informal activities	-	Formally controlled activities
	Security		Formally controlled activities	Formally controlled activities
Central actor	Green spaces	Informal activities	Formally controlled activities	-
Decentral actor	Health		Informal activities	
	NCE			Informal activities

Legend: Informal activities (dark grey); Formally controlled activities (light grey)

Source Own data and depiction 2015

concerned with aspects of climate adaptation when it comes to creating development plans. The town service 'Environment and Nature' has an advisory function in this, which is, however, limited to climate adaptation in general as the actions of this service are only peripherally affected. In Solingen, these two town services are the key actors (see Table 5). The town service 'Health' is heavily affected by the topic of thermal stress and does not differentiate between climate change mitigation and adaptation in its actions. It is dedicated to protecting people vulnerable to heat, monitors the immigration of animals that might carry diseases untypical for the region, and ensures the supply and quality of drinking water. Within the 'Department of Planning, Transport, the Environment, and Housing', the fire service is nearly unaffected, having neither specific knowledge nor heat-related responsibilities—with the exception of providing drinking water to drivers stuck in traffic jams during heat waves. The municipal enterprise 'Green Spaces' is affected in that one of its main tasks is the protection of urban green spaces, the relevance of which for urban climate is well known.

Among the political decision-makers, the 'Committee for Urban Development, the Environment, Climate Change Mitigation, Mobility' (abbr. UDECM) is a particularly relevant actor. Its members must consider climate change in their decisions. However, it lacks specific knowledge on the topic and the difference between climate change mitigation and climate adaptation is not always clear. However, there is some know-how regarding forest restructuring. The committee

Table 5 Overview of the actors in Solingen

		Urban planning	Environment
Key actor	Urban planning	-	
	Environment		-
	PTEH		
Central actor	UDECM		
Decentral actor			
		Informal activities	Formally controlled activities

Source Own data and depiction 2015

also functions as a connection to the administration and as a policymaker, making it a central actor (see Table 5).

The Biological Station 'Middle Wupper' is assigned to the associations and advocacy groups and has extensive specialist knowledge. The station is affected by climate change since the changes make its environment and nature conservation programs more difficult.

The public works of Solingen are only partially concerned with the effects of climate change. For example, the water level of the drinking water reservoirs operated by the public works depends on weather conditions. Periodic drought is a problem and may affect the quality of the drinking water. The actor does not distinguish sufficiently between climate change adaptation and climate change mitigation; its knowledge on the topic is deficient. The public works' heat- and drought-related actions are limited to protecting the reservoir water from algae.

Within the field of relevant actors, the town services 'Urban Planning' and 'Environment' are connected. Both are part of the 'Department of Planning, Transport, the Environment, and Housing', which is why there is an exchange of information between these two levels which can be traced back to the hierarchical structure of the city administration (see Table 5). As it is these town services that are responsible for implementing the resolutions of the 'Committee for Urban Development, the Environment, Climate Change Mitigation, Mobility', there is another connection here, which is independent of the institutional connections (see Table 5).

5.3 Actors in Wuppertal

In the branch 'Urban development, Building, Transport, Environment', the Departments of 'Urban Planning and Development' (abbr. Urban Planning) and 'Green Spaces and Forests' (abbr. Green Spaces) deal with thermal stress and therefore have profound expertise. Urban Planning is concerned with overheating in inner cities in terms of urban planning and development and functions as a key actor (see Table 6). Its neighboring department Green Spaces, defined as a central actor, furthermore records increasing heat and drought as well as the invasion of thermophile pests as damage to the green spaces. Furthermore, the owner-operated municipal enterprise 'Retirement Homes and Nursing Homes' (abbr. Social Affairs), a decentral actor, and the town service 'Social Affairs, Youth, School, Integration' are dedicated to protecting vulnerable groups during heat waves. In addition, the town service 'Health' is also dedicated to those new diseases which may occur in Wuppertal in the future due to global warming. Within the administration of Wuppertal, the fire department is also involved as heat waves affect firefighting.

In terms of associations and advocacy groups, the Chamber of Industry and Commerce and the parish of Wuppertal have limited knowledge on the topic of

Table 6 Overview of the actors in Wuppertal

Key actor	Urban planning
Central actor	Green spaces
Decentral actor	Wuppertal Institute
	Social affairs

Source Own data and depiction 2015

climate adaptation and are affected only marginally. At best, the parish considers protecting God's creation its responsibility.

Concerning economics, the Wuppertal Institute for Climate, Environment and Energy is involved. This actor is an expert in the field of climate adaptation and consults policymakers and the administration from the local to the national level. Public works are experts on heat-related climate adaptation but are not active in this regard. The Wupperverband has extensive knowledge especially on water-related aspects of climate adaptation. This is connected to heat in that heat waves affect the quality and level of water in reservoirs.

As shown in the following table, the relevant actors in Wuppertal exist in parallel —there are no interdependencies between them.

6 Conclusions

The analysis shed light on the background for implementing the chosen research approach of climate change governance in the Tri-City Area Bergisches Städtedreieck. According to the current state of research, climate adaptation projects are part of a climate change governance process provided that they are based on voluntary cooperation independently of administrative borders and are supported by public institutions. This means they conform to the idea of an "experimental regionalism" in which "a change of paradigms of spatial development in terms of content and processes (…) is meant to be achieved in the medium and long term through strategic selectivity in spatiotemporally specific, institutional experiments (…) by using soft model projects" (Wiechmann et al. 2012, p. 79; own translation). An inter-municipal approach for the Tri-City area Bergisches Städtedreieck regarding thermal stress is needed and must therefore be established. A fundamental similarity of the three cities is that they have all begun dealing with the topic of thermal stress, albeit independently. As all of the cities therefore have a similar starting point, an approach pursuing a joint development of contents and measures makes sense. As a climate change governance approach presupposes voluntary cooperation among the actors, the cooperation should be strengthened where it exists. The three cities differ widely in this regard: The scale stretches from an already well-developed informal connection (Remscheid) to a partially developed (Solingen) and a non-existent one (Wuppertal).

The offices in the Bergisches Städtedreieck which are relevant to urban planning are key actors in the context of thermal stress. Other key actors are the town service 'Environment' in Remscheid and the town service 'Environment and Nature' in Solingen. Central actors in the three cities are the technical service 'Green and Open Spaces' in Remscheid, the 'Green Spaces and Forests' department in Wuppertal, and the 'Committee for Urban Development, Environment, Climate Change Mitigation, Mobility' in Solingen. The key actors and central actors are therefore similar in their functions, which constitutes a beneficial basis for an inter-municipal exchange between the institutions. Since the decentral actors consist of the town service 'Health' and the 'Federal Association for Nature Conservation and the Environment' in Remscheid as well as the owner-operated municipal enterprise 'Retirement Homes and Nursing Homes' and the Wuppertal Institute for Climate, Environment and Energy in Wuppertal, the fields of action for this type of actor differ widely. Nevertheless, extending the inter-municipal cooperation is possible regarding health and particularly advisable regarding the topic of heat. There are two options that can be followed in the future: On the one hand, the decentral actors have to be strengthened in their radius of action concerning funding and responsibilities. Or on the other hand it has to be accepted that these actors are limited in their contribution concerning the climate change adaption process.

The analysis of the actors in Remscheid has shown that while the administrative structure does play a role in the connection between the actors, the connections which exist outside of this structure predominate. The fact that Remscheid—in comparison to the other two cities—has the most extensive knowledge and the highest density of measures shows that a structure that allows for both formal and informal activities is beneficial. This assessment is confirmed by the analysis of actors in Solingen as the state of knowledge and the density of measures is lower there and no formally controlled activities exist. In Solingen, actors are primarily connected through informal activities. The existence of both types of activities, formally controlled and informal, correlates with the state of knowledge and the density of measures. For example, no connection could be noted in Wuppertal, and the city also comes in last in terms of content. Concerning the role of informal and formal activities, it is proven that a high number of informal activities lead to a higher number of activities in the context of climate change adaptation within municipalities. It remains unclear whether a low level of connection between the actors causes a lower density of measures or whether the implementation of only a few measures causes a lower degree of connection between the actors.

In order to achieve an improved way of handling thermal stress in administrative action—, an initial evaluation of the state of measures that have already been implemented was conducted. This first step renders possible a comparison with the measures suggested in the previous project, which were designed according to the no-regrets principle. Of the nine heat-related measures suggested in the previous project that are relevant to BESTKLIMA, four have so far been implemented in each of the cities (see Table 7). As the measures were not weighted, each city sets its own focus in implementing the measures.

Comparing the three cities in terms of the action field 'Environment' shows that silvicultural strategies are already being employed to mitigate the effects of rising temperatures and heat waves. Within the action field 'Building', the measures of the subsection 'Architecture and Construction', e.g. ensuring climate comfort in buildings, are rather meagre. This may be because the cities' influence at the object level is small. This is where owners must act, which the city might achieve through PR work. Entirely untouched is the topic of thermal stress prevention in public transport and street spaces. These fields are so extensive and costly that implementing measures in road construction are only possible in isolated instances, rendering it inefficient.

Table 7 Comparison of the measures suggested in the previous project to the current status of heat-related measures

	Proposed measures from the previous project	Remscheid	Solingen	Wuppertal
Action field 'Environment'	Silvicultural strategies related to rising temperatures and heat waves	✓	✓	✓
	Assurance and improvement of urban green spaces with climatic relevance	✓		
Action field 'Building'	Assuring climate comfort indoors			
	Urban and climate modelling		✓	✓
	Assurance of cold-air generation and outflow zones		✓	✓
	Prevention against thermal stress in public transport and road space			
Action field 'Social Affairs'	Heat wave-preparation concept	✓		✓
	Setting up a local heat warning system	✓		
	Heat waves as a scenario in operational planning		(✓)	

Source Own data and depiction 2015

In contrast, the cities are most active in the field of land-use planning, which means that the focus here should not be on the transfer of knowledge but on the professional exchange among the practitioners This field should be the focus of further exertions for a successful climate change adaptation. The investigation can answer the question which measures were realized, but it cannot state clearly why these specific measures could not be realized apart from already mentioned assumptions. These assumptions need further investigation in a next step. The results show that the next step in the project must be about transferability on two levels: One level will be the further activation of the actors especially in the context of informal activities. In addition, the other level will be regarding the fact that climate change adaption not only comprises thermal stress but also storms, heavy rainfall events and insidious adaptation processes like the alteration of the biodiversity. BESTKLIMA is still working on identifying obstacles for the implementation of measures in the further course of the project. So far no final conclusions can be made concerning the aim of the research of BESTKLIMA, but a glance at the interim conclusion of our investigations can be given. The overall research goal, i.e. developing recommendations of action that will enable German municipalities to use their strategies and processes of climate adaptation effectively, will be reached in 2017.

Acknowledgements The project is funded by the 'Federal Ministry for the Environment, Nature Conservation, Building and Nuclear Safety' as part of the funding program for climate change adaptation measures within the funding priority 'Local lighthouse projects and establishing local and regional collaborations'. We would like to thank our student assistants María López Díaz, Anna Köhler and Martyna Nehring for supporting the research in the context of BESTKLIMA.

References

Baumüller, N. (2012). Hitze als Planungsfaktor: Klimaanpassung als kommunale Aufgabe der Stadtplanung. *PlanerIn.* (4), 17–19.

Baugesetzbuch 2015. Berlin.

Beierkuhnlein, C., & Foken, T. (2007). *Klimaanpassung Bayern 2020: Der Klimawandel und seine Auswirkungen - Kenntnisstand und Forschungsbedarf als Grundlage für Anpassungsmaßnahmen: Kurzfassung einer Studie der Universität Bayreuth. UmweltSpezial.* Augsburg: Bayerisches Landesamt für Umwelt.

Benden, J., Riegel, C., Trum, A., & Theißen, A. (2012). *Gewerbeflächen im Klimawandel (KlimAix): Leitfaden zum Umgang mit Klimatrends und Extremwettern*, Aachen: Institut für Stadtbauwesen und Stadtverkehr. Birkmann, J., Böhm, H. R., Buchholz, F., Büscher, D., Daschkeit, A., Ebert, S.,... Walkenhorst, O. (2013). *Glossar Klimawandel und Raumplanung* (2nd ed.). *E-Paper der ARL: Vol. 10.* Hanover: Verlag der ARL. http://nbn-resolving.de/urn:nbn:de:0156-73571. (Last Accessed 11 Dec 2015).

Birkmann, J., Seng, D. C., & Setiadi, N. (2013). Enhancing early warning in the light of migration and environmental shocks. *Environmental Science and Policy, 27*, S76–S88.

Bundesministerium für Verkehr, Bau und Stadtentwicklung, & Bundesinstitut für Bau-, Stadt- und Raumforschung (BMVBS & BBSR). (2013). *Wie kann Regionalplanung zur Anpassung an den Klimawandel beitragen? Ergebnisbericht des Modellvorhabens der Raumordnung "Raumentwicklungsstrategien zum Klimawandel".* In the series *Forschungen* (157). Berlin: BMVBS.

Bundesministeriums für Verkehr, Bau und Stadtentwicklung, & Bundesinstitut für Bau-, Stadt- und Raumforschung (BBSR). (2011). *Klimawandelgerechte Stadtentwicklung: Ursachen und Folgen des Klimawandels durch urbane Konzepte begegnen.* In the series *Forschungen* (149). Berlin: BMVBS.

Bundesregierung. (2008). *Deutsche Anpassungsstrategie an den Klimawandel: Vom Bundeskabinett am 17. Dezember 2008 beschlossen.* http://www.bmub.bund.de/fileadmin/bmu-import/files/pdfs/allgemein/application/pdf/das_gesamt_bf.pdf. (Last Accessed 15 Dec 2015).

Chmielewski, F.-M., Heyer, E., Hupfer, P., & Kuttler, W. (2005). *Witterung und Klima: Eine Einführung in die Meteorologie und Klimatologie.* Stuttgart: Teubner.

Deutscher Wetterdienst (DWD). (2015). *Wetterlexikon.* Available at: http://www.dwd.de/DE/service/lexikon/lexikon_node.html. (Last Accessed 07 May 2015).

Dosch, F. (2015). *Überflutungs- und Hitzevorsorge durch die Stadtentwicklung: Strategien und Maßnahmen zum Regenwassermanagement gegen urbane Sturzfluten und überhitze Städte; Ergebnisbericht der fallstudiengestützten Expertise " Klimaanpassungsstrategien zur Überflutungsvorsorge verschiedener Siedlungstypen als kommunale Gemeinschaftsaufgabe".* Bonn: BBSR.

Fröhlich, J., Knieling, J., & Kraft, T. (2014). *Informelle Klimawandel-Governance: Instrumente der Information, Beteiligung und Kooperation zur Anpassung an den Klimawandel.* Neopolis working papers (15). Hamburg: HafenCity Universität.

Hübner-Schmidt, K., Borries, B. v., & Hasemann, A. (2003). *Netzwerk- und Akteursanalyse: Ein methodischer Leitfaden.* Bonn: FriEnt. http://www.frient.de/publikationen/dokument/publikation/netzwerk-und-akteursanalyse/. (Last Accessed 22 Dec 2015).

Intergovernemental Panel on Climate Change (IPCC). (2014). *Klimawandel: Was er für Städte bedeutet: Kernergebnisse aus dem, fünften Sachstandsbericht des IPCC.* Berlin: BMZ. https://www.bmz.de/de/mediathek/publikationen/reihen/weitere_materialien/Cities_Briefing_Web_DE_GIZ.pdf. (Last Accessed 08 Jan 2016).

Pahl-Weber, E., Henckel, D., Klinge, W., Lau, P., Zwicker-Schwarm, D., Rütenik, B., & Besecke, A. (2015). *GERMANY: I. The Structure of Government and Administration in the Federal Republic of Germany.* BSR INTERREG III B project. http://commin.org/upload/Germany/DE_Country_Engl.pdf. (Last Accessed 15 Dec 2015).

Riegel, C., Trum, A., Maximini, C., & Vallée, D. (2013). *Klimaschutzteilkonzept "Anpassung an den Klimawandel für die Städte Solingen und Remscheid".* Aachen: ISB. http://www.bergisches-dreieck.de/fileadmin/user_upload/wirtschaftsregion/PDFs/Regionales_Klimaschutzteilkonzept_Klimawandel.pdf.pdf. (Last Accessed 04 Dec 2015).

Schönwiese, C.-D. (2003). *Klimatologie: 31 Tabellen.* Stuttgart: Ulmer.

Solga, J., & Blickle, G. (2009). Macht und Einfluss in Projekten. In M. Wastian, I. Braumandl, & L. Rosenstiel (Eds.), *Angewandte Psychologie für Projektmanager. Ein Praxisbuch für die erfolgreiche Projektleitung* (pp. 145–164). Berlin, Heidelberg: Springer Berlin Heidelberg.

Stadt Remscheid. (2015). *Organigramm der Stadtverwaltung Remscheid.* http://www.remscheid.de/rathaus-und-politik/medienpool/dokumente050/20150203_0.11.3_21.00_Organigramm-mit-Zusatz_20150201_Endversion.pdf. (Last Accessed 04 Jan 2016).

Stadt Solingen. (2015a). *Konzern Stadt Solingen.* http://www2.solingen.de/C125739700649BBD/html/2EB18BFD117C02E5C125766900459FA0?opendocument&nid1=14513. (Last Accessed 21 Dec 2015).

Stadt Solingen. (2015b). *Zentraler Betriebsausschuss.* http://www2.solingen.de/c125739700649bbd/0/1021_f_Zentraler_Betriebsausschuss?OpenDocument. (Last Accessed 04 Jan 2016).

Stadt Wuppertal. (2015). *Organigramm der Stadtverwaltung Wuppertal.* https://www.wuppertal.de/medien_zentral/bindata/organigramm-gesamt.pdf. (Last Accessed 06 Jan 2016).

Umweltbundesamt (UBA). (2013). *Wie können wir mögliche Klimaänderungen vorhersagen?* https://www.umweltbundesamt.de/themen/klima-energie/klimafolgen-anpassung/folgen-des-klimawandels/klimamodelle-szenarien/erwartete-klimaaenderungen. (Last Accessed 23 Dec 2015).

Unweltbundesamt (UBA). (2012). *KlimaExWoSt – Urbane Strategien zum Klimawandel.* https://www.umweltbundesamt.de/themen/klima-energie/klimafolgen-anpassung/werkzeuge-der-anpassung/projektkatalog/klimaexwost-urbane-strategien-klimawandel. (Last Accessed 07 Dec 2015).

Weischet, W., & Endlicher, W. (2008). *Einführung in die allgemeine Klimatologie: Mit 13 Tabellen und einer Tafel. Studienbücher der Geographie.* Berlin, Stuttgart: Borntraeger.

Wiechmann, T., Mörl, K., & Vock, A. (2012). Evaluation von Modellvorhaben der Raumordnung: Entwicklung einer Methodik am Beispiel der Modellvorhaben der Raumordnung zur Sicherung der Daseinsvorsorge. *Raumforschung und Raumordnung.* (1), 79–90.

Zimmermann, A. (2006). *Instrumente zur Akteursanalyse: 10 Bausteine für die partizipative Gestaltung von Kooperationssystemen.* Aus der Reihe: *Förderung partizipativer Entwicklung in der deutschen Entwicklungszusammenarbeit.* Eschborn: Deutsche Gesellschaft für technische Zusammenarbeit.

Author Biographies

Alice Neht is an economic geographer with a research focus on spatial planning, in particular the integration of climate mitigation processes and energy efficient urban planning. Before becoming a PhD student at the Institute of Urban and Transport Planning at the Faculty of Civil Engineering at the RWTH Aachen University, Alice contributed to research at the London School of Economics and Political Science and at the German University Oman. Current responsibilities include research project management, socio-scientific investigation analysis, and consultation with local authorities.

Kathrin Prenger-Berninghoff is an environmental scientist and spatial planner as well as co-manager of project BESTKLIMA, based at the Institute of Urban and Transport Planning.

Prof. Dr. Dirk Vallée is a civil engineer and Director of the Institute of Urban and Transport Planning at RWTH Aachen University. His research work concerns adaptation strategies to climate change and demographic change as well as land-use and transportation modelling.

Legacy of Authoritative Environmentalism and Path-Dependent Historic Institutionalism in the Climate Change Policy Dynamics of the Maldives

Ibrahim Mohamed and David King

Abstract The Maldives is considered one of the most vulnerable countries to climate change. Particularly, the Maldives is facing challenges of democratic governance and transparency, crucial for effective climate change policies and strategies. Limited financial and technical capacity also challenges integration of climate change in environmental policies. There is a dearth of studies on climate change politics and policy of small island nations like the Maldives. Hence, understanding policy dynamics of climate change in a political context is crucial. In this study, the theory of authoritative environmentalism was combined with insights from international relations and comparative politics of historic institutionalism to understand the climate change policy dynamics of the Maldives over the past twenty-five years. The study utilized a qualitative design using document analysis triangulated with semi-structured interviews. The results indicate the implementation and outcomes of climate change policy in the Maldives are influenced by a legacy of authoritative environmentalism and path-dependent historic institutionalism in governance. Consequently, political leadership at the highest level plays a crucial role in the implementation and outcomes of climate change policy in small island states.

Keywords Climate change policy · Historic institutionalism · Authoritative environmentalism · Adaptation · Democracy · Maldives

I. Mohamed (✉) · D. King
College of Marine and Environmental Sciences, James Cook University, Townsville, QLD 4811, Australia
e-mail: ibrahim.mohamed@my.jcu.edu.au

D. King
Centre For Disaster Management, James Cook University, Townsville, QLD 4811, Australia

W. Leal Filho (ed.), *Climate Change Research at Universities*,
DOI 10.1007/978-3-319-58214-6_13

1 Introduction

Climate change has become a crucial policy issue for many nations (Figueres 2013). In the majority of nations, environmental policy-making is influenced by democratic governance, while for others; environmental policies are influenced by various political and economic forces (Kraft 1999). In many instances, environmental policy-making is driven by diverse "actors, processes, decisions and institutions" with distinguishing characteristics (Kraft 1999), reinforced by path-dependent trajectories resulting from legacies of democratic as well as authoritarian rule and critical political changes (Steinberg and VanDeveer 2012). For instance, democratization in Latin American countries has resulted in effective environmental policies, even though institutional changes have been slow (Hochstetler 2012). Consequently, for many nations, the political regime determines the effectiveness of environmental governance and policy-making (Rivera-Batiz 2002), especially for "wicked" policy problems such as climate change (Howes et al. 2013; Huitema et al. 2011).

Environmental policy formulation approaches in climate change include administrative rationalism, democratic pragmatism and economic rationalism (Sosa-Nunez 2016). Administrative rationalism is based on the preference of scientific expertise over public opinion, while democratic pragmatism is based on wider public consultation to cater to public demands and interests. On the other hand, economic rationalism is designed to enhance cost-effectiveness and to satisfy the interest groups influencing the economy. However, the effectiveness of any of these approaches, once again, depends on the political and governance context and the extent of policy implementation of the country (Sosa-Nunez 2016).

Studies on democracy and environment show that liberal democracies have more effective environmental policies (Hochstetler 2012). However, critics argue that the environmental policies in democratic environmentalism are hampered by conflicts between different stakeholders as well as politicized public opinions, values and knowledge (McCright and Dunlap 2011). Consequently, Authoritative Environmentalism (EA) has emerged as a theory to understand the influence of authoritarian politics on climate change policy outcomes (Han 2015). Authoritarian environmental policies in countries with potential financial capacity often succeed as they pursue political patronage and legitimacy (Beeson 2010; Han 2015), as is the case in China (Beeson 2010; Gilley 2012). According to Gilley (2012), AE is based on the premise of the public being subjective and irrational compared to scientific views, hence justifying that only the technocrats, the state and their political elites must dictate policy.

Though AE has been used to study nations in the context of climate change in countries such as China and Korea, no literature exists on AE in the context of a small island state. Small Island Developing States (SIDS), while being at the forefront of climate change (De Souza 2016) are faced with political elitism dominating every aspect of their communities, owing to their small size and economic dependence (Benedict 1966). Consequently, AE is critical in the study of climate change adaptation in SIDS. The island nation of the Maldives presents an example of an island state that, though transitioned to a democratic state in 2008, remained under an authoritarian political regime for most of its history. The country has been a strong climate change advocate globally since as early as 1990s. However, for decades, attempts to address climate change in the country's environmental policies have remained unsuccessful. Limited financial and technical capacity, transparency, accountability as well as deficits in stakeholder participation have been considered as major factors that hinder successful implementation of the Maldives' climate change policies (Karthikheyan 2010; Transparency Maldives 2015).

Recognizing the need for effective implementation of climate change policies for the Maldives, this paper aims to explore the influence of the legacy of AE and historic institutionalism in shaping the context of the country's climate change policy. To do so, the paper presents an overview of the Maldivian political eras, and then undertakes a comparative policy analysis in the context of climate change policy of the Maldives over the past 25 years to understand the causes, effects and outcomes pertaining to climate change policy in the country. Studies on climate change policy and governance in the context of SIDS such as the Maldives is limited. Hence, the findings of this paper will be crucial in understanding the challenges faced by SIDS in implementing climate change policy and the role of politics and governance for climate change adaptation.

2 Context of the Study

The Maldives is an ideal case demonstrating strong leadership in the global climate change frontier over the last 25 years. This small island nation in the Indian Ocean has a population of 338,434 with 1.69% annual growth (MOFT 2015). The average elevation of the islands is 1.5 m above mean sea-level, making the entire country highly vulnerable to climate change (MHTE 2009). The Maldives was centrally governed as an Islamic kingdom from 1153 AD until 1968, when a single candidate system was established (Shafeeg 2000).

Ibrahim Nasir became the first president in 1968 and ruled until 1978 when Maumoon Abdul Gayoom took over and ruled for 30 years under the single candidate system. However, in 2008 the first multiparty election held ousted him (Rasheed 2014). During his regime, economic development focused on elitism, and clientelism was reinforced without the engagement of civil society in major Government policies (Carmen et al. 2015). Nevertheless, he was instrumental in bringing the vulnerability of the Maldives to global attention. The democratic transition of 2008 under president Nasheed faced major challenges in bringing the reforms envisaged in the new constitution, leading to a political crisis and the ultimate controversial change of the government midway through its five-year term (Rasheed 2014). The incumbent government came to power after winning the 2013 elections.

The first National Environmental Action Plan (NEAP-I) came into force in 1990. International assistance to develop the NEAP-I came after the devastating impacts of storm surges in the capital Malé in 1987 and the global concerns of sea level rise highlighted by the Maldives following this disaster. However, the NEAP-I as well as the preceding NEAP-II (1999) and NEAP-III (2009) did not integrate climate change policies effectively within governance and institutional structure. Hence, in 2015, realising these limitations and challenges in addressing climate change, in particular at a local policy level, the Maldives Climate Change Policy Framework (MCCPF) was developed.

3 Conceptual Framework

Comparative policy analysis provides understanding of causes and effects relating to climate change policy and politics (Purdon 2015). As the Maldives has undergone major political transformations over the past 25 years, comparative policy analysis can provide the context of AE and path-dependent historic institutionalism in climate change policy dynamics. Policy-making processes involve policy actors within the government recognizing the importance of key issues and developing policy instruments through political processes (Kingdon 1993). The "three streams: problems, proposals and politics" integrate, allowing policy entrepreneurs to reinforce their policies in critical times (Kingdon 1993).

A conceptual framework based on Kingdon's model and literature from three authors (Gilley 2012; Han 2015; McConnell 2010) was developed for this study. The model (Fig. 1) is used to explore the environmental and climate change policy of the Maldives and to frame our data analysis. Contextual factors describe the problems and drivers used in framing environmentalism, which then influence prioritization of environmental issues by the government. According to this framework, contextual factors influencing climate change perceptions of policy makers and government priorities influence the policy instruments. Consequently, the implementation and outcomes of the policies are determined by the political leadership. The conceptual framework demonstrates policy-making is an interlinked holistic process influenced by path-dependent historic institutionalism and AE.

4 Method

As this research involves an analysis of climate change policy of the Maldives over the past 25 years, we used a pragmatist qualitative approach involving document analysis and semi-structured interviews. Pragmatism utilizes human experiences to construct reality and hence is useful to explore multiple explanations of the reality of a phenomenon (Johnson et al. 2007).

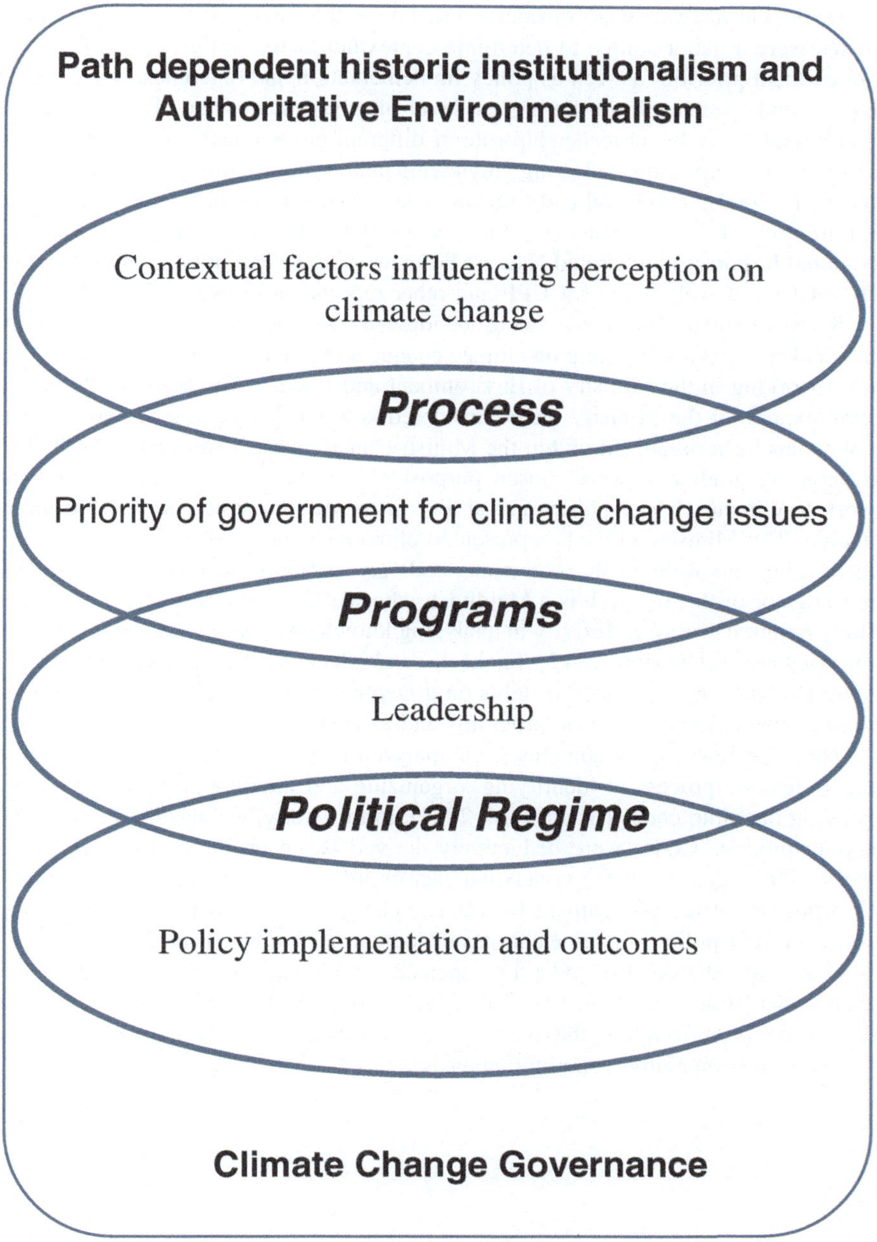

Fig. 1 Conceptual framework

Document analysis was considered crucial for this study, as government documents were a major source to determine contextual factors influencing policy and government priorities as well as policy implementation and outcomes. Some media reports and speeches from the highest level of government were also analyzed to understand the political leadership role of different governments. Documents were analyzed by exploring, selecting, reviewing and synthesizing policy documents (Bowen 2009). The official documents[1] selected included the National Environmental Action Plans (NEAPs), State of the Environment (SoE) Reports, National Reports to the United Nations Framework Convention on Climate Change (UNFCCC) as well as the MCCPF and other reports, as shown in Table 1.

Semi-structured interviews were conducted with two senior officials from national-level NGOs working on climate change and five civil servants at the policy level working in the Ministry of Environment and Energy. The high profile policy level experts of the Ministry were considered key policy actors, as environmental policy has been organized within the Ministry since its establishment in 1988. The experts, all graduates, were chosen purposively, based on the relevance of their work to climate change policy and ability to provide credible and reliable information. The Ministry officials represented climate change, energy, waste management and environmental protection and preservation sectors. The primary investigator of this research is a Maldivian who works for the Ministry, hence, this study required critical reflexivity in analyzing knowledge, experiences and views of interviewees (Davidson 2012; Shakeela and Becken 2015). The open-ended interview questions provided insights on government policy-making and successes in implementation of policies, programs and projects.

The interviews and documents were analyzed using thematic analysis, which is the systematic process of identifying, organizing and interpreting themes by categorizing data into codes (Braun et al. 2014). In this study, we used N-Vivo for the coding process. Codes were deductively derived based on the conceptual framework. The codes were: (1) contextual factors influencing climate change policy, (2) priorities of the government for climate change issues, (3) political leadership style, and (4) policy implementation and outcomes (Smith et al. 2009). Themes were developed from the codes by repeatedly reviewing the data extracts under each code (Braun et al. 2014). The themes were then defined and refined for discussion. Data from both the interviews and documents were triangulated (Bowen 2009) for a comparative temporal analysis.

[1]Documents were obtained from the official website of the Ministry of Environment and Energy at www.environment.gov.mv.

Table 1 Environmental philosophies and policy instruments in the Maldives from 1990 to 2015

Policy and governance	1990–2008	2008–February 2012	February 2012–2015
Philosophy of environmentalism	Authoritative environmentalism framed as sustainable development	Democratic environmentalism framed with economic rationalism	Authoritative Environmentalism framed with developmentalism
Constitutional status of environmental issues	Constitution did not recognize environmental protection	Articulated the duty of State to protect and preserve natural environment and all resources for future generations	Constitution of 2008 continues
Major environmental policies, strategies and action plans relevant to climate change	[a]First National Environment Action Plan (1990) [a]Second National Environment Action Plan (1999) National Development Plans (once every five years) National Solid Waste Management Policy Framework (2004) [a]National Implementation Strategy for Addressing Climate Change (2001) [a]National Adaptation Plan of Action (2007) [a]Sate of the Environment Reports (1994, 2001, 2002, 2004) National Biodiversity Strategy and Action Plan Vision 2020 (1999) and National Development Plan [a]First National Communication of Maldives to UNFCC (2001) Safer Island Strategy (2005)	[a]Third National Environment Action Plan (2009) "Aneh Dhivehi Raajje" The Strategic Action Plan 2008, [a]National Strategy for Sustainable Development, Maldives National Energy Policy and Strategy, [a]State of the Environment (2011)	Waste Management Policy (2015) Maldives Intended Nationally Determined Contribution (2015) [a]Maldives Climate Change Policy Framework (2015) Maldives as a Biosphere Reserve–Implementation Plan 2013–2017 (2012)

(continued)

Table 1 (continued)

Policy and governance	1990–2008	2008–February 2012	February 2012–2015
Institutions for environment policy implementation	Ministry of Environment 1988, and Environment Research Unit (1990) later changed to Environment Research Center. Established National Commission for the Protection of Environment in 1989, discontinued after 2007. Established the National Disaster Management Center in 2005.	Ministry of Environment integrated with other statutory bodies. Established the Environmental Protection Agency (2008). Established Climate Change and Energy Department (2009) within the Environment Ministry.	Ministry of Environment as an independent statutory body with the Climate Change Department (2012) and Energy Department (2012). Ministry of Fisheries and Agriculture. Ministry of Tourism. Ministry of Health.
Environmental laws and regulations enacted	Environment Protection and Preservation Act (1993). Environmental Impact Assessment (EIA) Regulations (2007). Regulation on Protection and Conservation of Environment in the Tourism Industry (1999). The Fisheries Act of the Maldives (1987). Tourism Act of Maldives (1999). Banning use of coral for construction. Law on Flora of inhabited islands (1998).	HCFC Regulation (Reg. no. 2010/R-19). Regulation on Environmental Liabilities (2012).	Waste Management Regulation (Regulation No: 2013/R-58). Environmental Impact Assessment Regulations (2012).
Priorities of government	To preserve and protect environmental resources for sustainable development.	To become carbon neutral and to build resilient communities through environmental stewardship and sustainable development.	To achieve economic prosperity and to protect and preserve environment.

[a] Refers to the documents analyzed in this study

5 Results and Discussion

Three different eras emerged in the political context and climate change policy agenda of the Maldives. The approach, environmentalism and milestones during different eras are summarized in Table 1.

6 Overview of the Three Different Eras

6.1 Autocratic Rule of Maumoon Abdul Gayoom: 1978–2008

The Maldives first created a statutory entity for the environment in 1984, by establishing the Environment Affairs Division within the Ministry of Home Affairs and Social Services (MEEW 2007). This Division was then upgraded to Ministry of Planning and Environment in 1988 (MEEW 2007). In 2005, the Ministry was reformed as the Ministry of Environment, Energy and Water and was directed by the National Commission for the Protection of the Environment (NCPE) established in 1989 (MEEW 2007). Although the Environment Ministry was mandated with the formulation of environmental policy to achieve sustainable development (SACEP 2002), policies were not based on any electoral incentives. Additionally, the Environmental Protection and Preservation Act (1993) restricted the environmental policy-making of the Ministry to those areas in which other ministries had no stake (SACEP 2002), until the amendment of the act in 2014. Consequently, until now, integration of environmental policy has been challenged by sectoral segregation, whereby environment related departments created within other ministries continue to formulate and implement their own policies. Democratization[2] which began with political turmoil in 2003 shifted policy-making for electoral incentives.

6.2 Democratic Rule: 2008–2012 February

For the first time in the history of the Maldives, the importance of the environment was recognized in the constitution enacted in 2008. According to the Article 22 of the new Constitution: “The State has a fundamental duty to protect and preserve the

[2]In September 2003 a prisoner was killed by prison guards of Maldives National Defense Force, leading to a prison riot followed by public protests in the Maldives. Maumoon A. Gayoom was elected for a sixth term in 2003 but was forced to bring reforms due to international criticism and political dissents (Shaheed and Upton 2008). Gayoom finally agreed on democratization, after 30 years of dictatorship, paving the way for formulation of a new constitution and democratic elections in 2008.

natural environment, biodiversity, resources and beauty of the country for the benefit of present and future generations" (Maldives Const. art. 22, 2008). After winning the first democratic election in 2008, President Nasheed reiterated that democracy is the most important climate change adaptation instrument, and regarded empowerment of people as a critical measure for a resilient community, owing to the democratic transition. Nasheed played an active role in bringing the mitigation potential of SIDS like the Maldives to global attention, by pledging to make the Maldives carbon neutral by 2020.

6.3 Regime Changes After February 2012 and Beyond

Due to political turmoil, a controversial regime change occurred in February 2012, forcing the resignation of Nasheed (Rasheed 2014). The interim government faced huge challenges nationally and globally, but continued with the inherited carbon neutral policy. Fresh elections were held in 2013, won by Yameen Abdul Gayoom. Critics argue that since his government came to power, a de-democratization has begun, with democratic values and freedoms being curtailed (Paddock 2016). For instance, the Press Freedom Index, issued by Reporters Without Borders, shows a sharp decline from 52nd in 2010 to 108th in 2013 (Fathih 2016a).

7 Comparative Analysis Based on the Conceptual Framework

7.1 Contextual Factors and Government Priorities

According to document analysis, contextual factors which influenced the prioritizing of climate change from 1990 to 2008 were the fragile nature of the islands and the crucial significance of the natural environment in sustaining the two major industries, namely tourism and fisheries. Estimates show that the Maldives generated US$4 billion from the tourism sector in the period between 2001 and 2009 (Rasheed 2014). Hence, a major contextual factor influencing government policies was the impact of climate change on tourism, and fisheries (MPHRE 1990). Consequently, major policies highlighted the importance of preserving the environment for sustainable development.

President Gayoom drew attention to the vulnerability of small islands on international platforms, following the 1987 flooding in Malé (MPHRE 1990). Following this, the Maldives ratified the UNFCCC and negotiated special recognition of the vulnerability of small islands at the UN Earth Summit of 1992 (SOE 2002). According to NEAP-I (MPHRE 1990), the dual policies: formulating and enacting environmental legislation and participation in international platforms, were regarded as major environmental goals. Priorities of the government also included modeling of impacts of climate change to the islands based on the findings of Inter-Governmental Panel on Climate Change (IPCC), given in their Assessment Reports (MPHRE 1990). In addition, institutional capacity building was prioritized, as human resources and technical capacity was limited. However, compared to international efforts, local level action was considered inadequate. According to one NGO interviewee:

> Gayoom was very vocal internationally, but locally there were few steps taken. (NGO1)

Pleading to the international community to assist in adaptation programs was heavily emphasized in the 1990s, as the Maldives had a negligible contribution to greenhouse gas emissions. For instance, Gayoom stated "Maldives urges industrialized nations to give financial and technological assistance to states which are under immediate threat" (Speech given in the Convention on the Sea and Environment, Italy, 1991). Consequently, adaptation was significantly emphasized during this era. For instance, a Safer Island Strategy (SIS) was developed in 2005, aimed to develop selected islands with proper hard engineering adaptation measures to relocate people living in vulnerable islands. However, this policy was futile, as relocating entire populations to other islands were highly controversial and undemocratic (Naseem 2016) and the government elected in 2008 did not support this policy.

Contextual factors during the 2008–2012 periods were based on utilizing the potential of the country and becoming carbon neutral by 2020. Hence, the National Sustainable Development Strategy of 2009 also reiterated a carbon neutral policy. Consequently, the government prioritized mitigation and developing the country's renewable energy sector. The government also re-evaluated hard engineering adaptation measures for coastal protection and began seeking alternatives, while population consolidation policy, popularized in the 1990–2008 period was considered incompatible with democracy. According to an NGO official interviewed, carbon neutral policy was an important instrument for public involvement and awareness of climate change, commenting:

> His carbon neutrality plan was a huge wake-up call for Maldivians, especially, because people who denied climate change started to question about it and started to talk about it. (NGO1)

The interim period between 2012 and 2013, after Nasheed was ousted, did not undergo major policy reversals, but a new policy to make the entire Maldives a marine reserve by 2017 was formulated. However, the contextual factors shifted when Yameen became president in 2013. A critical shift was in the reversal of the carbon neutral policy to a low carbon strategy and an emphasis on population consolidation (Naseem 2016). Additionally, a developmentalism approach, based on East Asia's developmental states such as Singapore, was aspired (Naseem 2016). This philosophy is based on the political context to pursue economic goals through governance of resources, markets and industries as a means to achieve national security and prestige (Kim 2016). Hence, the incumbent government has prioritized economic policies. According to one of the NGO informant's interviewed:

> When I was meeting, as part of young people [group] who met Yameen during his [presidential] campaign he told his supporters he is an economist and not an environmentalist. (NGO1)

Lack of commitment from the current government in climate change adaptation was also highlighted by a government official respondent who declared:

> In the international arena, Maldives is very loud and played key roles in climate change adaptation and resilience. But unfortunately, at home, at local level we don't see that much. (GOV4)

The findings from document analysis reveal that contextual factors influencing climate change policy between 1990 and 2008 were the vulnerability of the country to climate change and the threat to critical coastal marine resources. Due to the importance of the environment for development, environmentalism was framed as a means to achieve sustainable development in this period. Such framing may have been influenced by foreign experts advising the government during this period. Hence, authoritative environmentalism was the major policy philosophy during the 1990–2008 periods. Although vulnerability was contextualized during post 2008, until 2012 February, a shift from emphasizing adaptation to emphasizing mitigation was evident. Consequently, the potential of the nation to lead by example was the major policy from 2008 to 2012. Another contextual factor was the emphasis given to the role of public consultation and democratic governance. Due to the democratic values enshrined in the new Constitution, democratic principles and values became a huge priority, resulting in promoting democratic environmentalism. However, the global financial crisis of 2009 compelled economic rationalism as a major policy driver. Hence, reducing dependency on imported fuels became a major contextual factor which led to the carbon neutral policy.

7.2 Policy Implementation and Outcomes

Although addressing climate change at both international and national levels was prioritized, international advocacy achievements were more apparent during 1990–2008. Themes emerging from data extract on policy outcomes show some critical achievements from 1990 to 2008 (MPHRE 1990). The SOE (1994) highlighted the establishment of an Environment Research Unit in 1990, the enactment of the Environmental Protection and Preservation Act (1994) and installation of tide gauges to monitor the sea-level (MPHRE 1994). Despite the enactment of an Environment Law and a Research Center, lack of capacity was a major hindrance in achieving policy targets. Although an Environmental Impact Assessment (EIA) was mandated by the Environment Law, 74 coastal development projects were carried out between 2001 and 2004 without EIA (UNEP 2005). Additionally, there was a failure of environmental monitoring and evaluation, as the government lacked trained people to effectively monitor and implement policies (UNEP 2005).

Achievements in adaptation included integrated coastal zone management and measures to safeguard new settlements (MHAHE 1999). According to the SOE (2001), successes in the implementation included banning the use of coral in construction work in resorts and reducing the import duty for construction materials (MHAHE 2001). In addition, commencing solar energy projects and conservation of freshwater in islands was achieved. From 1990 to 2008, several policy documents were generated; however, the status of policy implementation has never been monitored or evaluated (UNEP 2005). In addition, segregated policies formulated by different ministries were never integrated, and lacked capacity, especially at island levels (UNEP 2005). Hence, policy implementation was not successful during the 1990–2008 period.

Policy implementation and outcomes during the period of 2008–2012 included major successes in mitigation through transformation of the energy sector and implementing renewable energy projects. Additionally, decentralization of environmental governance was initiated through local island councils. The government also established the Climate Change and Energy Department within the Environment Ministry in 2008. However, lack of capacity was a major issue in policy implementation. For instance, while the government amended regulations on EIA and Environmental Liabilities, compliance and enforcement were weak.

Post-2012 saw the historic Paris Agreement of 2105, where the Maldives played a crucial role as the chair of the Alliance of Small Island States (AOSIS). The government also began implementing several renewable energy projects. Also, coastal protection for thirteen islands started and seven islands were reclaimed by the end of 2015 (Fathih 2016b). One of the critical limitations in policy implementation was the integration of climate change adaptation and resilience in major development projects undertaken by the government, and was noted by one government official interviewed:

> Even in the critical projects like Safer Island Strategy, reclamation projects and coastal protection, climate change adaptation or resilience is not built in the design or built phases of projects of Ministry of Housing and Infrastructure. It is only implemented in the donor funded projects. (GOV4)

Hence, most development projects are not purposefully designed to build adaptive capacity. In addition, due to lack of evidence-based data and local research, most adaptation projects and programs are heavily influenced by donors. Most donor-funded projects face challenges in implementation as mentioned by one government official interviewed:

> Sometimes the way some of the donors plan to implement the projects are not efficient, but I think overall we benefit a lot from those projects. (GOV1)

Most officials from government and NGOs reiterated the lack of capacity in achieving the targets set in projects. According to one government official interviewed:

> I would say most of the projects we implement we are getting good results. But still we need improvements, I would say mainly the shortcomings are due to limited capacity. (GOV1)

In addition to the lack of capacity, representation of affected people in policy-making is also considered inadequate (Transparency Maldives 2015). Although most documents analyzed mentioned wide stakeholder consultation, such as the NAPA (MEEW 2007) and MCCPF (MEE 2016), representation of affected people living on islands and their views are not taken into account. According to an NGO official interviewed:

> I don't think it was done with consultation with the public... Public consultation was not enough. (NGO1)

In addition to this lack of representation, the inadequate knowledge of people regarding the most appropriate adaptation measures for their islands is also a major impediment to adaptive capacity. For instance, an NGO official interviewed commented:

> They are not aware. Usually what they will ask is for something they see on another island. ...They will want the same. ... Also during the EIA process, not much opportunity [for locals] to give suggestions. (NGO2)

These findings show that major successes in international advocacy and in participating in international climate regime were achieved during the past 25 years. However, achievements at the local level were inadequate. Policy implementation faced major challenges due to lack of monitoring and evaluation of policy outcomes. In addition, integration of climate change adaptation and resilience in most development projects has been hampered by the lack of proper environmental assessment mechanism, owing to lack of knowledge and capacity. After the democratization in 2008, democratic environmentalism was initiated, but the government was ousted before completing the five-year term. Policy reversals began with the regime changes from February 2012 onwards, and AE is more prominent in current government policies.

7.3 Political Leadership

During the autocratic rule before 2008, emphasis on human rights and good governance was weak in the Maldives (Bonnerjee 2014). During the presidency of Gayoom, climate change policy was largely governed by the strong political will and initiatives undertaken at the presidential level, with minimal involvement of the public (Carmen et al. 2015). Consequently, during the 1990–2008 era, AE was adopted. For instance, the policies of Population Consolidation and Safer Island Strategy were formulated by technocrats, dominated by direct political influence. Gayoom showcased highly liberal politics on international platforms, while at home, ruled with a heavy hand (BHRC 2015).

Gayoom was educated in Egypt and hence utilized his experience of the Islamic rule of Hosnei Mubarak (Mulberry 2012). Gayoom had gained political exposure and experience in working in the government of Nasir as a cabinet minister in 1977 (Shaheed and Upton 2008). Hence, he had in-depth knowledge of local politics, allowing him to utilize international support to reinforce his autocratic rule (Shaheed and Upton 2008). Gayoom has been acknowledged for utilizing the opportunities in the international fora to plead to rich countries on the challenges of climate change in the Maldives. Regarding past leadership, a government official respondent commented:

> I would say we had stronger leadership in the 90s and 2000s and from 2008 till 2012. (GOV2)

Consequently, during the 1990–2008 era, the Maldives was known to the West as a beautiful holiday destination whose existence was threatened due to global climate change impacts (Shaheed and Upton 2008).

The successor of Gayoom was Nasheed, a journalist educated in Britain (Gray and Foran 2015), with a wide range of experiences in Westminster Democracy. Nasheed not only became a champion of democracy after winning the election, but also became a champion of climate change, both nationally and internationally (Gray and Foran 2015). Nasheed brought a paradigm shift in the framing of climate change by changing the discourse of the Maldives from victimhood to an accomplishable nation (Gray and Foran 2015). In his speech given at the Climate Vulnerable Forum in November 2009, he stated, "In the Maldives, we want to focus less on our plight; and more on our potential" and for the first time, the Maldives began moving away from decades of advocacy based on victims with no potential. While demonstrating strong leadership internationally, Nasheed was also instrumental in local level advocacy and awareness through his famous under water cabinet meeting and carbon neutral pledge (Gray and Foran 2015). Nasheed's democratic principles also made him adopt democracy and good governance as key elements in climate change politics. In an interview to a documentary film on climate change, Nasheed stated, "Traditionally, we have always thought or highlighted saying that adaptation represents physical structures, revetments, embankments, breakwaters and so on. But we feel that one of the biggest; the most important adaptation issue [sic] is good governance. And therefore, consolidating democracy is very important for adaptation" (Television for Education Asia Pacific 2009).

Following the 2012 political turmoil, and the controversial resignation of Nasheed, his vice-president Waheed took over the country. Although policy reversals are common with such changes of regimes in developing countries (Steinberg 2012), this change did not undergo major policy reversals. According to one political figure of Waheed's administration, the institutional endurance of the Environment sector was maintained by avoiding political patronage, where discrimination of public servants based on politics or hearsay was disregarded (Mariyam Shakeela, Personal Communication, August 10, 2016). Steinberg (2012) believes insulation of professional bureaucracies from political patronage facilitates continuity of policies during political change and turmoil. Hence, during the aftermath of the political crisis of 2012, the Environment Ministry continued with existing policies.

The incumbent President, Yameen, is an economist with a strong background in economic policies and worked in the trade and economic sector for many years as a cabinet Minister. Hence, his government came with major policies for economic development. According to one government official interviewed:

> The current government is not very much thinking of climate change. Because current government's agenda is pretty much focused on economic growth. (GOV2)

The heavy focus of Yameen's government on economic development has undermined the climate change leadership the Maldives has enjoyed during the past decades. Hence the current government of the Maldives is regarded as a laggard in climate change leadership, especially at the local level. Regarding the lack of political commitment, an NGO official interviewed commented:

> I don't think we have that much commitment when it comes to policy makers.... I have not seen the President of the Maldives taking a personal stand, addressing climate change. (NGO1)

Although political leadership is weak, the professional bureaucracy within the Ministry is considered critical for climate change policy. An NGO official interviewed commented:

> There are technical people within the Ministry of Environment who would want to see a beautiful climate change [policy] paper and document and want to see something going forward. (NGO1)

Hence, the leadership style of the three past governments, before the incumbent government, had stronger commitments and political will to address climate change. While a more authoritative environmentalism was seen in the 1990–2008 period, democratic environmentalism was emphasized from 2008 to 2012. According to Steinberg (2012), strong environmental policies are observed under political leaders who have gained environmental awareness through education, media and civil society engagement. Hence, the findings show that leaders of the Maldives who had exposure to environmental issues and international politics were strong climate change pushers.

Future work that can complement this study needs to capture interviews of executive level politicians who have worked in different governments as well as in the incumbent government. Even though the environment Ministry is mandated with formulation of climate change policies and programs, senior public servants working in other institutions also need to be interviewed to better understand the influence of AE and historic institutionalism. Limitations of this study include varying levels of experiences of experts interviewed and the repetition of similar information across different government documents, along with a lack of substantive research-based evidence on policy implementation. In addition, due to the sensitivity of political issues, civil servants were constrained not to openly criticize government policies.

8 Conclusion

In this paper, we have analyzed the climate change policy dynamics of the Maldives over the last 25 years. This analysis shows that the Maldives is faced with huge challenges in environmental governance critical for climate change adaptation. Challenges resulting from undemocratic path-dependent historic institutionalism, lack of financial and technical capacity and human resources are major impediments for climate change adaptation. Small island nations like the Maldives have become strong advocates and leaders of global climate regime, resulting in the 2015 Paris Agreement coming into force (De Souza 2016). However, policy implementation and outcomes in small island nations are hampered by weak governance and high corruption (De Souza 2016).

Although the 2008–2012 periods saw a more democratic environmentalism in the Maldives, the legacy of path-dependent historic institutionalism has caused what Steinberg (2012) refers to as "policy churn". Conversely, similar to many East Asian countries, institutional resilience has been maintained during political changes by committed public servants (Steinberg 2012). This institutional endurance, however, is at stake now due to the recent amendment to the Civil Service Act (2007) of the Maldives, giving the Ministers the authority to select the top civil servant of the Ministries (Naish 2016). Additionally, the tourism industry as a crucial interest group in the political landscape, driving the socio-economic policies in the Maldives, has been regarded as a major attenuator of climate change risks and vulnerabilities (Shakeela and Becken 2015) and hence, has a limited contribution to climate change policy implementation.

Although a comprehensive climate change policy framework was developed by the Maldives in 2015, the policy lacks informed empirical knowledge on the adaptive capacity of small islands to enhance the resilience of island communities. The lack of representation of the people most affected by policy processes as well as high corruption and low enforcement mechanisms within the country are considered major impediments in implementing climate change policy. For instance, several coastal development projects have been conducted without proper assessments (UNEP 2005) or proper public consultation (Zuhair and Kurian 2016). Most coastal developments are hard engineering projects for partial protection from coastal erosion (UNEP 2005), often compromising the integrity of the coral reefs. The majority of these projects have exacerbated natural erosion by causing disruptions in sediment dynamics around the islands (Naylor 2015), impacting the resilience and adaptive capacity of many islands. The lack of planning laws (UNEP 2005) to address such measures is considered a major policy gap. For instance, the Safer Island Strategy (SIS) adopted in the 1990–2008 period was abandoned in 2010 with a new strategy called Integrating Climate Change Risks into Resilient Island Planning (ICCR) emphasizing sustainability (Naylor 2015). However, SIS was reinstated by the current government. According to Naylor (2015), an adamant hard engineering mindset and the SIS are embedded in current policies due to the legacy of historic institutionalism.

Our analysis shows that climate change policy in the Maldives has been challenged by the legacy of authoritative environmentalism and historic institutionalism inherited from the past undemocratic rule. Consequently, implementation of climate change policy is limited. Challenges of small island nations, as found by De Souza (2016), such as lack of representation of vulnerable and marginalized island communities, lack of knowledge and accessibility, high corruption, lack of effective compliance and enforcement mechanisms and inadequate rule of law are also significant in the Maldives. Additionally, the legacy of authoritative environmentalism has favored policies against the wider scientific views on the complexities and uncertainties of climate change, leading to distortion of scientific uncertainties to facilitate priorities of the governments. This study concludes that, for the Maldives to implement climate change policy effectively to enhance adaptive capacity, policy processes, programs and political decisions needs to adopt democratic environmentalism.

Acknowledgements We would like to thank the Department of Foreign Affairs and Trade of the Australian Government for funding the primary researcher's Ph.D. project at James Cook University. We are also grateful for the support and assistance provided by the College of Marine and Environmental Sciences of James Cook University. We also would like to acknowledge the heads of Live and Learn Environmental Education Maldives and Eco-Care Maldives for their assistance. Also special thanks to the staff of the Ministry of Environment and Energy and Dr. Mariyam Shakeela, for her contribution. We appreciate the editor in chief and the reviewers of this book, for their guidance.

References

Bar Human RightsCommittee (BHRC). (2015). *Trial Observation Report - Prosecution of Mohamed Nasheed*. Retrieved from London: http://www.barhumanrights.org.uk.

Beeson, M. (2010). The coming of environmental authoritarianism. *Environmental Politics, 19*(2), 276–294. doi:10.1080/09644010903576918.

Benedict, B. (1966). Sociological characteristics of small territories and their implications for economic development. *The Social Anthropology of Complex Societies. London: Tavistock Publications*, 23–36.

Bonnerjee, A. (2014). The Maldives: The shifting nature of a welfare state. In G. Koehler & D. Chopra (Eds.), *Development and Welfare Policy in South Asia*. Hoboken: Taylor and Francis.

Bowen, G. A. (2009). Document Analysis as a Qualitative Research Method. *Qualitative Research Journal, 9*(2), 27–40. doi:10.3316/QRJ0902027.

Braun, V., Clarke, V., & Terry, G. (2014). Thematic analysis. In Poul Rohleder & Antonia C. Lyons (Eds.), *Qualitative Research in Clinical Health and Psychology* (pp. 95–114). New York: Palgrave MacMillan.

Carmen, E.-B., Bruce, C. G., & Robert, K. (2015). A tale of two atoll nations: A comparison of risk, resilience, and adaptive response of Kiribati and the Maldives. In C. G. Bruce, P. M. Kelly, K. Robert, & T. Ailbhe (Eds.), *Climate Change and the Coast* (pp. 313–336). Boca Raton, FL: CRC Press.

Davidson, D. (2012). Analysing responses to climate change through the lens of reflexivity. *The British Journal of Sociology, 63*(4), 616–640. doi:10.1111/j.1468-4446.2012.01429.x.

De Souza, R.-M. (2016). Corruption, Climate Change, and Vulnerability in Small Island States. Retrieved from https://www.wilsoncenter.org/article/corruption-climate-change-and-vulnerability-small-island-states.

Fathih, M. S. (2016a, 4th February). Translation of President Yameen's 2016 Presidential Address. Retrieved from http://maldivesindependent.com/politics/translation-of-the-2016-presidential-address-at-the-peoples-majlis-opening-121984.

Fathih, M. S. (2016b, 21st April). Maldives 'remains very hostile for independent and opposition media'. Retrieved from http://maldivesindependent.com/politics/maldives-remains-very-hostile-for-independent-and-opposition-media-123694.

Gilley, B. (2012). Authoritarian environmentalism and China's response to climate change. *Environmental Politics, 21*(2), 287–307. doi:10.1080/09644016.2012.651904.

Gray, S., & Foran, J. (2015). Climate Injustice: The Real History of the Maldives. *Berkeley Journal of Sociology, 59*, 14–25.

Han, H. (2015). Authoritarian environmentalism under democracy: Korea's river restoration project. *Environmental Politics, 24*(5), 810–829. doi:10.1080/09644016.2015.1051324.

Hochstetler, K. (2012). Democracy and the Environemnt in Latin America and Eastern Europe. In P. F. Steinberg & S. D. VanDeveer (Eds.), *Comparative Environmental Politics: Theory, Practice and Prospects*. Cambridge, MA: MIT Press.

Howes, M., Grant-Smith, D., Reis, K., Bosomworth, K., Tangney, P., Heazle, M., Burton, P. (2013). *Public policy, disaster risk management and climate change adaptation*. Paper presented at the 2013 Public Policy Network Conference, Brisbane. http://eprints.qut.edu.au/64512/.

Huitema, D., Jordan, A., Massey, E., Rayner, T., van Asselt, H., Haug, C., Stripple, J. (2011). The evaluation of climate policy: theory and emerging practice in Europe. Policy Sciences, 44(2), 179–198. doi:10.1007/s11077-011-9125-7.

Johnson, R. B., Onwuegbuzie, A. J., & Turner, L. A. (2007). Toward a Definition of Mixed Methods Research. *Journal of Mixed Methods Research, 1*(2), 112–133. doi:10.1177/1558689806298224.

Karthikheyan, T. C. (2010). Environmental Challenges for Maldives. *South Asian Survey, 17*(2), 343–351. doi:10.1177/097152311201700210.

Kim, E.-s. (2016). The politics of climate change policy design in Korea. *Environmental Politics, 25*(3), 454–474. doi:10.1080/09644016.2015.1104804.

Kingdon, J. W. (1993). How do Issues Get on Public Policy Agenda. In W. J. Wilson (Ed.), *Sociology and the Public Policy Agenda* (pp. 40–50). Newbury Park, CA: SAGE Publications, Inc.

Kraft, M. E. (1999). Environmental policy Environmental Geology (pp. 216–221). Dordrecht: Springer Netherlands.

Maldives Constitution of 2008.

McConnell, A. (2010). Policy Success, Policy Failure and Grey Areas In-Between. *Journal of Public Policy, 30*(03), 345–362. doi:10.1017/S0143814X10000152.

McCright, A. M., & Dunlap, R. E. (2011). The Politicization of Climate Change and Polarization in the American Public's Views of Global Warming, 2001–2010. *Sociological Quarterly, 52* (2), 155–194. doi:10.1111/j.1533-8525.2011.01198.x.

MEE (Ministry of Environment and Energy). (2016). *Maldives Climate Change Policy Framework*. Malé, Maldives: Ministry of Environment and Energy.

MEEW (Ministry of Environment, Energy and Water). (2007). *National Adaptation Programme of Action*. Malé, Maldives: Ministry of Environment, Energy and Water.

MHAHE (Ministry of Home Affairs, Housing and Environment). (2001). *State of the Environment Report Maldives*. Malé, Maldives: Ministry of Home Affairs, Housing and Environment.

MHAHE (Ministry of Home Affairs, Housing and Environment). (1999). *Second National Environment Action Plan*. Malé: Ministry of Home Affairs, Housing and Environment.

MHTE (Ministry of Housing Transport and Environment). (2009). *Maldives National Strategy for Sustainable Development*. Malé, Maldives: Ministry of Housing, Transport and Environment.
MHTE (Ministry of Housing, Transport and Environment). (2009). *Third National Environment Action Plan*. Malé, Maldives: Ministry of Housing, Transport and Environment.
MOFT (Ministry of Finance and Treasury). (2015). Maldives Population and Housing Census 2014: Statistical Release:1 Population and Households. Maldives: Ministry of Finance and Treasury Retrieved from http://statisticsmaldives.gov.mv/population-and-households/.
MPHRE (Ministry of Planning, Human Resources and Environment). (1990). *National Environment Action Plan*. Malé, Maldives: Ministry of Planning, Human Resources and Environment.
MPHRE (Ministry of Planning, Human Resources and Environment). (1994). *State of the Environment Maldives*. Malé, Maldives: Ministry of Planning, Human Resources and Environment.
Mulberry, M. (2012, July). The Maldives - From Dictatorship to Democracy, and Back? Retrieved from https://www.nonviolent-conflict.org/maldives-dictatorship-democracy-back/.
Naish, A. (2016, 19th July). Majilis grants ministers authority over top civil servant. Retrieved from http://maldivesindependent.com.
Naseem, A. (2016, 11th May). Yameen's (?) Maldives. Retrieved from http://www.dhivehisitee.com/executive/yameen-maldives/.
Naylor, A. K. (2015). Island morphology, reef resources, and development paths in the Maldives. *Progress in Physical Geography, 39*(6), 728–749. doi:10.1177/0309133315598269.
Paddock, R. C. (2016, 27th August). Embattled Maldives President Faces Accusations of Corruption, Past and Present. Retrieved from http://www.nytimes.com/2016/08/28/world/asia/maldives-mohamed-nasheed-abdulla-yameen.html.
Purdon, M. (2015). Advancing Comparative Climate Change Politics: Theory and Method. *Global Environmental Politics, 15*(3), 1–26. doi:10.1162/GLEP_e_00309.
Rasheed, A. A. (2014). Historical institutionalism in the Maldives: A case of governance failure. *Maldives National Journal of Research, 2*(1), 7–28.
Shafeeg, A. (2000). History of Maldives. *Dhanfulhi, 1,* 1–139.
Shaheed, A., & Upton, J. (2008). *Maldives: Reform Deferred? Challenges and Lost Opportunities for Democratic Transition*. Paper presented at the 9th Annual Conference of Centre for the Study of Islam and Democracy,, Washington DC.
Shakeela, A., & Becken, S. (2015). Understanding tourism leaders' perceptions of risks from climate change: an assessment of policy-making processes in the Maldives using the social amplification of risk framework (SARF). *Journal of Sustainable Tourism, 23*(1), 65–84. doi:10.1080/09669582.2014.918135.
Smith, J. B., Vogel, J. M., & Ill, J. E. C. (2009). An architecture for government action on adaptation to climate change. An editorial comment. *Climatic Change, 95*(1), 53–61. doi:10.1007/s10584-009-9623-1.
South Asia Cooperation for Environment Programme. (2002). *Handbook on National Environmental Legislation and Institutions in the Maldives*. Retrieved from Colombo, Sri Lanka: http://www.sacep.org/pdf/Reports-Technical/2001-UNEP-SACEP-Law-Handbook%20-Maldives.pdf.
Steinberg, P. F. (2012). Welcome to the jungle: policy theory and political instability. In P. F. Steinberg & S. D. VanDeveer (Eds.), *Comparative environmental politics: theory practice and prospects* (pp. 255–284). Cambridge, Massachusetts, USA: MIT Press.
Television for Education Asia Pacific (Producer). (2009). Small Islands, Big Impact - featuring President Mohamed Nasheed of the Maldives. Retrieved from https://www.youtube.com/watch?v=10_RZl51-ko&feature=youtu.be.
Transparency Maldives. (2015). *Climate finance governance in Maldives: Pressing issues, Position Paper*. Retrieved from Maldives: http://transparency.mv/2015/08/climate-finance-governance-in-maldives-pressing-issues/.

United Nations Environment Programme. (2005). *Maldives: Post-Tsunami Environmental Assessment*. Retrieved from UNEP Nairobi, Kenya: http://www.unep.org/tsunami/reports/maldives.pdf.

Zuhair, M. H., & Kurian, P. A. (2016). Socio-economic and political barriers to public participation in EIA: implications for sustainable development in the Maldives. *Impact Assessment and Project Appraisal, 34*(2), 129–142. doi:10.1080/14615517.2016.1176404.

Part II
Research and Education on Climate Change Adaptation

Community-Based Adaptation to Climate Change in Egypt—Status Quo and Future Policies

Paulina Froehlich and Mohammad Al-Saidi

Abstract For many of Egypt's communities, climate change represent an increasing threat to development and livelihoods that remains underrepresented in terms of policies and strategies. The vulnerability of communities in agricultural land, coastal zones and urban spaces is investigated in this study using recent literature. This study aims at identifying policy options for community-based adaptation (CBA) to climate change in Egypt. Using key informants interviews and comparisons to similar case studies, it outlines three alternative policies sets: information-based, finance-based and capacity-based. It later introduces the obstacles and advantages of introducing the different CBA policies which represent potential focus areas of future climate change policy reforms. Further, it identifies critical factors for the success of the CBA approach like trust among all stakeholders and a clear allocation of responsibilities among governmental institutions.

Keywords Community-based adaptation · Climate change in Egypt · Policy analysis · Sea-level rise · Climate policy

1 Introduction

Climate change is a major concern for the countries of the Middle East and North Africa (MENA). Strongly affected by sea level rise (SLR), increasing temperatures, change of concentration and distribution of rainfall, most MENA countries also

P. Froehlich (✉)
Institute for Technology and Resources Management in the Tropics and Subtropics (ITT), Cologne University of Applied Science, Cologne, Germany
e-mail: paulina.froehlich@posteo.de

M. Al-Saidi
Center for Sustainable Development, College of Arts and Sciences, Qatar University, Doha, Qatar

W. Leal Filho (ed.), *Climate Change Research at Universities*,
DOI 10.1007/978-3-319-58214-6_14

struggle from a high population growth, fragile economies and political instability (Verner 2012). Consequently, the countries are increasingly directing their focus on adaptation and mitigation methods combating climate change. Adaptation describes ways to deal with the inevitable effects while mitigation aims to limit or reduce the roots of causes of climate change (ESCWA 2013). 'Hard' and 'soft' adaptation activities can be identified and used for the development of separate strategies. However, current studies suggest that comprehensive policy-making should combine the two sets of measures. While hard adaptation mostly focuses on infrastructure and economic production, soft adaptation aims to increase the communities' knowledge and awareness of environmental changes and impacts (Ambandi and Nicholles 2013).

Community-based adaptation is receiving growing attention from NGO's, climate change initiatives and researchers worldwide (Reid 2009). The presumption is that that climate change adaptation efforts are more effective when local knowledge is taken into consideration and communities are empowered to find out about the roots and effects of climate change. This is especially the case for low-income and vulnerable local communities. Due to the increased level of participation since the Arab Spring of 2011, environmental politics in Egypt have recently received a new impetus and popularity (Sowers 2013). This attention can also be seen within the context of the Sustainable Development Goals (SDGs), which were adopted in September 2015 by the UN member states. Goal 11 aims at making cities and human settlements inclusive, safe, resilient and sustainable while Goal 13 demands to take urgent action to combat climate change and its impacts. The role of communities is also highlighted by the SDGs. According to goal 16, communities have the right to participate in decision making that affects them directly (Reid 2009). Besides, these sustainable goals stress the importance inclusive decision-making and participatory approaches instead of classic top-down procedures (UN 2015).

In Egypt, there is a growing body of literature focusing on climate related threats, especially for the delta region about SLR and the increasing uncertainty over the Nile basin flow (citations). However, there is little focus on the urban level and hardly any comparisons between urban and rural CC related problems. Since the focus has been so far on quantifying threats and impacts, CC policy analyses are quite rare, apart from few reports by donor organizations. National studies from Egypt evaluating vulnerabilities to climate change effects and its emerging costs (NEEDS 2010) as well as adaptation options exist (UNDP and the Egyptian Cabinet 2011). However, they focus on infrastructural, government-led solutions only. There is also a study comparing two participatory urban planning programs (Hassan et al. 2011) and another study focusing on evaluating a community-based self-help initiative from Cairo in post-Mubarak times (Buckner and Buckner 2015). Both however do not focus on climate change. A larger study, conducted by the SEARCH program of IUCN in 2013 concentrated on resilience towards climate change and evaluated closely the adaptation measurements farmers have found in Egypt (IUCN 2014). The Participatory Development Project by GIZ currently carries out a project in combination with a local NGO to foster CC adaptation within urban informal districts (PDP 2015). No study or program, however, has focused yet on the combination of urban and rural adaptation issues and has bridged the gap towards policymaking.

2 Methods and Data

Data and comprehensive assessments are lacking on current climate change policies in Egypt while the community-level has largely not been addressed specifically. This is translates itself in a distinctive gap between the community reality and the policy level. Master plans, for urban planning for example, often failed due to lack of community involvement (Hassan et al. 2011). The study addresses this knowledge gap by providing a policy analysis for integrating communities into climate change adaptation and mitigation efforts. It first defines the problem and analyses recent literature on climate change vulnerability and risks for urban, costal and rural communities. Later, it lays out policy options by using examples from other countries and analyses of Egypt's special. Finally, the current policies and institutions concerning climate change adaptation in Egypt are identified while a policy measures for CBA in Egypt will be recommended Qualitative data used for this analysis is collected from secondary literature, stakeholder interviews and visits to four different urban and rural communities in early 2016. Discussions with inhabitants of local communities were organized together with interviews with local stakeholders. Further, semi-structured, detailed interviews were conducted with five experts from the Egyptian climate change sector to design CB adaptation options and discuss implementation issues.

While the study at hand follows a traditional policy analysis (problem analysis, goals, alternatives and recommendations), the policy alternatives are delineated conceptually and the assessment is largely based on stakeholders' perception. A quantitative verification is via, for example, calculating correspondent benefits and costs of each alternative is not provided nor is it feasible for such governance interventions. The policy alternatives rather represent different focus areas for future reforms rather than exclusive pathways. Although the reform policies and the adaptation measures might differ in each case depending factors like vulnerability level, economic and development livelihood as well as the community's characteristics like coherence and trust, the study provides valuable insights into status quo and overall reform problems and potential. It shows that communities might lack knowledge about relevant governmental plans and policies. On the other hand, institutional overlap or coordination failures represent key governance related problems. Such issues hinders policymakers and practical initiatives from implementing the concrete policy options and achieving the common goal of effective and sustainable adaptation to climate change.

3 Climate Change Adaptation in Egypt: Threats and Institutions

3.1 Climate Change Threats

Despite uncertainties associated with different climate change models for Egypt, the country is vulnerable to different kinds of expected effects of climate change. Such vulnerability results from several factors such as the great dependency upon one major water source, the Nile River, which accounts for 96% of the water resources available in Egypt (UNDP and the Egyptian Cabinet 2011). This water amount is estimated to be around. 55.5 billion cubic meter per year, based upon the agreement with Sudan of 1959, regulating the share of the Nile river among the two countries. This share is expected to change as Ethiopia and other upstream states are claiming more use rights for themselves (Jobbins 2012). Furthermore, Egypt's geographic location plays a role for its vulnerability. Next to its characteristics of an arid desert (96% of surface land) with difficult living- and development options, Egypt's coastal areas are also exposed to threats such as SLR (Sterman 2009). Additionally, poor people in terms of low-income and consequently low standard of health and education, have a higher sensitivity to climate change. The upper poverty line in Egypt ($2.50/day) was in 2013 at around 40% of the population (Smith and Deck 2013). Egypt ranks 108 out of 188 countries in the Human Development Index (HDI) (UNDP 2015). Egypt is however difficult to rank in terms of vulnerability in comparison to other MENA countries. It is better off when it comes to water resources as it depends largely on surface water and has additional groundwater resources, while Jordan, Tunisia or Algeria mostly depend on groundwater. However, Egypt is most vulnerable when taking population size and the future per capita use or threats such as sea level rise (SLR) into account. In comparison to other MENA countries, increasing temperatures will put a higher burden on Egypt considering the large agricultural production for example (Sowers 2013). Egypt is thus ranked among one of the most vulnerable countries towards climate change (EEAA 2010).

In addition, the economic vulnerability of Egypt towards climate change plays an important role. A study from 2013 by Smith and Deck examined the economic costs of climate change impacts on different sectors in Egypt under each two scenarios, a lower and a higher one. The results show a distinctive monetary loss looking at heat stress, water supply, air quality, and tourism: Accordingly, the total economic losses are estimated to range from 36–64 billion USD, around 2–6% of future gross domestic product by 2060 (Smith and Deck 2013). The agricultural sector is projected to have the greatest economic loss closely followed by the touristic one. The unfortunate combination of decreasing food production and consequently high food prices can increase malnutrition and unemployment alike. Further, an increased risk towards human health from higher air pollution and heat stress levels are expected, which in turn cause further costs (Smith and Deck 2013).

In summary, there are three spheres, which matter most when it comes to climate change effects in Egypt: agricultural land, coastal zones and urban spaces. A number of challenges arise for all three of them starting from flooding and salt-water intrusion by sea-level rise, minimizing crop productivity, animal disease and higher levels of evaporation due to rising temperatures reaching up to heat-stress, decreasing air quality and polluted water (Agrawla 2004; Smith and Deck 2013; Stermann 2009).

3.2 Institutions and Community-Based Initiatives

Figure 1 provides an overview of major stakeholders related to CC adaptation in Egypt. The relationships between these stakeholders are indicated while their overall importance in terms of influence and decision-making power is visualized by the size of each bubble. The size is a comparative approximation for illustration purposes. NGOs for example have no decision-making power but a relatively high influence on the CC scene in Egypt due to their activity all through the country.

On the macro-level, international institutions (e.g. UNFCCC) are often corresponding and coordinating with the national level institutions (e.g. Ministries). Egypt attends international meetings, like COP21 and intergovernmental meetings like those organized by the Arab Water Council concerning climate change. The ministries receive information as well as climate change mitigation and adaptation funding as a Non-Annex I country from such international institutions while they publish national level papers for UNFCCC, UNDP etc. Within national institutions,

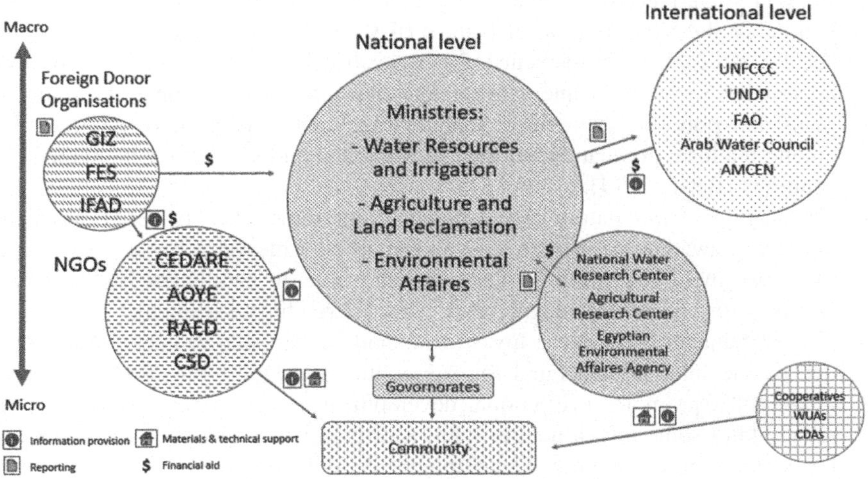

Fig. 1 Exemplified mapping of stakeholders and institutions involved in the climate change sector in Egypt

such as the ministries, research centers are hosted that receive a certain budget from the corresponding ministries in order to carry out the demanded national research on climate change. Other considerable research institutions on climate change in Egypt are based inside universities. Furthermore, the Ministry of Environment hosts a unit for climate change (the Egyptian Environmental Affairs Agency or EEAA) and is supposed to coordinate and integrate all national and international activities on the issue (NEEDS 2010). However, literature review and interviews with experts have shown that this unit does not a clear institutional role or the organizational capacity to perform its comprehensive coordination task.

National and International NGOs in Egypt form the link between the national level, the community level and foreign donors. NGOs practically carry out many of the projects planned or financed by foreign donor agencies but also collect their funding from private donations, the government or others. Mostly, NGOs heavily rely on collaboration with sub-national authorities such as governors to reach the local population (Jobbins 2012). Foreign donor agencies fund selected projects and programs, assist in management processes and share information. Civic participation consists of stakeholders in form of cooperatives, water-user associations (WUA), community-development associations (CDA) or individual activists. They work usually detached from the institutional levels. Their operation happens in defined limits, trained staff is rare and administrative capacity is short (Jobbins 2012).

Small number of vanguard community-based adaptation initiatives towards climate change exist in Egypt mostly by international donors and NGOs, and less by the government.

- Social, Ecological and Agricultural Resilience in the Face of Climate Change (SEARCH) is a project which was conducted between 2011–2013 in five Arab states (Jordan, Palestine, Lebanon, Egypt and Morocco) led by the International Union for the Conservation of Nature (IUCN). The objective was to increase social and ecological resilience for watershed ecosystems via participatory approaches in order to integrate climate risk into local, regional and national plans and policies. The project included a climate change perception survey among the local population, an intense workshop and support for implementing solutions (IUCN 2014).
- The ongoing Participatory Development Programme (PDP by GIZ) organizes dialogues, awareness raising events and small pilot measures towards CC. Such measures include rooftop gardens, which serve to familiarize urban target groups with options for making their areas more climate-resilient.
- The Sustainable Agriculture Investments and Livelihoods Project (SAIL), funded by the International Fund for Agriculture and Development (IFAD) from 2014–2023, promotes institutional development as the starting point and also rehabilitates some infrastructure. It has a strong focus on local communities via the support of Community Development Associations (CDAs) which should enhance.

- The Tadamun Initiative is a blog and urban movement from Cairo aiming at knowledge dissemination on citizens' rights over city development. Tadamun argues for more participatory and sustainable policies for urban areas and promotes environmental topics such as water supply, waste management or climate resilience but also art or economic development (see www.tadamun.info stand March 2016).
- Nawaya Egypt is an initiative aiming to empower environmental and social justice through sustainable agriculture in rural Egypt. Nawaya follows the pro-poor and bottom-up approach and understands itself as a technological hub working to spread positive and sustainable change through adaptation, production and spread of green technologies by experimental learning (see www.nawayaegypt.org, stand March 2016).

3.3 Current CC Adaptation Policies

Policy papers published between 2005–2015 addressing CC adaptation in Egypt rather focus on explorative ideas and options than concrete best practices, case study results or clearly defined future measures (similar observation also in Jobbins 2012 for that period). Further, in all publications, technical adaptations play a larger role than social and institutional ones. Note however that the period of extreme political changes between 2011 and 2013 in Egypt might have hindered CC policy updates. The latest publication, the Egyptian Intended Nationally Determined Contribution (INDC) from 2015 by EEAA, is supposed to summarize the state of art for Egypt's mitigation and adaptation plans towards the UNFCCC. It can therefore be analyzed in this place.

The two sections of the INDC paper, which matter for adaptation plans and policies, are Sect. 3.1.2. 'Egypt's Intended Actions to Promote Resilience' (including water resources, agricultural security, additional adaptation policies and measures) and Sect. 3.1.3 'Adaptation Action Packages' (including: coastal zones, water resources and irrigation, agricultural sector, health sector, rural areas, population and roads, tourism sector and the energy sector). The contents of the sections and the paper at large are descriptive and fall short from formulating concrete measures. For instance with regard to 'Agricultural Security' in Sect. 3.1.2, the paper states the following: "Concerning livestock, improving the current low productivity of cattle in addition to improving feeding programs are being considered. No clear adaptation options are defined for fishery wealth" (p. 7). Risk reduction, awareness building and institutional capacity on geographic data are mentioned as standing in the national adaptation focus but no reference indicates details on implementation progress, targets and action or investment plans. Further chapters like Sect. 3.1.3 provide merely a list of priorities and intended goals like the identification of health risks related to CC, the increase of irrigation efficiency, and land protection from degradation while maintaining current crop productivity.

With regard to community-focused actions, the INDC paper includes some action propositions in different sectors, yet without a clear implementation roadmap. Some of these actions which largely focus on awareness-building can be mentioned here:

- Coastal zones, 2. "Capacity building of the Egyptian society to adapt to climate change and associated risks and disasters" (p. 6).
- Water Resources and Irrigation 3. "Develop national policies to encourage citizens on water use rationalization" (p. 6).
- Agricultural Sector 7. "Develop (…) programs and policies to protect rural community and support its adaptive capacity to the expected trend in land use change, plant and animal production (…)" (p. 7).
- Health Sector 2. "Raise community awareness about climate change risks and means of adaptation." (p. 7).

Apart from the content analysis of INDC paper, one can frame this policy paper in comparison to some regional adaptation policies. Jordan for example has a put forward clearer CC adaptation policies. It lists defined technologies it plans to use in agriculture (e.g. drip, micro spray and night irrigation), names ministries in charge and the exact changes they must go through (e.g. installation of water testing labs in the ministry of health) and proposes the installation of Early-Warning systems and emergency rooms across the country (Hashemite Kingdom of Jordan 2015). Morocco begins its INDC with a coherent vision under which it takes all action measures. Further, it lists the total monetary spending on adaptation (64% of all CC related budget), formulates concrete approaches under which the named changes shall happen (e.g. sectoral, ecosystem-based, risk-prevention based) and indicates clear goals for 2020 and 2030 in numbers (e.g. reconstitution of forests on 2,00,000 ha, wastewater treatment to 100%) (Morocco 2015). A noticeable commonality of the CC adaptation policies of Egypt, Jordan and Morocco is the focus on 'hard adaptation measurements', which are mostly defined by infrastructural actions. 'Soft approaches' such as awareness training is named but not planned in detail and rather seem to supplement the other options. Furthermore, the national recommended adaptations tend to focus more on a coping-strategy than on a transformative approach. Note that Egypt is quite experienced in dealing with water scarcity or management and might thus have chosen to focus on water-related adaptation measures. In its INDC paper, it sets up a list of adaptation options towards increasing water resources such as desalination, rainwater harvesting, decreasing evaporation by reconstructed canal cross sections or use of deep groundwater reservoirs. All of which are infrastructural based. In general, the analyzed policies of Egypt can be judged to be in an early stage of evolvement, something that is admitted in the mentioned INDC paper. This paper recognized the need for further studies, clear and implementable adaptation strategies with financial commitment, solid knowledge base and strategic vision (p. 7).

4 Alternatives for Community-Based Adaptation

4.1 Policy Alternatives

Community-Based Adaptation (CBA) is "…a community-led process, based on communities' priorities, needs, knowledge and capacities, which should empower people to plan for and cope with the impacts of climate change" (Reid et al. 2009, p. 13). This includes the creation of space for the direct and practical involvement of communities, the devolution of rights and management authority from central government to local governments and communities, which is recognized by policy and law (Borrini-Feyerabend et al. 2007). Participation in planning is no stiff, clearly defined or simply implementable concept. Rather it is dynamic and constantly evolves as it is applied in different contexts. Participatory planning is believed to be a tool for shaping a system of rules and regulations, of managing conflicts, identifying and prioritizing needs and socially acceptable solutions and also for creating a sense of ownership (Hassan et al. 2011).

Alongside being a popular approach nowadays, the advantages of CBA interventions are also noticeable. The New Economics Foundation in collaboration with Care International conducted a study in Kenya and Niger, evaluating the economic outcomes of a CBA approach towards CC. The conclusion was that a properly planned and implemented CBA most likely leads to a distinctive increase of a community's income. Further, the report speaks of a number of positive 'soft outcomes' such as improved health, education, inclusive decision-making and less environmental damage (King 2014).

In essence, CBA policies reflect the needs of the community and thus effectively addresses local adaptation contexts to CC. Using academic literature, this study carried qualitative analysis to propose viable options for CBA in Egypt considering communities' needs and current CBA practices worldwide. Three CBA policies options are proposed, namely financial-based, information-based and capacity-based measures. The policy categories or options can be treated separately both conceptually and in practice. Information on CC form the base for threat analysis and action. Reid (2009) argues that CBA to CC needs to include climatic information from both, scientific and local, sides and that only the combination of the two can built a solid basis for risk evaluation and (2012) coping strategies. That for, the first policy option has an information-based focus. All CBA to CC literature stress the need for adequate financing. Simple financial services play a critical role and can have a great impact for especially poor people to increase their asset base, widen their economic possibilities and reduce their vulnerability to CC (Verner 2012). Therefore, monetary means cannot be neglected as a supportive strategy and thus form the second financial-based policy option. Verner (2012) further concludes for the Arab region: "Relying for years on powerful national governments has meant that local governments and municipalities have developed a very low capacity to adapt to climate change. This is a problem because local governments are in a much better position to understand the diverse relationships, resources, and vulnerabilities of local communities to climate change." (p. 223). The recognition of low capacities on the local scale, where it is most needed in order to adapt thus constitutes the third, the capacity-based policy option (Table 1).

(A) Financial, loan-based: The ability for individuals, households and communities to adapt to changing climate conditions and to cope with negative effects relies largely on their financial situation. Construction works at the house (e.g. isolation or storm resistance), technical devices (e.g. air conditioner, electric fan), medical treatment (e.g. for heat stress or cough) or for farmers a change in cropping (seeds, irrigation techniques) all require monetary means. Thus, it is more difficult for inhabitants of low-income districts to adapt to and withstand climatic threats. States subsidies on adaptation actions for low-income communities would help these inhabitants to cross the financial barriers. Loans encouraging households or communities to invest into adaptation are other options to lift the financial burden. "The promising role that communities can play in adaptation efforts highlights the need to ensure that they have sufficient access to financial resources. If adaptation money can reach this local level, instead of only flowing to the national level or large, top-down projects, it could facilitate adaptation activities that truly meet the needs of the most vulnerable people" (Christensen et al. 2012, p. 25). Financially supportive packages could include subsidies for water-saving technologies (e.g. irrigation systems) for farmers, loans or financial aid packages for community initiatives towards adaptation (in form of a competition/promotion) or prepare financial aid funds for recovering from climate related shocks such as floods or droughts or extremely cold winters. Also microfinance concepts can be linked to CBA as both ideas deal with the improvement of livelihoods and concentrate on the most vulnerable (Fenton 2011).

Table 1 Possible focus areas for a CBA policy in Egypt

Policy option	Distinctive features	Examples of adaptation measures
Information-based	• Focus on prevention, preemption and preparation • Broad target group • Relatively low-cost • Rather top-down despite considering local knowledge	• National databases • Awareness raising campaigns (TV spots, text messages, posters) • School syllabus
Financial-based	• Focus on impacts and crisis • Vulnerability-based implementation • Implementable on the short-run • Great instrument variety	• Community disaster funds • Adaptation loans • Subsidies on climate-adapting products
Capacity-based	• Focus on adaptive capacity and resilience • Selective targeting • Time-intensive • Higher participatory nature	• Trainings of trainers and multipliers in communities • Root-to-cause workshops • Teaching practical skills and new technology-based competences

(B) Knowledge, information-based: Some communities, especially rural ones, living under extreme climate conditions such as arctic or desert communities are familiar with certain climate circumstances and with the risks coming along. They have developed skillful methods of coping with extremes and gathered local knowledge on strategies throughout different generations. This pool of knowledge and skills is very valuable for the design of a suitable and sustainable policy supporting the adaptation efforts of these remote communities (Ayers and Forsyth 2009). Assessing, valuing and using the local knowledge when formulating a CBA policy is considered as an essential and indispensable step: "Climate information and communication systems informed by scientific and local knowledge have great potential for linking into and improving existing early warning systems." (CGIAR et al. 2012, p. 2). Studies highlight the value in using of the range of communication possibilities (mobile phones, radio and smart phones) for the spread of information. All information must be given in the local language, if necessary in an easy understandable dialect. Information about health conditions, possible housing improvement (construction works or purchase of cooling systems), and possible early warning systems for greater threats such as floods or sand storms would be to the benefit of urban communities. Short state sponsored TV spots, explaining climate change, its effects and protection possibilities, could help to spread awareness and make CC become a public topic. A clear and widely disseminated information base about climate change and its threats and possible responses can be nationally provided, informing the public and encouraging them to take adaptation measures (Verner 2012).

(C) Capacity, training and means-based: Measures in this category like trainings, workshops with producers, leaders and key change makers in the communities can help improve the so-called adaptive capacity of the respective community. The adaptive capacity of a system (or community) describes the ability to adjust to climate change to overcome potential damages and to take advantage of opportunities or to cope with the consequences (IPCC 2007). The ability of people, organizations, and systems, using available skills and resources, to face and manage disadvantage conditions or emergencies, is called the coping capacity (UNISDR 2009). Locally developed capacities have proved to be more effective and efficient than predetermined adaptation actions (CGIAR et al. 2012). For the present case study, community capacity building could be in form of a farmers training on the use of certified seeds (heat resistant, lower water demand), change of irrigation technique, integrated soil or pest management or animal vaccination against avian flu. Such training must offer a holistic approach towards the issue in order to make sure, the participants fully capture the issue from root to cause and find solutions themselves. Workshops with urban communities could include construction suggestions for creating shadow areas between the houses, how to establish rooftop gardens, general greenings of the neighborhood or finding joint disaster strategies in case of sandstorms or floods (Verner 2012). Building cooperatives among neighbors and exchanging information and skill could also be a form of capacity building. The Catalyst approach, a locally-led learning action, aims to let more experienced inhabitants assist and train others. The approach starts with the experiences and priorities of the individual and shifts to the community level through problem solving (Reid 2009). Additionally, capacity building for policy makers and local authorities plays a major role when designing a CBA policy. Furthermore, many CBA-experienced organizations recommend a train-the-trainers concept. Especially when climate change adaptation is relatively new on the national agenda, it is of a high importance to have people in the community with a better understanding of the role of climate information and the need for adequate responses and investments (CGIAR et al. 2012).

4.2 Context and Implementation of Policy Options

Experts interviewed for this research agreed in that no CBA policy would be successful without considering the community's willingness and readiness. These are mainly fostered by:

- Incentives (raising the public perception of an adaptation option to be something desirable/upgrading).
- Leading-by-examples (placing a beneficial adaptation measurement within the community, exhibiting the benefit).
- Sense of ownership (community's involvement whether by opinion, finance or physical work in order will increase sustainability of adaptation measures).
- Trust (between the community, its leader and the district's/regional official).

Further constrains were addressed during the interviews, of which the central ones shall be presented. When speaking of security or trust constraints, the sociopolitical context was highlighted. The security-first approach, which is recognizable in Egypt after the military, under the lead of president and General Abdel Fattah Al-Sisi, came into power in 2013, might have resulted in a neglect of socioeconomic and political issues under which trust can hardly emerge (Taha 2016). NGOs and governmental institutions alike assume the other party does not work transparently and is not legitimate to take the extent of influence it claims. "Civil society plays a weak role and is resistant to advancement (…) A possible reason could be that some civil society organizations often get engaged in exchanging criticism and monitoring errors and violations. Also, the financial transactions of some of these organizations are far from being above suspicion." These phrases are cited in the first paragraph of the Egypt's National Strategy for Adaptation to Climate Change and Disaster Risk Reduction within the chapter 'The role of civil society organizations and community participation" from 2011, p. 99 and display the described state well.

Egypt's CC sphere and CC policies lack a larger vision. The institutional side can be described as fragmented within a centralized setting, procedures are not managed coherently and responsibilities are not set clearly. Identification of responsibilities, monitoring and evaluation are needed, say the experts. This may be achievable by the 'climate change council', whose date of establishment and bare existence are hard to detect with contradictory facts in literature and among stakeholders. While Wodon (2014) states in a World Bank paper that a national committee for climate change was established in 1995, sources from the EEAA indicate that it was founded in 2007. The planning of a CC council was recently discussed in the summer of 2015, when a couple of Egyptian online newspapers such as madarmasr.com or egyptianstreets.com reported, that the prime minister, Ibrahim Mehleb, made a resolution to form a national council on climate change led by the minister of environment. This newly established council will be responsible to draft comprehensive national strategies on all CC issues, according to the reports. No official homepage exists and no paper was found published directly by the committee yet.

At large, key experts tend to favor policy options which include measures to enhance the adaptive capacity of local communities over other alternatives. They expressed that capacity training of the government officials and local authorities on CBA are desperately needed for a successful adaptation. Especially because Egypt's approach has always been the classical top-down path. Next to capacity building, experts stress the importance of the availability of knowledge. An accessible national information base on CC and its effects for politicians, investors and the public would form a consistent basis for further plans revealing options. All in all, adaptation options in Egypt must be tailor-made—according to the regional or local circumstances because 'adaptation readiness' differs among the regions. Moreover, gender has proven to play a major role in CBA (in other countries) and must be recognized as a valuable perspective for Egypt as well. Women bear the larger part of responsibility for the handling of natural resources within rural or low-income urban communities and tend to be less adaptation flexible due to lower capacities and chances (literacy, mobility, income etc.). "Women and men generally have different roles in rural societies in Arab countries. In some countries, women are often responsible for tending the land and livestock during the day and attending to household chores in the evening. Men typically handle the finances, trade in the market, and, if needed, migrate. Because natural resources are directly affected by the climate, women's roles in rural areas are often the most sensitive to climate change" (Verner 2012, p. 281). Additionally, women can also play an important role as multipliers of knowledge about natural resources, climate and adaptation option (Vincent et al. 2012).

5 Conclusions

Climate Change can undoubtedly be described as a major threat for Egypt as a state and its population on various sectors (health-wise, economically, socially, etc.). Adaptation options towards the changing climate are numerous and stretch from infrastructural to social and from individual to national. It is the state's responsibility to design and implement coherent adaptation policies in order to reduce the population's vulnerability and increase their resilience towards climate shocks. While this task is underperformed in Egypt, future climate change policy reforms need to be designed holistically and, importantly, with the participation of the local population. Community-based adaptation promises greater sustainability and justification. The identified policy options (financial-based, knowledge-based and capacity-based) represent potential focus areas of effective future policies. All three may, when carefully applied, bring along a number of so called 'soft benefits' effecting the economic, educational and health related status of the communities positively apart from only climate-related issues. Considering other countries' experiences, more attention is need for institutional constrains like overlaps and policy coherence and social constrains like participation, empowerment and ownership in designing CBA policies in Egypt. Institutionally, a clear allocation of responsibilities and effective management is a perquisite for successful coherent and effective policies. Socially, incentives, good examples and a sense of ownership can increase the community's readiness towards CBA. After all, lack of trust among the different stakeholders (state, NGO's and population) remains maybe the largest barrier to be crossed in order to beneficially cooperate. A trustful cooperation cannot simply arise from the will of the state but shall grow through the mutual work. Climate change adaptation is a serious threat for Egypt, especially affecting the ones with low-income, but ultimately manifesting itself in different forms across communities and classes. Thus, future risks from climate change to be tackled with wide participation at various levels with effective state-led guidance and support. Ultimately, key changes to ensure the success of any adaptation efforts will arise from reforming consumption, production and lifestyle patterns of local communities.

References

Agrawala, S. et.al. (2004). Development And Climate Change In Egypt: Focus On Coastal Resources And The Nile, OECD - Organisation for Economic Co-operation and Development, Paris.

Ambandi M., Nicholles N. (2013). POLICY BRIEF: Climate Change - Why Community Based Adaptation Makes Economic Sense, CARE - Cooperative for Assistance and Relief Everywhere, nef - new economics foundation, http://care.dk/wp-content/uploads/2013/02/PolicyBrief_ClimateChange_final_web.pdf (Last Accessed Februraby 2016).

Ayers J., Forsyth T. (2009). Community based adaptation to climate Change, Environment: Science and Policy for Sustainable Development, 51 (4), p. 22–31.

Buckner S., Buckner E. (2015). Post-Revolution Egypt: The Roy Adaptation Model in Community, Sage Journals, Nur Sci Q October 2015, F28, p. 300–307.

Borrini-Feyerabend G., Pimbert M., Farvar M. T., Kothari A., Renard Y., Jaireth H. and Warren P. (2007). Sharing power: A global guide to collaborative management of natural resources, London.

CGIAR - Consultative Group on International Agricultural Research, CARE - Cooperative for Assistance and Relief Everywhere, ENDA - Energie, Environnement et Développement (2012). Community-based adaptation: an approach to build resilience and sustainable development in West Africa, Communiqué Policy Brief, Research Program on Climate Change, Agriculture and Food Security.

Christensen, K., Raihan S., Ahsan R., Uddin A. M. N., Saleh Ahmed C., and Wright H. (2012). Financing Local Adaptation: Ensuring Access for the Climate Vulnerable in Bangladesh, ActionAid Bangladesh, Action Research for Community Adaptation in Bangladesh, Bangladesh Centre for Advanced Studies, and International Centre for Climate Change and Development, Dhaka.

EEAA (2010). National Environmental, Economic and Development Study (NEEDS) for Climate Change, Cairo, https://unfccc.int/files/cooperation_and_support/financial_mechanism/application/pdf/egypt_final_report_needs.pdf (Last Accessed January 2016).

The Egyptian Cabinet and UNDP (2011). United Nations Development Programme, Egypt's National Strategy for Adaptation to Climate Change And Disaster Risk Reduction, http://www.climasouth.eu/drupal/docs/Adaptation011%20StrategyEgypt.pdf (Last Accessed Februrary 2016).

ESCWA - Economic And Social Commission For Western Asia (2013). Guide On Climate Change Negotiations For Representatives And Negotiators From Arab Countries, New York https://www.unescwa.org/sites/www.unescwa.org/files/publications/files/e_escwa_sdpd_13_tp-10_e.pdf (Last Accessed January 2016).

Fenton, A. (2011). Community-based adaptation and microfinance: a win-win partnership? Guest post on IIED - International Institute for Environment and Development, http://www.iied.org/community-based-adaptationmicrofinance-win-win-partnership (Lasst Accessed February 2016).

Hashemite Kingdom of Jordan (2015). Intended Nationally Determined Contributions, As per United Nation Framework Convention on Climate Change, www4.unfccc.int/submissions/INDC/Published%20Documents/Jordan in March 2016 (Last Accessed January 2016).

Hassan G.F., Hefnawi A., Refaie M. (2011). Efficiency of participation in planning, In: Alexandria Engineering Journal, Vol.50, p. 203–2012, Alexandria.

IUCN (2014). International Union for Conservation of Nature - Regional Office for West Asia, A Guiding Toolkit for Increasing Climate, https://cmsdata.iucn.org/downloads/iucn_report__3_.pdf (Last Accessed February 2016).

IPCC (2007). Climate Change 2007 - Impact, Adaptation and Vulnerability. https://www.ipcc.ch/pdf/assessment-report/ar4/wg2/ar4_wg2_full_report.pdf (Last Accessed April 2017).

IPCC (2013). Intergovernmental Panel on Climate Change, Data Distribution Center, http://www.ipcc-data.org/guidelines/pages/gcm_guide.html (Last Accessed January 2016).

Jobbins G. (2012). Climate Change and Water in the Arab Region. Mapping Knowledge and Institutions, Regional Technical Cooperation Programme, Adaptation to Climate Change in the Water sector in the MENA Region (ACCWam), GIZ, Cairo.

King S. (2014). COMMUNITY-BASED ADAPTATION IN PRACTICE: A global overview of CARE International's practice of Community-Based Adaptation (CBA) to climate change, Copenhagen.

Morocco (2015). Intended Nationally Determined Contributions, – As per United Nation Framework Convention on Climate Change, http://www4.unfccc.int/Submissions/INDC/Published%20Documents/Morocco/1/Morocco%20INDC%20submitted%20to%20UNFCCC%20-%205%20june%202015.pdf (Last Accessed March 2016).

PDP (2015). Participatory Development Project in Urban Areas, GIZ, Deutsche Gesellschaft für Internationale Zusammenarbeit, Cairo, https://www.giz.de/en/worldwide/16278.html (Last Accessed Feburary 2016).

Reid H., Canon T., Berger R., Alam M. and Milligan A. (2009). Participatory Learning and Action. Communitybased adaptation to climate change, Nottingham.

Smith J. Deck L., McCarl B., Kirshen P., Malley J. and Abdrabo M. (2013). Potential impacts of climate change on the Egyptian Economy. UNDP - United Nations Development Programme, Cairo.

Sowers J. (2013). Environmental Politics in Egypt, Activists, Experts and the State, Routledge Studies in Middle Eastern Politics, Oxon.

Sterman D. (2009). Climate Change in Egypt: Rising Sea Level, Dwindling Water Supplies, International Action, Climate Institute, Washington DC. http://climate.org/archive/topics/international-action/egypt.html (Last Accessed January 2016).

Taha, H. (2016). Militarisation and security challenges in Egypt, In: Hokayem, E., Taha, H. Egypt After the Spring: Revolt and Reaction, the Adelphi Series, Vol 55 Issue 453–454, p. 177–200, London.

UN (2015) United Nations. Sustainable Development Goals, http://www.un.org/sustainable development/sustainable-development-goals (Last Accessed January 2016).

UNDP (2015) United Nations Development Programme. Sustainable Development Goals, About Egypt: www.eg.undp.org/content/egypt/en/home/countryinfo/ (Last Accessed January 2016).

UNISDR (2009) United Nations Office for Disaster Risk Reduction. Terminology, www.unisdr.org/we/inform/terminology (Last Accessed May 2016).

Verner D. (2012). Adaptation to a Changing Climate in the Arab Countries: A Case for Adaptation Governance and Leadership in Building Climate Resilience. The World Bank, Washington DC.

Vincent K., Wanjiru L., Aubry A., Mershon A., Nyandiga C., Cull T., and Banda K. (2012) Gender, Climate Change and Community-Based Adaptation, UNDP - United Nations Development Programme. New York.

Wodon Q., Burger N., Grant A. and Liverani, A. (2014). Climate Change, Migration, and Adaptation in the MENA Region, MPRA Paper No. 56927, The World Bank, Washington DC.

Implications of the Paris Climate Change Agreement for Adaptation Research and Universities

Edward A. Morgan, Willow Hallgren, Fernanda Helfer, Oz Sahin, Johanna Nalau, Esther Onyango, Wade Hadwen and Brendan Mackey

Abstract The recent Paris climate change agreement has resulted in a consensus between 196 countries on the need for action, including a stated aim to try to limit warming to 1.5 °C. Despite its weaknesses, it will provide the policy basis for climate change action and research within signature countries. Scientific research is highlighted in the agreement as having a key role to play, and higher education institutions will be expected to lead the way in this research. Much of the focus will be on mitigation but the agreement places substantial emphasis on climate change adaptation. The purpose of this article is to discuss what the agreement might mean for climate change adaptation research, and how it might impact upon research agendas in higher education institutions. In considering how the agreement might impact upon different areas of adaptation research, the paper highlights how

E.A. Morgan (✉) · W. Hallgren · F. Helfer · O. Sahin · J. Nalau · E. Onyango · W. Hadwen · B. Mackey
Griffith Climate Change Response Program, Gold Coast Campus, Griffith University, Parklands Drive, Southport, QLD 4222, Australia
e-mail: ed.morgan@griffith.edu.au

W. Hallgren
e-mail: w.hallgren@griffith.edu.au

F. Helfer
e-mail: f.helfer@griffith.edu.au

O. Sahin
e-mail: o.sahin@griffith.edu.au

J. Nalau
e-mail: j.nalau@griffith.edu.au

E. Onyango
e-mail: e.onyango@griffith.edu.au

W. Hadwen
e-mail: w.hadwen@griffith.edu.au

B. Mackey
e-mail: b.mackey@griffith.edu.au

W. Leal Filho (ed.), *Climate Change Research at Universities*,
DOI 10.1007/978-3-319-58214-6_15

effective interdisciplinary collaboration and research is going to be needed to support the lofty aims of the Paris agreement. It discusses the challenges of this research approach and how universities might be able to support the Paris agreement.

Keywords Climate change adaptation · Interdisciplinary research · Vulnerability · Resilience

1 Introduction

The Paris Climate Change Agreement resulted in 196 countries agreed to limit atmospheric warming to 2 °C and to "pursue efforts" to limit warming to 1.5 °C as compared with pre-industrial times (i.e. before 1750). In addition, the parties agreed to "aim to reach global peaking of greenhouse gas emissions as soon as possible" (Article 4, point 1) and achieve net zero emissions in the second half of the 21st Century. The Agreement was hailed as a success after pessimistic expectations prior to the meeting, but there remain significant questions over whether the aims can be achieved and achieved quickly and equitably (Clémençon 2016).

The Agreement makes it clear that science and research are the key to providing a robust basis for its implementation. Higher education institutions, and especially universities, are likely to be leaders and partners in developing the scientific understanding to inform responses to climate change. They have already played a key role in providing much of the research that is the basis of the IPCC Assessment Reports. This role is expected to be even stronger after the Paris Agreement.

Notably, the Paris Agreement elevates adaptation as a significant policy issue to be considered. Article 7 of the Agreement specifically addresses adaptation, and presents the possibility of a global adaptation goal (Magnan 2016) and national level assessments of adaptation progress and reporting mechanisms. As an interdisciplinary climate change research team we were interested in the implications of this adaptation focus for universities and researchers. Here, we draw on our wide range of expertise in different aspects of adaptation to analyse the Paris agreement in the context of the current literature around adaptation. We use a qualitative document analysis combined with a synthesis of the current adaptation literature to assess the implications of the Agreement, and particularly the call for national adaptation assessments.

The purpose of this article is to discuss what might be needed from researchers and universities in light of the Paris Agreement and identify the challenges this research poses. It is also the intention of this study to suggest some key aspects of adaptation research for researchers and universities to focus on. First, we briefly discuss what is new for adaptation in the Paris Agreement and areas that are missing for adaptation, in order to provide key background context. This document analysis compares the Paris Agreement with previous agreements, and then compares what the Agreement says about adaptation in the context of existing literature. Then we discuss the implications for adaptation research, its links to decision-making in the implementation of the Paris Agreement commitments and the role of universities in supporting the commitments of the Agreement. This analysis draws on a review of the literature to compare the state-of-the-art for adaptation assessment with the needs implied by the Paris Agreement. It then places the results of the comparison into the context of the challenges and opportunities for universities. The analysis only focuses on one aspect of what an adaptation assessment might require—assessing vulnerability and adaptation options—as it is here that research has a vital role to play. Choosing options and monitoring them will follow from this, and research can have a role there too, but we do not discuss this here. Similarly, the analysis is not exhaustive—there are no doubt a range of other implications for universities and potential research priorities—the aim to highlight what we see as the key challenges and opportunities, based on the current state of adaptation research.

2 Adaptation in the Paris Agreement

In light of growing evidence of climatic changes that are irreversible (Solomon et al. 2009), adaptation has received increasing attention in policy, practice and research. Adaptation, however, has steadily gained more prominence and importance in UNFCCC agreements from Conferences of the Parties (COP), as climate change and the scope of necessary actions has become better understood. The Copenhagen Accord that emerged from COP15 in 2009 attracted substantial criticism for its failures on mitigation agreements (Rogelj and Meinshausen 2010; Rogelj et al. 2010) and included very little discussion of adaptation. In contrast, the following COP at Cancun (COP16, 2010) resulted in several agreements, one of which focused on adaptation. The Cancun Agreements stated that adaptation should be given the same priority as mitigation, and they established the Cancun Adaptation Framework, “with the objective of enhancing action on adaptation” (UNFCCC 2010, point 13).

After the limited scope of COPs 17–20, COP21 in Paris became the focal point for a legally binding and universal agreement on climate. Although mitigation remains the focus of the Paris Agreement, adaptation goals are prioritised more than in previous agreements. The Agreement also expands the aims and goals for adaptation, begins to crystallise some of the elements of adaptation of previous agreements, and also encourages the development of a framework for adaptation assessment and reporting.

One of the successes of the Paris Agreement is a global adaptation goal (Article 7, point 1). Magnan (2016, p. 160) points out that the Agreement “aims to build a collective understanding of adaptation through metrics and tools that capture each country’s efforts”. The aim of a more collective understanding of and approach to adaptation, as well as the idea of collating and comparing the efforts of the Parties suggests a more ambitious and collaborative attitude towards adaptation. The Paris Agreement includes a call for National Adaptation Plans to be submitted to the UNFCCC, highlighting a shift towards greater action on adaptation. Similarly, the Parties agree to include the "state of adaptation efforts, support, experiences and priorities" (Decision 100, (a), (ii), p. 14) into the global stocktake called for in the Agreement. However, the language around the adaptation communication is less definite than for mitigation reporting: "Each Party should, *as appropriate*, submit and update periodically an adaptation communication, which *may include* its priorities, implementation and support needs, plans and actions, without creating any additional burden for developing country Parties" (Article 7, point 10, emphasis added). It is also not clear how such a communication would be assessed.

The language used in the adaptation elements of the Paris Agreement does imply a growing consensus on how adaptation should be framed, discussed and implemented. For example, the terms "vulnerability, resilience and adaptive capacity" appear several times throughout the Agreement. These are the key terms that are increasingly providing the conceptual basis for climate change adaptation research (e.g. Gallopín 2006; Ford et al. 2015; Adger 2006; Folke 2006). The use of these terms, in addition to the focus on sharing information, strengthening institutions and scientific knowledge, and moving towards the assessment of adaptation, reflects efforts to create a more connected approach to adaptation within the UNFCCC. Hence, the Paris Agreement can be interpreted as a step towards crystallising adaptation as a concept, in order to support efforts towards implementing adaptation actions. However, the relationship between these three concepts is still somewhat unclear and the subject of debate and research (Gallopín 2006). This is important because an adaptation approach focused on resilience might not be the same as an approach focused on vulnerability. For example, recent focus on social-ecological resilience emphasises the ability of systems to transform (Folke 2006) and it is unclear how transformation might fit with vulnerability, which is more focused on sensitivity, exposure and capacity to respond (IPCC 2014a, b).

The Paris Agreement also emphasises reporting and assessing adaptation efforts. It includes adaptation in the global stocktake and calls for adaptation communications. Further, there are calls for "assessment of climate change impacts and vulnerability" (Article 7, point 9c), "monitoring and evaluating and learning from adaptation plans, policies, programmes and actions" (Article 7, point 9d) and to "review the adequacy and effectiveness of adaptation and support provided for adaptation" (Article 7, point 14c). The inclusion of a section on climate-related loss and damage (building from the Warsaw Agreement) would also appear to reflect moves towards assessing and monitoring adaptation at a global level. Assessing loss and damage remains a challenge (Huggel et al. 2013) but it does provide a mechanism for considering adaptation in terms of impacts of climate change and mitigation efforts (Huq et al. 2013; James et al. 2014).

There appears to be an effort within the Paris Agreement to place adaptation on a similar footing to mitigation in terms of global effort and cooperation. The Paris Agreement is well balanced between the two concepts, and the attempt to both clarify and encourage assessment of adaptation indicates a move towards a global approach to adaptation, just as there is for mitigation. However, whereas for mitigation, clear targets on emissions can be reported and assessed (as seen with the NDCs), measuring adaptation is far more complex, and the research into adaptation is less well developed (Tol 2005; Füssel 2007; IPCC 2014a). Monitoring and evaluation of adaptation is difficult because it deals with complex social-ecological processes that need to be considered over long time-frames and at a variety of different scales, and so viewpoints differ on success and failure (IPCC 2014a, Chap. 14; Adger et al. 2009). The language of the Paris Agreement reflects this complexity and uncertainty and results in the adaptation text being much less definite than that for mitigation.

Hence, although adaptation receives significant attentions in the Paris Agreement, we argue that the language around it is less definite and clear than for mitigation. This likely reflects the complexity of adaptation and our less well-established understanding of adaptation, how to achieve it, and how to assess it. In this respect, the Paris Agreement highlights the challenges and the opportunities for adaptation research going forward, which is the focus of the next section.

3 Adaptation Research to Support the Paris Agreement

This section discusses the research challenges specifically related to climate change adaptation support policy and the Paris Agreement, and how universities can support adaptation research efforts.

The Paris Agreement suggests that there is an urgent need for the development of more concrete, politically and economically feasible adaptation policies to allow for National Adaptation Plans (NAP) to be submitted to and assessed by the Parties of the UNFCCC. This implies a requirement for more specific and detailed options for adaptation, that build upon more general adaptation measures that have been set in existing plans (e.g. the national climate change action plans developed by European and Latin American countries, see Metternicht et al. 2014). Currently, there are NAP guidelines to assist countries to undertake the steps and activities to facilitate 'country-owned, country-driven' medium and long-term adaptation to climate change (UNFCCC 2015). These guidelines are non-prescriptive and are designed to create an opportunity for countries to learn how to manage "multiple stress factors that combine in complex ways across scales" (Least Developed Countries Expert Group 2012). The nature of these guidelines reflects the challenge of adaptation from both a research and policy perspective—the complexity of interrelated factors across different scales resulting in highly context-specific adaptation needs. Nonetheless, there is agreement that the first stages of adaptation need to be about identifying adaptation needs and options, and providing some assessment of these options (Least Developed Countries Expert Group 2012; Hinkel et al. 2013).

We suggest that these steps necessitate an integrated Vulnerability, Impact and Adaptation (VIA) Assessment. VIA assessments would seek to provide a clear understanding of the causes, areas of vulnerability and actual impacts, based on site-specific knowledge and scientific data, as well as taking into account political, social and cultural processes that may facilitate or impede the implementation of the suggested adaptation measures (PROVIA 2013; Bizikova et al. 2009). Here, we draw on the state of the art of the adaptation literature to discuss how adaptation research can help formulate 'best practice' VIA assessment to enhance the effectiveness of climate change adaptation, support the development of NAPs. The discussion highlights some of the challenges that need to be addressed and the role universities can play in supporting climate change adaptation.

3.1 Interdisciplinary and Integrated Research

Both adaptation research in general, and VIA assessments specifically, should be designed to be interdisciplinary and integrated. The first challenge for an effective VIA assessment is to do away with the traditional silo mentality, and to ensure such research projects are performed by a multidisciplinary team of technical experts (e.g., specialists in climate science, ecosystems and biodiversity, socio-economic scientists, experts in stakeholder engagement, policy development and communications), in order to incorporate biophysical, policy, socioeconomic and cultural perspectives. Calls for interdisciplinary and integrated research in climate change are not new, of course, and the difficulties are well documented (Adger 2006; Adger et al. 2009; Thornton et al. 2014).

Uncertainty can be a particular challenge for such research. Uncertainty has been a major challenge for climate change research and policy (Dessai 2007) and the challenge is greater for adaptation (Fussel 2007). Hence, recognition and evaluation of uncertainties is essential. In an adaptive process, uncertainty analysis also enables the identification of which variables are uncertain and determinant in explaining vulnerability to climate change (Vignola et al. 2009). Uncertainty also needs to be communicated effectively, and researchers and policymakers will need to develop a shared understanding of the uncertainty to ensure that action is not delayed by calls for more research, and that failures are understood as part of the uncertain process.

Universities will need to support this interdisciplinary approach by allowing time and resources for collaboration and recognising that skills such as engagement and communication are currently not well-evaluated and often under-valued. Interdisciplinary research centres are becoming more common, but genuinely interdisciplinary research can still face institutional challenges.

3.2 Scale and Context

VIA assessment should be site-specific and conducted at the local scale. The challenge here is that many assessments show results on regional and global scales (e.g. Vörösmarty et al. 2010; Houghton and English 2014; McIver et al. 2015), providing a broad view that seldom helps interested parties to design site specific local adaptation (Vignola et al. 2009). We suggest that VIA assessments should continue being undertaken at local level to produce more tangible adaptation initiatives. This comes with the policy challenge that NAPs are to be reported at a country scale, and thus an effective global-scale adaptation tracking is necessary. There are likely complex interactions between vulnerabilities, impacts and adaptations at the local and national scale. If the focus of VIA assessment is at the local scale, this mismatch needs to be addressed.

Similarly, tools and methods employed in the assessment should be tailored to the specific context in which the assessment outputs are to be applied. Outputs should be site-specific and applicable to the study area. The role of researchers includes the identification of the best tools and methods to use, given the scope of the VIA assessment. This underlines the challenge that adaptation research faces. Whereas mitigation reporting will tend towards generalizable, comparable methods, adaptation assessment seems to require the opposite (Fussel 2007). This is a problem both for researchers—who are searching for general conclusions to report to the scientific community—and policymakers—who want to compare and contrast policies and actions to learn from them (or ideally transplant methods from one area to another). We suggest that adaptation research will increasingly need to learn from, compare and draw more general conclusions from context-specific adaptation.

Universities are well-placed to support research for the regions in which they are based. However, there is pressure to make research internationally relevant, or to make it of interest to the dominant US and European audience. Universities can balance this by recognising that local and regional outreach and collaboration will be vital to support adaptation research. They do of course face the pressure of needing to compete for research funding, and research funding bodies also need to recognise how local and context specific research is important.

3.3 Stakeholder Engagement

Stakeholders should be involved during the entire assessment process. Stakeholders can provide important contextual knowledge so researchers are able to understand local processes of adaptation and ecosystem management. The Paris Agreement notes the importance of a “gender-responsive, participatory and fully transparent approach”, that considers “vulnerable groups, communities and ecosystems”, and is guided by “the best available science and, as appropriate, traditional knowledge,

knowledge of indigenous peoples and local knowledge systems". In the first stages, for example, stakeholders will assist with the definition of the scope and determining the objectives of the assessment; in the final stages, they will validate and ensure the relevance and use of the final outputs. They will also help prioritise the adaptation options set in the assessment. The role of researchers should include finding methods for better stakeholder engagement in climate change adaptation assessments and to effectively communicate the results to them (Harrison et al. 2013; PROVIA 2013). Importantly, this engagement has the potential to improve planning and policy as well (Cameron and Grant-Smith 2014; Collins and Ison 2009). However, engagement is no guarantee of success, and stakeholder engagement in both research and policy can be resource-intensive.

Importantly, researchers will need to ensure that results are communicated in a way that is relevant to policymakers (Vignola et al. 2009). At the local and national levels, a participatory process should ideally take place in the policy design process, including problem identification, strategy formulation, selection of policy options, and monitoring and evaluation. At the global level, scientists can play an important advisory role to national policymakers who participate in international negotiations related to climate change adaptation. Such interactions are not without difficulties, however—adaptation and adaptation research can easily be politicised and used to justify actions taken from a purely political motive or to support a particular political agenda and maintain power structures (Eriksen et al. 2015). Similarly, policy processes can become overly scientific, ignoring the concerns and values of those impacted (whether by climate change or adaptation) (Lackey 2007). Researchers will have to explore ways of interacting with policy to create shared understandings of the issue(s) at stake (Hoppe 2013).

Researchers will need to find new ways of working with a wide range of stakeholders (Gardner et al. 2009; Harrison et al. 2013; PROVIA 2013; IPCC 2014b), much in the same vein as they will be required to work with colleagues from different disciplines, and universities will need to support this. At the same time universities will need to support such engagement and value it, and provide training and support to develop the necessary skills. There is an increased focus on impact in assessing research in universities, which would seem to parallel this need for greater interaction with stakeholders and policy-makers. However, there are still concerns over how impact will be measured. A focus on impact within the scientific community or on short-term impact will not support greater engagement.

3.4 Climate Change Communication

A dialogue about climate change needs to continue between climate change scientists of all relevant disciplines, decision makers and the civil society. Scientists, or science research managers, will need to devote substantial efforts in communicating research results to non-scientists, such as local NGOs, public administration and the media, in order to increase their capacity to influence the implementation adaptation (Vignola et al. 2009). However, climate change remains a politically divisive topic. Although the Paris Agreement might have created a consensus on the need for action, translating that on the ground will be challenging where climate scepticism and conflicting political issues can stall progress. Universities can support these efforts by that they continue to support climate research, focus on outreach and provide training and support to researchers that try to get involved in these issues. Of course, key will be continuing to educate current and future students, teaching them to think about and understand the issue of climate change in all its complexity.

3.5 Equity

A final issue that must always be considered, and that cuts across all of the above, is that of equity. Research remains an Anglo-centric, Western-focused and male-dominated field, yet the impacts of climate change will be felt most heavily in the global south and among the poorest and most vulnerable communities (IPCC 2014a). At the same time, adaptation must not be about protecting the wealth of those that have it at the expense of the poor. Researchers and universities have a responsibility to ensure those that cannot afford their own research are not left out, and that data, information and ideas are relevant, and available to all. Similarly, universities need to ensure that they challenge inequity and inequality within their institutions. This is as much a challenge for society as a whole as it is for research, but it must be part of the adaptation agenda at every level.

4 Conclusion

This article has discussed the adaptation aspects of the Paris Agreement, their implications for adaptation research and the role of universities in implementing the commitments of the Agreement. We argue that there is a clear attempt to place adaptation alongside mitigation in the Agreement, highlighting its growing importance as the climate changes. However, our analysis also demonstrates that adaptation remains complex and uncertain, and significant challenges remain if it is to become a part of mainstream policy and planning. To support this, we argue that adaptation research will need to (a) address the complexity and uncertainty of social-ecological systems, (b) be context-specific, (c) connect the local scale to national interests, and (d) be equitable and inclusive of stakeholders. Universities will have to support these efforts through training and recognition of aspects of research beyond simply academic novelty and impact. This may need them to challenge existing institutional barriers to the sort of research needed. No doubt this is a massive challenge, but we suggest that there are also opportunities for genuinely interdisciplinary, inclusive research to engage with stakeholders to help implement the Paris Agreement and support climate change adaptation. Just as the climate is changing and forcing societies to adapt, so are the expectations of what researchers and universities can provide in terms of knowledge and understanding. Researchers played a key role in putting climate change and the need for mitigation on the public agenda and, although that battle is still being fought, universities and researchers need to adapt their institutions to better support the adaptation needs highlighted in the Paris Agreement.

Acknowledgements We acknowledge the support provided by the Griffith Climate Change Response Programme for this study. We also acknowledge the support provided by CNPq (Brazil) to Dr Fernanda Helfer, one of the authors of the paper.

References

Adger, W. N. (2006). Vulnerability. *Global Environmental Change,* 16(3), 268–81.

Adger, W. N., Eakin, H., & Winkels, A. (2009). Nested and teleconnected vulnerabilities to environmental change. *Frontiers in Ecology and the Environment*, 7(3), 150–157.

Bizikova, L., Bellali, J., Habtezion, Z., Diakhite, M., & Pinter, L. (2009), *IEA Training Module – Volume 2 – Vulnerability and Impact Assessments for Adaptation to Climate Change*. Nairobi: United Nations Environment Programme, 59 p.

Cameron, J., & Grant-Smith, D. (2014). Putting people in planning: participatory planning, inclusion and power. In J. Byrne, J. Dodson, & N. Sipe (Eds.), *Australian Environmental Planning: Challenges and Future Prospects* (pp. 197–205). London: Routledge.

Clémençon, R. (2016). The Two Sides of the Paris Climate Agreement: Dismal Failure or Historic Breakthrough. *Journal of Environment and Development*, 25(1), 3–24.

Collins, K., & Ison, R. (2009). Jumping off Arnstein's ladder: social learning as a new policy paradigm for climate change adaptation. *Environmental Policy and Governance*, *19*(6), 358–373.

Dessai, S., O'Brien, K., & Hulme, M. (2007). Editorial: On uncertainty and climate change. *Global Environmental Change*, 17(1), 1–3.

Eriksen, S. H., Nightingale, A. J., & Eakin, H. (2015). Reframing adaptation: The political nature of climate change adaptation. *Global Environmental Change*, 35, 523–533.

Folke, C. (2006). Resilience: The emergence of a perspective for social–ecological systems analyses. *Global Environmental Change*, 16(3), 253–267.

Ford, J. D., Berrang-Ford, L., Biesbroek, R., Araos, M., Austin, S. E., & Lesnikowski, A. (2015). Adaptation tracking for a post-2015 climate agreement. *Nature Climate Change*, 5(11), 967–969.

Füssel, H. M. (2007). Adaptation planning for climate change: Concepts, assessment approaches, and key lessons. *Sustainability Science*, 2(2), 265–275.

Gallopín, G. C. (2006). Linkages between vulnerability, resilience, and adaptive capacity. *Global Environmental Change*, 16(3), 293–303.

Gardner, J., Dowd, A-M., Mason, C. & Ashworth, P. (2009). *A Framework for Stakeholder Engagement on Climate Adaptation*. CSIRO Climate Adaptation Flagship Working Paper 3. Kenmore: CSIRO, 32 p.

Harrison, P., Holman, I.P., Cojocaru, G., Kok, K., Kontogianni, A., Metzger, M.J., & Gramberger, M. (2013). Combining Qualitative and Quantitative Understanding for Exploring Cross-Sectoral Climate Change Impacts, Adaptation and Vulnerability in Europe. *Regional Environmental Change,* 13(4), 761–80.

Hinkel, J., Bharwani, S., Bisaro, A., Carter, T., Cull, T., Davis, M., Klein, R., Lonsdale,K., Rosentrater, L., & Vincent K., (2013) *PROVIA Guidance on Assessing Vulnerability, Impacts and Adaptation to Climate Change – SUMMARY*. Nairobi: United Nations Environment Programme, 20 p.

Hoppe, R., Wesselink, A., & Cairns, R. (2013). Lost in the problem: The role of boundary organisations in the governance of climate change. *Wiley Interdisciplinary Reviews: Climate Change*, 4(4), 283–300.

Houghton, A. & English, P. (2014). An Approach to Developing Local Climate Change Environmental Public Health Indicators, Vulnerability Assessments, and Projections of Future Impacts. *Journal of Environmental and Public Health*, 132057.

Huggel, C., Stone, D., Auffhammer, M., & Hansen, G. (2013). Loss and damage attribution. *Nature Climate Change*, 3(8), 694–696.

Huq, S., Roberts, E., & Fenton, A. (2013). Loss and damage. *Nature Climate Change*, 3(11), 947–949.

IPCC (2014a). *Climate Change 2014: Impacts, Adaptation, and Vulnerability. Part A: Global and Sectoral Aspects*. Contribution of Working Group II to the Fifth Assessment Report of the Intergovernmental Panel on Climate Change. Cambridge: Cambridge University Press, 1132 p.

IPCC (2014b). *Climate Change 2014: Impacts, Adaptation, and Vulnerability. Part B: Regional Aspects*. Contribution of Working Group II to the Fifth Assessment Report of the Intergovernmental Panel on Climate Change. Cambridge: Cambridge University Press, 688 p.

James, R., Otto, F., Parker, H., Boyd, E., Cornforth, R., Mitchell, D., & Allen, M. (2014). Characterizing loss and damage from climate change. *Nature Climate Change*, 4(11), 938–939.

Least Developed Countries Expert Group (2012). *National Adaptation Plans. Technical guidelines for the national adaptation plan process*. Bonn: United Nations Climate Change Secretariat, 152 p.

Lackey, R. T. (2007). Science, scientists, and policy advocacy. *Conservation Biology: The Journal of the Society for Conservation Biology*, 21(1), 12–17.

Magnan, A. K. (2016). Climate change: Metrics needed to track adaptation. *Nature*, 530(7589), 160.

Metternicht, G., Sabelli, A., and Spensley, J. (2014) Climate Change Vulnerability, Impact and Adaptation Assessment: lessons from Latin America. International Journal of Climate Change Strategies and Management, 6(4), 442–476.

McIver L., Kim R., Woodward A., Hales S., Spickett J., Katscherian D., Hashizume M., Honda Y., Kim H., Iddings S., Naicker J., Bambrick H., McMichael A.J., Ebi K.L. (2015). Health Impacts of Climate Change in Pacific Island Countries: A Regional Assessment of Vulnerabilities and Adaptation Priorities. *Environmental Health Perspectives*, 1509756.

PROVIA (2013). *PROVIA Guidance on Assessing Vulnerability, Impacts and Adaptation to Climate Change*. Nairobi: United Nations Environment Programme, 174 p.

Rogelj, J., & Meinshausen, M. (2010). Copenhagen accord pledges are paltry. *Nature*, 464(7292), 1126–1128.

Rogelj, J., Chen, C., Nabel, J., Macey, K., Hare, W., Schaeffer, M., Markmann, K., Hohne, N., Andersen, K.K., & Meinshausen, M. (2010). Analysis of the Copenhagen Accord pledges and its global climatic impacts—a snapshot of dissonant ambitions". *Environmental Research Letters*, 5(3), 034013.

Solomon, S., Plattner, G.-K., Knutti, R., & Friedlingstein, P. (2009). Irreversible climate change due to carbon dioxide emissions. *Proceedings of the National Academy of Sciences of the United States of America*, 106(6), 1704–1709.

Thornton, P. K., Ericksen, P.J., Herrero, M., & Challinor, A.J. (2014). Climate Variability and Vulnerability to Climate Change: A Review. *Global Change Biology*, March, 1–16.

Tol, R. (2005). Adaptation and mitigation: trade-offs in substance and methods. *Environmental Science & Policy*, 8(6), 572–578.

UNFCCC (2010). Cancun Agreement to the United Nations Framework Convention on Climate Change, II, 13. Dec. 11. Available from: http://unfccc.int/resource/docs/2010/cop16/eng/07a01.pdf#page=4.
UNFCCC (2015). *Navigating the Landscape of Support for the Process to Formulate and Implement National Adaptation Plans*. Bonn: United Nations Climate Change Secretariat, 48 p.
Vignola, R., Locatelli, B., Martinez, C., & Imbach, P. (2009). Ecosystem-based adaptation to climate change: What role for policy-makers, society and scientists? *Mitigation and Adaptation Strategies for Global Change,* 14 (8), 691–696.
Vörösmarty, C. J., McIntyre P. B., Gessner, M. O., Dudgeon, D., Prusevich, A., Green, P. A., Glidden, S., Bunn, S. E., Sullivan, C. A., Reidy, C. A., & Davies, P. M. (2010). Global threats to human water security and river biodiversity. *Nature,* 467, 555–561.

Climate Change and Extreme Events over Dabaa Region, Egypt

Magdy Kamal Badir and Hamdy Abd AlRahman

Abstract Mediterranean coast of Egypt is narrow band of earth on the Mediterranean Sea extend to approximately 1500 km length and around 20 km wide. Dabaa region affected with the climate of Mediterranean Sea, as Temperature, Relative Humidity and Precipitation. The climate of the Mediterranean coast moderate temperature all the year, high Relative Humidity and highest precipitation comparing to others area in Egypt. Around the world measured climatic conditions for many years, so it is possible to define what is considered normal and extreme event. Extreme meteorological events may be good markers of climatic change or variability. Extreme weather and climate events are due to the often large loss of human life and exponentially increasing costs associated with them. In this study, focuses on changes in extreme temperature and precipitation during the period 1981–2010 and the surface wind (direction & speed) during the period 1981–2014 over Dabaa region. The results indicated that the mean annual air temperature over the studied region increased by about 0.1 °C during the period of investigation. The yearly precipitation over Dabaa region varied between a minimum amount of 25.8 mm in 2010 and a maximum amount of 237.9 mm in 1991, with an average annual precipitation of about 125 mm. The mean precipitation rate over the study period was 3.2 mm/rainy day and the mean wind directions are North West direction. Three extreme events have been recorded during the study period; scarce

M.K. Badir (✉)
Tanta University, Tanta, Egypt
e-mail: mbadir64@gmail.com

M.K. Badir · H.A. AlRahman
Cairo University, Cairo, Egypt

M.K. Badir
WMO Regional Training Center, Egyptian Meteorological Authority, Cairo, Egypt

H.A. AlRahman (✉)
Zagazig University, Zagazig, Egypt
e-mail: gd68@hotmail.com

H.A. AlRahman
Climate Department, Egyptian Meteorological Authority, Cairo, Egypt

W. Leal Filho (ed.), *Climate Change Research at Universities*,
DOI 10.1007/978-3-319-58214-6_16

precipitation (2.2 mm/rainy in 12 day) in 2010, extreme high precipitation (3.8 mm/rainy in 63 day) in 1991 and highest daily temperature 21.6 °C at 2010.

Keywords Extreme events · Precipitation · Gust wind · Mediterranean coast · Dabaa region · Egypt

1 Introduction

Increases in extreme climate events have greater negative impacts on human society and natural environments than changes in climate means. Several studies have reported how a single catastrophic precipitation event can devastate long-term accomplishments of human society within a short period (e.g. Zong and Chen 2000). It is also well documented how greatly an extreme hot or cold temperature event can increase human mortality, as well as energy utilization (e.g. Huynen et al. 2001; UNEP 2004). There have been many studies in recent decades that have examined changes in monthly, seasonal, and annual climate means for the purpose of capturing the fingerprints of climate change due to anthropogenic modification of atmospheric chemistry. In contrast, systematic and extensive studies about changes in extreme climate events covering broad regions (e.g. continental or hemispheric scale) have been a much more recent development, primarily due to difficulties in collecting high spatial and temporal resolution climatic data across international boundaries.

Since the late 1990s, many efforts have been made to assess changes in spatial and temporal patterns of extreme climate events using observational climatic data. Analyses of long-term climatic characteristics of extreme events, including their intensity, duration, and frequency, are needed for developing mitigation and adaptation plans. In these efforts, data from a greater number of long-term weather stations across more extensive regions are indispensable to the detection of global-scale changes in climate extremes. However, there are many difficulties in obtaining reliable long-term climatic data due to financial and institutional barriers, including limitations imposed by many countries on exchanging data sets with daily or higher resolutions, as well as poor maintenance of data observation and archive practices.

In general, there are, in developing countries, greater difficulties in maintaining observation networks and archiving quality controlled data in an appropriate format due to lack of funding, technology, and human resources, as well as political instability or armed conflict in some regions.

As a result, many stations have only short records or long periods of missing data, and many valuable long term climatic data, including those archived throughout colonial periods. In some countries that only became independent during the 20th century, data remain in a non-digital format and not readily available for use in analyses (Page et al. 2004). A further difficulty was that no standard definitions of extreme climate indices were agreed by the international community until the late 1990s, which has also delayed the required monitoring efforts of extreme climate events across large regions.

Many regional workshops for detecting extreme climate events have been organized successfully in areas such as North Africa, Central and South America, and the Middle East (Peterson and Manton 2008). Among many previous papers, Alexander et al. (2006) provided the latest, most comprehensive analysis regarding global-scale changes in extreme climate events through combining results obtained by many regional meetings on extreme climates.

In particular, as the Intergovernmental Panel on Climate Change (IPCC) Fourth Assessment Report (2007) stated, the direction and magnitude of changes in extreme climate are not necessarily identical to those in climate means. Thus, the exploration of the associations between extreme climate events and climate means is an important task for projecting future changes in extreme events using climate means, which are currently simulated more reliably by climate models than extreme events.

The region has shown large climate shifts in the past and has been identified as an important area for future climate change projections. The rate of winter precipitation over the Mediterranean basin is affected by the climatic pattern over the Atlantic Ocean, while the dry climate summer is controlled by both the Asian and African Monsoons. The Mediterranean is expected to become generally drier with prolonged droughts in summer and reduced rainfall in winter, while a higher proportion of rainfall will fall on very wet days. Many researchers have addressed the investigation of extreme precipitation over the Mediterranean basin. This was intensively carried out for the Western Mediterranean basin and along the European Mediterranean coasts, e.g.

In October 2007, the Supreme Council for Energy adopted a strategy for energy supply and use, which comprehensively integrates the main policies and measures that could meet the longer term challenges facing the national energy industry. The strategy confirms the ongoing activities, adding to them nuclear power generation from Dabaa station.

The present study is an initial fair trial to get closer to the weather extremes over Dabaa region in the north coast of Egypt. It focuses on the extreme events, which have been recorded for the precipitation and air temperature.

2 Area of Investigation

Dabaa region (Fig. 1), which is the study area, extends for about 50 km and boarded by the Mediterranean Sea to the north, western desert to the south, Alexandria to the east and Mersa Matruh to the west.

Dabaa synoptic station with WMO No. 62,309 lie at Latitude 30° 56′ 58″ N and longitude 28° 26′ 41″ E working 24 h per day from 7th Dec.1947 at now.

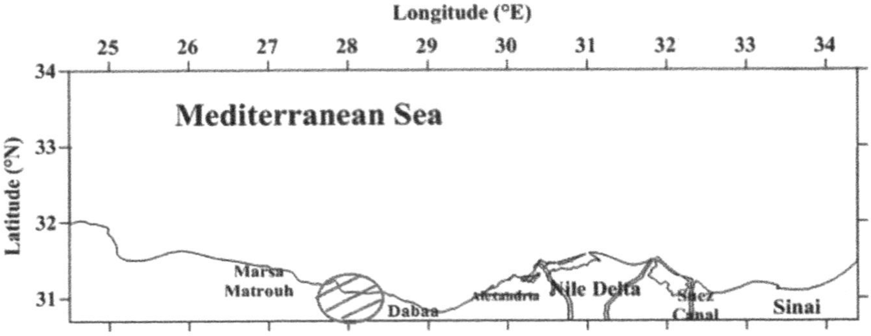

Fig. 1 Egyptian Mediterranean Coast, showing the study area (Dabaa Region)

3 Data and Methods of Analysis

Precipitation amounts and air temperature over Dabaa region, derived from the data archive of Egyptian Meteorological Authority EMA, are used, covering the period 1981–2011.

An extreme precipitation event is defined when the mean amount of precipitation/rainy day exceeds or precedes the usual mean of precipitation/rainy day.

The definition goes for the air temperature when the anomaly highly exceeds or precedes the annual mean air temperature for the region of investigation.

4 Results

4.1 Precipitation

The yearly amount of precipitation over Dabaa region during the study period (1981–2010) varies between a minimum amount of 25.8 mm in 2010 and a maximum amount of 237.9 mm in 1991 (Fig. 2). The annual mean amount of precipitation is 125 mm with a precipitation rate of 3.2 mm/rainy day.

In 2010 the total amount of precipitation was 25.8 mm in only 12 days, i.e. 2.2 mm/rainy day. This is extremely low than the normal rate which is 3.2 mm/rainy day. Three years are shown to have the highest amounts of precipitation. These years are 1991 with a total precipitation of 237.9 mm in 63 days, the year 1992 of 217.9 mm in 43 days and the year 2000 of 229.1 mm in 54 days. These results are also confirmed using the anomaly calculations as shown in Fig. 3.

Climatically, Dabaa region is characterized by moderate temperature all the year, high Relative Humidity and highest precipitation comparing to others area in Egypt.

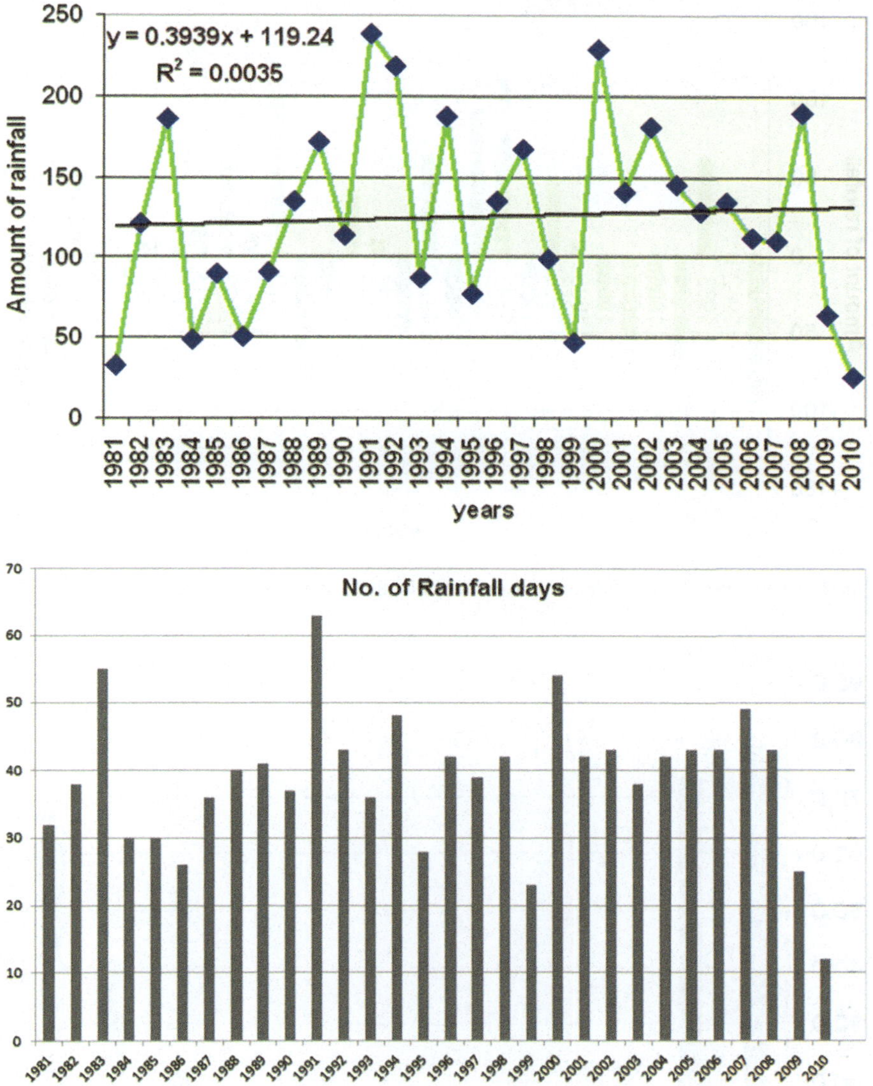

Fig. 2 Annual mean precipitation mm and Number of rainy days/year over Dabaa region (1981–2010)

It has only two periods: a rainy period from November to April and dry period from May to October. The winter months (January, February and December) over the study period is shown in Fig. 4. On the monthly basis, January was the month of the highest winter precipitation event over the period of investigation. January 1991 the total amount of precipitation was 41.9 mm in 18 days, i.e. 2.3 mm/rainy days.

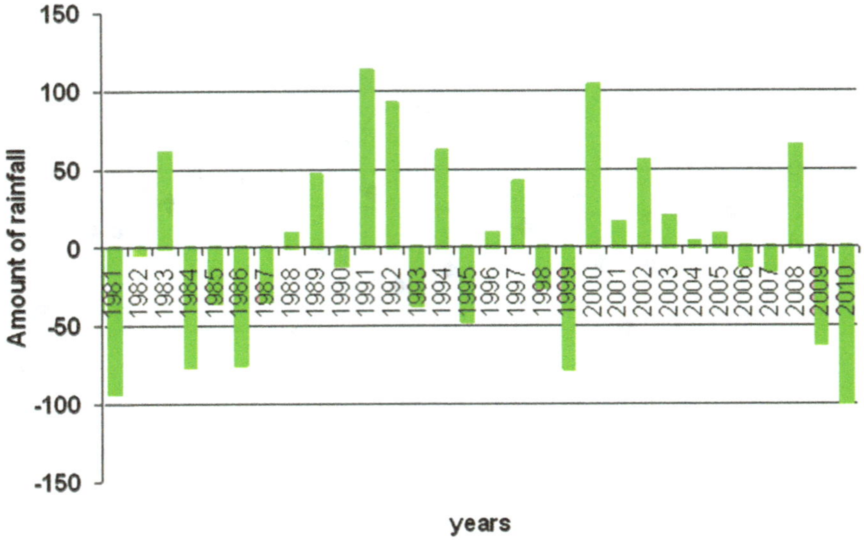

Fig. 3 Annual precipitation anomaly over Dabaa region (1981–2010)

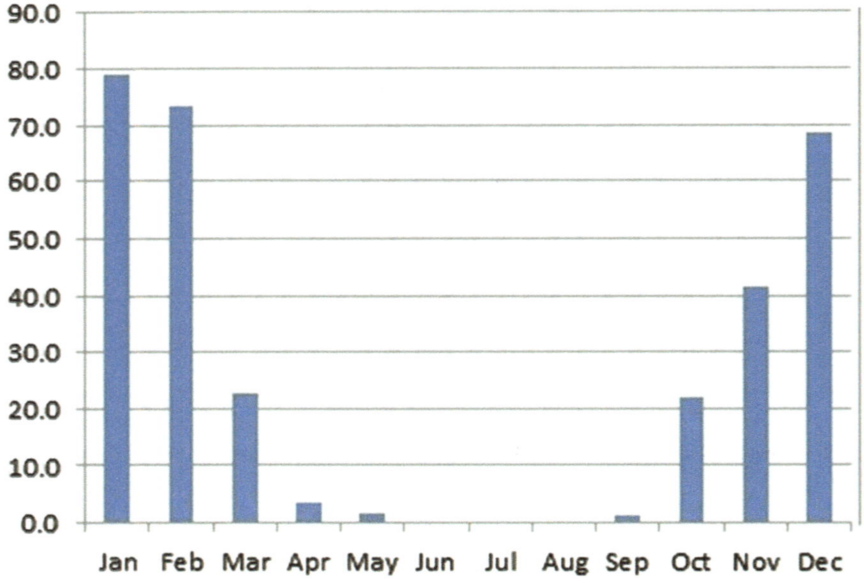

Fig. 4 The winter months over Dabaa region (1981–2010)

In on the other hand, January 2000, had a total precipitation amount of 102.8 mm in 17 days, i.e. 6.1 mm/rainy day, which is extremely high.

4.2 Air Temperature

Climatically, Dabaa region is characterized by moderate temperature all the year. The average monthly air temperature ranges between minimum of 18.8 °C in February and maximum of 21.7 °C in August with an annual average of 20.25 °C. The rate of increasing air temperature over Dabaa region during the 30 years study period was about 0.1 °C. The same trend was given for the period 1981–2010. This trend can be mathematically represented by the following equation:

$$Y = 0.074X + 18.8 \quad (1)$$

In order to investigate the extreme events in the air temperature during the 30 years of investigation, the anomaly for the whole data set is calculated and given in Fig. 5. The anomaly varies between −1.2 and +1.7 °C.

Examining the year 2010, the year of scarce rain, the highest annual mean air temperature periods were observed (Fig. 6). The air temperature during this period was high and above the normal means, when the annual mean air temperature exceeded 21.7 °C.

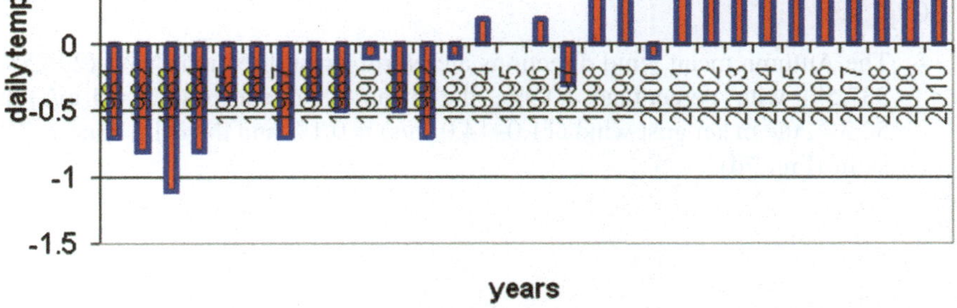

Fig. 5 Annual air temperature anomaly over Dabaa region (1981–2010)

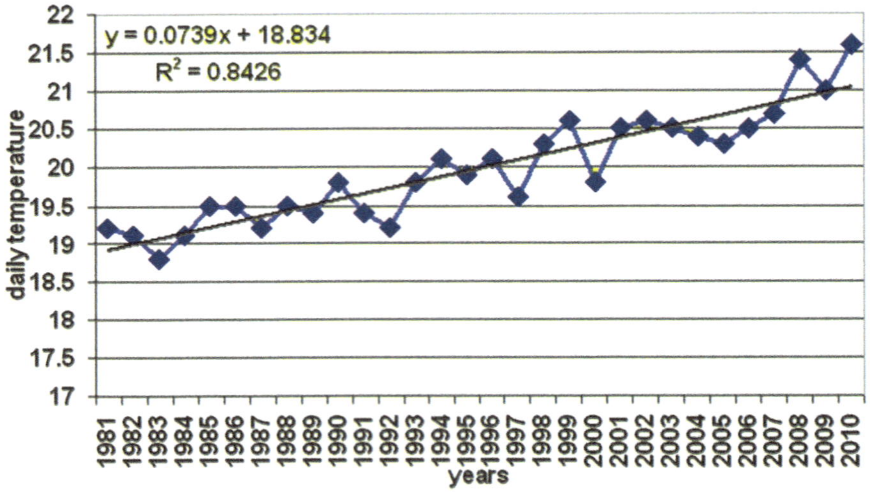

Fig. 6 Annual mean air temperature (°C) at Dabaa region (1981–2010)

4.3 Seasonally Surface Wind Directions and Speed

- The winter mean wind directions are West direction (W) (247.5°–292.5°) with percentage 20.3%, the mean wind speed (1.5–4.5 m/s) is 48.3%, the mean gust wind (14.0–17.0 m/s) is 0.5% and the calm wind is 4% as in (Fig. 7a).
- The spring mean wind directions are North West direction (NW) (292.5°–337.5°) with percentage 24.0%, the mean wind speed (1.5–4.5 m/s) is 48.2%, the mean gust wind (14.0–17.0 m/s) is 0.1% and the calm wind is 3% as in (Fig. 7b).
- The summer mean wind directions are North West direction (NW) (292.5°–337.5°) with percentage 40.3%, the mean wind speed (1.5–4.5 m/s) is 51.1%. The mean gust wind (11.0–14.0 m/s) is 0.1%, and the calm wind is 3% as in (Fig. 7c).
    - The Autumn mean wind directions are North West direction (NW) (292.5°–337.5°) with percentage 30.9%, the mean wind speed (1.5–4.5 m/s) is 52.3%, the mean gust wind (11.0–14.0 m/s) is 0.1% and the calm wind is 7% as in (Fig. 7d).

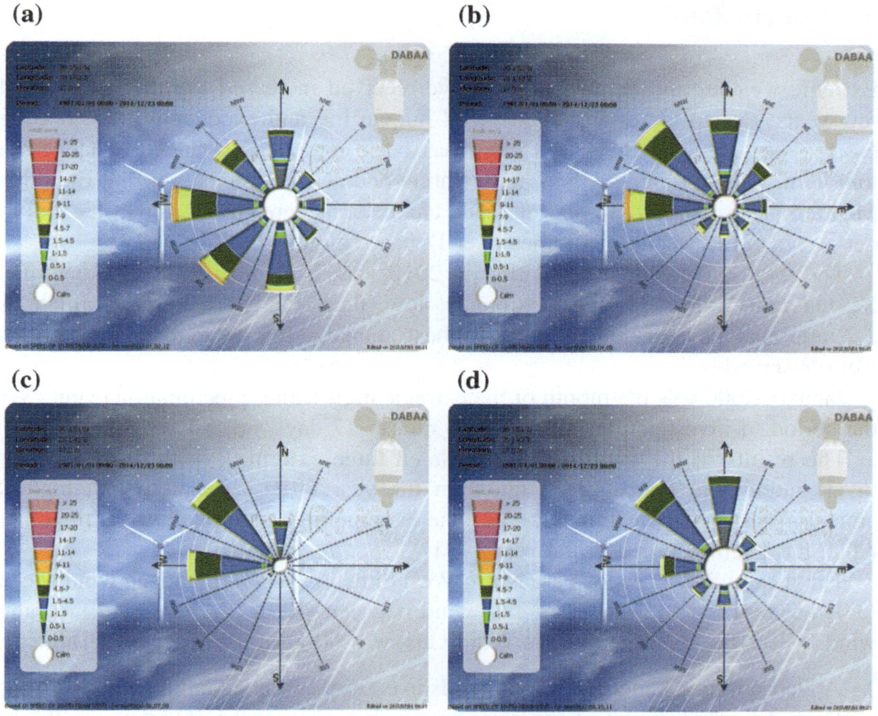

Fig. 7 Seasonally mean wind directions and speed over Dabaa region (1981–2014). **a** Winter, **b** Spring, **c** Summer, **d** Autumn

Table 1 Extreme meteorological events at Dabaa region from 1947 to 2015

Climactic event	Value	Date
Highest temperature	47.1 °C	30/05/1961
Lowest temperature	0 °C	15/01/1948
Maximum daily amount rainfall	57 mm	13/12/1977
Highest sea level air pressure	1033.3 hPa	05/01/1992
Lowest sea level air pressure	994.6 hPa	18/12/1962
Maximum surface wind gust	60 kt/235°	20/03/1967

4.4 Some Facts on Dabaa Region Climate Extremes

Extreme meteorological events may be good markers of climatic change or variability, so it is important to keep good records of such extremes. A Egyptian collection of such events has been collected by Egyptian Meteorological Authority (EMA). Data has been collected on all types of events such as temperature, pressure, rainfall, wind, … at Dabaa synoptic station from 1947 at now. Some of the information can be seen in Table 1.

5 Conclusions

Extreme weather and climate events have received increased attention in the last few years; due to the often large loss of human life and exponentially increasing costs associated with them. No work has been completed related to changes in long-term extreme precipitation, temperature and sea level events in the Egyptian Mediterranean region. The present study can be considered as an initial fair trial to get closer to the weather extremes over Dabaa region in Egypt.

The total amount of precipitation in 2010 was 25.8 mm in only 12 days, i.e. 2.2 mm/rainy day. This is extremely low than the normal rate which is 3.2 mm/rainy day. In this year, the highest annual mean air temperature periods were observed.

January 2000 was the month of the extreme high winter precipitation event over the period of investigation having 102.8 mm in 17 days, i.e. 6.1 mm/rainy day.

The results of the present study declared three extreme events in the area of investigation; the scarce precipitation in the year 2010 associated with highest annual mean air temperature 21.7 °C. The extreme high precipitation, on the other hand, has been recorded in the year 1991.

The mean wind directions are North West direction (NW), the mean wind speed (3.0–4.5 m/s) is 50%, the mean gust wind between (14.0–17.0 m/s) and the calm wind is 3%. The extreme Surface Wind Gust was 60 kt in SW direction (Table 1).

1 A Dabaa region pattern in recent trends in extreme weather and their impacts has been discerned. Some of the extreme weather phenomena associated with climate change is increasing in frequency and intensity within Dabaa. In some cases the impacts of these changes have had a significant effect on societies and economies throughout Egypt.
2 Extreme precipitation, often associated with floods and damage to infrastructure and crops, appears to be increasing in severity and frequency.
3 Low-lying coastal zones are considered to be particularly vulnerable to climate change, especially through sea level rise, changes in wave climate and in storminess.
4 Insurance industry data clearly show that the number of loss-relevant weather extremes has increased in Mediterranean coast. There is increasing evidence that at least part of these increases is driven by global warming. Some of the hazard-driven increases in loss events may even have been moderated by human activities through loss prevention measures.
5 Human factors play a part in moderating the impacts of heat waves. Extreme heat has had a considerable impact on human health in Egypt with significant mortality, notably during the heat waves of 2010.
6 For many crops in Egypt, weather extremes are the major factor in climate-change impacts on production. An increased frequency of extreme weather events is likely to be unfavorable for crop production.

References

Alexander LV, Zhang X, Peterson TC, Caesar J, Gleason B, Klein Tank AMG, Haylock M, Collins D, Trewin B, Rahimzadeh F, Tagipour A, Rupa Kumar K, Revadekar J, Griffiths G, Vincent L, Stephenson DB, Burn J, Aguilar E, Brunet M, Taylor M, New M, Zhai P, Rusticucci M, Vazquez-Aguirre JL. 2006. Global observed changes in daily climate extremes of temperature and precipitation. *Journal of Geophysical Research* 111: D05109 DOI:10.1029/2005JD006290.

Huynen MMTE, Martens P, Schram D, Weijenberg MP, Kunst AE. 2001. The impact of heat waves and cold spells on mortality rates in the Dutch population. *Environmental Health Perspectives* 109: 463–470.

IPCC (Intergovernmental Panel on Climate Change). 2007. *Climate Change 2007: The Physical Science Basis. Contribution of Working Group I to the Fourth Assessment Report of the Intergovernmental Panel on Climate Change*, Solomon S, Qin D Manning M Chen Z, Marquis M, Averyt KB, Tignor M, Miller HL (eds.). Cambridge University Press: Cambridge, UK; New York, NY, USA.

Page CM, Nicholls N, Plummer N, Trewin B, Manton M, Alexander L, Chambers LE, Choi Y, Collins DA, Gosai A, DellaMarta P, Haylock MR, Inape K, Laurent V, Maitrepierre L, Makmur EEP, Nakamigawa H, Ouprasitwong N, McGree S, Pahalad J, Salinger MJ, Tibig L, Tran TD, Vediapan K, Zhai P. 2004. Data rescue in the southeast Asia and south Pacific region: challenges and opportunities. *Bulletin of the American Meteorological Society* 85: 1483–1489.

Peterson TC, Manton MJ. 2008. Monitoring changes in climate extremes: A tale of international collaboration. *Bulletin of the American Meteorological Society* 89: 1266–1271.

Zong Y, Chen X. 2000. The 1998 flood on the Yangtze, China. *Natural Hazards* 22: 165–184.

Climate Change Adaptation in Yam and Cassava Production, Cross River State, Nigeria: The Role of Higher Educational Institutions

Hilda Chia Eta and Austine Oko Angba

Abstract This paper has as aim to discuss the role of higher educational institutions in climate change adaptation in the production of yam and cassava in Cross River State, Nigeria with emphasis on the type and level of skills acquired by lecturers of the Faculty of Agriculture, Forestry and Wildlife Resources Management, the roles played and challenges encountered. The paper is based on an empirical study. The University of Calabar was purposively selected to represent higher educational institutions because there is no root and tuber research institute in Cross River State. However, the study can be extended to cover the entire south-south geopolitical zone of Nigeria to incorporate the views of the root and tuber research institute in the zone. Results of the study revealed that key technical skills directly linked to climate change adaptation e.g. ability to develop technologies for climate change adaptation; ability to design frameworks for monitoring climate change adaptation projects were acquired by only 40 and 46.7% of the respondents respectively. Regarding the level of skills acquired, results show that 13.3 and 40% of respondents had low and moderate skill levels respectively. The paper reveals that the prominent roles played by respondents were: teaching multidisciplinary professionals and students on climate change issues (4.6); providing opportunities for staff skill development and training on climate change issues (4.4). This paper will provide useful information to all stakeholders in sub Saharan Africa in particular that are concerned with climate change adaptation in the root and tuber crop production sub-sector.

Keywords Climate change adaptation · Educational institutions

---

H.C. Eta (✉) · A.O. Angba
Department of Agricultural Economics and Extension, Faculty of Agriculture, Forestry and Wildlife Resources Management, University of Calabar, PMB 1115 Calabar, Cross River, Nigeria
e-mail: hildajens@yahoo.com

A.O. Angba
e-mail: drangba@yahoo.com

W. Leal Filho (ed.), *Climate Change Research at Universities*,
DOI 10.1007/978-3-319-58214-6_17

1 Introduction

Increasing the productivity of food crops such as yam, cassava and maize in Africa, especially in the face of climate change, cannot be achieved without the benefits of agricultural research and educational systems. This is so because these systems play a central role in shaping innovation processes for climate change adaptation. In Cross River State, yam and cassava in particular make up the major dietary composition of the population as they constitute the basis of carbohydrate energy for both the rural and urban masses. They are used as substitute for wheat flour in confectioneries (e.g. Cassava flour for biscuits, bread, cake etc.), some are used for compounding animal feed, organo-mineral fertilizer formulation, raw materials in agro-industries (e.g. alcohol), and raw materials in textile and pharmaceutical industries (Iyagba 2010). Onwubuya and Ajani (2012) posited that Nigeria is the world's largest producer of cassava (with estimated annual production of about 40 million metric tonnes); yam (35.01 million metric tonnes, representing 70% of world's production); sweet potato (1.56 metric tonnes); and cocoyam- taro (3.3 million metric tonnes). Even though the country is a leading producer of some of these crops, it should be mentioned that the crop production sub-sector at large is highly susceptible to changes in climate due to dependence on the weather and climate for its activities. For instance, in Nigeria it is estimated that approximately 90% of the rural dwellers who serve as the major producers of these crops rely heavily on rain-fed agriculture as a means of livelihood and food security (Ministry of Environment of the Federal Republic of Nigeria—MOE FRN 2003). In recent times, the weather and climatic conditions for production of these crops have become highly variable with the variability attributed to changes in climate. It is predicted that crop yields would be halved by 2020 and by the year 2100 net revenues from these crops would fall by 90% (Food and Agriculture Organization—FAO 2008). Hence, in the advent of climate change educational and training institutes have become particularly significant in that they assist in developing manpower and also serve as sources of knowledge and technology that will ensure adaptation to the adverse effects caused by changes in climate (Lynam 2012). According to Madukwe (2008), the educational system plays three primary roles in the agricultural sector, namely: training of manpower (agricultural researchers, extension personnel, policy makers, farmers, and other actors in the sector); carrying out research to generate agricultural technology; and transferring technology directly to farmers. FAO (2013) however purported that response to climate change requires additional roles from researchers and scientists that will enhance the development of balanced technical and institutional innovations; hence research and educational professionals have to reorient their traditional roles toward the wider dimensions of climate change adaptation. According to Chambwera and Anderson (2011), some of the practical ways by which agricultural research and educational systems can address climate change are via:

(i) linking indigenous knowledge with modern technologies;
(ii) sharing experiences and enhancing collaboration between farmers and research organizations, as well as developing joint innovation for adaptation and mitigation;
(iii) collaborating through training, and via human and technological capacity development; and
(iv) fostering multidisciplinary networks with research and knowledge institutions with the aim of spreading good practices around the globe.

Chambwera and Anderson (2011) also emphasized that agricultural research can play the role of identifying and testing climate change adaptation and mitigation measures, technologies and policies that target the poor and which aim at improving resilience, food systems, and sources of income. Furthermore, these authors added that research can mainstream agriculture in climate change policies at all levels and ensure that climate issues are taken into consideration in agricultural policies in ways that are valuable to the vulnerable. The American College and University Presidents' Climate Commitment [ACUPCC] (in Briefing 2011) on their part advocated that research and higher education institutions can perform the following roles as far as climate change adaptation is concerned:

(a) increase the number of core courses and electives offered in school curricular on climate change adaptation so as to increase specialized information on the issue;
(b) provide openings on campus for the society to experiment and/or role-model adaptation solutions;
(c) create, test and disseminate information to local communities on adaptation strategies, regional climate forecasts as well as work with these communities to implement solutions and explain the science; and
(d) identify adaptation strategies that also contribute to mitigation efforts.

Kotecha (2010) purported that in Nigeria, universities can actively participate in climate change adaptation by developing the conceptual, methodological and analytical tools necessary for understanding the diverse issues of climate change, integrating climate change and adaptation concerns in curriculum development, developing new or specialized programmes and establishing joint programmes as well as student exchange programmes and innovation teaching. Makungwa (2010) corroborated this view by opining that scientists should incorporate activities on climate change science, impact analysis and adaptation strategies in research priorities and programmes, and that research and educational institutions should be linked with expertise in climate modeling, climate forecasting and downscaling of forecasts. Achike and Okpara (2012) on their part noted that trainings should lay emphasis on the tools and methods for measuring vulnerability and adaptation as well as modeling climatic scenarios and socioeconomic conditions. These researchers expound further that trained students could be encouraged to participate actively in field work and that efficient and effective collaboration and partnership could be established in the planning and delivery of academic and research programmes in order for adaptation to be meaningful.

It is therefore against this backdrop that this paper has as aim to discuss the role of higher educational institutions in climate change adaptation in the production of yam and cassava in Cross River State, Nigeria, with emphasis on the type and level of skills acquired by lecturers of the Faculty of Agriculture, Forestry and Wildlife Resources Management for climate change adaptation as well as the roles played and challenges encountered.

2 Agricultural Research in Nigeria

According to Adekunle et al. (2012), even though investment in agricultural research and knowledge generation in developing countries has generated innovations/technologies with the potential to address some of the challenges brought by climate change, their impact has been disappointing largely due to the way research has been undertaken. Agricultural research system in Nigeria has been described as top-down, centralized, monolithic and with isolated structures (Agwu et al. 2008). These authors added that there is generally a weak or non-existent interrelationship among actors, hence absence of interactive learning mechanisms. They noted further that researches carried out in the numerous universities and research institutes (both national and international) are not well coordinated and are often poorly linked to the production sector; hence, they do not seem to target the needs of beneficiaries. Considering that climate change has only recently become a global challenge, most stakeholders especially those involved in agricultural research and educational systems have received little formal training pertaining to the subject matter, thus they are handicapped in the execution of the roles expected of them. These shortcomings have had a negative influence on respondents' capacities to contribute positively to climate change adaptation both at the local and global levels.

Studies (Urama and Ozor 2011) have revealed that a multidisciplinary, multidimensional and multi-institutional approach is required to tackle the adverse effects of climate change; hence this paper helps to explain the contribution of the Faculty of Agriculture, Forestry and Wildlife Resources Management of the University of Calabar, Nigeria, to the climate change challenge especially in the root and tuber crop sub-sector. The results of the study will serve as a baseline data on available capacities for climate change adaptation among lecturers who constitute one of the key actors in yam and cassava production in the study area. It will help development practitioners, and scientists to design and implement interventions for strengthening weaknesses identified thereby enhancing these actors' efforts to addressing the adverse effects of climate change. Finally, the findings of the study will form a good reference material for scholars and researchers interested in climate change adaptation especially at the higher educational level and in the Nigerian context in particular.

3 Methodology

This paper is based on the review of related literature and empirical study. The study was carried out in Cross River State (CRS), located in the South-South geopolitical zone of Nigeria. The state has eighteen local government areas (LGAs) divided into three agricultural zones as follows:

1. Ogoja agricultural zone comprising of Ogoja, Yala, Bekwarra, Obudu and Obanliku, LGAs;
2. Ikom agricultural zone comprising of Ikom, Abi, Yakurr, Obubra, Etung, and Boki LGAs;
3. Calabar agricultural zone comprising of Calabar Municipal, Calabar South, Bakassi, Akampa, Odukpani, Akpabuyo, and Biase, LGAs.

The population of the study comprised of all staff of agricultural research and educational institutions in Cross River State. These are: staff of public sector agricultural research institutes, staff of university-based research institutes, staff of private research institutes, staff of research departments of agricultural companies, staff of institutions that provide post-secondary agricultural education and vocational training in agriculture. A multi-stage sampling technique was employed to select the respondents for the study. In the first stage, purposive sampling technique was used to select the University of Calabar with her Faculty of Agriculture, Forestry and Wildlife Resources Management to represent the research and higher educational institutions in the state. This is because the institution has most departments that make up the Faculty of Agriculture, Forestry and Wildlife Resources Management and also because there is no yam and cassava research institute in Cross River State. In the second stage, simple random sampling technique was used to select thirty (30) out of sixty one (61) academic staff of the rank of senior lecturer and above, across the departments of the faculty, giving a sample size of 30. Primary data for the study was obtained by the use of structured questionnaires that were validated by some staff of the Department of Agricultural Economics and Extension, University of Calabar and administered to respondents. Hence, a total of 30 questionnaires were administered and all were returned for analysis. Even though this study is limited by the fact that the views of research institutes dealing with root and tuber crops were not captured, the study can be extended to cover the entire south-south geopolitical zone of Nigeria, in which case more tertiary and research institutions will be involved, hence ensuring that their views are also taken into consideration. The section that follows presents the analysis of the questionnaires, results and implications.

4 Analysis and Interpretation of Results and Implications

4.1 Socio-Economic Characteristics of Respondents

(a) Gender

Table 1 shows that majority (86.7%,) of the lecturers interviewed were males. This finding suggests that few senior female staff constitute the workforce of this institution. The finding also upholds the view of Maguire (2012) who noted that women are underrepresented at all levels of Agricultural Education and Training institutions, from postsecondary to tertiary education.

(b) Age

Table 1 shows that, majority (56.7%) of the lecturers of the faculty of Agriculture were between 41 and 50 years old, 30% were between 51 and 60 years old, while 6.7% were above 60 years old. Considering that majority of the staff fell within the age range of 41 to 50 years, it can be inferred that the workforce is relatively young. Hence, they can engage in rigorous research activities that would contribute to the alleviation of the adverse effects of climate change in the study area.

(c) Marital status

All the lecturers (30) interviewed indicated that they were married (Table 1). This has the implication that respondents will be willing to seek solutions to the challenges posed by climate change so as to ensure agricultural sustainability for the sake of posterity.

(d) Highest educational qualification

Majority (96.7%) of lecturers interviewed had PhD degrees while only 1 (3.3%) lecturer had an M.Sc. degree. This finding suggests that the agricultural educational system in the study area is capable of carrying out researches that will generate innovations for climate change adaptation as well as build the innovation capacities of stakeholders involved in yam and cassava production.

(e) Working experience

Table 1 shows that the majority (86.6%) of lecturers had worked for the university for between 11 and 30 years (i.e. 73.3% had worked for between 11 and 20 years, 13.3% between 21 and 30 years). Only 13.3% of the respondents had worked for between 1 and 10 years. This result shows that most of the respondents have worked for the institution for a considerably period of time, implying that they have acquired a wealth of experience in their various disciplines. This experience is valuable for performing important roles as far as climate change adaptation is concerned.

Table 1 Distribution of lecturers by socio-economic variables

Socio-economic variable	Lecturers: n = 30	Percentage
	Frequency	
Gender		
Male	26	86.7
Female	4	13.3
Age (years)		
20–30	0	0
31–40	2	6. 7
41–50	17	56.7
51–60	9	30.0
61–70	2	6.7
Marital status		
Married	30	100
Widow	0	0
Divorced	0	0
Single	0	0
Highest educational qualification		
No Formal Education	0	0
First school leaving certificate	0	0
Ordinary level certificate	0	0
OND/NCE	0	0
BSc/HND	0	0
MSc	1	3. 3
PhD	29	96.7
Others	0	0
Working experience (years)		
1–10	4	13. 3
11–20	22	73.3
21–30	4	13.3
Above 31	0	0
Membership into organization/group		
Yes	30	100
No	0	0
Type of membership organization		
Religious group	28	93.3
Thrift/Esusu	8	26.7
Cooperative society	17	56.7
Social club	11	36.7
Others	0	0

Source Field Survey (2014)

(f) Membership into organization/group

Results on Table 1 show that all lecturers belonged to at least one organization/group. Membership into organizations or groups that engage in educational and research activities in particular may facilitate information sharing on climate change issues.

(g) Type of membership organization

Majority (93.3%) of lecturers belonged to religious groups, 56.7% were members of cooperative societies, and 36.7% were members of social clubs while 26.7% belonged to thrift/esusu groups. Membership of respondents into these organizations or groups may provide the forum for influencing the implementation of development projects especially those related to climate change adaptation in their respective communities.

4.2 Types of Climate Change Adaptation Skills Acquired by Lecturers

In order to assess the type of climate change adaptation skill (i.e. technical, managerial or organizational) acquired by the respondents, a list of relevant climate change adaptation skill options (as obtained from literature) was presented to respondents to indicate ‘Yes’ if they had been acquired and ‘No’ if they had not.

Table 2 shows that the lecturers had acquired technical competencies such as: ability to teach/educate students on climate change concerns (90%); ability to disseminate research findings on climate change stakeholders through journals, newsletters (90%); ability to supervise students’ field research on climate change issues (90%); and ability to conceptualize research ideas and apply up-to-date research methods and design techniques for climate change adaptation projects (70%). Only 46.7% of the respondents were able to design frameworks for monitoring climate change adaptation projects while 40% were able to develop technologies for climate change adaptation. With respect to the managerial/ organizational skills, 73.3% were able to build partnerships and networks with other stakeholders to enhance action on climate change adaptation issues, 63.3% could monitor and evaluate climate change adaptation projects and another 63.3% had the ability to lobby and advocate for policies that favor climate change adaptation. Only 46.7% could mobilize financial resources for research activities on climate change adaptation activities.

It can therefore be inferred from these results that most of the respondents had acquired technical skills that were directly related to the traditional roles of the university, which are to: train manpower (agricultural researchers, extension personnel, policy makers, farmers, and other actors); carry out research to generate agricultural technology; and transfer technology directly to farmers (Madukwe

Table 2 Distribution of lecturers per climate change adaptation skill acquired

Skill acquired	Freq.	%
Technical		
Ability to teach/educate students on climate change concerns	27	90
Ability to develop technologies for climate change adaptation	12	40
Ability to conceptualize research ideas and apply up-to-date research methods and design techniques for climate change adaptation projects	21	70
Ability to disseminate research findings on climate change stakeholders through journals, newsletters.	27	90
Ability to supervise students' field research on climate change issues	27	90
Ability to conduct impact analysis of climate change effects on stakeholders	19	63.3
Ability to design frameworks for monitoring climate change adaptation projects	14	46.7
Ability to design frameworks for evaluating climate change adaptation projects	17	56.7
Managerial/organizational		
Ability to monitor and evaluate climate change adaptation projects.	19	63.3
Ability to mobilize financial resources for research activities on climate change adaptation activities	14	46.7
Ability to build partnerships and networks with other stakeholders to enhance action on climate change adaptation issues	22	73.3
Ability to lobby and advocate for policies that favour climate change adaptation	19	63.3
Total No. of respondents	30	

Source Field Survey (2014)
Respondents mentioned the acquisition of >1 skill

2008). Key technical skills that are directly linked to climate change adaptation e.g. ability to develop technologies for climate change adaptation; ability to design frameworks for monitoring climate change adaptation projects etc. were not acquired by a significant proportion of the lecturers. The implication is that few climate change technologies would be developed by the university researchers for dissemination to farmers and other stakeholders, and that the adoption/ implementation of developed technologies would not be effectively monitored. This may be the reason for failure to implement climate change adaptation technologies by yam and cassava farmers. According to FAO (2008), research provides the backbone for climate change adaptation and mitigation methodologies and needs to be linked with social science research on how to introduce new methodologies, crop varieties, etc. to communities, to ensure that the innovations are taken up, and are benefitting the different members of vulnerable communities. This notion is supported by Achike and Okpara (2012) who advocated that incorporating climate change science and adaptation strategies in Nigerian agricultural research priorities and programmes is essential in developing new skills as well as generating new climate knowledge that can be transferred to farmers and other stakeholders.

Table 3 Distribution of lecturers based on level of acquisition of each climate change adaptation skill

Skill acquisition	Frequency	Percent
No skill (0)	0	0
Low skill (1–4)	4	13.3
Moderate skill (5–8)	12	40
High skill (9–12)	14	46.7
Total	30	100

Source Field Survey (2014)

4.3 Level of Climate Change Adaptation Skills Acquired by Lecturers

To ascertain the level of skill acquisition, respondents were categorized into four groups based on the number of skills acquired. Those who had acquired zero skill were of the category No skill. Similarly, those who had acquired from 1 to 4, 5 to 8 and 9 to 12 skills fell under the category of low skill, moderate skill and high skill respectively.

Table 3 indicates that 46.7% of the lecturers had a high level of skills for climate change adaptation, 40% had a moderate level of skills while 13.3% had a low level of skills. None of the lecturers lacked the skill to enable them play a vital role in climate change adaptation. Even though the bulk of the lecturers had acquired a high level of skills, the result shows that the majority (53.3%) of the respondents had skills ranging between low and moderate. This can be attributed to a low proportion of lecturers who have had specialized knowledge or expertise in climate change issues. This finding confirms the assertion of Achike and Okpara (2012) that most Nigerian teaching staff were trained before climate change became a recognized problem, hence lack of expertise in the subject matter.

4.4 Roles of Lecturers

In order to identify the roles played by respondents with respect to climate change adaptation in the production of yam and cassava in the study area, a set of roles that respondents were expected to play as obtained from literature were presented. Respondents were asked to rate these roles using a 5 point Likert scale. The response categories and the corresponding weighted values of the scale were as follows: Strongly agree = 5, agree = 4, Undecided = 3, disagree = 2, strongly disagree = 1. Roles with mean score values above or equal to 3 represented the principal roles played by the respondents while those with mean score values less than 3 were not regarded as key roles.

Table 4 Distribution of lecturers on perceived roles

Perceived role	Total score	Mean score	SD	Rank
Teaching multidisciplinary professionals (e.g. researchers, agricultural extension staff, etc.) and students on climate change issues	137	4.6*	0.24	1st
Conducting research to develop new agricultural technologies to facilitate adaptation to adverse effects of climate change	125	4.2*	0.23	3rd
Providing facilities and opportunities for experimenting adaptation solutions to climate change	108	3.6*	0.21	8th
Role-modelling adaptation solutions to climate change	101	3.4*	0.20	9th
Disseminating knowledge through publications about climate change to local communities	125	4.2*	0.23	3rd
Working with local communities to identify opportunities for the institution to provide education, research, and pilot projects on adaptation	118	3.9*	0.22	5th
Collaborating and/or partnering with other stakeholders to communicate relevant knowledge or anticipated local/regional climate changes	113	3.8*	0.22	6th
Providing opportunities through seminars, conferences etc. for staff skill development and training on climate change issues	132	4.4*	0.23	2nd
Providing incentives for faculty and graduate students to develop expertise on projects relating to climate change issues	115	3.8*	0.21	6th
Grand mean		3.9		

Source Field Survey (2014)
*Critical roles

Table 4 shows the first four major roles perceived to be performed by the respondents or their institutions together with their corresponding mean score values. These are: teaching multidisciplinary professionals (e.g. researchers, agricultural extension staff, students, etc.) on climate change issues (4.6); providing opportunities through seminars, conferences etc. for staff skill development and training on climate change issues (4.4); conducting research to develop new agricultural technologies to facilitate adaptation to adverse effects of climate change (4.2); and disseminating knowledge through publications about climate change to local communities (4.2). Contrarily, a comparison of these mean score values with the grand mean score value (3.9) shows that respondents perceived roles such as: collaborating and/or partnering with other stakeholders to communicate relevant knowledge or anticipated local/regional climate changes (3.8); providing incentives for faculty and graduate students to develop expertise on projects relating to climate change issues (3.8); providing facilities and opportunities for experimenting adaptation solutions to climate change (3.6); and role-modelling adaptation solutions to climate change (3.4) as the least roles played.

Generally, the finding shows that the main roles performed by the university were those closely related to the traditional roles of academic teaching, research, and knowledge dissemination, while the least performed roles were those that required an in-depth knowledge and understanding of the climate change subject. The implication is therefore that students' innovation capacities for climate change adaptation may not be sufficiently enhanced. According to FAO (2013), engaging in activities that provide an in-depth knowledge on climate change such as modelling of future climate impacts on complex food security systems is imperative for addressing local impacts in agro-ecological zones at the local and national levels. It will ensure the creation of detailed national and international knowledge and provide data bases of climate change impacts on food security.

4.5 Constraints to Effective Performance of Climate Change Adaptation Roles by Lecturers

To assess factors constraining respondents from effectively performing their roles, a list of constraints from literature that could hinder effective contribution to climate change adaptation was presented to respondents. They were asked to rate them using a five point Likert scale with response categories: Strongly agree (5), Agree (4), Undecided (3), Disagree (2) and Strongly disagree (1) was used for this purpose. Factors with mean score values above or equal to 3 implied critical or major constraints, while those with mean score values less than 3 implied minor constraints.

In Table 5, the results shows that lecturers perceived inadequate funds for research and training (4.93); inadequate funds for implementation of climate change technologies (4.73); inadequate infrastructure/equipment (4.63); inadequate funds for dissemination of adoptable technologies (4.57); and inadequate monitoring and evaluation of disseminated technologies (4.27) as major constraints. These constraints can be attributed to poor investment in agricultural research and educational systems. According to Enete and Amusa (2010), Nigerian government's funding of agricultural research is well below the average for Africa as a whole (0.85% of GDP). Lack or inadequate investment in agricultural research and educational systems means that there will be a limited or absence of adoptable scientific climate change adaptation technologies for dissemination to yam farmers and other stakeholders in the value chain. It also implies that the few technologies that have been developed and disseminated to farmers and other stakeholders are not effectively monitored and evaluated.

Table 5 Distribution of staff of SMANR, lecturers and staff of CRADP by constraints to innovation capacity development

Constraint	Lecturers: n = 30		
	Mean score	SD	Rank
Inadequate funds for research and training	4.93*	0.09	1st
Inadequate funds for implementation of climate change technologies	4.73*	0.09	2nd
Inadequate funds for dissemination of adoptable technologies	4.57*	0.09	4th
Inadequate infrastructure/equipment	4.63*	0.09	3rd
Inadequate staff strength and capacity	4.13*	0.08	7th
Insufficient technical expertise for research	3.9*	0.08	10th
Lack of access to recent relevant publications and exposure	3.53*	0.08	18th
Weak coordination and cooperation with stakeholders on climate change issues	3.83*	0.08	11th
Inconsistent or insufficient government policy on climate change	4.17*	0.08	6th
Inadequate monitoring and evaluation of disseminated technologies	4.27*	0.09	5th
Lack of commitment to research on climate change issues by government	3.97*	0.08	9th
Secretiveness among colleagues	2.87	0.07	23rd
Inadequate climate change technologies for dissemination	3.63*	0.08	15th
Poor access to knowledge and information on new climate change adaptation technologies/innovations	3.77*	0.08	12th
Inaccurate information on weather forecasts	3.43*	0.08	20th
Poor staff remuneration	3.73*	0.08	13th
Influence of donor agencies	3.13*	0.07	22nd
Weak legal framework	3.63*	0.08	15th
Brain drain and high staff turnover	3.63*	0.08	15th
Administrative bottlenecks/bureaucracy	4.07*	0.08	8th
Poor research integrity	3.53*	0.08	18th
Poor accountability	3.67*	0.08	14th
Mistrust of other organizations	3.27*	0.08	21st
Lack of self-confidence	2.47	0.06	24th
Grand mean	3.42		

Source Field Survey (2014)
*Critical constraints

5 Conclusion

Increasing the productivity of food crops such as yam, cassava and maize in Africa, especially in the face of climate change, cannot be achieved without the benefits of agricultural research and educational systems. These systems help to facilitate innovation processes innovations that can aid climate change adaptation. However, in Nigeria as a whole and in Cross River State in particular lecturers in the Faculty of Agriculture, Forestry and Wildlife Resources Management do not have adequate skills to enable them contribute meaningfully to the climate change challenge. In order for these actors to contribute effectively to climate change adaptation especially in the root and tuber crop sub-sector, technical skills such as ability to develop technologies for climate change adaptation; ability to design frameworks for monitoring climate change adaptation projects etc. are a necessity. This is because farmers rely on these researchers to develop, disseminate and monitor climate change adaptation innovations and projects, so as to boost crop yields. Furthermore, these researchers are ill-equipped to perform roles that require an in-depth knowledge and understanding of the climate change subject. This has been attributed to inadequate funds for implementation of climate change projects, inadequate infrastructure and limited collaboration with other stakeholders on climate change issues. To ensure that these respondents contribute meaningfully to climate change adaptation specifically in the root and tuber crop sub-sector, future research should be carried out to assess the climate change adaptation capabilities of all stakeholders in the sub-sector with the view to identifying weaknesses so as to design and implement interventions to strengthen them.

References

Achike A I, Okpara U T (2012) The Role of University Education in Adapting Nigerian Agriculture to Climate Change. In Critical Issues in Agricultural Adaptation to Climate Change in Nigeria. Anselm, A & Uguru, M. I. (Eds.). Chenglo Limited. Pp. 13–28

Adekunle A A, Ellis-Jones J, Ajibefun I, Nyikal R A, Bangali S, Fatunbi O, Ange A (2012) Agricultural innovation in sub-Saharan Africa: Experiences from multiple-stakeholder approaches. Forum for Agricultural Research in Africa (FARA), Accra, Ghana. Pp. 4–7

Agwu A E, Madukwe M C, Dimelu M U (2008) Innovation system approach to agricultural development: Policy implications for agricultural extension delivery in Nigeria. African Journal of Biotechnology. Vol 7 No 11, pp. 1604–1611

Briefing (2011) Integrating climate change into agricultural research for development in Africa. The International Institute for Environment and Development (IIED) briefing Pp. 1–4

Chambwera M, Anderson S (2011). Integrating climate change into agricultural research for development in Africa. The International Institute for Environment and Development (IIED) briefing Pp. 1–4

Enete A A, Amusa T A (2010) Challenges of Agricultural Adaptation to Climate Change in Nigeria: A Synthesis from the Literature. Field Actions Science Reports. Journal of Field Actions. Vol. 4, Pp. 1–12

Food and Agriculture Organization [FAO] (2008). Climate change adaptation and mitigation in the food and agriculture sector. Technical background document from the expert consultation held on 5th to 7th March 2008. FAO, Rome. Pp. 2–10

Food and Agriculture Organization [FAO] (2013). Facing the challenges of climate change and Food security: The role of Research, Extension and Communication for Development. Occasional papers on Innovation in Family Farming. FAO, Rome. Pp. 7–20

Iyagba A G (2010) A Review on Root and Tuber Crop Production and their Weed Management among Small Scale Farmers in Nigeria. ARPN Journal of Agricultural and Biological Science. Vol. 5 No 4 Pp. 52–55

Kotecha P (2010) Climate Change, Adaptation and Higher Education: Securing our Future. Southern African Regional Universities Association (SARUA). Leadership dialogue series. SARUA. Wit. South Africa. Vol 2 No. 4

Lynam J (2012) Agricultural Research within an Agricultural Innovation System. In Agricultural Innovation Systems: An Investment Sourcebook. The World Bank. Washington, D.C. Pp. 261–269

Madukwe M C (2008) Practice Without Policy: The Nigerian Agricultural Extension Service. An Inaugural Lecture of the University of Nigeria, Nsukka, Delivered on April 29th, 2008

Maguire C J (2012) Agricultural Education and Training to Support Agricultural Innovation Systems. In Agricultural Innovation Systems: An Investment Sourcebook. The World Bank. Washington, D.C. Pp 107–121

Makungwa S (2010) Adaptation, agriculture and Food Security. Climate Change, Adaptation and Higher Education: Securing our Future. Southern African Regional Universities Association (SARUA) Leadership dialogue series SARUA. Wit. South Africa Vol. 2. No. 4. Pp. 22–46

Ministry of Environment of the Federal Republic of Nigeria [MOE-FRN] (2003). Nigeria's First National Communication under the United Nations Framework Convention on climate change. Abuja, Nigeria. Pp. 15–56

Onwubuya E A, Ajani E N (2012) Strategies for improving production and processing of cocoyam among women farmers in Anambra State, Nigeria. Universal Journal of Education and General Studies Vol. 1. No. 6. Pp. 169–173

Urama K, Ozor N (2011). Agricultural Innovations for Climate Change Adaptation and Food Security in Western and Central Africa. Journal of Tropical Agriculture, Food, Environment and Extension. Vol.10 No 1. Pp. 1–15

Change and Analysis of Extreme Rainfall Indices During 1960–2010 and 2011–2100 in Abidjan District (Côte d'Ivoire)

Jean Homian Danumah, Samuel Nii Odai, Mahaman Bachir Saley, Lucette You Akpa, Joerg Szarzynski and Fernand Koffi Kouame

Abstract Flood risk occurrence is very often related to heavy precipitation. The availability of analysis of weather data is a potential source for long term flood risk prediction and management. The aim of this paper was to determine and analyse trends of observed and future rainfall indices from 1961 to 2010 and 2011 to 2100 using rclimdex model in Abidjan District. This work was based on the integration of daily weather data within rclimdex model throughout quality control test, homogeneity test and indices calculation of ten (10) rainfall indices. The results showed an overall decrease trend of the rainfall indices namely through a negative trend in the annual total rainfall, maximum number of consecutive wet days, and number of extremely wet days during the period from 1961 to 2010. Exception was made from 1995 to 2010 where the same indices showed a positive trend. However, the results have showed also an increase trend of consecutive wet days (*CWD*), Simple daily intensity index (*SDII*) and Number of heavy precipitation days (*R10*) indices from 2011 to 2100. Thus these findings explain the nowadays flood occurrence and indicate that rainfall extreme under flood risk events will continue in the future. Therefore it call decision makers for preparedness and mitigation strategies in Abidjan District.

Keywords Climate events · Rainfall indices · Rclimdex · Abidjan · Côte d'Ivoire

S.N. Odai · J. Szarzynski
WASCAL Program (West African Science Service Center on Climate Change and Adapted Land Use), GRP on Climate Change and Land Use, Kwame Nkrumah University of Science and Technology, Kumasi, Ghana

J.H. Danumah (✉) · M.B. Saley · L.Y. Akpa · F.K. Kouame
Centre Universitaire de Recherche et d'Application en Télédétection (CURAT), Laboratoire Associé Francophone (LAF n° 401), 22 BP 801, Abidjan 22, Côte d'Ivoire
e-mail: danumahjean@yahoo.fr

W. Leal Filho (ed.), *Climate Change Research at Universities*,
DOI 10.1007/978-3-319-58214-6_18

1 Introduction

Climate change is considered to be the biggest challenge faced by mankind in the twenty first century (Osman et al. 2013). Natural disaster is a global issue and needs to be examined at global, regional and local scale. Climate changes may increase the frequency, magnitude and the seasonality of extreme events such as flood and may occur more frequently in the future (Huong and Pathirana 2013; Pederson et al. 2012). Population demand, green gas emission from industrialization and land use lead to severe climate variability such as increases of extreme events (flood and drought) occurrence in the world. In Africa, the situation is very likely to worsen as intergovernmental panel on climate change (IPCC) projected higher frequencies and intensities of floods and droughts (IPCC 2007) for the continent as a consequence of climate change. Disasters can catalyse structural and irreversible change by creating new conditions and relationships within environmental, socioeconomic and political structures, institutions and organizations (Birkmann 2010). Increase of water vapour in the atmosphere affects precipitation events and the risk of flooding and the changing climate is thus perceived through extreme events which tend to modify the magnitude of the predicted climate (Mouhamed et al. 2013). Flood risk occurrence is very often related to heavy precipitation, hence the need of knowledge on weather variability for most reliable flood management and mitigation; climate indices becomes a potential source for short term and long term extreme rainfall prediction.

Climate indices determination thus, has become important to analyse weather data trends in present conditions and in the future; to predict extreme rainfall under climate change. The World Meteorological Organization (WMO), CCI/CLIVAR Expert Team and ETCCDMI team for Climate Change Detection Monitoring Indices recommend 27 different climate indices. These indices have been developed to characterize the intensity, duration and frequency of extreme events via daily information, and/or defined relevant derived variables. The “rclimdex” package provides an appropriate template to calculate these indices efficiently.

rclimdex model is a statistical package developed by WMO CCI/CLIVAR Expert Team and ETCCDMI team. This model requires quality control and homogeneity tests of data before all indices calculation. These indices gives more comprehension in analysis of variations and trends in extreme climate events (drought and floods) which are more sensitive to climate change than are the mean values, hence have received much attention (IPCC 2007). The rclimdex model in the case of temperature and precipitation extremes change detection has been used in many studies (Alexander et al. 2006; Cinco et al. 2014; Haylock et al. 2006; Yan et al. 2014; Stephenson et al. 2014; Keggenhoff et al. 2014; Yu and li 2014; New et al. 2006; Revadekar et al. 2011; Babatolu et al. 2014; Gbode et al. 2015; Zhao et al. 2012). In West Africa, rclimdex has been used to detect extreme events mainly drought in sahelian zone (Mouhamed et al. 2013) added to other extreme indices calculation approaches (Sarr 2011; Ali 2011; L’hote et al. 2002).

However, very few studies have been undertaken in climate indices analysis in Africa because of several difficulties in accessing daily meteorological data (Lamptey 2009).

More recently in Côte d'Ivoire especially in Abidjan, we have registered increasingly important phenomena of floods, with its effects. Extreme rainfall is the main natural disaster which causes loss of many lives, destruction of infrastructures, and displacement of people during the rainy season.

However, the analysis trend of daily weather data for understanding in present condition and managing future extreme events in Côte d'Ivoire is practically non-existent. Weather data analysis for many years were based on determining break on the times series using annual and monthly rainfall and some statistical methods such as Pettit and Buishand test, application of Nicholson indices to bring out the wet and dry period in case of rainfall variability and shows general trend, inter-annual behaviour (Brou 2005; Savane et al. 2003; Goula et al. 2006; Hauhouot 2008; Kouassi et al. 2008).

The significant lack of research gap identified by this study is that, recent scientific work undertaken in the district of Abidjan concentrated on monthly and annual rainfall variability during past and current conditions (Savane et al. 2003). This is a piece-meal approach and does not provide a deep understanding of weather trend and link climate change to natural disaster within the district. Currently, no studies have yet been undertaken to analyse rainfall indices trends under present and future climatic conditions implications in flood occurrence using rclimdex model at Abidjan.

The aim of this paper was to determine and analyse trends of observed and future rainfall indices (10) from 1961 to 2010 and 2011 to 2100 using rclimdex model in Abidjan. For this purpose, assessment process of observation and future daily rainfall data are conducted following three steps: quality control test, Homogenization and climate indices in rclimdex model.

2 Methodology

2.1 Study Area

The district of Abidjan is located in the south of Côte d'Ivoire between latitudes 5° 10 and 5° 38 North and longitudes 3° 4 and 5° 21 West (Fig. 1). It consists of thirteen (13) municipalities since 2001, ten (10) municipalities in Abidjan and three (3) others town namely Bingerville, Songon and Anyama and covers an area of approximately 2119 km^2. The population is about 4,739,752 inhabitants in the metropolis, and 4,460,355 inhabitants in the main city (INS, 2013), which represents 20.3% of the national population as it is in 2013. In addition, the population grows constantly mainly as a result of high industrialization, urbanization and the later political crisis.

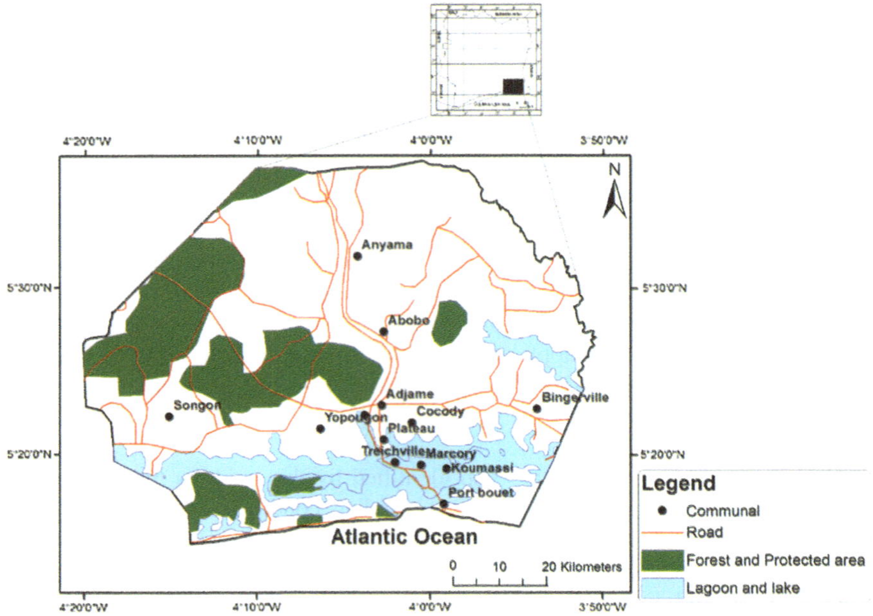

Fig. 1 Location of the study area (*Source* Author)

2.2 Climatic Variability

The study area has an equatorial climate transition (Attieen Climat), characterized by four seasons: two dry seasons (December to April and August to September) and two rainy seasons (May to July and October to November) within the annual cycle:

- Long dry season from December to April;
- Long rainy season from May to July;
- Short dry season from July to September;
- Small rainy season from October to November.

2.3 Precipitation

The annual average rainfall of Abidjan is around 1873 mm from 1960 to 2012 (Fig. 2).

The high annual rainfall recorded in Abidjan district during the period 1960–2012 ranged from 2800 mm in 1963 to 1020 mm in 1990 with an average of 1910 mm. Generally in the 1960s, the annual rainfall ranged between 2000 and 3000 mm. After 1987, there has been a drop in rainfall and this has oscillated

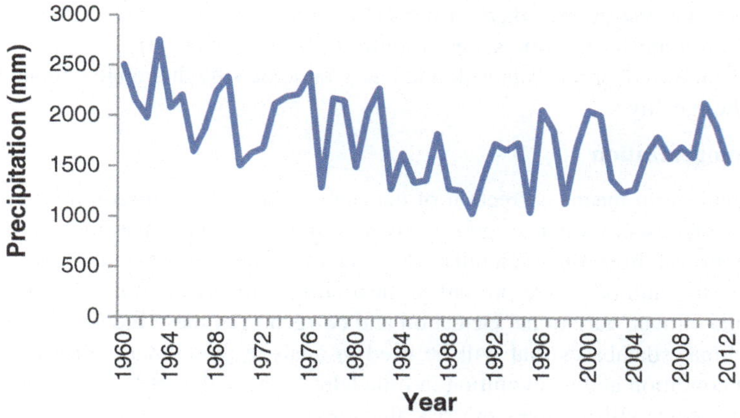

Fig. 2 Total annual rainfall variability of Abidjan from 1961 to 2012 (*Source* Author)

between 1500 and 2200 mm, a reduction of more than 500 mm compared to the 1960s.

2.4 *Methodology Approach*

2.4.1 Data and rclimdex Model Processing

The software "rclimdex" provides an appropriate template to calculate the ten (10) rainfall indices efficiently. The originality of rclimdex is to make quality control tests even on daily data that do not follow a normal statistical distribution. The calculation of climate indices requires first making quality control test and secondly homogenization test on the datasets used.

In this study 50 years of observed daily weather data (precipitation, maximum and minimum temperature) from 1960 to 2010 and generated daily rainfall data with LARS-WG model from 2011 to 2100 have been used in rclimdex model and considered for the different process to calculate the indices of climate extremes of Abidjan District.

- **Quality Control test**

The method of quality control is to help the user to identify registry errors that may exist on the daily data. The main steps for the quality control of the data in rclimdex is summarized in the following steps:

- replace missing data by −99.9 values which is code that rclimdex model required for missing data;
- have not more than 365–366 daily observations per year;

- Repeat on consecutive days in the same variable value (Tmin, Tmax or rain);
- sudden jump in the time series (rupture or average change);
- Detection of aberrant values defined as a value outside the range climate known in the locality.

- **Homogenization**

Homogenization means correction of data sets with artificial discontinuities due to changes in observation networks (moving station, change instrument measuring change in the immediate vicinity of a station, changing observer, etc.). These artificial discontinuities are present in most climate records and may interfere with the actual climate variations. Detection and correction of these breaks are needed to build climate databases that will be used to analyze the climate variability under present condition and its evolution in time later because the artificial discontinuities that can occur will involves the risk that future projections based on past observations are biased. The advantage of homogeneity test was to correct data series as homogeneous climate data series and systematically involved a diagnosis on average analysis of extremes because between that date and the end of the series, the series becomes daily reference series (QAS).

- **Climate indices**

This step is to determine, the frequency, amount and duration over a period of events extremes by calculating rainfall indices based on daily times series. The Expert Team on Climate Change Detection Monitoring Indices (ETCCDMI) have undertaken a set a regional analyses for understanding climate extremes events and trends. In this study, among 27 core indices computed on the daily basis by rclimdex to analyze extreme rainfall events, ten (10) rainfall indices were calculated and considered most relevant (Table 1).

3 Results

From daily weather data collection ten (10) rainfall indices were calculated based on observed data from 1960 to 2010 and future data from 2011 to 2100 through rclimdex model in this study.

3.1 Consecutive Wet Days (CWD)

Figure 3a, b shows the evolution of consecutive wet days (*CWD*) in Abidjan between the period 1960–2010 and 2011–2100. Observation of Fig. 3a indicates that there was no an important inter-annual variability of *CWD* from 1960 to 2010 based on the general trend represented by the straight line. The range on *CWD* is

Table 1 Daily extreme precipitation indices

Precipitations indices			
Identification	Indicators name	Definitions	Units
PRCPTOT	Annual total wet-day precipitation	Annual total PRCP in wet days (RR $\geq$ 1 mm)	mm
RX1day	Max 1-day precipitation amount	Monthly maximum 1-day precipitation	Mm
Rx5day	Max 5-day precipitation amount	Monthly maximum consecutive 5-day precipitation	Mm
SDII	Simple daily intensity index	Annual total precipitation divided by the number of wet days (defined as PRCP $\geq$ 1.0 mm) in the year	Mm/day
R10	Number of heavy precipitation days	Annual count of days when PRCP $\geq$ 10 mm	Days
R20	Number of very heavy precipitation days	Annual count of days when PRCP $\geq$ 20 mm	Days
Rnn	Number of days above nn mm days	Annual count of days when PRCP $\geq$ nn mm, nn is user defined threshold	Days
CWD	Consecutive wet days	Maximum number of consecutive days with RR $\geq$1 mm	Days
R95p	Very wet days	Annual total PRCP when RR > 95th percentile	mm
R99p	Extremely wet days	Annual total PRCP when RR > 99th percentile	mm

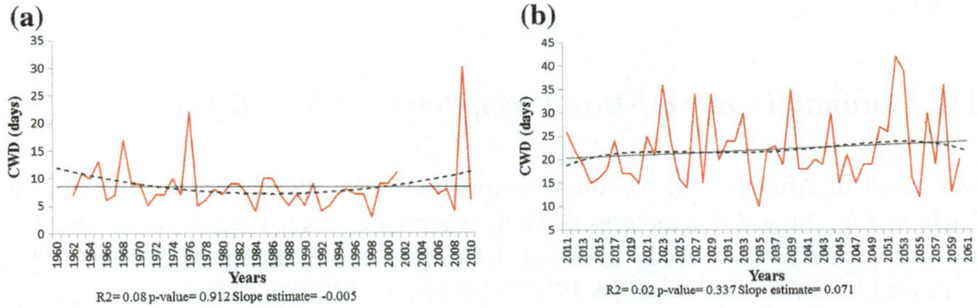

Fig. 3 Consecutive wet days (CWD). **a** 1960–2010. **b** 2011–2100)

comprised between 2 and 30 wet days. Except in 1968, 1975 and 2010 where there was a peak of 17, 23 and 30 of wet days respectively. This directly means important amount of rainfall during these years.

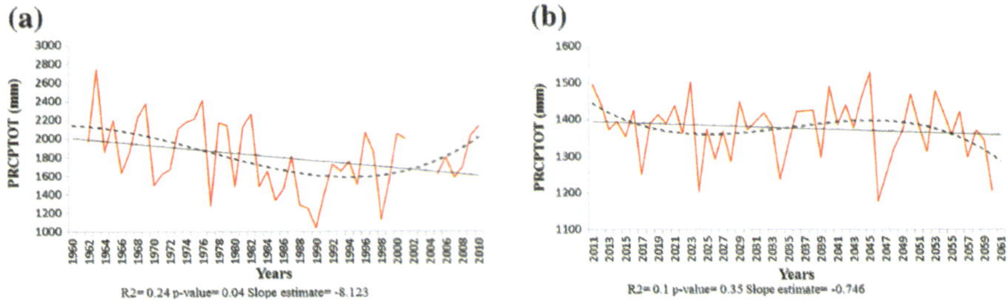

Fig. 4 Annual total wet-day precipitation (PRCPTOT). **a** 1960–2010. **b** 2011–2100)

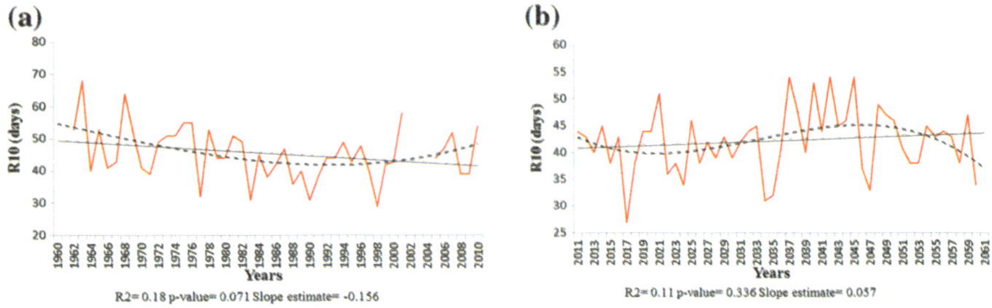

Fig. 5 Number of heavy precipitation days (R10). **a** 1960–2010. **b** 2011–2100

In Fig. 3b, *CWD* from 2011 to 2100 present a fluctuation between 10 and 40 wet days. This Figure shows an important variability of wet days under several observed peak (20–45 days) which is opposite to results in Fig. 4a. Average trend of this period indicate an increase of wet days which assume that rainfall intensity will be more persistent in future.

3.2 Annual Total Wet-Day Precipitation (PRCPTOT)

The result of Annual total wet-day precipitation (*PRCPTOT*) index from 1960 to 2010 in Fig. 4a was characterised by a general downward trend with a decline amount of *PRCPTOT* from 2900 to 1000 mm during the period 1960–1990. Followed by an upward trend of *PRCPTOT* (1000–2200 mm) from 1990 to 2010 (Fig. 5a). *PRCPTOT* result is very significant (p-value = 0.04 < 0.05) from 1960 to 2010. But in Fig. 4b representing the same index from 2011 to 2100, there was a difference compared to previous in Fig. 4a. The overall range of *PRCPTOT* (1000–1500 mm) was slightly low than *PRCPTOT* from 1960 to 2010. However, *PRCPTOT* fluctuation in 2011–2100 is more pronounced and reveals a short decrease of *PRCPTOT* from 2011 to 2030 followed by an increase of the same index from 2030 to 2045 and finally a decrease from 2045 to 2100 (Fig. 4b).

The overall trend of this period shows an insignificant downward trend of *PRCPTOT* while in Fig. 4a (1960–2010) there was an important decrease of *PRCPTOT*.

3.3 Number of Heavy Precipitation Days (*R10*)

The number of heavy precipitation days over 10 mm (*R10*) index in Abidjan District show opposite trends from 1960 to 2010 (Fig. 5a) and 2011 to 2100 (Fig. 5b). In Fig. 5a, general tendency trend of *R10* present a decrease trend from 1960 to 2010. First of all, a medium decline of *R10* from 1960 to 1990 between 60 and 30 days was observed. Then, secondly from 1995 to 2010, the curve shows a resumption of *R10* between 42 and 50 days. Inversely, *R10* index from 2011 to 2100 increase based on the general observe trend (Fig. 5b). Hence, *R10* index reached the most important peak from 2030 to 2045 with 52–54 days.

This index *R10* under present and future condition indicated that rainy days over 10 mm started increasing from 1995 to 2010 and will continue until 2100. This can be an explanation for the current heavy precipitation days.

3.4 Number of Very Heavy Precipitation Days (*R20*)

The number of very heavy precipitation days over 20 mm (*R20*) index in Abidjan District also shows opposite trends between 1960–2010 (Fig. 6a) and 2011–2100 (Fig. 6b). From 1961 to 2010, the number of very heavy precipitation days over 20 mm (*R20*) show a decrease trend (Fig. 6a). But, two (2) periods were distinguished: a decrease of *R20* from 1961 to 1995 between 43 and 18 days, followed by an increase of the same index between 18 and 25 days especially from 1995 to 2010. *R20* index result was very significant (p-value = 0.043 < 0.05) from 1960 to 2010. While in Fig. 6b, this *R20* index, average trend is constant means no decrease

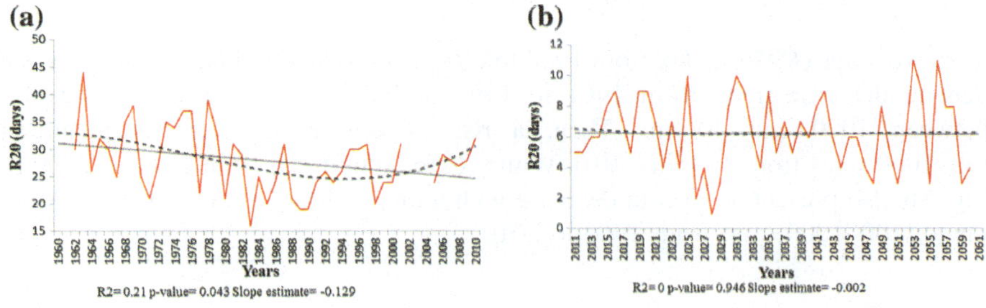

Fig. 6 Number of very heavy precipitation days (R20). **a** 1960–2010. **b** 2011–2100

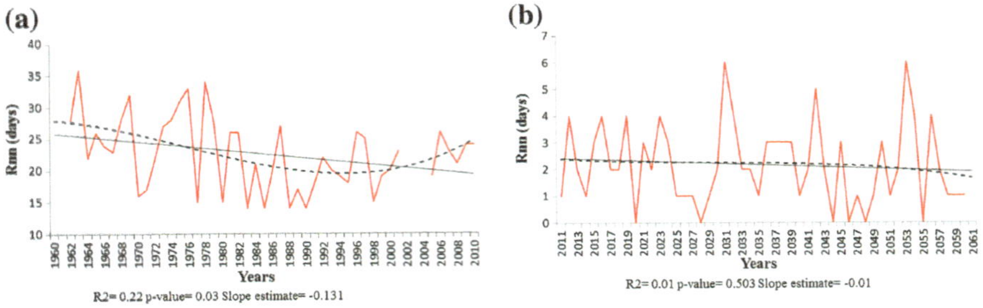

Fig. 7 Number of days above 25 mm (Rnn). **a** 1960–2010. **b** 2011–2100

or increase from 2011 to 2100. However, the amount of *R20* during this period will be too small (0–10 days) compared to baseline *R20* (10–45 days) and means that *R20* index will be minimal in future (2011–2100).

3.5 Number of Days above 25 mm (Rnn)

From 1961 to 2010, the trend of the number of days above index 25 mm (*Rnn*) generally declined in Fig. 7a. But as *R20* two big periods were characterised in *Rnn* index: a decrease number of days above 25 mm (*Rnn*) from 1961 to 1995 between 5 and 40 days, followed by a rise of *Rnn* between 15 and 25 days especially from 1995 to 2010. *Rnn* index like *PRCPTOT* and *R20* result is very significant (p-value = 0.03 < 0.05) from 1960 to 2010. This same index *Rnn* in Fig. 7b showed a slight decrease from 2011 to 2100 and its range will be too small (0–6 days) compared to baseline *Rnn* range (10–45 days). The significance of *Rnn* index results could support heavy precipitation observed nowadays based on *Rnn* increase from 1995 to 2010.

3.6 Very Wet Days (R95p)

Very wet days (*R95p*) index from 1960 to 2010 is presented in Fig. 8a. There was a general decrease trend from 1961 to 1995 of *R95p* with important fluctuation between 1000 and 40 mm. Then, a rise of this *R95p* index is identified (30–1000 mm) from 1995 to 2010. While *R95p* index trend from 2011 to 2100 in Fig. 8b also present an overall decrease with a range (400–50 mm) less than *R95p* range (100–1000 mm) from 1960 to 2010.

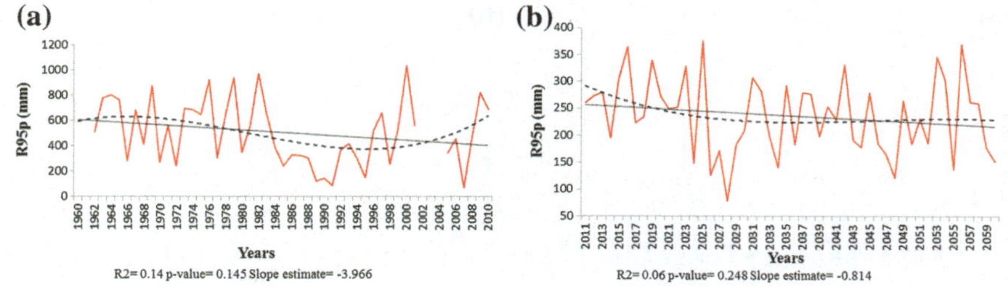

Fig. 8 Very wet days (R95p). **a** 1960–2010. **b** 2011–2100

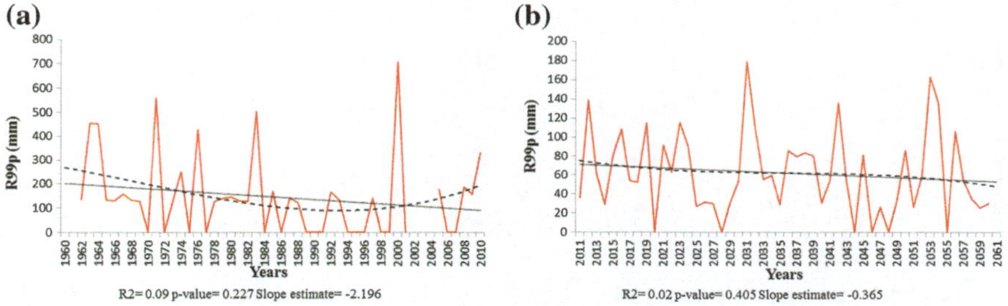

Fig. 9 Extremely wet days (R99p). **a** 1960–2010. **b** 2011–2100

3.7 *Extremely Wet Days (R99p)*

In Fig. 9a, the amount of *R95p* was different compared to extremely wet days (*R99p*) though there was some similarity from 1960 to 2010. Indeed, there was a negative trend of R99p from 1961 to 1995 with fluctuation between 0 and 560 mm, followed by a positive trend from 0 to 700 mm during the period 1995–2010. Regarding Fig. 9b, *R99p* index from 2011 to 2100 present a negative trend and was not so important (0–200 mm) compared to *R99p* (0–700 mm) from 1960 to 2010. Both Figures showed a negative trend of extremely wet days (*R99p*).

3.8 *Maximum 1-Day Precipitation Amount (RX1day)*

Maximum 1-day precipitation amount (*RX1day*) index result showed a general negative trend between 1960 and 2010 (Fig. 10a) and 2011–2100 (Fig. 10b). There was an insignificant decrease of Maximum 1-day precipitation amount (*RX1day*) from 1960 to 1995, following a short rise of *RX1day* from 1995 until 2010 (Fig. 10a). This slight general decrease trend from 1960 to 2010 fluctuated between 0 and 200 mm of water/decade with two exceptions made in 1972 and 1983 with the highest *RX1day* peak of 230 mm and 310 mm respectively.

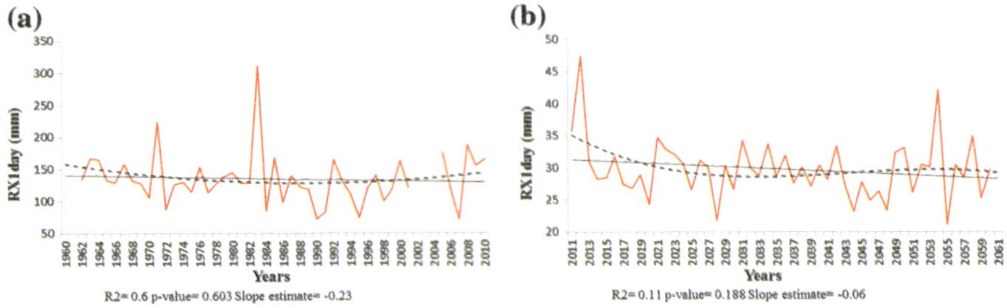

Fig. 10 Maximum 1-day precipitation amount (RX1day). **a** 1960–2010. **b** 2011–2100

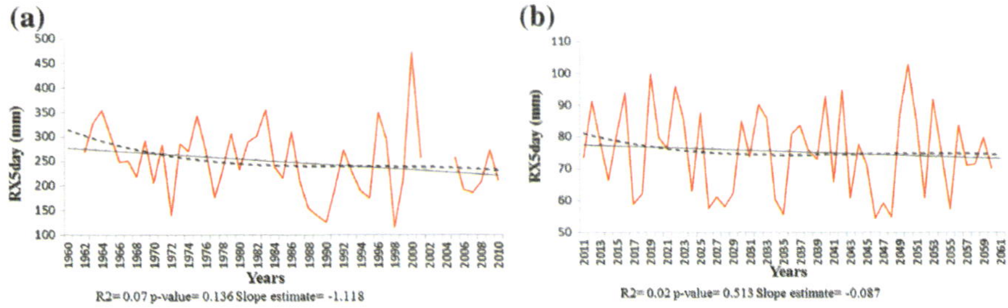

Fig. 11 Maximum 5-day precipitation amount (RX5day). **a** 1960–2010. **b** 2011–2100

This *RX1day* index result from 2011 to 2100 is shown in Fig. 10b. There was an overall negative trend with fluctuation between 0 and 0 mm of water/decade from 2011 to 2100. However, it was observed that during the period of 2011–2100, *RX1day* reached the highest peak in 2014 with 48 mm of water/decade which is an indicator of extreme rainfall occurrence.

3.9 Maximum 5-Day Precipitation Amount (RX5day)

Maximum 5-day precipitation amount (*RX5day*) index results showed a general negative trend between 1960 and 2010 (Fig. 11a) and 2011–2100 (Fig. 11b). The plotted result in Fig. 11a reflects changes of Maximum 5-day precipitation amount from 1960 to 2010. There was a downward trend of *RX5day* characterised by a reduction of *RX5day* from 350 to 100 mm of water/decade excepted in 2000 where it was noted the highest *RX5day* rate 460 mm.

RX5day index from 2011 to 2100 (Fig. 11b) has a similar decrease trend in Fig. 12a. However, difference was identified in fluctuation range which was between 50 and 150 mm of water/decade.

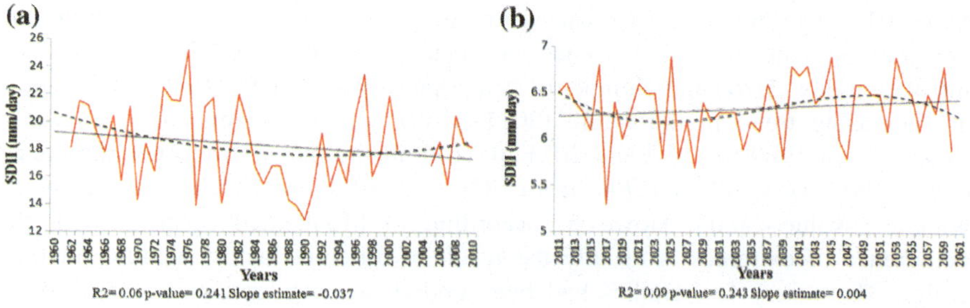

Fig. 12 Simple daily intensity index (SDII). **a** 1960–2010. **b** 2011–2100

3.10 Simple Daily Intensity Index (SDII)

Simple daily rainfall intensity (*SDII*) index result showed inverse general trend between 1960–2010 (Fig. 12a) and 2011–2100 (Fig. 12b). Simple daily rainfall intensity (SDII) index decreased from 1960 to 1990 with an exception of 1975 which recorded the highest intensity index 25 mm/day of that period and after that, there was an increase of *SDII* from 1990 to 2010 and the range was between 18 and 24 mm/day. Afterwards, *SDII* index present in Fig. 12b an increase based on general trend from 2011 to 2100. Both Fig. 12a, b of *SDII* index started by a decrease of intensity and later there was a rise of *SDII* from 1990 to 2010 (Fig. 12a) and 2030 to 2100 (Fig. 12b).

In all, ten rainfall indices (10) (PRCPTOT, CWD, R10, R20, Rnn, R95p, R99p, RX1day, RX-5day and SDII) were calculated and compared with observed (1960–2010) and future (2011–2100) data. These rainfall indices showed a general decreased trend of most rainfall indices in present and in future.

4 Discussion

Weather and extreme events analysis mainly floods were investigated through rainfall indices calculation of Abidjan in present and future condition using rclimdex model. The results based on the 10 rainfall indices calculated revealed variation between negative and positive trends from one index to another. Regarding the decreasing trend of these rainfall indices: Annual total wet-day precipitation (*PRCPTOT*), Number of heavy precipitation days (*R10*), Number of very heavy precipitation days (*R20*), Number of days above 25 mm (*Rnn*), Very wet days (*R95p*) and Extremely wet days (*R99p*) from 1960 to 2010 summarized a reduction of rainfall in Abidjan following observation made by L'Hote et al. 2002 and Ali 2011 which revealed a decrease in annual total rainfall in west Africa mainly in the sahelian zone. However, the same results shows positive trend for consecutive wet days (*CWD*) index from 1960 to 2010 which is in accordance with

Sarr 2011 who found an increasing trend of consecutive wet days from the late1980s in sahelian region. Hence, consecutive wet days (*CWD*), Simple daily intensity index (*SDII*) and Number of heavy precipitation days (*R10*) indices show an increasing trend in the future (2011–2100) with no significance (P-value > 0.05). From 1960 to 2010 and 2011 to 2100 all the indices were not significant except *PRCPTOT, R20* and *Rnn* from 1960 to 2100 trends which were significant with a P-value < 0.05. However, according to Mouhamed et al. (2013); the occurrence of extremes is usually the result of different factors at different time scales. Then, very few studies had been undertaken concerning extreme indices calculation in humid zone of West Africa to compare with results of this study. But availability of daily data from other stations within the study area could improve the results consistency by comparison between rainfall indices from various stations.

5 Conclusion

rclimdex model was used to calculate ten (10) rainfall indices to assess flood occurrence as an extreme event from 1960 to 2010 and 2011 to 2100 throughout quality control and homogenisation tests followed by indices calculation of the data. The ten (10) rainfall indices (*PRCPTOT, CWD, R10, R20, Rnn, R95p, R99p, RX1day, RX-5day* and *SDII*) results and analysis between observed and future indices revealed a general decreasing trend in rainfall from 1960 to 2010 and 2011 to 2100. However, specific observation made indicates that there was a positive trend from 1995 to 2010 for all observed rainfall indices. This positive trend confirm the extreme rainfall events more frequent nowadays in Abidjan and some indices including *PRCPTOT, R20* and *Rnn* results were very significant (p-value < 0.05). Then *CWD, SDII* and R10 indices from 2011 to 2100 results showed an increase trend. These rainfall indices tendencies clearly indicate that heavy rainfall will occur in the future within Abidjan district. It is therefore important for authorities of the district and government of the country to cope with the finding and invest in it and develop preparedness and adaptation measures.

Acknowledgements We are grateful for the financial support provided by the German Federal Ministry of Education and Research (BMBF) under the auspices of the West African Science Service Centre for Climate Change and Adapted Land Use (WASCAL) project. We are also grateful to the WASCAL GRP-CCLU, Kumasi, Ghana.

Competing Interests Section

I declare and certify that this research article is for pure academic purpose. In fact, it is one specific objective of my Ph.D. research. Therefore, there is a non-financial competing interest.

References

Alexander, L V., Zhang, X., Peterson, T C., Caesar, J., Gleason, B, Klein Tank, A, Haylock, M, Collins, D, Trevin, B, Rahimdah, F, et al., (2006) Global Observed Changes in Daily Climate Extremes of Temperature and Precipitation. Journal of Geophysical Research, III D05109. http://dx.doi.org/10.1029/2005JD006290.

Ali, A (2011) Climate variability and change in the Sahel. Understanding the current situation by observing Climate change in the Sahel. A challenge for sustainable development. AGRHYMET Monthly Bulletin (Special Issue), 17–20. http://www.agrhymet.ne/PDF/Bulletin%20mensuel/specialChCang.pdf.

Babatolu, J, Akinnubi, R, Folagimi, A and Bukola, O (2014) Variability and Trends of Daily Heavy Rainfall Events over Niger River Basin Development Authority Area in Nigeria. American Journal of Climate Change, 3: 1–7. doi:10.4236/ajcc.2014.31001.

Birkmann, J, Buckle, P, Jaeger, J, Pelling, M, Setiadi, N, Garschagen, M,... and Kropp, J (2010) Extreme events and disasters: a window of opportunity for change? Analysis of organizational, institutional and political changes, formal and informal responses after mega-disasters. Natural Hazards, 55(3):637–655.

Brou, Y T (2005) Climat, mutations socio-économiques et paysages en Côte d'Ivoire. Mém. HDR, Univ. Sci. Techn. Lille, France, 212.

Cinco, T A, de Guzman, R G, Hilario, F D, & Wilson, D M (2014) Long-term trends and extremes in observed daily precipitation and near surface air temperature in the Philippines for the period 1951–2010. Atmospheric Research, 145–146: 12–26. doi:http://dx.doi.org/10.1016/j.atmosres.2014.03.025.

Gbode, I E, Akinsanola, A A and Ajayi, V O (2015) Recent Changes of Some Observed Climate Extreme Events in Kano. International Journal of Atmospheric Sciences, Volume 2015, Article ID 298046, 6. http://dx.doi.org/10.1155/2015/298046.

Goula, B T A, Kouassi, V J & Savané, I (2006) Impacts du changement climatique sur les ressources en eau en zone tropicale humide: cas du bassin versant du Bandama en Côte d'Ivoire. Agron. Afr., 18(1): 1–11.

Hauhouot, C (2008) Analysis of the rainfall risks in the precarious districts of Abidjan: Survey of cases in Attécoubé, *Geo-Eco-Trop*, 2008, 32: 75–82.

Haylock, M R, Peterson, T C, Alves, L M, Ambrizzi, T, Anunciação, Y M T., Baez, J, Barros, V R, Berlato, M A, Bidegain, M, Coronel, G, Corradi, V, Garcia, V J, Grimm, A M, Karoly, D, Marengo, J A, Marino, M B, Moncunill, D F, Nechet, D, Quintana, J, Rebello, E, Rusticucci, M, Santos, J L, Trebejo, I, Vincent, L A (2006) Trends into taland extreme South American rainfall in1 960–2000 and links with sea surface temperature. Journal of Climate 19: 1490–1512.

Huong, H T L and Pathirana, A (2013) Urbanization and climate change impacts on future urban flooding in Can Tho city, Vietnam, Hydrol. Earth Syst. Sci., 17: 379–394. www.hydrol-earth-syst-sci.net/17/379/2013/ doi:10.5194/hess-17-379-2013.

IPCC, (2007) Climate Change 2007: Synthesis Report. Contribution of Working Groups I, II and III to the Fourth Assessment Report of the Intergovernmental Panel on Climate Change [Core Writing Team, Pachauri, R.K and Reisinger, A. (eds.)]. IPCC, Geneva, Switzerland, 104.

Keggenhoff I, Elizbarashvili M, Amiri-Farahani A, & King L (2014) Trends in daily temperature and precipitation extremes over Georgia, 1971–2010. Weather and Climate Extremes, 4: 75–85. doi:http://dx.doi.org/10.1016/j.wace.2014.05.001.

Kouassi, A M, Kouamé, K F, Goula, B T A, Lasm, T, Paturel, J E & Biémi, J (2008) Influence de la variabilité climatique et de la modification de l'occupation du sol sur la relation pluie-débit a partir d'une modélisation globale du bassin versant du N'zi (Bandama) en Côte d'Ivoire. Rev Ivoir. Sci. Technol. 11: 207–229.

Lamptey B (2009) An analytical framework for estimating the urban effect on climate. International Journal of Climatology, 30: 72–88.

L'Hote, Y, Mahé, G, Some, B, Triboulet, J P (2002) Analysis of a Sahelian annual rainfall index from 1896 to 2000; the drought continues. Hydrological Sciences—Journal 47(4):563–572.
Mouhamed, L, Traore, S B, Alhassane, A and Sarr, B (2013) Evolution of some observed climate extremes in the West African Sahel. Journal Weather and Climate Extremes 1:19–25.
New, M, Hewitson, B, Stephenson, D A, Tsiga, A, Kruger, A, Manhique, A, Gomez, B, Coelho, C A S, Masisi, D N, Kululanga, E, Mbambalala, E, Adesina, F, Saleh, H, Kanyanga, J, Adosi, J, Bulane, L, Fortunata, L, Mdoka, M L, Lajoie, R (2006). Evidence of trends in daily climate extremes over Southern and West Africa. J. Geophys. Res. 111, D14102. http://dx.doi.org/10.1029/2005JD006289.
Osman, Y, Al-Ansari, N, Abdellatif, M, Aljawad, S B and Knutsson, S (2013) Expected Future Precipitation in Central Iraq Using LARS-WG Stochastic Weather Generator. Engineering, 6: 948–959. http://dx.doi.org/10.4236/eng.2014.613086.
Pedersen, A N, Mikkelsen, P S and Arnbjerg-Nielsen, K (2012). Climate change-induced impacts on urban flood risk influenced by concurrent hazards. Journal of Flood Risk Management, 5: 203–214.
Revadekar, J V, Patwardhan, S K, and Rupa Kumar, K (2011). Characteristic Features of Precipitation Extremes over India in the Warming Scenarios. Advances in Meteorology, Volume 2011, Article ID 138425, 11. doi:10.1155/2011/138425.
Sarr, B (2011). Return of heavy downpours and floods in a context of changing climate. Climate change in the Sahel. A challenge for sustainable development. AGRHYMET Monthly Bulletin (Special Issue), 9–11. http://www.agrhymet.ne/PDF/Bulletin%20mensuel/specialChCang.pdf.
Savane, I., Coulibaly, N. and Soro, N., 2003. Analyse multicritère spatiale pour l'identification de la vulnérabilité aux risques naturels d'Abobo (banlieue de la ville d'Abidjan, Côte d'Ivoire). *Rev. Ivoir. Sci. Technol*, 4: 94–107.
Stephenson, T S, Vincent, L A, Allen, T, Van Meerbeeck, C J, McLean, N, Peterson, T C,… & Trotman, A R (2014). Changes in extreme temperature and precipitation in the Caribbean region, 1961–2010. International Journal of Climatology, 34 (9): 2957–2971.
Yan, G, Qi, F, Wei, L, Aigang, L, Yu, W, Jing, Y, … & Qianqian, M (2014). Changes of daily climate extremes in Loess Plateau during 1960–2013. Quaternary International 1–17. http://dx.doi.org/10.1016/j.quaint.2014.08.052.
Zhao, C, Wang, W, & Xing, W (2012). Regional Analysis of Extreme Temperature Indices for the Haihe River Basin from 1960 to 2009. Procedia Engineering, 28(0), 604–607. doi:http://dx.doi.org/10.1016/j.proeng.2012.01.776.

Climate Change Mitigation and Adaptation in Higher Education Institutions: The Case Study of the Faculty of Physical and Mathematical Sciences at the University of Chile

Claudia Mac-Lean, Maisa Rojas, Luis Vargas and Natalia Vicencio

Abstract This study presents climate change related actions at the Faculty of Physical and Mathematical Sciences at the University of Chile on climate change mitigation and adaptation. Specifically, actions in research and development, education, extension, and operational measures on campus are discussed in light of the experience of other Higher Education Institutions and the available international evidence. Selected initiatives from internationally recognized Universities are revised in order to highlight initiatives which promote climate change awareness and develop low-carbon sustainable lifestyles. The paper addresses the role of the University of Chile in disseminating climate change stewardship in education and demonstration projects. For that purpose, we describe the following climate change related actions at the Faculty: (i) carbon footprint calculation, (ii) sustainable campus initiative, (iii) courses in climate change related topics, (iv) minor on

C. Mac-Lean (✉)
Office of Engineering for Sustainable Development, Faculty of Physical and Mathematical Sciences, University of Chile, Beauchef 850, Po Box 8370451 Santiago, Chile
e-mail: cmaclean@ing.uchile.cl

M. Rojas
Department of Geophysics, Faculty of Physical and Mathematical Sciences, University of Chile, Beauchef 850, Santiago, Chile
e-mail: maisa@dgf.uchile.cl

L. Vargas
Department of Electrical Engineering, Faculty of Physical and Mathematical Sciences, University of Chile, Beauchef 850, Po Box 8370451 Santiago, Chile
e-mail: lvargasd@ing.uchile.cl

N. Vicencio
Department of Chemistry and Biotechnology, Faculty of Physical and Mathematical Sciences, University of Chile, Beauchef 850, Santiago, Chile
e-mail: naty.vicencio@gmail.com

W. Leal Filho (ed.), *Climate Change Research at Universities*,
DOI 10.1007/978-3-319-58214-6_19

sustainability, (v) postgraduate diploma on organizational sustainable management, (vi) research in climate change, and (vii) generation of a climate change mitigation and adaptation plan.

Keywords Climate change · Higher education institutions · Mitigation and adaptation

1 Introduction

Human induced climate change is without doubt one of the greatest challenges that humankind has to address this century. After decades of political negotiations on this subject, the outcome of the last "Conference of the Parties", that took place in Paris in November 2015, has finally set both measurable ambition, with a concrete timeline, and the mechanisms for accountability and financing to tackle climate change. Because 195 countries adopted the agreement unanimously, it is to date the strongest political signal that the world as a whole has adopted to solve this problem.

The Paris agreement has set the course towards a decarbonisation of the world economy. The transformational challenge that this change implies is immense, and Universities will be key players in that transformation. It is the role of Higher Education Institutions (HEIs) to train the next generation of leaders and decision makers, create new knowledge and do this in innovate ways.

What the new international context, with the Paris agreement, means for HEIs is that all these efforts need to be aligned with the overall goal of transforming our societies towards decarbonisation, and this, in the coming following decades. This requires important innovation in learning, discovery and innovation. Engineering faculties have a special role to play in this transformation through developing the engineering solutions necessary for a low carbon society.

There is a medium-long history of Universities and HEIs concerned with sustainable development as demonstrated by the existence of devoted associations—for instance The Sustainability Tracking, Assessment & Rating System STARS—and national and international initiatives. These initiatives are well aligned with the UN Sustainable Development Goals (USDG), and the recently finished UN decade of Education for Sustainable Development (2005–2014) highlighted progress in many areas and regions of the world.

There is a tight relationship between the USDG and Climate Change (CC). Therefore all the progress in institutionalising sustainable development views in HEIs will be instrumental for developing climate change action in a timely manner. This conviction has also reached at the Office of Engineering for Sustainable Development (OESD) of the Faculty of Mathematical and Physical Sciences. The OESD is in charge of promoting a culture of engineering for sustainable development at the Faculty and implementing corresponding actions. More and more these actions have included the climate change challenge. Hence the purpose of this paper is to present climate change related actions at the Faculty of Physical and Mathematical Sciences at the University of Chile on climate change mitigation and adaptation. By following the international experience in this field (Cornell 2009; Massachusetts Institute of Technology 2015; UCLA 2008), the actions are grouped in four categories: research and development, education, extension, and operational measures on campus. These four categories are discussed and a comparative analysis is performed in light of the experience of other Higher Education Institutions and the available international evidence.

It must be kept in mind that the review process was focused on some selected HEI and does not have the purpose of encompassing all actions that have been devised for facing climate change mitigation and adaptation worldwide. Another limitation of this work is that the revision is constrained to American universities, which although are ranked among the best of HEI in the world, they do not reflect all the diversity from other cultures and backgrounds.

In this paper we first review international experiences of Climate Change action plans in Universities (Sect. 2). Then, after a short introduction of the Faculty of Physical and Mathematical Sciences at University of Chile, its main climate change related actions are presented in Sect. 3. Finally, in Sect. 4 the conclusions of this work are summarized.

2 Review of Climate Action Plans in Higher Education Institutions

This chapter presents a brief review of Climate Action Plans (CAPs) at certain American universities that are part of the American College & University President's Climate Commitment.

Over 650 schools have joined the American College & University Presidents Climate Commitment (ACUPCC 2016). This breakthrough program is reshaping institutions and communities while training the future political, business, and scientific leaders who will help solve climate change. ACUPCC signatories commit to measure and report their greenhouse gas emissions, take immediate actions to reduce them, and develop and implement a plan to go climate neutral.

From that list of universities, in this paper three of them have been selected to study their plans and strategies to introduce CC: Massachusetts Institute of Technology (MIT), University of California Los Angeles (UCLA), and Cornell University. This selection was done based on data availability, CAPs robustness, and degree of advance in their implementation. We have analysed their CAPs and, from our perspective, we have selected the most compelling initiatives, which could play a role model for the development of the CAP at the Faculty of Physical and Mathematical Sciences at University of Chile.

Another university attached to ACUPCC with an interesting CAP is the University of Arizona, which after identifying the direct and indirect causes of carbon emissions, has classified the reduction targets in the context of: energy, transport, agriculture, refrigerants, recycling and waste; and it has also included Education, Research and Community Outreach (Global Institute of Sustainability at Arizona State University 2009). In addition, the Sustainability Strategic Plan of Yale University is divided into the following strategic areas: Campus Systems (Campus Planning, Building Design, and Construction, Waste Management, Transportation, Food and Dining, Environmental Health and Safety), Earth Systems (Energy and Greenhouse Gas Emissions, Water use and Land Management), Administrative Systems (Finance and Business Operations, Procurement, Cleaning and Maintenance) and Education and Engagement (Yale Office of Sustainability 2010). Another interesting case is Duke University, which after conducting a full inventory of emissions of greenhouse gases, has developed a significant methodology based on the calculation of a cost/offset ratio to discern which mitigation strategies should be included in the Greenhouse Management Plan (Hummel 2005).

The actions undertaken by universities regarding CC may be classified in the following four categories: Research and Development, Education, Extension and Outreach, and In Campus initiatives. In the following subsections the main ideas and initiatives of the three selected universities are presented.

2.1 *Massachusetts Institute of Technology*

(A) Research and Development

The main goal is to accelerate progress towards low- and zero-carbon energy technologies. To accomplish this MIT is developing:

- Eight New Low-Carbon Energy Centers: Five now and the remainders to follow.
- New research to transform major energy systems.
- A new study: The 2 °C Challenge Accelerating the Transition to a Zero-Carbon Future.

(B) Education

The main objective is to educate a new generation of climate, energy and environmental innovators. To achieve this, MIT is proposing students:

- To develop an Environment and Sustainability degree option.
- To develop an online Climate Change and Sustainability credential.
- To explore broad adoption of principles of "benign and sustainable design".

(C) Extension

Outreach has the following main initiatives:

- Educate leaders in industry and government.
- Expand the capacity of MIT's Climate CoLab.

- Mobilize the strength of alumni.
- Create a web portal on climate change.

(D) Operational measures on campus:

The objective in this area is to actively pursue new carbon-cutting strategies across campus and activate the campus as a living lab. Main activities in this area are:

- Reduce campus greenhouse gas emissions 32% by 2030.
- Eliminate the use of fuel oil in campus power generation by 2019.
- Actively pursue new carbon-cutting strategies across campus.
- Enact "carbon shadow pricing."
- Deploy an open data platform for campus energy use.

2.2 University of California, Los Angeles (UCLA)

(A) Research

UCLA currently has over 160 faculty engaged in climate and sustainability-related research. This research activity at UCLA has created 22 centres at the matters of Natural & Physical Sciences, Technology, Policy & Law, and Human Health and Environmental Justice.

(B) Education

- Develop Green Office Workshops.
- Use the Daily Bruin, UCLA Today, and other campus print and web-based media to create a series of educational articles explaining the components of sustainability and providing information on professional careers in sustainability.
- UCLA offered over 200 courses focused on or directly related to climate change and sustainability.
- 9 ungraduated programs.
- 7 graduated programs included Masters and PhD.
- 9 Student Groups and Organizations.

(C) Extension

- Educational events.
- City of Los Angeles and Los Angeles County Engagement.
- Internships.
- Mentorship.
- Campus Sustainability Tour.
- Community Engagement.

(D) Operational measures on campus

- Building Ventilation System Energy Conservation.
- Lighting Efficiency.
- Solar Power Production.
- Fume Hood Occupancy Sensors.
- Appliance Replacement Incentive Program.
- Dedicated Server Room Consolidations.
- Policy Changes: During the winter break the campus shuts down the HVAC in unoccupied buildings and those buildings that do not require temperature stability.
- Fleet GHG Emissions Reduction Initiatives.
- Commute Emissions Reduction Initiatives.
- Air Travel Emission Reduction Initiatives.
- Lab Energy Efficiency Program (LEEP): has the goal to promote energy conservation and efficiency specifically in campus laboratories.
- Housing and Residential Life Initiatives: are designed to encourage the students to modify their behavior in the areas of recycling, energy use, reduced use/waste management, and reuse.

2.3 Cornell University

(A) Research

Cornell is a premier hub for sustainability, advancing research and cultivating collaboration. It has seven research centers with sustainability-related topics:

- Atkinson Center for a Sustainable Future
- Energy, Environment, & Economic Development
- Cornell Agricultural Experiment Station
- Energy Materials Center at Cornell
- Food, Agriculture, and Development
- Institute for Computational Sustainability
- KAUST-Cornell Center for Energy and Sustainability

(B) Education

- Work to ensure that climate literacy and sustainability are part of the curriculum and educational experience at Cornell.
- Expand Cornell's Energy Conservation Engagement Program (ECEP) to achieve a 1% reduction in annual electrical energy usage and to educate the campus community, including thousands of future graduates, on the value of sustainable energy use.

(C) Extension

- Participate actively in climate literacy outreach to ensure that the broader community receives Cornell's assistance in making critical decisions for a sustainable future.

(D) Operational measures on campus

- Eliminate on-site coal combustion.
- Utilize combined heat and power (CHP), or the simultaneous production of electricity and the utilization of "waste" heat for campus heating requirements, to optimize the efficiency of the Central Energy Plant.
- Integrate Building Energy Standards and energy modeling into the building design, review, and approval process in order to maximize energy efficiency.
- Optimize the campus Heat Distribution System to increase efficiency and cost-effectiveness and to facilitate the integration of Cornell's future energy sources.
- Capitalize on more than 50 campus waste streams and other university-owned biomass resources to generate renewable energy through CURBI (Cornell University Renewable Bioenergy Initiative).
- Eliminate the combustion of fossil fuel for campus heating by developing an Enhanced Geothermal System (EGS) hybridised with biogas. Prepare a preliminary design and phased implementation plan for a Hybrid Enhanced Geothermal System and build a demonstration project.
- Support the expansion of regional wind generation capacity and integrate wind power into Cornell's renewable energy portfolio.
- Implement broad-based, mission-linked carbon management strategies such as forest management, carbon capture and sequestration, and community projects to offset unavoidable university emissions.
- Integrate Sustainability into the Work Place.
- Identify realistic actions and responsible parties to lead the effort to increase Cornell's STARS (Sustainability Tracking, Assessment, & Rating System) rating to become one of the first Platinum-rated STARS campuses in the nation.

3 Climate Change Related Actions at the Faculty of Physical and Mathematical Sciences at the University of Chile

Founded in 1842, the University of Chile is the main and oldest institution of higher education owned by the State in Chile, with a national and public character. It is a research-oriented institution, and it is organised in 14 Faculties.

The Faculty of Physical and Mathematical Sciences (FCFM), in which our work is focused, host the School of Engineering. It has 425 professors, 1.100 graduate students and 4.860 undergraduate students. The engineering studies at University of Chile last six years. The main specialisations comprise mechanical, industrial, computer science, civil construction, electrical, chemical, mining, and biotechnology engineering.

The Office of Engineering for Sustainable Development (OESD) at the FCFM was founded in 2014. The OESD seeks to promote and integrate a culture of engineering for sustainable development at the Faculty, enhancing the excellence of the formation of the students. In line with the international evidence presented in the previous section, the Office works in four aspects: teaching, research, extension and operations of the Campus. The OESD at the FCFM understands that there is no Sustainable Development without taking Climate Change into account. Therefore it has taken as a core mission to integrate Climate Change into the mainstream educational activities of the Faculty. In this line, there have been a number of initiatives that have been developed driven by two parallel forces.

The OESD itself was created in order comply with a "Cleaner Production Agreement", that was signed between the Head of the University and the Chilean government. Many of the actions that the office carried out during the first two years of its existence were directly related to that agreement, without an the development of an own strategic plan. On one hand the personal interested of Academics involved in the OESD have helped focussing on CC in the OESD actions. After having fulfilled compromises of the Clean Production Agreement, has the Office had the time to reflect on a more medium and long-term plan of actions, and these include integrating Climate Change into all activities of the Faculty. A full overview of this process includes a description of more isolated actions, and then more long-term development Climate Change Mitigation and Adaptation Plan (Sect. 4). In this section, the OESD climate change related actions at the Faculty are described.

(A) Research

- *Research in Climate Change.* The Faculty hosts the Centre for Climate and Resilience Research (CR2), which focuses directly on Climate Change. In addition, it hosts the Centre of Energy, the Chilean Solar Energy Research Centre, and the Andes Geothermal Centre of Excellence, which are related to climate change as their research focuses on energy generation and renewable energies.

The CR2 research centre was founded in 2012, with 75 academics and professionals. It focus is on Earth System Science, with an interdisciplinary approach and in close relation to stakeholders, its main goal is to contribute to a better understanding of the Earth System and to promote actions to achieve societal resilience in Chile. Their research proposal addresses relevant aspects in the fields of biogeochemistry, climate dynamics, ecological services, human dimension, and modelling.

(B) Education

- *Courses in Climate Change related topics*. The matter of climate change teaching and research at the FCFM is revised in (Rojas, M. et al., 2015). A summary of the courses that address Climate Change at any level of depth is given in Table 1.

Twenty courses where identified to cover topics related to Climate Change at the FCFM, of which four courses have complete dedication to Climate Change, four courses include a unit, ten courses cover the topic in a single class, and five incorporate it with an exercise or homework. Nine of these courses are compulsory – out of a universe of about 240 compulsory courses at the FCFM, and eleven are offered in an elective scheme.

The Faculty of Physical and Mathematical Sciences of the University of Chile offers 28 Minors, of which three are climate change related minors: Minor in

Table 1 Summary of courses at FCFM that address climate change

Course characterisation	Summary
Course by department	6 DIC 4 Workshops 4 Other departments 3 IQBT 2 DGF
Level of depth	4 courses with full dedication 4 courses with a unit 10 courses with an individual class 5 courses with exercises/homework
Course type	9 compulsory 11 electives
Semester	3 in 3rd semester 5 in 5th or 6th semester 11 no defined semester
Topics covered	All aspects of climate change: physics basis, impacts and mitigation
Teaching methodology	9 courses utilise lectures 13 courses utilise group work and discussion
Evaluation methodology	Project, Exam, Reports, and Presentations
Average number of students per course	2 courses >= 60 students 3 courses = 40 students 15 courses = 10–20 students

Acronyms of departments: *DGF* Department of Geophysics, *DIC* Department of Civil Engineering, *DIE* Department of Electrical Engineering, *DIMEC* Department of Mechanical Engineering, *DMIN* Department of Mining Engineering, *EH* Humanities Studies, *EI* Engineering School, *IQBT* Department of Chemistry and Biotechnology

Meteorology and Climate, Minor in Renewable Energies, and Minor in Engineering for Sustainable Development.

– *Minor in Engineering for Sustainable Development.* The competences students should have by the end of this Minor program are:

  - The student links sustainability to complex systems analyses, in its social, environmental and economic dimensions;
  - The student assumes sustainability as a professional attribute related to critical thinking and responsibility;
  - The student integrates complexity and sustainability educational tools in the practice of its own field of specialisation;
  - The student discusses, promotes, and proposes sustainable solutions to its local environment.

The previously described competences have determined the definition of the structure of the Minor – which has 45 credits. The structure of the Minor aims not only to introduce basic sustainability contents and topics, but also to permit students to specialise in their own fields of interest, and to experience the complexities linked to sustainability through a practical workshop.

The structure of the Minor comprises one mandatory course and three elective courses. Firstly, there is a mandatory introductory course to Engineering for Sustainable Development, which purpose is to introduce a common background in terms of engineering and sustainability to the students. There is a wide range of new forms of learning in ESD programmes (Wals 2013), including trans- and inter-disciplinary learning, social learning, project-based learning, gaming, computer simulations, distance learning, backcasting, case-studies, policy-laboratories, problem-based learning, bootstrapping, values education, ecological footprint analysis, experiential approaches, reflective journal writing. In our case, we have adopted among others interdisciplinary learning, project-based learning, values education, and the ecological footprint analysis.

Secondly, three elective courses must be chosen from a pool of approximately 15 courses offered by the various Engineering Departments within the Faculty. These elective courses comprise subjects such as climate systems, environmental engineering, environmental economics, sustainability in construction, innovation for sustainability, sustainability in mining, social project evaluation, and renewable energies. This initiative also led to the creation of two new elective courses in the undergraduate programme. The Minor ends with a practical project where students propose a sustainability intervention to their surroundings. The design of this workshop followed the ideas of (Segalas et al., 2010) in the sense that sustainability courses at technological universities should focus their content on the social and institutional aspects of sustainable development, and apply a constructive and community-oriented pedagogical approach.

To date the Minor in Engineering for Sustainable Development, has had nearly 80 students since it was launched the second semester of the year 2014.

(C) Extension

- *Postgraduate Diploma on Organisational Sustainable Management.* This programme is oriented to professionals in the areas of engineering, science, and architecture, in charge or interested in leading change towards sustainability inside their organisations

The programme was launched in 2015 and seeks to contribute from the engineering and science fields to the formation of professionals and leaders capable of evaluating, managing, and promoting sustainability as an attribute of value generation within their organisations. This is achieved through specific skills and knowledge, in terms of strategies and methodological tools of sustainability and the circular economy.

(D) Operational measures on campus

- *Carbon Footprint Calculation.* The carbon footprint calculation at the FCFM has followed the Greenhouse Gas (GHG) Protocol (World Resources Institute 2004) and the consolidation methodology used to determine organizational boundaries, which is known as operational control approach. The methodologies, along with all the GHG reports of the American College & University Presidents Climate Commitment are available at http://rs.acupcc.org/stats/complete-ghg/.

The carbon footprint has been calculated as part of the commitments contained in the Cleaner Production Agreement signed by the Faculty with the Government of Chile. The calculation triggered a few changes in the manner the organisation registers information, in order to allow for data availability.

The information is organised in three scopes. The first scope (Scope 1) corresponds to direct emissions (vehicles owned by the university, heating, and machines), the second (Scope 2) includes the indirect emissions from the energy grid, and the third one (Scope 3) includes transport of the community, paper use, and waste. (Table 2)

The main results for the year 2014 are the following:

The final share of CO_2 emissions at the Engineering Faculty is shown in Fig. 1.

Reference total emissions (Scopes1 + 2 + 3) at UCLA are 398.972 CO_2 eq ton, and for Cornell 224.650 CO_2 eq ton. These numbers are nearly ten times the case of the Engineering Faculty in terms of per full time enrolment.

Table 2 Scopes of the carbon footprint calculation for the FCFM

Scope	Total annual emissions [CO_2 eq ton]
Scope 1	310
Scope 2	3.091
Scope 3	4.957
Total	8.358
Emissions per person	1,3

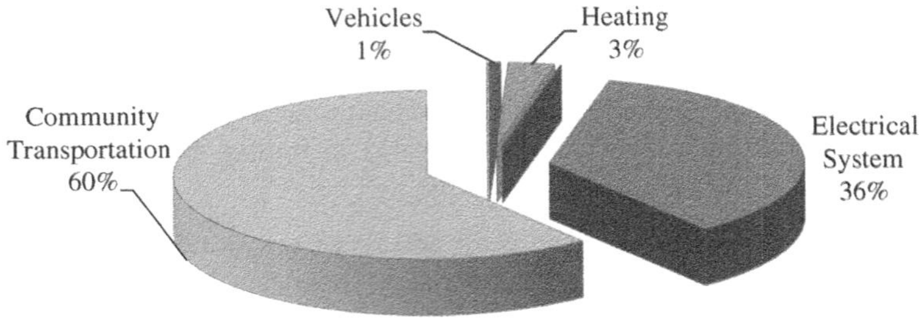

Fig. 1 Carbon footprint calculation distribution

- *Sustainable Campus Initiative.* There have been a number of initiatives to transform the operation and infrastructure of the Engineering Campus into a more sustainable organisation. Among those projects the most important are a new building complex and the compliance of a Cleaner Production Agreement. This agreement was signed between the Head of the University and the Chilean government.

The size of the new building complex, named Beauchef 851, has 50.000 m^2, and has been designed and built with high sustainability standards, including technologies such as a grey water recycling system, a solar PV plant of 16 kW, solar thermal for heating, efficient lighting and air conditioning, and a CO_2 monitoring system. Evidence of greening was given by the LEED Gold rating (full compliance in December 2015). In terms of costs, the LEED certification for Beauchef 851 was around US$ 2.100.000.

The OESD also has active collaboration with the Architecture Office of the Campus, consequently most of the retrofit or new buildings are incorporating sustainability concepts and LEED considerations. Finally, a recycling system for the entire Campus was launched during the first semester of 2016, efficient lighting is the new standard at the Faculty, and HVAC is being shut down in unoccupied classrooms and offices.

4 Generation of a Climate Change Mitigation and Adaptation Plan

After three years of work of the OESD at the Faculty of Physical and Mathematical Sciences at the University of Chile, there is a strong conviction that the path of the future engineering education is linked to the sustainability concept, which in turn is tied with adaptation and mitigation climate change actions. In Chile, due to the place engineers have in the society, they are perhaps the most influential professionals changing the physical environment nowadays, and they must carry the stewardship of sustainability accordingly.

By taking MIT's Plan for Action on Climate Change as a reference (Massachusetts Institute of Technology 2015) and, in order to be able to actively pursue new carbon-cutting strategies across campus, the FCFM along with the Climate and Resilience Research Center (CR2, www.cr2.cl), has committed to develop a Climate Change Mitigation and Adaptation Plan during the year 2016. The participatory approach of the process has been diagnosed as crucial, in order to raise awareness and education in the community in regards to climate change.

The plan development has two stages: one open activity where the pillars of climate mitigation and adaptation at the FCFM are identified; and a second open event where specific climate actions for each pillar are recognised. The plan includes mitigation, adaptation, and capacity building measures.

5 Conclusions

From the international experience, Climate Actions Plans of the Massachusetts Institute of Technology (MIT), University of California Los Angeles (UCLA), and Cornell University, are considered a useful experience to foster the Climate Change Mitigation and Adaptation Plan of the Faculty of Physical and Mathematical Sciences at the University of Chile.

Climate Change related actions of the Faculty of Physical and Mathematical Sciences at the University of Chile embraces both Mitigation and Adaptation, which are organized in four areas: Education, Research, Extension and Operational measures on campus. Main initiatives in these areas are the completion of the cleaner production agreement, the formation of the Office of Engineering for Sustainable Development, the calculation of the carbon footprint, and the already developed campus greening initiatives.

From the current work, it is possible to observe that the Climate Change related actions at the FCFM are aligned with the American College & University Presidents Climate Commitment, and ACUPCC experience will definitely serve as useful input for our Faculty.

Thus, one of the main lessons from the paper is the alignment of FCFM action plan with the international experience, which grouped the efforts in a participatory approach in order to raise awareness and education in the community in regards to climate change.

Thus, the future work of devising and implementing the Climate Change Mitigation and Adaptation Plan at FCFM will be built on solid basis that could respond to the needs of our present and future expectations.

References

ACUPCC, http://rs.acupcc.org/, last revised in May 2016

Cornell University. Cornell Climate Action Plan: Making Climate Neutrality a Reality. Ithaca: s.n., 2009.

Global Institute of Sustainability at Arizona State University. (2009). Carbon Neutrality Action Plan . Tempe, Arizona: Arizona State University.

Hummel, S. (2005). CHARTING A PATH TO GREENHOUSE GAS REDUCTIONS. Greening Campus VI Conference Proceedings, (págs. 1-19). Muncie, Indiana.

Massachusetts Institute of Technology. A Plan for Action on Climate Change. Massachusetts: s.n., 2015. http://climateaction.mit.edu/

Rojas, M. et al. Climate Change Education and Literacy at the Faculty of Physical and Mathematical Sciences of the University of Chile (2015). International Journal of Global Warming.

Segalas, J., Ferrer-Balas, D., and Mulder, K.F. (2010), "What do engineering students learn in sustainability courses? The effect of the pedagogical approach", Journal of Cleaner Production, Vol. 18, 275-284.

UCLA. Climate Action Plan. Los Angeles, California : s.n., 2008. http://www.sustain.ucla.edu/wp-content/uploads/2013/04/UCLA-Climate-Action-Plan.pdf

Wals, A. E. J. (2013), "Sustainability in Higher Education in the Context of the UN DESD: A Review of Learning and Institutionalization Processes", (article in press) Journal of Cleaner Production, http://dx.doi.org/10.1016/j.jclepro.2013.06.007.

World Resources Institute (WRI), The Greenhouse Gas Protocol - Corporate Accounting and Reporting Standard, Revised Edition, 2004.

Yale Office of Sustainability. (2010). Sustainability Strategic Plan. New Haven: Yale University. http://sustainability.yale.edu/sites/default/files/strategicplanupdatejune2011.pdf

Towards Climate Change Capacity Development in Universities: Climate Change Training Needs of Agriculture Lecturers in Niger Delta Region of Nigeria

C.C. Ifeanyi-Obi, C.C. Wigwe, U.R. Etuk and O.M. Adesope

Abstract Based on the climate change curriculum for universities developed by Ozor and Madukwe (Journal of Agricultural Extension 16, 2012), Chakeredza et al. (ICRAF Working Paper no. 82. World Agroforestry Centre, Nairobi, Kenya, 2009) and the climate change curricula development meeting held at Entebbe, Uganda, 2011, this paper aimed at assessing the climate change training needs of agriculture lecturers in the Niger Delta region of Nigeria with the view to provide universities management a relevant guide in planning training courses for the agriculture lecturers. Specifically, the study focused on examining lecturers understanding of the concept of climate change, their sources of climate change information, the extent of awareness on climate change made by the universities and the climate change training needs of the lecturers. One hundred and forty-four agriculture lecturers were selected using cluster sampling technique. Data were collected using structured questionnaire and analyzed with descriptive statistical tools. Mean age of the lecturers was 43 years with 63% having a doctorate degree while the rest have either masters (33%) or bachelors (4%) degree. Average number of years spent in the job was 8 years. The agriculture lecturers mainly perceive climate change to be increased intensity of solar radiation (M = 3.60) and increase in atmospheric temperature (M = 3.42). Their major source of climate change information was journal papers (74%) and internet (71%). Almost all (94%) agreed that climate change concepts is covered in some courses offered at their university at undergraduate level but none of the four universities sampled have climate change as a course of study both at the undergraduate and post graduate level. Of the five professional meetings examined, agriculture lecturers attend mainly conferences (35%) and seminars (34%).

C.C. Ifeanyi-Obi (✉) · C.C. Wigwe · O.M. Adesope
Department of Agricultural Economics and Extension,
University of Port Harcourt, Port Harcourt, Rivers State, Nigeria
e-mail: clara.ifeanyi-obi@uniport.edu.ng

U.R. Etuk
Department of Agricultural Economics and Extension,
University of Uyo, Uyo, Akwa Ibom State, Nigeria

W. Leal Filho (ed.), *Climate Change Research at Universities*,
DOI 10.1007/978-3-319-58214-6_20

Only one university out of four used for the study has climate change posters displayed in strategic places in the university. Majority (59%) graded the general effort of their university management on climate change as low. All the lecturers agreed that they need training in the seventeen areas captured as climate change training needs with future climate models, projections, predictions and forecast (M = 3.63) and climate impacts on different areas of agriculture (M = 3.58) as the most needed area of training. Information gathered from this study will be useful to university management in planning training programme for their lecturers.

Keywords Climate change · Training needs · Agriculture lecturers · Capacity development

1 Introduction

Climate change is no longer news as the impacts are globally felt and experienced by almost all sectors of the economy especially the agricultural sector. It has become one of the topical 21st century global threat, hence the serious attention given to it at all levels. The agricultural sector, especially in the developing world like Nigeria, are facing increasing challenges resulting from climate change mainly due to their poor infrastructural capacity heightened by poor knowledge and skill on climate change issues.

Though there is increased awareness of climate change issues in the country, not only as a result of governmental efforts, but mostly occasioned by the incidence of climate change experienced by the people. The efforts towards equipping the citizenry to effectively adapt to this change is yet to be effective. Different key players in climate change are yet to be recognized and pooled together for a more effective and sustainable integration of climate change knowledge. Oladipo (2010) noted that meaningful success in climate change response efforts could be achieved through the establishment of platform or framework where all the climate actors can operate in a coordinated manner.

Almost all parts of the country are experiencing the impacts of climate change in varying dimensions and intensity, but more pronounced in some areas as a result of their geographical location, natural environment as well as human activities. For instance, the Niger Delta region of Nigeria is the only known oil producing area in the country and host over 123 gas flaring sites making Nigeria one of the highest emitters of greenhouse gases in Africa, (Akinro et al. 2008; Nzeadibe et al. 2011). The flares contribute greatly to greenhouse gas emission, hence more climate change scenarios in the region resulting in increased health challenges and reduced productivity in the agricultural sector. Despite the high activity of oil producing industries in the region, agriculture still remains the major livelihood activity of the indigenous people in the region characterized by use of local tools and high dependence on weather signals. Farming activities of the inhabitants are exposed to both climate change extreme events and other environmental stressors like frequent oil spillage on their farms, air and water pollution, coastal surges, floods and erosion.

Chakeredza et al. (2009) opined a strong positive correlation between levels of education and economic development in any country and stresses that good education should provide solutions to development problems. According to them, tertiary agricultural educational institutions in Sub-Saharan Africa are expected to advice on solutions to the climate change challenges facing the continent, unfortunately in their present state, these institutions are ill-equipped to move with speed to address these issues. This could be in the knowledge base capacity as well as current learning structure. Current pedagogies are centrally defined by university governance structures which are very hierarchical and rigid, and in most cases, discourage the culture of shared thinking and collaboration required for addressing complex system-related challenges such as global environmental change, including climate change (Urama 2009). Similarly, Bloom et al. (2005) noted that higher education plays a critical role in preparing and providing the leadership to meet her challenges and to stimulate sustainable development. Effective adaptation to complex global challenges such as climate change requires well-coordinated and collaborative efforts to galvanise ad hoc response activities at individual and institutional levels across Africa, to create new curricula, new teaching methods, new pedagogies and university governance structures suitable for building necessary capacities for climate change adaptation, (Ozor and Madukwe 2012).

Increasingly, climate change is becoming a challenge to agricultural production because it increases risks and uncertainties for farmers (Chakeredza et al. 2009). They further posited a need for graduating agricultural students to be well-versed with the challenges posed by climate change if they are going to advise the communities they will be working with appropriately. According to them, the graduating students also need to understand the implications of climate change to economic development and international trade, this will help to shape and sustain future policy-making in climate change issues. The challenges therefore are to develop good curricula, produce relevant learning resources and capacitate educators. Different efforts have been made by various researchers and institutions (Chakeredza et al. 2009; Climate change curriculum development meeting, Entebbe, Uganda 2011; Ozor and Madukwe 2012) to develop climate change curricula that will effectively mainstream climate change education into the agricultural education. In addition to developing relevant and useful curricula for climate change, there is also a need to capacitate the agricultural lecturers who are the educators of these students. The climate change curricula development meeting in Entebbe 2011 in view of effectively mainstreaming the developed climate change curricula for higher institutions recommended in-depth training and capacity development of university staff who are going to be in charge of training the students. According to their recommendation, these training could be in form of short courses, seminars, workshops or even specific training at graduate level in order to teach the courses effectively.

In view of the on-going, there is a need not only to develop and mainstream relevant climate change curricula into the agricultural educational system, the lecturers who are the main instructors of these students need to be properly trained in order to teach the students effectively. Proper knowledge on the different modules in the climate change curriculum need to be acquired by these lecturers. Based on the climate change curriculum for universities developed by Ozor and Madukwe (2012), Chakeredza et al. (2009) and the climate change curricula development meeting, Entebbe, Uganda, 2011, this paper assessed the climate change training needs of agriculture lecturers in Niger Delta region of Nigeria with the view to provide university management relevant guide in planning training courses for the agriculture lecturers.

Objectives of the study were to:

1. describe the demographic characteristics of agriculture lectures in study area
2. examine agriculture lecturer's perception of climate change in the study area
3. identify the agriculture lecturer's source of climate change information
4. assess the level of climate change awareness created by universities in the study area
5. identify climate change training needs of agriculture lecturers in the study area

2 Methodology

Cluster sampling was used to select sample for the study. Each of the nine states namely Rivers, Bayelsa, Abia, Akwa Ibom, Imo, Cross River, Ondo, Delta and Edo, formed a cluster from which four states were randomly selected for the study namely: Rivers, Imo, Cross River and Akwa Ibom. Thirty six agriculture lecturers were randomly selected from the faculty of agriculture in each of the Federal university in the four states selected for the study, giving a total of 144 agriculture lecturers for the study. These are University of Port Harcourt, University of Uyo, University of Calabar and Federal University of Technology, Owerri. Data was collected using structured questionnaire. Analysis of data was done using descriptive statistics such as mean, frequency counts and percentages with the aid of SPSS software (version 21). The limitation of selecting federal universities is the non-inclusion of state owned universities which did not allow for comparison.

2.1 Demographic Characteristics of Agriculture Lecturers

Table 1 presents the distribution of demographic characteristics of agriculture lecturers in the Niger Delta region of Nigeria. As regards sex of respondents, majority (66%) were males while only 34% were females. The large number of male lecturers indicate that the universities management are yet to align with the

Table 1 Demographic characteristics of agriculture lecturers in the study area

S/N	Variable	Frequency	Percentage	Mean
1	Gender			
	Male	95	66	
	Female	49	34	
2	Age (years)			43
	≤ 30	12	8	
	31–40	47	33	
	41–50	65	45	
	>50	20	14	
3	Marital status			
	Married	119	83	
	Single	25	17	
4	Educational qualification			
	B.Sc	6	4	
	M.Sc	48	33	
	Ph.D	90	63	
5	Position held			
	Graduate assistant	8	6	
	Assistant lecturer	23	16	
	Lecturer II	20	14	
	Lecturer I	42	29	
	Senior lecturer	35	24	
	Associate professor	10	7	
	Professor	6	4	
6	No of years spent in the job			
	<5	40	28	
	6–10	61	42	
	11–15	38	26	
	Above 16	5	4	
7	Monthly income (₦)			
	<100,000	4	3	
	101,000–150,000	58	40	
	151,000–200,000	40	28	
	201,000–250,000	16	11	
	Above 250,000	26	18	
8	Area of specialization			
	Animal science	17	12	
	Agricultural economics	28	19	
	Crop science	19	13	
	Fishery	15	10	
	Forestry and wildlife	13	9	
	Agricultural extension	24	17	
	Soil science	21	14.9	
	Human ecology and nutrition	7	4.9	

Source Field survey 2016

global best practice on women affirmative allocation into public offices and government parastatal. The mean age (M = 43 years) of the agriculture lecturers in the Niger Delta region of Nigeria indicate that most of the lecturers are still at their productive age hence strong enough to withstand the stress and rigour in the academia especially, in tackling the issues of climate change mitigation and adaptation. Majority (83%) were married and possessed either M.Sc (33%) and or Ph.D (63%) degree. Only 29% earn monthly income above ₦200,000. This points to the fact that many of the lecturers cannot sponsor themselves to professional trips such as international conferences and workshop in the area of climate change considering their salary level.

Result showed that universities in Niger Delta region have lecturers in virtually all areas of specialisation in agriculture. These include animal science, crop science, soil science, fisheries, agricultural economics, agricultural extension, forestry and wild life management and, human ecology and nutrition. Of all the specialties, the applied aspects of agriculture which is the social science aspect of agriculture (agricultural extension = 19% and agricultural economics = 17%) ranked the highest. However, the pure agriculture specialisation areas such as crop, soil and animal sciences did not rank far from the social science aspect of the discipline-agriculture.

3 Agriculture Lecturer's Perception of Climate Change

Table 2 indicates the agriculture lecturer's perception of climate change. The results revealed that the lecturers were in the affirmative in respect to majority of the factors used to capture their perception of climate change. Out of the 21 statements used to capture their perception, 11 recorded positive affirmations with increased intensity of solar radiation having the highest mean. Perception is an important factor influencing people's climate change adaptation decision. Individual's perception of climate change goes a long way in determining the adaptation choices they make. Having the right perception to climate change facilitates the readiness to take up effective adaptation measures by farmers.

This result is an indication that the lecturers have some basic knowledge of climate change. This may not be surprising as they are in the academia and as such should be enlightened on issues of global concern like climate change. Also, the findings from Table 2 is in consonance with some earlier research (Ndamani and Watanabe 2015; Viscusi and Zeckhauser 2006) which identified increase in rainfall, temperature and solar radiation as major perceptions of climate change. Also, in a similar study by Barimah et al. (2015), 52.2% of the respondents attested to the fact that they have heard of the concept of climate change, and over 71% were able to describe the concept of climate change as change in climate parameters, rainfall and sunshine.

Table 2 Agriculture lecturer's perception of climate change

Perception of climate change	Strongly agree	Agree	Disagree	Strongly disagree	Mean	Remark
Increasing fluctuation in onset of rains	65 (45)	65 (45)	12 (8)	2 (1)	3.34	Agreed
Decreasing fluctuation in onset of rains	15 (10.4)	19 (13.2)	80 (55.6)	30 (20.8)	2.11	Disagreed
Increasing fluctuation in cessation of rains	43 (29.9)	87 (60.4)	14 (9.7)	–	3.20	Agreed
Decreasing fluctuation in cessation of rain	13 (9.0)	26 (18.1)	77 (53.5)	28 (19.4)	2.17	Disagreed
Increased intensity of rainfall	48 (33.3)	74 (51.4)	21 (14.6)	1 (0.7)	3.17	Agreed
Reduced intensity of rainfall	11 (7.6)	35 (24.3)	72 (50)	26 (18.1)	2.22	Disagreed
Increased duration of rainfall	36 (25)	75 (52.1)	26 (18.1)	7 (4.9)	2.97	Agreed
Reduced duration of rainfall	17 (11.8)	28 (19.4)	84 (58.3)	15 (10.4)	2.33	Disagreed
Increase in atmospheric temperature.	70 (48.6)	65 (45.1)	8 (5.6)	1 (0.7)	3.42	Agreed
Decrease in atmospheric temperature	11 (7.6)	22 (15.3)	81 (56.2)	30 (20.8)	2.10	Disagreed
Increased intensity of flooding	49 (34)	86 (59.7)	8 (5.6)	1 (0.7)	3.27	Agreed
Decreased intensity of flooding	11 (7.6)	20 (13.9)	81 (56.2)	32 (22.2)	2.07	Disagreed
Increased frequency of flooding	50 (34.7)	88 (61.1)	4 (2.8)	2 (1.4)	3.29	Agreed
Decreased frequency of flooding	10 (6.9)	14 (9.7)	83 (57.6)	37 (25.7)	1.98	Disagreed
Increased intensity of solar radiation (sunshine)	67 (46.5)	68 (47.2)	9 (6.2)	–	3.60	Agreed
Decreased intensity of solar radiation	10 (6.9)	15 (10.4)	88 (61.1)	10 (6.9)	2.03	Disagreed
Longer dry season (drought)	20 (13.9)	75 (52.1)	45 (31.2)	4 (2.8)	2.77	Agreed
Shorter dry season	14 (9.7)	38 (26.4)	80 (55.6)	12 (8.3)	2.38	Disagreed
Increased longer days than night	12 (8.3)	56 (38)	68 (47)	8 (5.6)	2.50	Agreed
Reduced longer days than night	9 (6.2)	20 (13.9)	97 (67.4)	18 (12.5)	2.14	Disagreed
General undefined weather	73 (50.7)	54 (37.5)	14 (9.7)	3 (2.1)	3.37	Agreed

Source Field survey 2016

4 Agriculture Lecturer's Source of Information on Climate Change

Lecturer's sources of information in the area of climate change is shown in Table 3. The results from the Table revealed that lecturers seek for information on climate change from myriad of sources. Television, internet based tools and journal papers recorded over 70% each, indicating that they form strong hold of information source of the respondents. Newspapers, personal experiences, radio and mobile phones also formed a considerably major source of lecturers' information source on climate change (with over 50%). The results are in concordance with Barimah et al. (2015) who found newspaper, radio, television, books, internet and journals to be among the various sources of information on climate change.

5 Climate Change Awareness Created by Universities

Climate change awareness created by universities in the Niger Delta region of Nigeria is presented in Table 4. It was shown that no university in the study area have any programme (course of study) in climate change both at the undergraduate and post-graduate level but they climate change issues are included in the course content of the universities. Also, there are no short courses (diploma) in climate change offered in the universities used for the study. Most of the respondents (74%) affirmed that no posters on climate change were displayed in strategic positions in the university. This showcases the low emphasis laid on climate change by many universities and the cogent need for universities management to beef up their efforts in sensitization and awareness creation on climate change. It is not surprising that a

Table 3 Climate change information sources of agriculture lecturers

S/N	Variable: major source of information on climate change	Frequency	Percentage
1	Newspaper	83	57.6
2	Radio	85	59
3	Mobile phones	75	52.1
4	Other colleagues	59	41
5	Text books	62	43.1
6	Meterological centres	51	35.4
7	Television	111	77.1
8	Personal experience	99	68.8
9	Internet based tools	103	71.5
10	Journal papers	106	73.6
11	Family and friends	36	25

Source Field survey data 2016

Table 4 Climate change awareness created by universities

S/N	Variable	Frequency	Percentage
1	Is climate change covered in any course content offered at the undergraduate level in your faculty		
	Yes	144	100
2	Does your university offer any course on climate change at the undergraduate level in your faculty		
	No	144	100
3	Does your university offer any course on climate change at the post graduate level in your faculty		
	No	144	100
4	Does your university offer short courses (Diploma) on climate change in your faculty		
	No	144	100
5	Are there posters on climate change displayed in strategic places in your university?		
	Yes	36	24
	No	108	76
6	How would you grade the general effort of your university administration on climate change sensitization		
	High	5	4
	Moderate	54	37
	Low	85	59

Source Field survey 2016

greater percentage (59%) of the respondents rated their efforts as regards climate change sensitization low.

6 Climate Change Training Needs of Agriculture Lecturers

Table 5 shows the climate change training needs of agriculture lecturers in the study area. The results revealed that the lecturers are conscious of the need to improve their capacity and knowledge in climate change as they indicated the need for training in the seventeen statements used to capture climate change training need.

This findings from Table 5 suggest that, irrespective of the high educational qualification of the lecturer's, they have a dire need for training on climate change. This could done in the form of conferences, exchange programmes, workshops and seminars. Doing this will no doubt, improve the competence of the lecturers to responsively teach climate change topics if the different curricula developed are eventually mainstreamed into the school system. Presently, it will help them to

Table 5 Climate change training needs of agriculture lecturers in the study area

Areas of climate change	Training mostly needed	Training moderately needed	Little training needed	No training needed	Mean	Remark
Introduction to climate change (definitions of the major concepts)	67 (46.5)	43 (29.9)	25 (17.4)	9 (6.2)	3.17	Agreed
Global warming and environmental changes	84 (58.3)	38 (26.4)	20 (13.9)	2 (1.4)	3.42	Agreed
Agro biodiversity	67 (46.5)	65 (45.1)	11 (7.6)	1 (.7)	3.38	Agreed
Biofuels	70 (48.6)	60 (41.7)	12 (8.3)	2 (1.4)	3.38	Agreed
Climate system (components of climate system)	66 (45.8)	45 (31.2)	32 (22.2)	1 (.7)	3.22	Agreed
Past climate events (paleoclimatology and geological evidence	72 (50)	54 (37.5)	16 (11.1)	2 (1.4)	3.36	Agreed
Recent climate change(evidence for climate change, indigenous knowledge and perceptions)	84 (58.3)	53 (36.8)	6 (4.2)	1 (.7)	3.53	Agreed
Future climate models and projections, predictions and forecast	101 (70.1)	33 (22.9)	10 (6.9)	–	3.63	Agreed
Climate change vulnerability	75 (52.1)	55 (38.2)	14 (9.7)	–	3.42	Agreed
Climate change impacts on different areas of agriculture	92 (63.9)	43 (29.9)	9 (6.2)	–	3.58	Agreed
Climate change mitigations and adaptation	81 (56.2)	55 (38.2)	8 (5.6)	–	3.51	Agreed
Economics of climate change	86 (59.7)	44 (30.6)	14 (9.7)	–	3.50	Agreed
Climate smart agriculture	84 (58.3)	53 (36.8)	6 (4.2)	–	3.55	Agreed
Climate change extension	78 (54.2)	56 (38.9)	9 (6.2)	1 (.7)	3.47	Agreed
Climate change and food technology	78 (54.2)	59 (41.0)	6 (4.2)	1 (.7)	3.49	Agreed
Handling and use of weather equipment	65 (45.1)	65 (45.1)	12 (8.3)	2 (1.4)	3.34	Agreed
Policy issues on climate change	90 (62.5)	46 (31.9)	7 (4.9)	1 (.7)	3.56	Agreed

Source Field survey 2016

better teach the existing topics in climate change covered in some courses offered in the universities. This will go a long way in equipping the graduate with climate change knowledge and better dispose them to contribute significantly to climate change adaptation in the country upon graduation especially in the Niger Delta region of Nigeria where gas flaring is a topical issue, and majority of its populace are agrarians that depend on subsistence and crude way of farming.

Ifeanyi-obi et al. (2012) outlining the implications of climate change to agricultural system in Nigeria stressed the need for change and expansion in the capacity and role of extension agents. This expansion in capacity may not be for only the extension agents, all stakeholders like the agriculture lectures must also enhance their capacity in order to contribute meaningfully to the increasing demand of climate change.

7 Conclusions and Recommendations

In conclusion, the findings revealed that lecturers have basic understanding of the concept of climate change. They are making conscious efforts to access information in the area of climate change and are desirous of undertaking training in different areas of climate change so as to boost their capacity.

It is therefore recommended that management of universities in the Niger Delta region of Nigeria should make concerted efforts to have climate change courses in their curriculum. Also, increased sensitization of the university community on climate change through posters and banners in the university compound is important as this will help create awareness and consciousness of climate change among the staff and students.

In addition, there is a need to improve the knowledge of the lecturers in the different areas of climate change. This could be done through sponsoring them to short courses, workshops, trainings, seminars, symposium etc. that are organized in the area of climate change. University management could also on their own organize trainings for the lecturers in climate change. All this will help to improve their overall capacity on climate change issues thereby equipping them to effectively teach the different areas of climate change covered in the proposed curriculum.

References

Akinro, A.O., Opeyemi, D. A., & Ologunagba, I.B. (2008). Climate Change and environmental degradation in the Niger Delta region of Nigeria: Its vulnerability, impacts and possible mitigations. *Research Journal of Applied Sciences*, 3 (3):167–173.

Barimah, P. T., Kwadwo, S. O. & David, O. (2015). Assessment of people's knowledge and perception on climate change: A case study of Asunafo North District, Ghana. *International*

Journal of Innovative Research in Science, Engineering and Technology, 4(1). 18417–18424. DOI:10.15680/IJIRSET.2015.0401003

Bloom, D. D., Canning &. Chan, K. (2005). Higher Education and Economic Development in Africa. World Bank Report No 85. Available at: http://siteresources.worldbank.org/EDUCATION/Resources/278200-1099079877269/547664-1099079956815/HigherEd_Econ_Growth_Africa.pdf

Chakeredza, S. Temu, A. B., Yaye, A., Mukingwa, S., Saka, J. D. K. (2009). Mainstreaming climate change into agricultural education: Challenges and perspectives. ICRAF Working Paper no. 82. Nairobi, Kenya: World Agroforestry Centre.

Climate Change University Curriculum Development Meeting Entebbe, Uganda. Held 25–26 November, 2011. Pp.: 1–14.

Ifeanyi-obi, C.C., Etuk, U.R. and Jike-wai, O. (2012) Climate change, effects and adaptation strategies; Implication for Agricultural extension system in Nigeria. *Greener journal of Agricultural sciences,* 2(2); Pp. 053–060.

Ndamani, F. & Watanabe, T. (2015). Farmers' perceptions about adaptation practices to climate change and barriers to adaptation: A micro-level study in Ghana. *Waters, 7.* 4593–4604. doi:10.3390/w7094593

Nzeadibe, T. C., Egbule, C. L., Chukwuone, N. A. & Agu, V.C. (2011). Climate change awarenesss and adaptation in Niger Delta Region of Nigeria. *African Technology policy studies network, working paper series*, No 57.

Oladipo, E. (2010). Towards enhancing the adaptive capacity of Nigeria: A review of the country's state of preparedness for climate change adaptation. Submitted to Heinrich Boll Foundation Nigeria.

Ozor, N. and Madukwe, M.C. (2012) Influencing Curriculum Development and Knowledge of Climate Change Issues in Universities: The Case of University Of Nigeria Nsukka. *Journal of Agricultural Extension,* 16 (1).

Urama, K.C. (2009) Higher Education for Sustainable Development in Africa. Nairobi, ATPS; p. 9

Viscusi, W.K. & Zeckhauser, R. J. (2006). The perception and valuation of the risks of climate change: a rational and behavioral blend. *Springer, 77.* 151–177. DOI:10.1007/s10584-006-9075-9

Greek Universities Addressing the Issue of Climate Change

Constantina Skanavis, Aristea Kounani and Ioannis Ntountounakis

Abstract Nowadays, a great challenge, that most countries are struggling with, is the rapid change of Earth's climate. Meanwhile, the sea level is constantly rising, floods and droughts are increasing and global greenhouse gas emissions continue to rise. Climate change (CC) is a wide world worriment, which requires international confederates, such as industries, educational institutions, individual civilians to come in agreement, in order an environmental and energy revolution process to be somehow initiated. Universities, as having the important mission of educating the future decision makers, indisputably can contribute to mitigating and adapting CC. The scope of this survey is to assess whether or not Greek universities have incorporated into their curricula and research a significant CC content. Additionally, it is aimed to identify the outreach of the Universities to the Greek society, through civic ecology procedures. Greece is a country that generates electricity by burning coal, in a high quota, to cover inhabitants' diurnal energy needs. The findings of this research would enlighten on whether the higher educational institutions in Greece have given the pertinent significance to the issue of CC.

Keywords Greek universities · Greece · Climate change · Higher education

C. Skanavis (✉) · A. Kounani · I. Ntountounakis
Department of Environment, University of the Aegean,
University Hill, 81100 Mytilene, Greece
e-mail: cskanav@aegean.gr

A. Kounani
e-mail: akounani@yahoo.gr

I. Ntountounakis
e-mail: giannisntountounakis@gmail.com

W. Leal Filho (ed.), *Climate Change Research at Universities*,
DOI 10.1007/978-3-319-58214-6_21

1 Introduction

Today, climate change is one of the greatest menaces humanity is confronted with, due to its multiple implications on human survival. Human beings need a decent environment for effective and fruitful life (Offorrma 2014). The issues and challenges of climate change affect all of us, regardless of nationality, religion and profession. Furthermore, CC is interrelated with economy, infrastructure, health, natural resources, transport and industry, making environmental decision a real task. Distinctly, the prospect of climate change adverse doesn't seem that feasible in the near future. The dominant effect of CC is experienced heavily in tropical and subtropical regions (Downing et al. 1997). Scientists have long perceived that the emissions of greenhouse gases and alterations in land use in an industrialized society bring relentless changes in climates and affect ecosystems (Young et al. 2015). It becomes readily apparent that human activities are those that influence the climate. Anthropogenic climate change is likely to present new and largely unforeseen challenges for societies (Naess et al. 2005). Thus, the challenge posed by climate change cannot be overcome because it constitutes a significant threat to global economic growth. The impact of climate change on the economy is not a recently developed phenomenon; nonetheless its effects are turning into more noticeable. Climate change has added its toll on low agricultural production, and has also affected tourism and the pace of industrialization (Huho 2015). A simple example of this fact is the droughts, which apart from agricultural devastation have also affected drastically tourism and industry.

Climate change in the context of sustainable development requires a holistic approach that promotes a better quality of life in a long time frame, rather than a quality of life that aims at short-term profits (Bangay and Blum 2010). Climate change could be possibly addressed with several intervention strategies. The consequences of climate change have been recognized internationally as areas of great concern, with two main lines of action: mitigation and adaptation (Fussel and Klein 2006). The mitigation, most often, refers to changes in anthropogenic conduct to reduce the extent of CC. Adaptation strategies aim to decrease the severity of the adverse effects of climate change on vulnerable communities or infrastructure (Fussel 2005; Owen et al. 2013).

Universities have a significant role in implementing adaptation and mitigation of climate change (Adger et al. 2005). The aim of this survey is to assess whether or not Greek universities have seriously incorporated the climate change issue into their curricula and research efforts. Also its scope is to examine the extent of the implementation of adaptation and mitigation, by giving emphasis in identifying the outreach of the Universities to the society, meaning communicating through civic ecology procedures, on global CC and its impacts on the region.

1.1 Adaptation to Climate Change

Adaptation to climate change is defined, as the adjustment in ecological, social or economic systems in response to observed or expected changes in climatic stimuli. It also includes the effects and impacts, as well as efforts to alleviate adverse CC (Adger et al. 2005).

Adaptation increases the ability of individuals, groups, or organizations to adapt to changes, implement decisions and participate in actions. Adaptation can be implemented in "preparation for" or "in response to" impacts, generated by a changing climate. Hence, adaptation is a continuous stream of activities, actions, decisions and attitudes that appraise decisions regarding all aspects of life, and reflect existing social norms and processes. Therefore, it might be impracticable to separate climate change adaptation decisions from actions triggered by other social or economic events. This verity makes understandable that, attributing adaptations to climate change is not a simple process (Adger et al. 2005).

The objectives correlated with implementing adaptation decisions are more likely to focus on reducing the cumulative impacts of climate change, ensuring that adaptive measures taken by one organization do not adversely impact upon others, avoiding anticipated adverse impacts of climate change, and ensuring that the distributional impacts of adaptation are minimized (Adger et al. 2005).

Even though climate change education is important to all education levels, from primary schools to universities, it is in the higher education that there is a need to tackle it in a systematic way. This is due to the fact that university students will soon pursue careers in science, education, law or engineering among others, and hence, need to be conscious of the impact their professions have both on the environment as a whole and on the climate in particular (Adger et al. 2005).

1.2 The Civic Ecology Approach

Civic Ecology is a framework for community resilience. It can be defined as the integrated web of energy, resources, financial, information and cultural flows and interactions that re-envisioned, created and managed by citizens, acting for the common good within a geographically-defined community and its city-region. It is a human ecology of place, intimately integrating both natural and social/culture systems. In this framework, citizens are empowered to "own" their place and they know that they don't have be an expert to impact their community's future. Thus, civic ecology is a structure that facilitates creation of a shared vision. Individuals without a technical background seem to understand the systems approach and acquire skills on how to design community patterns without any trouble and with minimal education. By, creating and managing local community systems, the community's DNA is created, the basis for perpetuating the community from generation to generation. Citizens can establish systems that address current problems, but they also have to create a living culture, that is emergent and learns forward to address future complications (IMCL 2016).

1.3 The Role of Education

If the role of education is to help learners of all ages to build up the knowledge, skills and capacities, which enable them to think critically, to deal with problems, and to address uncertainty, then the focus of climate change interventions should be placed on more holistic ways of confronting climate change reality (Bangay and Blum 2010).

Environmental awareness and especially the part that concerns climate change issues have to be incorporated in all levels of education, from pre-school to university (Leal Filho 2010). Education plays a momentous role in sustainable development and climate change. It is an implement that aims to increase public awareness and comprehension of global environmental issues, but can also contribute to a better and more sustainable future for humanity. The younger generation reflects the future citizens of society who will be called to confront the issues and impacts of climate change. Environmental degradation, increasing global penury and climate change are just a few of the problems that mankind, and principally the youth, is called to provide solutions for. The first step in recognizing the importance of climate change in our lives is to imagine the influence of humans on environment (Feher et al. 2010).

Education, including formal education, public awareness and training, should be recognized as a process by which human beings and societies can fully realize their potential to quality of life. Education is critical for promoting sustainable development and ameliorating the capacity of people to address the environmental issues (Bangay and Blum 2010). The role of education is identified, as a vital process to reverse the problem of climate change. Environmental education (EE) contributes to adaptation and mitigation measures. The objectives of EE are the augmentation of environmental knowledge, awareness and active participation (Skanavis and Petreniti 2006). Education for sustainable development seeks to achieve similar results, by allowing students to make decisions and selections promoting sustainable development (Bangay and Blum 2010). The literature reveals lack of scientific understanding of the processes of climate change within the wider public. The consequences, due to absence of understanding, result into ineffective mitigation actions, and possibly weakened support for government initiative on the issue (Huxter et al. 2015).

Climate change, today, is no longer the exclusive domain of scientific experts. It calls for action from all citizens. Higher education, in particular, has an important constructing role in educating students about climate change. Previous researches have indicated that environmental studies at university level may amend and upgrade students' education or culture concerning the environment (McMillan et al. 2004).

1.4 *The Role of Universities*

Although education on climate change issues is significant at all levels of education, it is legitimate to focus on higher education. This is because students will soon follow careers in science, education, technology and other sectors and should be aware of the consequences attached by their professional activities and decision-making both on the environment and particularly on the global climate. As institutions of higher education can promote social change (Virtanen 2010), they have the obligation to take initiatives, in order to address and provide answers to such complications of modern society, which are related to the sustainability of the planet and the survival of human life. Universities not only should they contribute to dissemination of knowledge but they should also set the standards. Universities themselves, have to convert into greener institutions (Virtanen 2010). The overall objective of a university could be to achieve energy autonomy and be self-sustained. Therefore, activities dealing with students should include the understanding of climate change and related consequences. Higher education institutions should aim to further research on climate issues. Promotion of students' environmental participation in regional or international workshops, seminars and conferences should also be high in a university's agenda (Carlin 2010).

2 Case Study: Greek Universities

2.1 *Methodology*

2.1.1 Research Area

The study takes place in Greece, which has a population of 10,815,197 inhabitants. Greece with an extent of 131.957 km^2 has a relief characterized by the accumulation of many mountains. Only the 1/5 of the land consists of plains, usually surrounded by high mountains. Greece has a coastline about 13,676 km and more than 2500 islands, but only 227 of which are inhabited (European Union 2016).

The generation of electricity in Greece comes from thermal power plants. In Western Macedonia region, the 50% of total electricity is where it is being produced. Lignite is the significant domestic energy source, accounting for 53.15% of domestic production in 2011. Natural gas contributes a 28.3%. At the same time the aim to increase the electricity needs input from renewable energy resources (RES) has been set to reach 34% by 2020. In the same context, emphasis is placed on busting the use of gas in the energy consumption. In 2011, based on data for the Interconnected System, the 66.5% of the installed capacity of power plants are thermal power stations, including coal-fired 4930 MW, with 730 MW oil and gas 4579 MW. The 19.6% is composed of large hydroelectric plants and the 13.9% are units RES (RAE 2016).

2.1.2 Research Instruments

Greek Universities' web sites were checked in order to extract information whether or not they have incorporated CC issue into their curricula and research efforts. This investigation aimed to identify the Universities' community outreach, meaning communicating through civic ecology procedures on global CC issues and related local impacts.

All Greek universities were initially assessed through a qualitative research approach in order to have the three more environmentally active selected for further study. A group of 4 questions were sent by email to all Greek Universities, asking them whether or not the university has in their curricula a course or a class on CC, whether or not it does research on CC, whether or not it is organizing CC seminars, meetings, conferences. The selected universities, based on the prior questions, were the National and Kapodistrian University of Athens, the University of the Aegean and the Democritus University of Thrace. A questionnaire of 13 questions was sent by email, concerning the following: (1) the total number of undergraduate students, (2) the total number of post graduate students, (3) the total number of Ph.D. candidates, (4) the whole number of undergraduate courses related with the CC issue, (5) the exact number of post graduate courses related with the CC issue, (6) the total number of post graduate programs related with the CC issue, (7) the precise number of researchers' publications related to CC issue, (8) the total number of seminars, events etc., that the university is organizing concerning the CC issue.

2.1.3 Research Sample

In Greece Higher Education, is provided by Universities, Polytechnic schools and Technological Institutes. Moreover, in the academic year 1997–1998 an Open University was enacted, offering distance-learning opportunities. Higher University Education has been charged with the task of high scientific training of the future experts (HNARIC 2016). Today, in Greece 23 Universities are in operation and the total number of active students is about 221,942 for the year 2014–2015 (HSA 2016).

The National and Kapodistrian University of Athens, officially founded in 1837, is the first University not only of Greece but of the Balkan Peninsula and the Eastern Mediterranean region. It has got 8 Schools (School of Economics and Political Sciences, School of Education, School of Health Sciences, School of Law, School of Philosophy, School of Science, School of Theology and School of Physical Education and Sport Science) and 33 Departments (e.g. Law, Biology, Chemistry, Geology and Geo-environment etc.). It is considered the largest university in Greece, with a total of 65,010 students (45,553 undergraduates, 11,479 postgraduates and 7978 Ph.D. candidates) in the current school year (NKUA 2016).

The University of the Aegean (UAegean) was founded in 1984 aiming to introduce new approaches in higher education in Greece and to promote regional development. Situated in 6 picturesque islands in the Aegean Archipelago, the

UAegean offers a unique natural, cultural and human environment for modern studies in the ancient cradle of knowledge. In a span of thirty years, the UAegean has evolved into an international research–oriented University offering 18 undergraduate (BA or B.Sc) and 28 postgraduate (MA or M.Sc) programs in modern interdisciplinary thematic areas such as environment, communication systems, cultural informatics, product design, food and nutritional sciences, education design and Mediterranean studies. In addition, the UAegean has established joined international postgraduate programs (i.e. in Biodiversity, Environmental Policy and Management, European Integration) as well as joined Ph.D. degree programs in a wide range of thematic areas. The total number of students is 16,265 (13,947 undergraduates, 1662 postgraduates and 656 Ph.D. candidates) in the current school year (UAegean 2016).

Democritus University of Thrace (DUTH) was established in July 1973. The University plays an important role in strengthening the national and cultural identity of the region of Thrace, and contributes to the high level of education in Greece. The administration of DUTH is headquartered in Komotini, which is the capital city of the Administrative Region of Eastern Macedonia and Thrace. The DUTH is currently operating eight Faculties and nineteen Schools in four cities of Thrace—eight in Komotini, five in Xanthi, four in Alexandroupolis and two in Orestiada. Overall, 25,280 students are studying at DUTH (21,520 undergraduates, 2046 postgraduates and 1712 Ph.D. candidates) in the current school year (DUTH 2016).

2.2 Results

As the research has shown, Greece is a country where Higher Education is quite strong in a variety of disciplines. A research of Eurostat showed that, 1/3 of young people, in Greece, up to 24 years old, were university students in 2009. This is a percentage of 29.9%, while the highest in general in Europe was 27% (EUROSTAT 2014). The recent years a further increase has been noted in Greece (Fig. 1) (HSA 2016).

Post-graduate studies lead to the award of a diploma of specialization. Greek universities offer a total of 213 post-graduate courses. Only the Open University offers ten post-graduate courses (HNARIC 2016). Figure 2 presents the total number of students at the Greek Universities.

In Fig. 3 the answers to the questions that were emailed to the participating Universities are being presented.

The Case of the National and Kapodistrian University of Athens (NKUA)

The university with the largest student body (65,010 students) is NKUA. The data have shown that out of the total number of 1979 undergraduate courses only 4 of them were related with the CC issue, which means a percentage of 0.2. At the post graduate level the percent becomes 0.58. The total number of the post-graduate

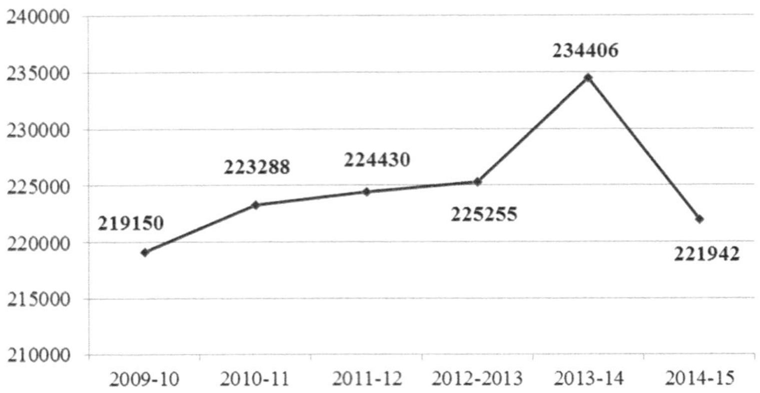

Fig. 1 Total Greek universite's active students *based on data received by HSA*

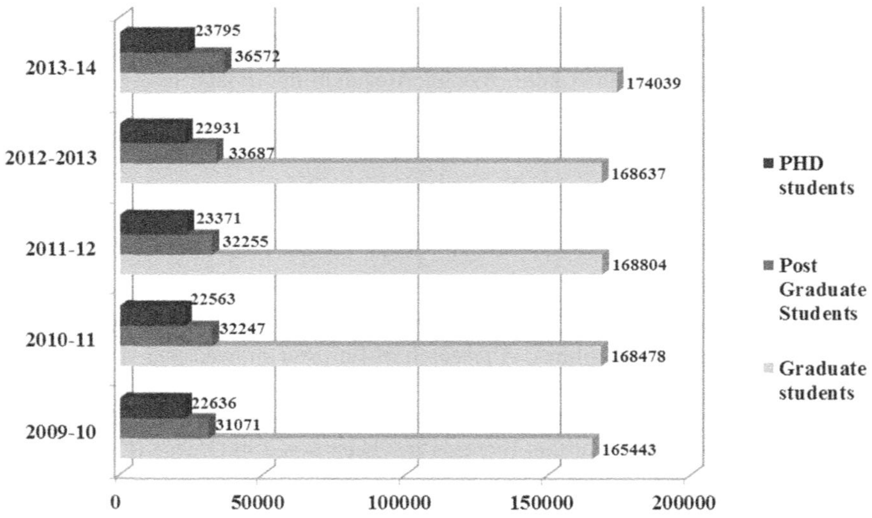

Fig. 2 The total number of Greek University Students *based on data received by HSA*

programs is 110, and none of them is on the CC specifically, and only 5 (4.55%) have even courses related with the issue of CC.

Data based on Scopus, showed that in 2015, NKUA from the produced publications a 0.2% was related to the CC issue (SCOPUS 2016). NKUA organized a total number of 90 seminars, conferences and events, with 3 (3.33%) of them being

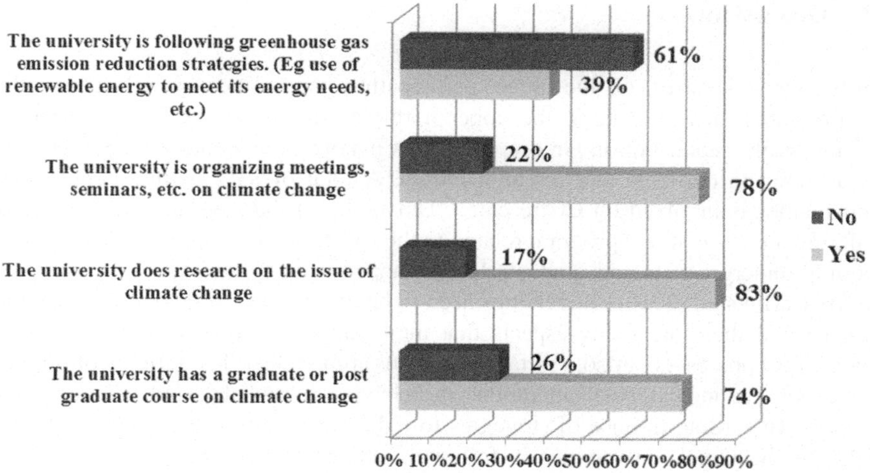

Fig. 3 Greek Universities' response *based on survey's data*

related to CC, for the year 2015. One of the major outreach efforts was the 8th International Scientific Conference on Energy and Climate Change, which was held in Athens, in 2015.

The Case of the University of the Aegean (UAegean)

University of the Aegean has a fairly large student population (16,265) and offers 1278 undergraduate courses. Only 2 of the university's courses were related with the CC issue, which means a percentage of 0.156. At the post graduate programs' courses the percent of CC ones increased to 0.769. The total number of the post-graduate programs is 40, and 1 of them is on the CC, which means a 2.5%. Out of the UAegaean's publications only a 0.7% was related to the CC issue (SCOPUS 2016). The University's outreach efforts through seminars, conferences and other events had 1 out of the 117 that took place, concentrating on CC topic, a percentage therefore of 0.85, for the year 2015.

The Case of the Democritus University of Thrace (DUTH)

This University with a student body of 25,280 and 1485 undergraduate courses has only 5 of them being related with the CC issue, which means a percentage of 0.33. At the post-graduate programs the percentage increases to 0.78. The total number of the post-graduate programs is 28, and none of them is on the CC, and only a 3.57% has courses related with the issue of CC (DUTH 2016). Related to DUTH produced publications a percentage of 1.18 was related to the CC issue (SCOPUS 2016). Their outreach attempts based on civic ecology procedures on global CC, were on the level of 1.11% of the total seminars, conferences and other events organized by the University (DUTH 2016).

3 Discussion

From the responses of the three participating universities, which cover the metropolitan area of Athens, the upper northern part of Greece and the islands' region, higher education in Greece, in its vast majority (74%) has incorporated CC issue into the curricula and research efforts. A total of 83% of the participating universities is involved on CC research. University of the Aegean is the only one offering a post-graduate program related to the CC issue. The courses related to CC both at undergraduate and graduate levels happen to be quite low.

As Carlin said (2010), universities need to be more aware of their environmental impact and there are many aspects that they can do in order to strengthen sustainability processes. Also, Virtanen (2010) argued that institutions of higher education are indicators of alterations in behavior, knowledge and practices in a society. Regarding the use of strategies to reduce greenhouse emissions, only the 39% of the Greek universities is following strategies that reduce its ecological footprint policy (Papaioannidou et al. 2016). The universities should be the leaders on efforts leading to reduction of greenhouse emissions. In addition, universities are charged with community outreach responsibility through civic ecology efforts. This would empower citizens to increase their level of resilience (Papaspiliou et al. 2014). A high enough participation (78%), of the assessed universities, is recorded in the spectrum of organization of seminars, events, conferences etc. on CC, in order to spread knowledge and raise scientific awareness. However, the results indicate that out of their individual outreach efforts only a percentage of 3.33% is devoted by the NKUA on CC. This is lower (1.11%) at DUTH and even less (0.85%) at UAegean.

As these issues of intergenerational ethics, climate justice, and "deep" sustainability become more important in higher education, there is a critical need to find a place for them on Greek campuses' curricula and way of life and collaboration with other universities around the world. University research centers in environmental and sustainability education are particularly suited to carry out the task of envisioning and practicing the appropriate socio-ecological changes. It is suggested that environmental educational approaches supported by the national mechanism of each country could lead in reinforcement of their local initiatives (Ryan et al. 2010). Nomura and Abe (2010) revealed the significance of the governmental initiatives in enhancing education about sustainability. Policies and funding support by the government are considered critical points for strengthening sustainable development education in academia. It is also suggested that sustainability leadership support among senior management in university campuses is a key factor for continuing and enforcing environmental efforts (Normura and Abe 2010). Finally although most (83%) of the Greek Universities have in their curriculum sustainable development content and most of them have green campus procedures, only a 26% of them have a related university policy (Papaioannidou et al. 2016). This possibly explains why CC is not a top priority of the Greek universities' educational and research agenda, which is what our research concludes at.

It is articulated that universities could offer various innovative initiatives in education about climate change, so as to strengthen understanding about the dimensions of sustainability (Ryan et al. 2010). As Greek universities have a great percentage of highly educated students, not only studying but also being involved in scientific researches, they could play a crucial role to the adaptation and mitigation of CC. Since this research has given data about how the Greek Universities address the issue of CC, it could be a tool in the hands of the Greek policy makers to change the tactics of the implementation of this great global issue.

References

Adger NW, Arnell NW, Tompkins EL (2005) Successful adaptation to climate change across scales. Global Environmental Change 1 (5): 77–86

Bangay C, Blum N (2010) Education responses to climate change and quality: Two parts of the same agenda? International Journal of Educational Development 30:359–368

Carlin S (2010) Getting to the Heart of Climate ChangeThrough Stories. Chapter in Universities and Climate Change: Introducing Climate Change to University Programmes (Climate Change Management), Walter Leal Filho. Springer (DE). vol. 1. Berlin. pp 89–97

Democritus University of Thrace (DUTH) (2016) internal evaluation report DUTH: 2015, Available at: http://modip.duth.gr/index.php/2011–08-26-07-51-24/internalreportsduth. Accessed 20 April 2016

Dowing TE, Ringius L, Hulme M, Waughray D (1997) Adapting to Climate Change in Africa. Mitigation and Adaptation Strategies for Global Change 2:19–46

European Union. European countries—Greece (2016), Available at: http://europa.eu/about-eu/countries/member-countries/greece/index_el.htm/ Accessed 8 April 2016

EUROSTAT (2014) Eurostat regional yearbook 2014: Education, pp. 92 Available at: http://ec.europa.eu/eurostat/documents/3217494/5786237/KS-HA-14-001-03-EN.PDF/43efce65-4b6c-4105-8c93-a6fce178c5ee?version=1.0. Accessed 8 April 2016

Feher A, Hauptvogl M, Trangosova P, and Svetlanska L (2010) How to Educate for a Healthy Climate at a University? An Intergenerational Cooperation (A Case Study from Slovakia). Chapter 11 in Universities and Climate Change: Introducing Climate Change to University Programmes (Climate Change Management), Walter Leal Filho. Springer (DE). vol. 1. Berlin. pp 137–149

Fussel HM (2005) Vulnerability in climate change research: a comprehensive conceptual framework, paper presented at the University of California International and Area Studies Breslauer Symposium, University of California, Berkeley. Available at: www.pikpotsdam.de/, fuessel/download/bresl05_published.pdf. Accessed 9 March 2016

Fussel HM, Klein RJT (2006) Climate change vulnerability assessments: an evolution of conceptual thinking. Climatic Change 75:301–29.

Hellenic National Academic Recognition and Information Center (HNARIC) (2016) Tertiary Education in Greece. Available at: http://www.doatap.gr/en/index.php. Accessed 9 April 2016

Hellenic Statistical Authority (HAS) (2016) Greece in figures: January–March 2016, Available at: http://www.statistics.gr/en/greece-in-figures. Accessed 9 April 2016

Huho JM (2015) Climate Change Knowledge Gap in Education System in Kenya. International Journal of Innovation and Research in Educational Sciences 2(3):227–233, Available at: https://www.researchgate.net/publication/279179079. Accessed 1 March 2016

Huxster JK, Uribe-Zarain X, Kempton W (2015) Undergraduate Understanding of Climate Change: The Influences of College Major and Environmental Group Membership on Survey Knowledge Scores. The Journal of Environmental Education 46(3): 149-165.

International Making Cities Livable (IMCL) (2016) Citizen Engagement via Civic Ecology: Interview with Tim Smith. Available at: http://www.livablecities.org/articles/citizen-engagement-civic-ecology-interview-tim smith. Accessed 30 March 2016

Leal Filho W (2010) Climate Change at Universities: Results of a World Survey. Chapter in Universities and Climate Change: Introducing Climate Change to University Programmes (Climate Change Management), Walter Leal Filho. Springer (DE). vol. 1. Berlin. pp 1-19

MacMillan K, Money K, Downing S, Hillenbrand C (2004) Giving your organisation SPIRIT: An overview and call to action for directors on issues of corporate governance, corporate reputation and corporate responsibility. Journal of General Management 30 (2): 15–42.

Naess LO, Bang G, Eriksen S, Vevatne J (2005) Institutional adaptation to climate change: Flood responses at the municipal level in Norway. Global Environmental Change 15 (2): 125–138.

National and Kapodistrian University of Athens (NKUA) (2016) The university. Available at: http://en.uoa.gr/the-university.html. Accessed 24 April 2016

Nomura K, Abe O (2010) Higher education for sustainable development in Japan: policy and progress. International Journal of Sustainability in Higher Education 11(2): 120–129.

Offorma GC (2014) Climate Change and the Need for New Curriculum Development in Nigerian Universities Nigeria. Climate Change, Agriculture & Higher Education: Multidisciplinary Issues & Perspectives 2(1): 158–170, Available at:https://www.researchgate.net/publication/280554166. Accessed 9 April 2016

Owen R, Fisher E, McKenzie K (2013) beyond reduction: Climate change adaptation planning for universities and colleges. International Journal of Sustainability in Higher Education 14(2):146–160, Available at: https://www.researchgate.net/publication/259533045. Accessed 9 April 2016

Papaioannidou D, Sardi C, Skanavis C (2016) How close to being "Green" are the Greek universities. 13th International Conference on Protection and Restoration of the Environment, Mykonos Island, Greece, 3rd to 8th of July, 2016

Papaspiliou K, Skanavis C, Giannoulis C (2014) Civic Ecology Education and Resilient Societies: a Survey of Forest Fires in Greece. Journal of Education and Training Studies 2(2): 48–57

Regulatory Authority for Energy (RAE) (2016) Power generation Available at: http://www.rae.gr/site/categories_new/consumers/know_about/electricity/production.csp. Accessed 10 April 2016

Ryan A, Tilbury D, Corcoran PB, Abe O, Nomura K (2010) Sustainability in higher education in the Asia Pacific: developments, challenges, and prospects. International Journal of Sustainability in Higher Education 11(2): 106–119

SCOPUS (2016). Available at: https://www.scopus.com Accessed 25 April 2016

Skanavis C, Petreniti V (2006) Non-Formal Environmental Education in Formal Education: Centers on Environmental Education in Greece, International Conference on Protection and Restoration of the Environment VIII, Chania, Greece https://www.researchgate.net/publication/260421601_Non-formal_Environmental_Education_Centers_of_Environmental_Education_in_Greece. Accessed 2 March 2016

University of the Aegean (UAegean) (2016), The University: Statistics. Available at: http://www.aegean.gr/aegean2/statistics.html. Accessed 2 April 2016

Virtanen A (2010) Learning for Climate Responsibility: Via Consciousness to Action. Chapter in Universities and Climate Change: Introducing Climate Change to University Programmes (Climate Change Management), Walter Leal Filho. Springer (DE). vol. 1. Berlin. pp 231–240

Young BE, Dubois NS, Rowland EL (2015) Using the climate change vulnerability index to inform adaptation planning: Lessons, innovations, and next steps. Wildlife Society Bulletin 39(1): 174–181

Using Improved Varieties of Pearl Millet in Rainfed Agriculture in Response to Climate Change: A Case Study in the Tillabéri Region in Niger

Abdourahamane Tankari Dan-badjo, Halima Oumarou Diadie, Sabrina Maria Rita Bonetto, Carlo Semita, Elena Isotta Cristofori and Anna Facello

Abstract The seasonal effects of global warming and water shortages begin to be observed on agricultural production and forecast trends encourage studies on adaptation to climate change. In Niger, West Africa, farmers have always had to cope with irregularity and poor distribution of rainfall. In recent years, a variation in the frequency and duration of rainy season were observed, suddenly affecting a drop in agricultural productions with the resulting food crisis. Therefore, it is necessary to find measures to adapt to the climate variability. This study focus on the Tillabéri region (Niger) where pearl millet is one of the main agricultural product. In the last few years, variations in rainfall distribution and quantity have negatively influenced the yield of the millet crops. A climatic assessment of the region has been verified collecting information from both previous studies and satellite data. Two early improved varieties of pearl millet (SOSAT-C88 and HKP) drought resistant have been distributed to local farmers in 16 pilot areas of the Tillabéri region and the crop yields were compared to those of the local traditional variety cultivated in the same area. The results have identified a significant increase

A.T. Dan-badjo
Soil Sciences Department, Faculty of Agronomy,
Abdou Moumouni University, BP 10960, Niamey, Niger

H.O. Diadie
Crop Production Department, Faculty of Agronomy,
Abdou Moumouni University, BP 10960, Niamey, Niger

S.M.R. Bonetto (✉) · C. Semita
Earth Sciences Department, CISAO, University of Turin,
Via Valperga Caluso 35, 10125 Turin, Italy
e-mail: sabrina.bonetto@unito.it

E.I. Cristofori · A. Facello
TriM- Translate into Meaning, Corso Sommelier 24, Turin, Italy

W. Leal Filho (ed.), *Climate Change Research at Universities*,
DOI 10.1007/978-3-319-58214-6_22

in production, up to 62%, with the improved varieties compared to the local one. These results suggests the possibility of a potential extension, in this region, of improved varieties to mitigate the effects derived by climate change in the agricultural productivity in order to avoid famine and guarantee food security.

Keywords Pearl millet · Early varieties · Crop yelds · Climate change · Tillabéri · Niger

1 Introduction

The effects of global warming and water shortages are beginning to be observed on food production, particularly crop and forecast trends encourage studies on adaptation.

In last decades, the agricultural production in Niger shows a general decreasing trend. This is strongly related to the combination of several environmental elements including in particular the climate variability and the extremes. Water resources availability and weather hazards such as droughts and floods have a strong impact on crops yield. Indeed, the irregularity in the precipitation trends and consequential hazard events affect the decrease in the agricultural production (FAO 2009; Tubiello et al. 2008).

In particular, in Niger, in addition to the irregularity and poor distribution of rainfall in space and time, in recent years it has been observed both a late start and early cessation of rainy season causing situations of food insecurity (Lawan et al. 2014). Therefore, an adequate assessment of precipitation trends and anomalies provides an essential element to be integrated into strategic plans for agriculture, especially for developing countries. Adaptation measures have been recently offered to farmers, including the use of improved varieties (varieties with short or medium cycle). However farmers may be reluctant to replace traditional or customary varieties from other varieties even if the improved performances have been demonstrated.

In this work, first results of a specific project with an NGO (COOPI) developed in the Tillabéri region, near to Niamey, are shown. The climate assessment has been pointed out using satellite data in order to verify the precipitation trends and anomalies of the rainy seasons. Due to recent scarce harvest of pearl millet, which is one of the main agricultural products in the region, two improved varieties of pearl millet have been tested comparing their crops yield in respect to traditional local variety. Local farmers have been involved in the project in order to raise awareness and train them on the possibility of the use of early varieties in the context of climate change.

The goal of this study is to demonstrate the contribution of short-cycle varieties in yield increase in a context of climate change to increase food security and to improve life conditions of the concerned population.

2 National and Local Climate Assessment

Niger is a Sahelian country, the climate is tropical arid and semi-arid. It is in fact one of the hottest areas of the globe and is characterized by four types of seasons (Lawan et al. 2014):

- cold season (mid-December to mid-February)
- hot dry season (March–May)
- rainy season (June–September)
- warm rainless season (October to mid-December).

This study is focused on the Tillabéri region, where suitable agronomic strategies and the development of specific projects for food security are planned and promoted (Fig. 1). Agricultural plans require the knowledge of climatic local assessment, particularly in relation to temperatures and precipitation trend and anomalies.

Climatic information on the Tillabéri area has been deduced by previous studies, particularly in the framework of the Anadia Niger project, commissioned by the World Meteorological Organization and operated by a partnership between National Directorate of Meteorology (DMN) of Niger, Institute of Biometeorology (IBIMET) of the National Research Council (CNR) and the Inter-University Department of Science, Planning and Policies territory (DIST) of the Politecnico di Torino. In general, the action has been focused on the assessment of natural disasters and their impacts on agriculture and food security. In detail, the study has

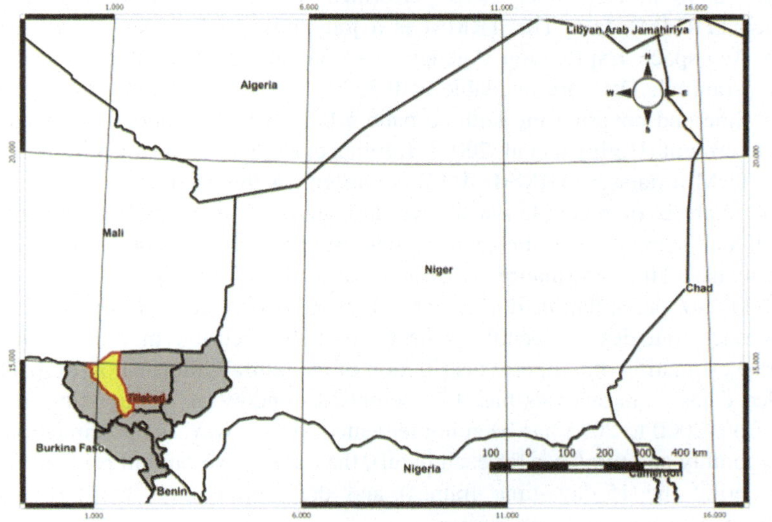

Fig. 1 Location map of the studt area (*grey* = the Tillabéri region, *yellow* = the Tillabéri municipality

been conducted in order to provide a climate characterization at regional level, with time period from 1981 to 2010, and to identify future climate keys parameters (Lawan et al. 2014).

In the Anadia Niger project, the analysis was provided using the DMN ground based network, considering the parameters of rainfall and temperature. Only the stations with a dataset covering the last 30 years since 1980 have been selected (12 stations in total). In this case the ground based stations are limited by research constrains and, in many cases, data are often not able to provide adequate or continuous observations.

In this work, the assessment provided into the Anadia Niger project has been integrated by means of satellite rainfall estimates which provide a more uniform spatial coverage compared with ground measurements. Also the comparison of results obtained from satellite data are compared with results obtained from Anadia ground stations in order to perform a validation of the two analyses.

In recent years, remote sensing instruments provide a huge amount of data that have become extensively used for detecting and extracting information about the spatial and temporal distribution, trends and anomalies of precipitation in order to obtain the basic information for the research of precipitation variation and for the long term water resource management. These data are means for quantifying the rainfall in developing countries or remote locations, where conventional rain station is sparse (Shaban et al. 2009; Scheel et al. 2011). Moreover, remote sensing techniques can contribute to assure a complete and systematic database of cartographic, climatological records, in addition to in situ, historical and current data, to produce meaningful and useful information for environmental management.

In particular, our research presents an analysis of rainfall climatology from January 2000 to December 2015 performed using Tropical Rainfall Mapping Mission (TRMM) data. The TRMM is a joint mission between NASA and the Japan Aerospace Exploration Agency (JAXA), designed to study and monitor tropical rainfall. Data are available at $0.25° \times 0.25°$ spatial resolution, with 3 h revisit time and covering the latitude band 50° N–S with temporal coverage from 1998 to present (Huffman et al. 2007; Huffman et al. 2010). The most recent version of the TRMM dataset (TRMM 3B42) is chosen for this study.

The analysis of precipitation allows derivation of the rainfall seasonality and rainfall day's trend over the area of interest, both the Tillabéri region and its municipality. This information is essential in order (i) to assess the inter-annual variability of precipitation during the analysed period and (ii) to plan suitable agronomic strategies or specific projects for food security. In order to have an accurate estimation of temporal distribution of the rainfall information from TRMM gridded dataset, an analysis based on temporal window of 15 days-time for each pixel from 2000 to 2015 has been implemented. In this way, the precipitation trend (mm/monthly and mm/year for each pixel), the cumulative rainfall gridded datasets (considering the 15 days-time frames) and the rainfall days trend (n° rainfall days/year for each pixel) are derived.

After the rainfall deficit in the 1980s, in the last two decades the trend presents a strong inter-annual variability of rainfall; however the observed trend is an increase

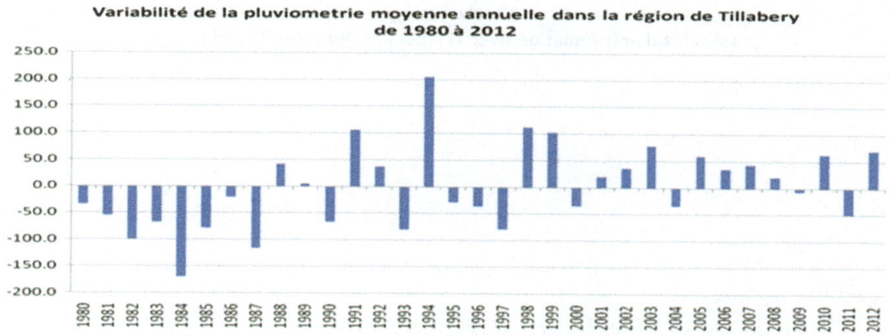

Fig. 2 Evolution of the average annual rainfall of 1980–2012. (*Source* Lawan et al. 2014)

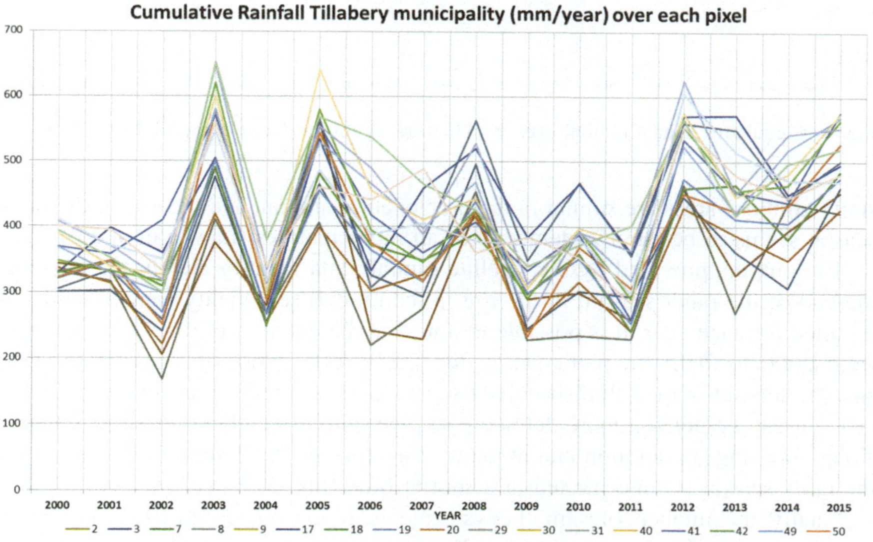

Fig. 3 Cumulative annual rainfall over the Tillabéri municipality for the time period 2000–2015

in the precipitation (amount) in the last two decades. (Figure 2) (Lawan et al. 2014). The Figs. 3 and 4 show the annual precipitation trend derived by satellite based products over the study area. The results match well with the results presented in the Anadia Niger project showing a strong inter-annual variability, characterized by a decreasing trend between 2009–2011 and by a recovery from 2012.

Concerning the rainy season, the outputs provided in the Anadia Niger project shows in generally a delay in the beginning of the rainy season on the Tillabéri region, even if there is a small part of the region, including the municipalities in the Northen sectors and someone in the South easternmost portions, with slightly early

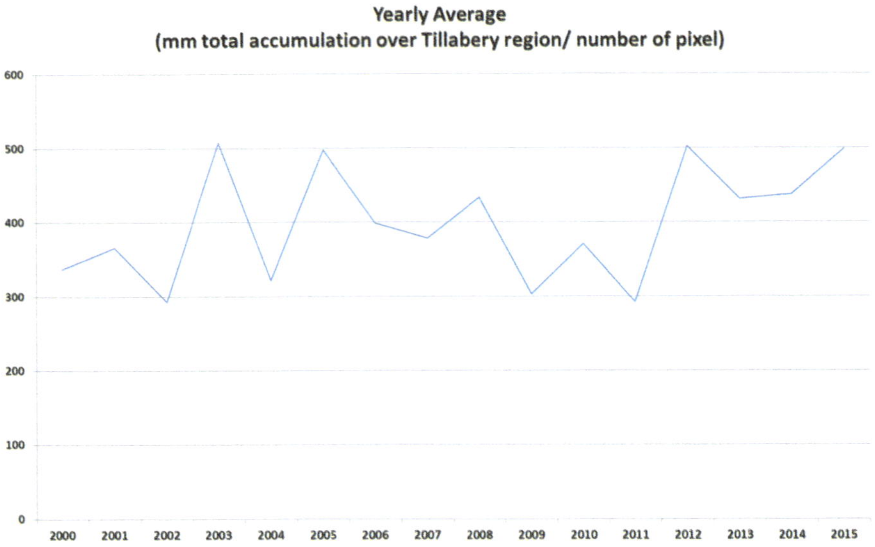

Fig. 4 Cumulative annual rainfall over the Tillabéri region for the time period 2000–2015

start. All those areas are similar in the precipitation amount and the end of the rainy season seems to be late (September) for all municipalities (Lawan et al. 2014).

The outputs provided by the satellite rainfall data identify clearly the start and the end of the rainy season. In the Fig. 5 the rainfall seasonality (mm/monthly) is reported for each year. It is possible to note that the duration of the season changes from 2000 to 2015. In particular in the 2000–2005 period the season shows a smaller amount of rainfall distributed over a longer interval. In recent years (2012–2015) it can be noted a more defined peak, the rainfall trend curves show a higher slope, focusing on duration and amount. Therefore the total amount of rainfall for the rainy season is concentrated in a shorter time (Fig. 5).

Finally, the numbers of rainfall days is increased in the decades 1990s and 2000s compared to the 1980s and 1990s, while generally the lower number of no-rainfall days confirm the upward trend of rainfall (Lawan et al. 2014). The results of number of rainfall days derived from satellite rainfall data (threshold 0.01 mm rainfall/days) are showed in the Fig. 6. It is possible to notice that number of rainfall days decreased between 2009 and 2011, while they slightly increased from 2012.

Based on previous observations on rainy season, the rainfall trend is increased from 2012, but at the same time precipitations are showed a defined peak on duration and amount. Probably, in accordance with the precipitation trend, rainfall days are concentrated in a limited period.

Generally by the analysis of rainfall satellite data, combined with the existing data, it is possible to note a slight increase in precipitation, as a number of days and amount, focused in a limited timeframe.

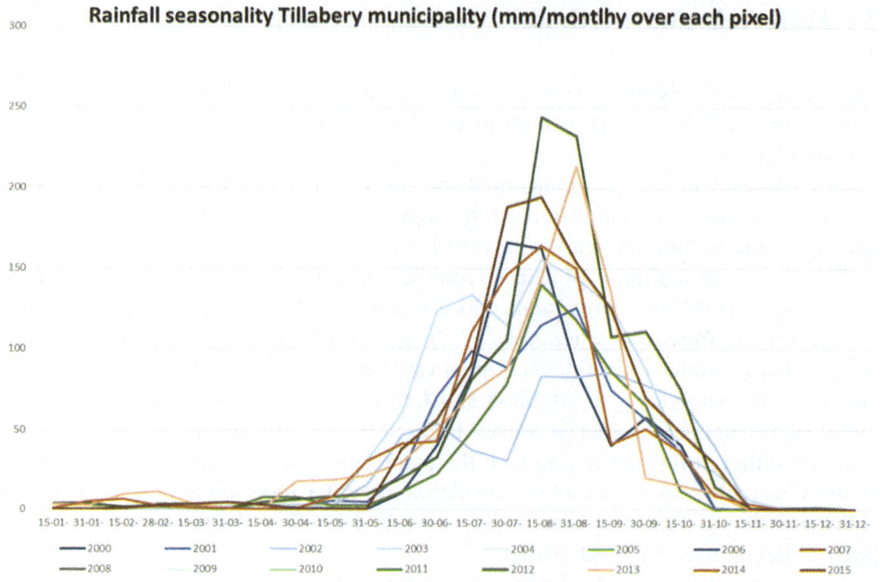

Fig. 5 Rainfall seasonality over the Tillabéri region for the time period 2000–2015

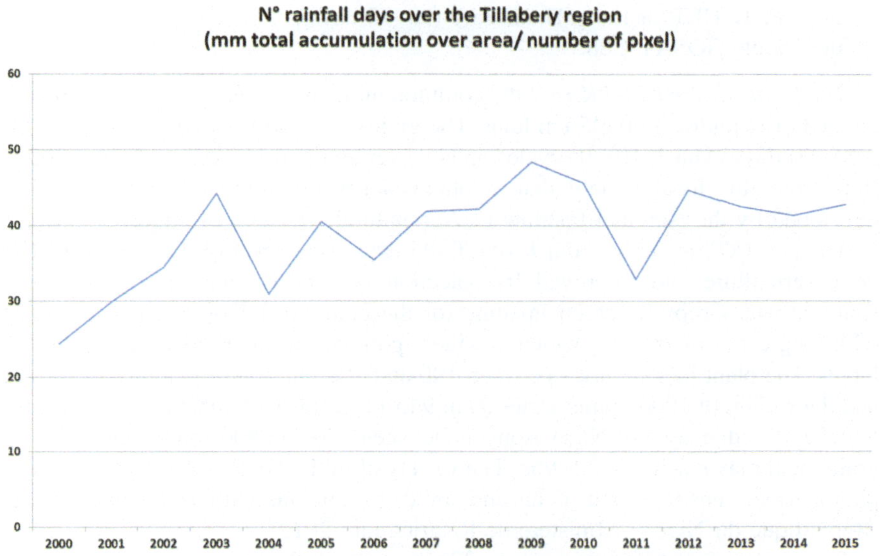

Fig. 6 Number of rainfall days (threshold 0 mm rainfall/days) over the Tillabéri region for the time period 2000–2015

3 Materials and Methods

The study illustrates the agronomic strategies and the development of a specific project for food security promoted in the Tillabéri region in relation to the local climate change.

In particular, the project concerned the use of early pearl millet varieties in a context of climate variability in order to improve the productivity due to the scarce harvest of the traditional pearl millet cultivated in the areas in the last few years.

The local agricultural practices consist on associated system of cultures: milled—cowpea (*Vigna unguiculata*, locally defined "Niébé") or millet—sorghum.

The millet (*Pennisetum glaucum [L.]*) is a cereal plant commonly known as pearl millet or millet whose development cycle is very variable depending on their variety: early varieties (75–90 days) and late varieties (120–150 days). The term millet represents pearl millet in West Africa; however, it means pearl millet as well as finger millet in East Africa. In fact, there are a considerable number of regions in West Africa where pearl millet is considered a regular food in the diet. Pearl millet is the mainstay crop in the Sahel region. In this document, the term millet refers to pearl millet as used in West Africa.

Once defined the climatic assessment and recent precipitation trend, three millet varieties were test in this project in the Tillabéri region:

- the local variety Haïni Kiré,
- the variety HKP or Early Haïni Kiré (90–95 days),
- the variety SOSAT-C88.

The local variety Haïni Kiré is the common millet in Western Niger. The spike is tapered or cylindrical 60–95 cm long. The grains are yellow in color and the cycle is 85–90 days (Mai 2010). Its cycle can however go up to 110 days. (Azuma 2009). It is from this local variety that is obtained several improved varieties and/or developed by the National Institute of Agricultural Research of Niger.

Variety HKP or Early Haïni Kiré (90–95 days) was developed by the Ministère de l'Agriculture and improved by selection of four generations by ICRISAT (International Crops Research Institute for the Semi-Arid Tropics). It is a variety with long ears (55–60 cm), which is widely popularized in all areas "milicoles" of Niger. The plant height varies between 190 and 200 cm; tillering ability is average and the weight of 1000 grains varies from 9 to 11 g. It is particularly suitable for the Sahel and Sudan zone of Niger isohyets between 350 and 800 mm where potential grain yield may reach 1.5–2.5 t/ha. The variety of millet HKP is slightly sensitive to photoperiod sensitive and coal and mildew, but insensitive to head miner (République du Niger—Ministère de l'Agriculture 2012).

The variety SOSAT-C88 is obtained by recurrent selection from the crossing in Mali between local cultivars Souna and Sanio Mali. Flowering starts from 58th day after sowing, maturity from 85 to 90 day after sowing. The average plant height is 200 cm. The ability tillering is estimated at 3 tillers per plant. The potential grain yield is 1.5–2 t/ha. His harvest index is 30%. The spike is dumbbell, compact, medium length of 28 cm, circumference of 12 cm. The seed is yellow-olive color, oval. The 1000-seed weight is 10 g. This variety of millet is susceptible to striga, the borer blight and stem borer; resistant to mildew and ergot, tolerant to lodging, drought and slightly sensitive to photoperiod. The recommended cultivation area is the area between 350 and 600 mm annual rainfall on sandy soils and semi-clay (République du Niger—Ministère de l'Agriculture 2012).

Seeds of each variety were distributed to local farmers and monitoring on cycle duration and crop evolution and yield measurement of the millet improved variety in respect to the traditional local variety was carried out. For each variety tested, the analysis concerned all his growth cycle, from sowing to harvesting grain. The determined parameters are the number of tillers ears with grains, the number of ears Grain/plot; the gross and net weight (after drying in the sun); weight husk and the yield of dry grains by accession.

Data were processed using SPSS and Excel. The obtained data were submitted to variance analysis to identify the presence or absence of significance (at the 5% probability) between the average yields of millet crops depending on the variety used. In the presence of significant differences, the Student Newman Keuls test (at 5% level) was conducted to identify significantly different treatments depending on the variable (Roberts et al. 1999; Zar 1999).

The experimental phase

The experimental phase of this study involved four municipalities of the Tillabéri department that are Anzourou, Dessa, Sakoira and Tillabéri. In each municipality, 2 villages were selected and in each village, 2 pilot farmers were chosen. So in total, the experiments were conducted on 16 pilot areas ($4 \times 2 \times 2$). In the test sites the climatic parameters are quite similar with a precipitation trend as referred in the previous paragraph.

Farmers involved were 16 with an age ranging between 30 and more than 60 years old. Statistical analysis on social, economic and education conditions were conducted In particular, 81.25% of them are in the age group from 30 to 60 years and 18.75% have an age upper to 60 years.

Outcomes of the investigation revealed that 70% of the farmers do not have a scholar instruction or agricultural basic training. Schooled farmers are 30% considering primary or secondary school level.

The number of fields varies according to the households: 68.75% of the investigated households owns two fields, 18.75% have one field and 12.50% have 3 fields.

Each farmer received 10 kg of seeds of improved millet between June 27th and July 10th, respecting the local traditional sowing period for millet (since mid-June to the beginning of July).

Most of the farmers (93.25%) sowed all the improved seeds, while the 6.25% sowed only a part of the seeds.

31.25% of farmers sowed in June and 68.75% of them in July. None of them had to make the re-sowing.

81.25% of the involved farmers own 2 or 3 fields, so if they sow at least a field with the varieties improved, it would guarantee them at least a part of the annual production of millet in case of irregularity of rainy season.

At the end of the experimental phase, the opinion of the farmers who tested the different millet varieties was collected in order to verify the satisfaction and social agreement in respect to the introduction of new varieties of millet.

4 Results

Table 1 reports the opinions of the farmers concerning the springing up of the plants, the development of the crop and the bioticks constraints according to the varieties of millet. It reveals a good germination and a good crop development with both improved varieties in particular for the HKP variety. With the local variety, the germination is medium and the crop development is bad. Concerning the constraints biotic, we observed many more enemies attacks of cultures on the local variety than to both improved varieties.

The Table 2 reports the date of sowing and the durations of the phenological stages according to the varieties. For the improved varieties, the longest period is observed between the tillering and the earing with 39 days and 36 days respectively for the SOSAT-C88 and the variety HPK. For the local variety, the longest period consists of the flowering—maturity stage (36 days).

Generally, the improved varieties reached the maturity more quickly than the local variety: 80 and 82 days respectively for the variety SOSAT-C88 and the variety HPK, instead of the 92 days of the local variety.

Table 1 Crop development according to the varieties

Varieties Crop stages	SOSAT-C88	Early Haini Kiri (HKP)	Local variety Haini Kiré
Germination	Good	Good	Medium
Crop development	Good	Very good	Bad
Biotic constraints (attacks of the enemies of the cultures)	Weak	Weak	Medium

Table 2 Sowing periods and duration of the phonological stages in days after sowing according to the varieties

Varieties	Sowing	Springing up	Tillering	Earing	Flowering	Maturity
SOSAT-C88	05 Juillet 2015	–	21	60	63	80
HKP	10 Juillet 2015	–	23	59	62	82
Local variety Haini Kiré	27 Juin 2015	–	22	53	56	92

Table 3 Weight in grains according to the tested varieties of millet

Varieties of millet	Weight in grains (g/m^2)
SOSAT-C88	120
HKP	140
Local variety Haini Kiré	86

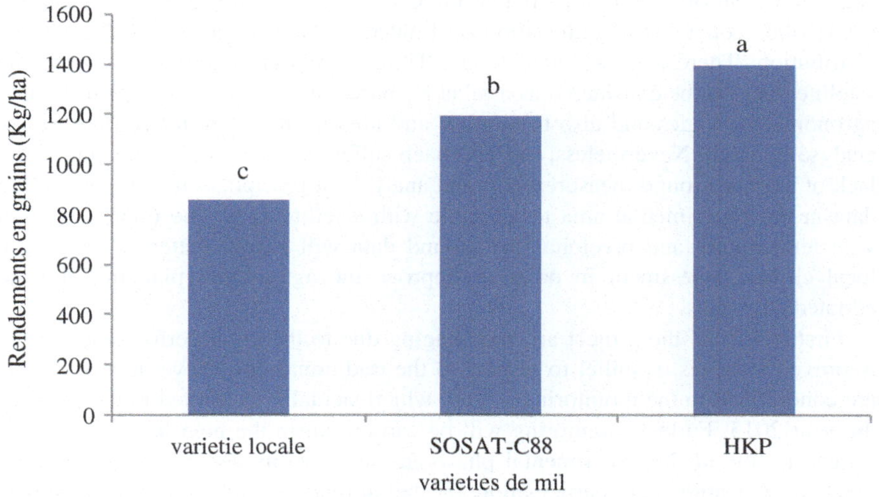

Fig. 7 Crop yield (Kg/ha in grain) according to the varieties of the millet

Table 3 and the Fig. 7 report the average weight in grain and the crop yields in grains of the varieties of millet tested by the experimental farmers. The weight of the grains is bigger for both improved varieties. The results of the statistical analyses revealed highly significant differences of the crop yields according to the varieties. The highest yield (1400 kg/ha) is obtained with the improved variety HPK, followed by the improved variety SOSAT-C88 (1200 kg/ha) and finally by the local variety HP (860 kg/ha). Indeed, compared with the local variety, we observe a significant increase of the yield crop of about 62% with the HKP variety and 40% with the SOSAT-C88 variety.

Histograms are significant different at the threshold of 5%.

5 Discussion

In this study, the climatological analysis is performed using remotely sensed data in order to integrate and compare the results obtained from satellite data with the assessment provided into the Anadia Niger project. By the comparison, it is possible to observe that the general trend of the climatology of the Tillabéri region is highlighted by both of them and are coherent, in particular the precipitation trends derived by satellite data match well with the results presented in the Anadia Niger project, even if no ground station are present nearer to the experimental areas. Traditionally, ground based data provide a more accurate and detailed observations, such as temperatures and rainfall. However, satellite data provide a more uniform spatial coverage, a huge amount of data and an easier, not expensive and faster collection of information, in particular in developing countries or remote locations, where often conventional rain station are limited in number and/or spatial-temporal distribution. Therefore, as regard to the Tillabéri region agricultural project, the satellite data can be considered as a reliable source of data in order to plan suitable agronomic strategies and also to monitor and identify the more interesting area to analyse in detail. Nevertheless, this approach suffers of some limitations due to the lack of regular ground measurements and analysis of precipitation and temperature data in the experimental area to integrate with satellite data. The implementation with temperature and precipitations ground data will allows bettering describing local climate assessment in order to improve the agricultural planning and the adopted strategies.

First results of the project are encouraging due to the good performance of the improved varieties of millet to respect to the traditional one. Nevertheless, results are conditioned to the monitoring period, which, actually, is limited to the crops of the year 2015. Further monitoring will be carried out in the next few years.

The results of the experimental phase are also conditioned by the agricultural practise, education and participation of the farmers involved. Personal choice, habits and experience of the farmers should influence the crop yield obtained (quantity of seed really sowed, period of the sowing, irrigation type and quantity, equipment, etc.), as well as schooling and comprehension of the problem.

Unfortunately, the academic level of the experimental farmers is very low: more than two thirds of the farmers (70%) are illiterate and are unable to read and understand the information registered on the packagings of the improved seeds. That compromises a good use of the improved varieties. Therefore, it would be necessary to plan sessions of information and awareness in language "Zarma", which is the most spoken in the region of Tillabéri.

Most of the farmers (87.5%) declared that the seeds quality were good and 12.5% assert that seed were of medium quality, but all the experimental farmers who tried the improved varieties would like to continue to use them to face the climatic variations. It suggests that the distribution of the improved varieties through "fields schools "or" the experimental farmers" is an effective method allowing to reduce the reluctance of the farmers towards the adoption of the new techniques or the news strategies to reduce the impacts of the climate change.

The project is actually limited to the analysis and comparison between different varieties of millet, but, in order to guarantee food security and improve agricultural production, agricultural strategies have to include similar experimental studies also for the the other species traditionally cultivated in the area.

6 Conclusion

A decrease of agricultural yields in Niger due to climate variability (the strong inter-annual variability characterized by an alternation of decreasing and increasing trends, and a slightly change in duration in the rainy season) encourages research projects to find measures to adapt to this climate variability, assuring food security particularly in poor areas.

The study illustrates the agronomic strategies and the development of a specific project for food security promoted in the Tillabéri region in relation to the local climate change.

In particular, the project concerned the introduction of two improved pearl millet varieties in a context of climate variability in order to improve the productivity due to the scarce harvest of the traditional pearl millet cultivated in the areas in the last few years.

The varieties of improved millet were tested in 16 pilot areas involving local farmers and the phenological stages, from sowing to harvesting, were monitored and compared to those of the traditional millet variety. The determined parameters are the number of tillers ears with grains, the number of ears Grain/plot; the gross and net weight (after drying in the sun); weight husk and the yield of dry grains by accession.

The results of this experimental study showed a good germination and a good crop development of millet with both improved varieties in particular for the variety HKP.

Both the improved varieties SOSAT-C88 and HPK finish their cycle respectively in 80 and 82 days, whereas the local variety finishes its cycle in 92 days.

The yields crop in grain of millet are respectively 1400, 1200 and 860 kg/ha for the improved varieties HKP, SOSAT-C88 and the local variety. In particular, the yield of the HPK variety (up to 1400 kg/ha) is rather close to those obtained with the same variety in 2013 in a region of Niamey (1388–1863 kg/ha) where weather conditions are rather close to that of the region of Tillabéri.

The yield crop showed a significant increase of the order of 62% with the HKP variety and 40% with the SOSAT-C88 variety with regard to that of the local variety. It suggests that both improved varieties adapt themselves well to the irregularity of the rainy season in the region of Tillabéri

The varieties HKP and SOSAT C 88 are resistant in the drought on the contrary to the local variety and they can pursue their development in case of temporary hydric deficit. This result suggests that these earlier varieties can be good alternatives (strategies of adaptation) to the climate change marked by the late starting of the rainy season or its early stop in this region.

Future development of the research will increase the quantity of seeds (15 kg/farmer) with an earlier distribution at the beginning of June in order to have enough time to plan agricultural works, according to the local periods of sowing for cereal crops in particular for millet (middle of June up to the beginning of July).

Moreover, the area will be implemented with a weather station to allow correlating the data obtained with the local climate parameters such as the height and rainfall distribution over time.

Similar study on other improved varieties of cereals, as the cowpea—*Vigna unguiculata* ("niébé") and the sorghum will carried out in future projects.

In future, the result of this study will be shared with other areas in Niger in order to educate and train the farmers on the usefulness of the improved varieties in the context of climate change.

References

Azuma H. (2009) Les céréales au Niger: accent sur le mil et le sorgho, Rapport d'études, 117p. Association pour la Collaboration Internationale en matière d'Agriculture et des Forêts du Japon (JAICAF) Ed. Sozosha, LTD

FAO (2009) Food Security and Agricultural Mitigation in Developing Countries: Options for capturing synergies. 84 p. http://www.fao.org/docrep/012/i1318e/i1318e00.pdf (Last Accessed 18 Mai 2016)

Huffman, G.J., Adler, R.F., Bolvin, D.T., Nelkin, E.J. (2007) The TRMM Multi-satellite Precipitation Analysis (TMPA), In Satellite Rainfall Applications for Surface Hydrology, [Hossain F. and Gebremichael M., Eds.], Springer Verlag, chapter 1, pp. 3–22

Huffman, G.J., R.F. Adler, D.T. Bolvin, E.J. Nelkin (2010) The TRMM Multi-satellite Precipitation Analysis (TMPA). Chapter 1 in Satellite Rainfall Applications for Surface Hydrology, F. Hossain and M. Gebremichael, Eds. Springer Verlag: 3–22, ISBN: 978-90-481-2914-0

Lawan K G, Bacci M. and Mouhaimini M. (2014) Caracterisation Climatique De La Region De Tillabéri. Report N°1 Anadia Niger Project. 36 p. http://www.meteo-niger.org/data/files/ANADIA_RT1_Climat_300614.pdf (Last Accessed 18 Mai 2016)

Mai O. A. (2010) Pratiques paysannes de gestion des mils précoces et tardifs et suivi de la floraison à Sina Kora (Zone de Kouré), mémoire de fin d'études ITA, Option Production Végétale, FA, UAM de Niamey, 46 p.

République du Niger – Ministère de l'Agriculture (2012) Catalogue national des espèces et variétés végétales (CNEV), 276 p. http://www.fao.org/fileadmin/user_upload/spid/docs/Niger/CatalogueNationaldesEspecesetVarietesVegetales-Niger.pdf (Last Accessed 18 Mai 2016)

Roberts, Maxwell; Russo, Riccardo (1999) Following up a one-factor between-subjects ANOVA". A Student's Guide to Analysis of Variance. Filey, United Kingdom: J&L Composition Ltd. pp. 82–109. ISBN 0-415-16564-4

Scheel, M.L.M., Rohrer, M., Huggel, C., Santos Villar, D., Silvestre, E., Huffman, G.J. (2011) Evaluation of TRMM Multi-satellite Precipitation Analysis (TMPA) performance in the Central Andes region and its dependency on spatial and temporal resolution. Hydrol. Earth Syst. Sci., 15, pp. 2649–2663

Shaban, A., Robinson, C., El-Baz, F. (2009) Using MODIS Images and TRMM Data to Correlate Rainfall Peaks and Water Discharges from the Lebanese Coastal Rivers. Journal of Water Resource and Protection, 1(4), pp. 227–236

Tubiello F., Schmidhuber J., Howden M., Neofotis P. G., Park S., Fernandes E., Thapa D. (2008) Climate Change response strategies for agriculture: Challenges and opportunities for the 21st century. Agriculture and Rural Development Discussion Paper 42 The World Bank 75 p. http://www-wds.worldbank.org/external/default/WDSContentServer/WDSP/IB/2009/06/16/000333038_20090616005440/Rendered/PDF/489210NWP0dp4210Box338934B01PUBLIC1.pdf (Last Accessed 18 Mai 2016)

Zar, Jerrold H. (1999) Biostatistical Analysis (4th ed.). Newburry Park, CA: Prentice Hall. pp. 208–230. ISBN 0-130-81542-X

Preserving Cultural Heritage from the Impacts of Climate Change. The Case of Musical Instruments and Scores

Artemisa M. Reyes Gallegos and Víctor Manuel Lopez-Lopez

Abstract Climate change impacts physical, social, and cultural aspects that directly threaten forms of life, worship and socialization in communities (and their values). The purpose of this work is to promote awareness of the impacts, not only economically but also those to cultural properties. Public and private organizations show little interest in introduce projects that consider variable of risks impacts of climate change on cultural heritage. Even when the intensity of extreme climate events impacts are increasing every year, both the national infrastructure and the cultural heritage. This work intends to highlights the importance of adaptation and prevention of the effects of climate, particularly those examples of heritage that are related to cultural musical activity, such as documentary collections of sheet music, and musical instruments. Many properties of cultural heritage are at risk of deterioration due to environmental degradation and global warming. Consider that impacts in isolation are not enough to know their effects in the cultural field. Better understanding and management of risks demands an integral vision for preserving both historical buildings such as concert halls and the conditions for the musical instruments that kept within their walls. In conclusion, climate impacts to society, for better or for worse, and climate change can definitely affect our own political, economic and social future. Changes taking place adaptation is an option to safeguard the historical cultural heritage and mean an opportunity for culture.

Keywords Cultural heritage · Climate change · Musical instruments · Preserving cultural heritage · Scores

---

A.M. Reyes Gallegos (✉)
Faculty of Music, National Autonomous University of México (UNAM), Xicoténcatl 126, Col Del Carmen, 04100 Coyoacán, México City, Mexico
e-mail: artemis.rg@comunidad.unam.mx

V.M. Lopez-Lopez
National Polytechnic Institute, Building 12, Unidad Profesional "Adolfo López Mateos" del IPN. Delegación G.A.M., Mexico City, Mexico
e-mail: victorlopez@ai.org.mx

W. Leal Filho (ed.), *Climate Change Research at Universities*,
DOI 10.1007/978-3-319-58214-6_23

1 Introduction

In general terms, it can be asserted that all kinds of natural and cultural heritage are exposed to multiple hazards of damage, loss, destruction, exhaustion or transformation. This can threaten their physical integrity and compromise their social and economic values. In addition, in the case of cultural heritage, it means a loss of a source of information from the past and from a symbol of identity. The reason to raise awareness to stakeholders, decision makers and to managements of cultural policies is to highlight the importance to deal and protect the cultural properties of these climatic impacts, and in this way begin safeguard actions.

There are many types of factors affecting the preservation of cultural heritage, which have their origin in the impacts on the environment: those produced by man and those produced by the live cycle of the materials with which these are made. They can be affected by human contact such as mass tourism, socio-economic pressures and changes in lifestyles.

By correlating the natural risks and those of human origin, it is common to speak of negative consequences in national and communal economies, loss of infrastructure and implications direct and immediate to the population. However, cultural heritage generally omitted or ignored is still a factor of sustainable development (RC 2012). Its conservation status is related to the state of the infrastructure and involvement population.

Perhaps, the most important challenge for world heritage in our time is to keep their values to society while at the same time, to accommodate changes imposed by large global phenomena such as growing inequality, globalization, climate change and mass urbanization.

Music goes far beyond the sound emission of instruments and voices. Music and traditional musical instruments deeply convey spiritual and cultural principles and aesthetic values to groups and communities, as well as expertise in many fields, which are necessary to know from different points of view (Dournon 2000).

2 Background

Since 2005 the *Center for World Heritage* (WHC) of the *United Nations Educational, Scientific and Cultural Organization* (UNESCO),[1] warned about the impacts of climate change on cultural heritage, highlighting the importance of acting for its adaptation, mitigation and protection. Because of this, the Vilnius Report[2] (WHC 2006) insisted that the global community must to implement

[1]With the IUCN, ICOMOS and ICCROM.

[2]Document result of a joint report from the World Heritage Centre, its Advisory Bodies, and a broad group of experts to the 30th session of the World Heritage Committee at Vilnius, Lithuania, 8–16 July 2006.

Table 1 Main factors of threat to the outstanding values of the world heritage properties

Construction and development	Transport infrastructure
Infrastructure services	Pollution
Local conditions that affect physical tissue	Use social and cultural heritage
Other human activities	Management and factors institutional
Climate change and severe weather events	

strategies to protect the values, integrity and authenticity of the sites and properties of cultural heritage, and develop safeguard projects.

Before, in 1982 the WHC published its first report concerning cultural heritage impacts due to climate change and severe weather events. Although the studies since then have been irregular, the broad increase of impacts to such properties (2% in 2001 reaching 10% in 2013) reflects the statistical analysis advertised in 2014 (Veillon 2014).

Three important documents alerted the international community to pay attention to the impacts of climate change and cultural heritage: two reports were prepared by the WHC: *Predicting and managing the impacts of Climate Change on World Heritage* (WHC 2006) and the *Strategy to assist States Parties to implement management responses*, (Collete 2007). Lastly, *Policy Document on Impacts of Climate Change and World Heritage* (UNESCO/WHC 2008).

Is important for to have an approximation about the current situation of cultural heritage, the analysis made by the WHC, product of a study that spanned from 1979 to 2013, in which 77% of the 2642 cases considered, located at 108 states parties, and are inscribed in the World Heritage list.

Thu, in 2015, the system of information of the State of Conservation (SOC) of the WHC, highlighted 14 primary involvement factors that represent a threat to the outstanding values of the World Heritage property (WHC/UNESCO 2015). For cultural heritage sites, in the context of contemporary life, stand out among others (Table 1).

3 Methodology

Is important to say that, this work takes into account several sources not only those official sources. On the one hand, some academic works written by researchers, as well as studies by luthiers: specialists in the maintenance, care, and repair of musical instruments, and by manufacturers of musical instruments. In this instance there are opinions and experience of specialists in construction and repair, maintenance and tuning of musical instruments from the Faculty of Music of the National Autonomous University of México (UNAM).[3] On the other hand,

[3]A luthier, specialist in construction and repair of percussion, bowed string instruments and guitar, and responsible of maintenance and tuning of pianos, both part of the team of technical managers responsible for instruments of the Faculty of Music of the UNAM.

interviews to professional musicians in the field of classical music. (Burke and Macdonald 2014; Vaughan 1981), particularly those who are part of symphony orchestras. In addition, empirical evidence of the experiences musical performers was used.

Among the revised information, it is worth mentioning the reference to climate change and its impacts in different areas of human activities. The literature pertaining to the threats that affect different types of cultural property has been also revised (De La Torre 2014). In particular, it focuses on the impacts of climate change on the documentary archives, in this case: scores, and on the effects of temperature and humidity to musical instruments (Arnoult 2003; Brandt and Foucaud 1998; Cervantes 2008; Anon., s.f.; Gómez Gonzalez et al. 2008; Matthews and Feather 2003; Mcilwaine 2006; Tacón Clavaín 2008).

Specific information is not abundant, but it refers to the risk control plans of cultural heritage and the management and preserving them (Cassar 2011; Chapuis 2009; France 2009). In all perspectives, the musical point of view is privileged by correlation to the management of cultural heritage.

4 Cultural Heritage and Climate Change

The main factors related to climate change and sever weather that impacts cultural heritage, are storms and flooding, followed by temperature changes, droughts, and finally humidity changes.

Given that cultural heritage is an individual and community social construct linked to social interactions and the notions of identity and cultural cohesion, the responses to the impacts of climate change should not separated from physical, social and cultural dimensions. A multidimensional understanding of impacts and decisions for mitigation and adaptation of its effects is imperative where they converge on the synergy of disciplines, including arts, humanities, as well as social, physical and biological sciences, technology and engineering (Reyes Gallegos 2014).

Both the United Nations Organization (UN) and UNESCO have taken letters in the issue, since climate change, directly and indirectly is affecting both tangible and intangible[4] cultural manifestations and diverse heritage collections. In extreme cases, even these can become a 'national security issue'.

Multidisciplinary studies have been carried out by specialists, such as those published by the Getty Conservation Institute, the Environmental Technologies and Pollution Prevention European Commission, the WHC of UNESCO, the International Federation of Library Associations and Institutions (IFLA), Core Activity on Preservation and Conservation (PAC), the Intergovernmental Panel on Climate Change (IPCC) the International Centre for the Study of the Preservation and Restoration of Cultural Property (ICCROM), the International Council on

[4]Also nominated as 'material' and 'immaterial' heritage.

Monuments and Sites (ICOMOS), the International Union for conservation of nature (IUCN), to name just a few of the most representative.

All of them conclude that the extreme weather events resulting from climate change are a shock to global society and present risks and effects that can become irreversible on the physical, social, and cultural heritage, as well as on natural heritage.

5 Climatic Impacts on Cultural Heritage

Several factors can affect cultural buildings such as concert halls, museums, conservatories and music schools (Worland 2011). Some are of outdoor origin and others originate inside them. The main external cause is environmental conditions. Internal causes are poor control of temperature, humidity and poor air circulation (Brandt and Foucaud 1998). In both cases, factors such as humidity, temperature and ventilation relative, are involved as well as increased ultraviolet ray levels (UV). Specifically, humidity and temperature, environmental factors, are determinant in the performance of materials and the conservation of musical instruments and scores (Tacón Clavaín 2008) but these affect human behavior as well. Whatever the point of interest, "the assessment of the impacts of climate change on cultural World Heritage must thus account for the complex interactions within and between natural, cultural and societal aspects" (Collete 2007).

The adverse impacts of climate change will have consequences on all of humanity, including the products of human creativity. Interactions are two-way: on the one hand, they are subject to climatic changes, but some of them are also, propitiatory of climatic variations. Changes in climate have physical and social impacts on cultural heritage. Important direct physical impacts can be expected on movable and immovable assets such as: historical buildings, wood and organic material items, stability of structures; and on all types of heritage, such as real and personal properties (among those, cultural heritage and musical items) run the risk of being exposed to high levels of humidity, temperature and UV rays. (WHC 2006; CPM 2006).

Impacts on cultural heritage, in extreme cases, can lead to the degradation or destruction of the properties thus causing a loss of elements of cultural memory. The *Report on predicting and managing the impacts of climate change on World Heritage*, published in 2007 by the UNESCO presents a breakdown of the relationship between climate change and the effect-impact produced by the results of climate change to different climatological phenomena.

In Mexico, a methodical study of the situation of cultural heritage in general and its vulnerability to the effects of climate change has not been made, but the *National Climate Change Strategy 2013–2018, 10-20-40 Vision* (2013),[5] presents a breakdown by climate events in municipalities with a high risk of disruption (see Fig. 1).

[5]NCCS for its acronym in English. Document of the current Federal Government of México by 2012–2018.

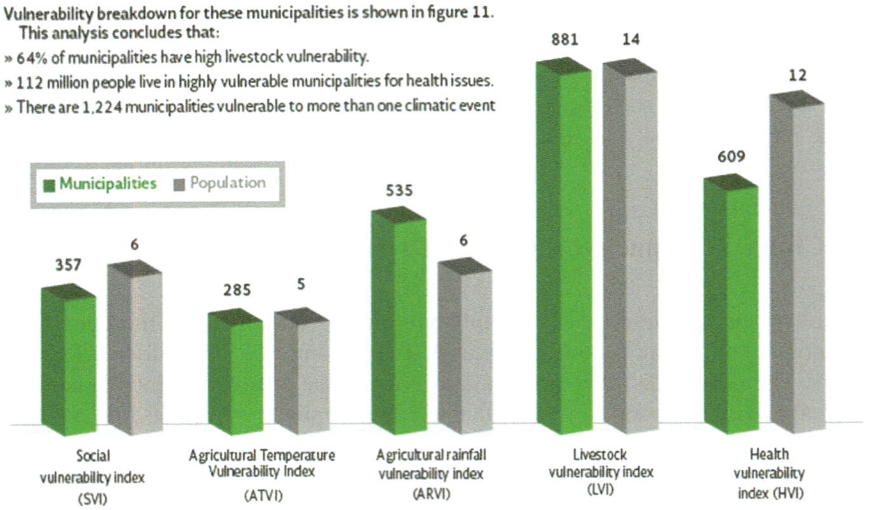

Fig. 1 Vulnerability indexes (Semarnat 2013)

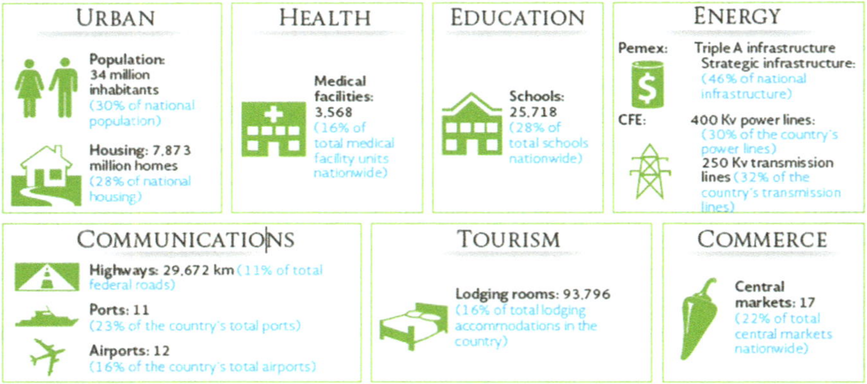

Fig. 2 Analysis of strategic infrastructure in municipalities with a high risk of floods and landslides (Semarnat 2013)

This document identifies various types of strategic social and economic infrastructures exposed to catastrophic events (see Fig. 2) Although cultural heritage is not mentioned, we must take note that in municipalities there are many different types of heritage and maybe we can deduce that these will also be affected at different levels.

6 Musical Heritage Assets

Culture is often associated with art and aesthetics themes, and is seen as a process of creativity and an application of scientific and technological resources, joined with the field of sociology, and more specifically with humanities. The concept of culture associated with aesthetics, it relates to both artistic and scientific creative procedures. In such cases, each community has certain elements that "create" and "preserve" culture in its various forms, producing, executing or interpreting cultural works. States, the private initiative and civil society organizations have museums, concert halls, theatres, libraries, conservatories, universities and other venues. The "cultural creator symbolizes the freedom of thought and expression, which has been one of the motivating forces of the struggles for human rights throughout history" (Stavenhagen 2001).

Art, Philosophy, and Philology have constantly been considered as accessories and have been given no priority. The concept of culture has been based on indices of economic and technological development. Fortunately, now it is recognized that culture is also involved in science, technical, and economic fields, then, sometimes take in account for maker decisions.

Music in the academic field is considered 'art', and thus, is an element of culture. It was not until a distinction was made between the types of heritage as being tangible or intangible, that culture became classified as heritage and music was able to be declared as a cultural heritage of humanity. By its nature, it has a place both in the conception of tangible and intangible heritage, since it can manifest itself in rituals and traditional events that revolve around traditional values and spiritual conceptions. Also recurring to musical instruments and material objects to express it.

It is surprising to note the absence of the mention of music properties in national legislation, of production and musical activities when they really are a significant part of daily, official, civic life, as well as of recreate activities. There are abundant references and descriptions of practice and musical creation in documents of different fields, (e.g. law, education, history, literature, philosophy, academic and artistic research, migration, performance, press, judicial documentation, sociology, etc.). Reviews of concerts are considerably abundant, but attention to what happens to the objects and infrastructure with which music is performed is hardly existent. Musical activity addresses both physical and virtual community spaces and public and private entities aimed at musical recreation, such as conservatories, schools, bands, orchestras, broadcasters and others (Baz 2008).

The fact that direct mention is not made in Mexican legislation about musical goods as part of the cultural heritage of the nation does not mean that they are not recognized. Mexico values the importance and significance of projects and programs for their protection, conservation, and dissemination both nationally and internationally and they identify the musical goods as necessary elements for the projection of their nation. (National Development Plan 2012–2018).

Mexico has a wide range of historical and cultural heritage goods. A major achievement of the emphasis of cultural heritage preservation is that they can be used for sustainable development and as a projection of local, national and international identity. Important building considered as historical and artistic heritage are more attended, thus they retain its value.

Since the establishment in 1988 of the National Council for Culture and the Arts (CONACULTA[6]), management and safeguarding of cultural heritage was delegated at an institutional level to the General Directorate of sites and monuments. That is responsible for protecting, restoring, preserving and cataloguing the federal sites and monuments for their artistic and historical value.

Recently, under a presidential initiative, Congress decreed the creation of the Ministry of Culture (Congreso de la Unión 2015), assigning them among other issues to "preserve, protect and maintain the archaeological, historical and artistic monuments that make up the cultural heritage of the nation, [as well as] scientific research in anthropology and history, [...] the conservation and restoration of cultural, archaeological and historical heritage, [...] the protection, conservation, restoration and recovery of that heritage and the promotion and dissemination of such materials". (art. 41 Bis). Here again, there is no specific mention of music or musical goods.

Music is directly related to musical instruments, but it also concerns scores: the printed documents used for reading the notes resulting in sound execution (Reyes Gallegos 2014).

7 Effects of Climate Change on Musical Heritage

The transcendent year for the preservation of library materials, including scores, was 1986. On this date for the first time, an International Conference was organized around this topic. Highlights from this conference include the consideration of the preservation of collections of sheet music, musical instruments, buildings, people, and the environment (Sánchez Hernampérez 1999).

Returning to the issue of climate: temperature, humidity, UV levels and ventilation, are all fundamental aspects for the musical heritage. Great care has been taken to adapt buildings and services areas with heating and air conditioning to provide comfort to the public. However, the same attention has not been provided to keep microclimates protecting instruments and musicians, nor to the spaces for the archived scores (Fig. 3).

Musicians playing instruments know good that tuning is a problem if the air temperature is above or below normal room temperature, as may occur during outdoor performances of concert bands or symphonic concerts (Worland 2011). Instruments from both the brass and woodwind families tend to modify their pitch in relation to the reaction of the materials with which they are made and when there are changes in humidity and temperature, their pitch is high when the air is hot and

[6]By its acronym in Spanish (official language in Mexico).

Fig. 3 Classroom for individual instrumental lessons at the faculty of music—UNAM. http://oferta.unam.mx/escuela-facultad/9/facultad-de-musica (May 20th, 2016)

low when the air is cold, while stringed instruments react inversely to changes in temperature (Fig. 4).

Several musicians have commented on the changes in the intonation of their instruments during performances. The pitch fluctuates according to the temperature and weather conditions. Every instrument changes in different ways depending on the features of the musical family they belong to, therefore string, woodwind, and brass instruments modify pitch in different manners. It is always evident that temperature affects the pitch and the tuning of musical instruments.

Another instrument present in musical activity is the piano, which is made mainly of wood but has a big metal mechanism. (Key Arts Piano Houston s.f.). Humidity greatly affects the piano. Seasonal changes, and even daily changes in humidity, cause wood parts and keys and action parts to swell or shrink, resulting in sticking and sluggish notes. The harp, other instrument made with wood, metal and, natural (animal), nylon and wire metal strings also is affected by these variations of temperature and humidity. In the case of this instrument, in addition to tuning alterations by wood and metal structure, also strings contract with cold and dilate with heat, affecting in pitch stability and moisture alters their touch. In conclusion, for every musical instruments, excessive humidity and extreme temperature changes are dangerous enemies.

Collections of scores (other element of musical heritage) are exposed to two types of threats that can be catastrophic: those that are caused by natural phenomena and those which have their origin in human acts or omissions. They are especially vulnerable to dust, in addition to extreme temperature and humidity.

On the other hand, in modern competitions for concert halls (Statham, s.f.) architects take great care to state that the acoustical conditions have been carefully considered. Now a day, it is equally relevant to consider indoor air quality and humidity (Fig. 5).

Fig. 4 Rehearsal of symphonic orchestra of the National Polytechnic Institute (IPN), before the concert. Symphonic orchestra of the IPN, Mexico

Fig. 5 "Nezahaulcoyotl" concert hall. México. Concert hall of the philharmonic orchestra of the National Autonomous University of Mexico

8 Conclusion/Reflection

Although there are a considerable number of buildings for musical activity in Mexico, the state of instruments and collections of sheet music have not really had enough focus of attention to know exactly what is happening to them.

Among the cultural venues related to music (museums, concert halls, theatres, conservatories, schools, universities, archives, libraries) perhaps those most conditioned to keep instruments and sheet music, are those under the administration of CONACULTA and the National Institute of Fine Arts, as well as the public and private universities and radio, and television stations. A comprehensive listing is located in "Documentary musical heritage in the context of sustainable cultural development" (Reyes Gallegos 2014).[7]

The Faculty of Music of the UNAM has formed and maintained a team of specialists for the conservation and restoration of its instrumental collection (wood, stringed instruments, guitars, harps, woodwinds, brass, percussion, pianos, harpsichords, and organs). It also has a control for the document collections in their library. Actions for the protection of these instruments and printed material focuses mainly on the deterioration that may have occurred due to use and constant consultation, as well as reactions to temperature, humidity, lighting and material wear largely related to climate. It will be necessary yet, starting from the diagnosis presented, to undertake specific research on the behavior of each of the families of instruments to determine their reactions to climate impacts for improved preservation and maintenance plans and the best accurate in musical performance.

Awareness of the value of the cultural heritage of everyday life will lead to the implementation of policies and management projects for conservation that integrates considerations of landuse planning and above all, that take into account context, economically and socially integrated, after all, as said Koïchiro Matsuura.[8] "Cultural diversity is linked to the dynamic process whereby cultures change but retain their identity" (WCCD-UNESCO 2010). Sites and assets should not be protected in an isolated manner nor as museum objects, but rather as living entities at social, economic and cultural levels.

References

Anon., s.f. *Conservation préventive du patrimoine documentaire. Introduction.* [En línea] Available at: http://webworld.unesco.org/safeguarding/fr/introduction.htm Último acceso: 15 May 2016].

ARNOULT, J.-M., 2003. Les bibliothèques irakies en 2003: un nouveau chapitre de l'histoire interminable des désastres. *International Preservation New,* September, Issue 30, pp. 20–29.

BAZ, R. V., 2008. Los archivos musicales. estado de la cuestión. En: *El archivo de los sonidos: la gestión de fondos musicales.* Primera ed. Salamanca: ACAL, pp. 21–122.

BRANDT, A.-C. & FOUCAUD, J.-F., 1998. Environnement et conservation des documents dans les bibliothèques et les centres d'archives. En: *Protection et mise en valeur du patrimoine des bibliothèques. Recommendations techniques.* Paris: Ministére de la culture et de la communication. Direction du livre et de la lecture, pp. 15–24.

BURKE, S. & MACDONALD, S., 2014. Creativity and Conservation: Managing Significance al the Sydney Opera House. *APT Bulletin,* 45(2/3. SPECIAL ISSUE ON VALUES-BASED PRESERVATION), pp. 31–37.

---

[7]See Appendix 12.8 p. 796. "Main institutions and bodies responsible for preserving cultural heritage" ("Principales instituciones y organismos encargados de preservar el patrimonio cultural").

[8]Koïchiro Matsura was Director-General of UNESCO, Chairperson of UNESCO's World Heritag.

CASSAR, M., 2011. Impact of climate change on cultural heritage. From international policy to action. En: *Conservation perspectives. The GCI Newsletter.* London: GCI Newsletter, pp. 10–11.

CERVANTES, G. D., 2008. *Los documentos especiales en el contexto de la archivística.* México: s/e.

COLLETE, A., 2007. *Report on predicting and managing the impacts of climate change on World Heritage and Strategy to assist States Parties to implement appropriate management responses. Climate Change and World Heritage.* s.l.: Ministerio de cultura de España.

CPM, 2006. *Prévision et gestion des effets du changement climatique sur le patrimoine mondial.* Vilnius: CPM.

CHAPUIS (ed), M., 2009. *Preserving our heritage, improving our environment. 20 years of EU research into cultural heritage.* Brussels: Publications Office of the European Union. European Commission.

Congreso de la Unión, 2015. *Decreto por el que se crea la Secretaría de Cultura. México, Diario Oficial de la Federación.*

DE LA TORRE, M., 2014. Values in Heritage Conservation: A Project of The Getty Conservation Institute. *APT Bulletin,* 45(2/3, SPECIAL ISSUE ON VALUES-BASED PRESERVATION), pp. 19–24.

DOURNON, G., 2000. *Guía para la recolección de músicas e instrumentos tradicionales.* Francia: UNESCO.

Federal Government of Mexico, 2013. *NCCS (National Climate Change Strategy. 10-20-40 Vision),* Mexico: Federal Government of Mexico.

FRANCE, F. G., 2009. *Best Practice and Standards in Environmental Preservation for Cultural Heritage Institutions: Goals, Knowledge, Gaps.* [En línea] Available at: http://www.bl.uk/aboutus/stratpolprog/ccare/events/France.pdf Último acceso: 20 June 2012].

GÓMEZ GONZALEZ, P. J., MONTERO GARCÍA, J., BAZ, R. V. & HERNÁNDEZ OLIVERA, L., 2008. *El archivo de los sonidos: la gestión de fondos musicales.* Salamanca: ACAL Asociación de Archiverois de Castilla y León.

KEY ARTS PIANO HOUSTON, s.f. *Humidity-Temperature and clime conditions affecting pianos.A compilation from Piano Technicians Guild and other sources.* [En línea] Available at: http://www.keyartspianohouston.com/climate.html Último acceso: 30 April 2016].

MATTHEWS, G. & FEATHER, J., 2003. *Disaster management for libraries and archieves.* Hampshire: Ashgate.

MCILWAINE, J., 2006. *IFLA Disaster Preparedness and Planning: a Brief Manual.* Paris: IFLA-PAC.

RC, 2012. *El Patrimonio cultural y la gestión del riesgo de desastres.* [En línea] Available at: http://www.colombiahumanitaria.gov.co/Prensa/Paginas/121204.aspx Último acceso: 16 mayo 2016].

REYES GALLEGOS, A., 2014. *El patrimonio musical documental en el contexto del desarrollo cultural sustentable.* Tesis para obtener el grado de Doctorado en Música (Musicología) ed. México: Universidad Nacional Autónoma de México. Programa de Maestría y Doctorado en Música.

SÁNCHEZ HERNAMPÉREZ, A., 1999. *Politicas de conservación en bibliotecas.* Colección Instrumenta Bibliológica ed. Madrid: Arco/Libros.

SEMARNAT, 2013. *National Climate Change Strategy. 10-20-40 Vision.* Mexico: Federal Government of Mexico.

STATHAM, H. H., s.f. The Structure and Arrangement of Concert Halls. *The Structure and Arrangement of Concert Halls,* Volumen 38th Sess. (1911–1912), pp. 903–905.

STAVENHAGEN, R., 2001. Derechos Culturales: el punto de vista de las ciencias sociales. En: *¿A favor o en contra de los derechos culturales?.* París, Francia: Ediciones UNESCO Los derechos humanos en perspectiva, pp. 19–48.

TACÓN CLAVAÍN, J., 2008. *La conservación preventiva en archivos y bibliotecas. Prevención y protección.* Madrid: Ollero & Ramos. Ministerio de Cultura de España.

UNESCO /WHC, 2008. *Policy Document on the Impacts of Climate Change on World Heritage Properties.* Paris: UNESCO.

VAUGHAN, D., 1981. Orchestral Sound in Concert Halls-1. *The Musical Times,* January, 122 (1655), pp. 14–17.

VEILLON, R., 2014. *State of Conservation of World Heritage Properties. A statistical analysis (1979–2013).* Paris: UNESCO / WHC.

WCCD-UNESCO, 2010. *Invertir en la diversidad cultural y el diálogo intercultural. Informe Mundial de la UNESCO.* Luxemburgo: UNESCO.

WHC/UNESCO, 2015. *State of Conservation Information System (SOC).* [En línea] Available at: http://whc.unesco.org/en/soc [Último acceso: 20 May 2016].

WHC, 2006. *Predicting and Managing the Effects of Climate Change. The Impacts of Climate Change on World Heritage Properties,* Paris: UNESCO-HQ.

WORLAND, R., 2011. *Demonstrating the Effect of Air Temperature on Wind Instrument Tuning..* [En línea] Available at: http://acoustics.org/pressroom/httpdocs/161st/Worland.html [Último acceso: 30 April 2016].

Development and Application of Corn Model for Climate Change Impact Assessment and Decision Support System: Enabling Philippine Farmers Adapt to Climate Variability

Orlando F. Balderama, Lanie A. Alejo, Edgardo E. Tongson and Rhia T. Pantola

Abstract This purpose of this paper is to present results of a university led research and extension undertaking in providing solution to corn farming in coping with climate variability. Methods employed were science tools such as simulation and climate modelling, integration of automated weather station for real-time weather data inputs and Short Messaging System (SMS) as decision support to government workers and farmers. Specifically it aimed to develop a localized corn model; assess future corn production under climate change scenarios and; develop decision support system for corn production. A local model was developed for climate change assessments and development of decision support for corn farmers. The model was able to predict the observed data on yield and timing of phenological events from the actual experiments and actual farmer's field with high goodness of fit ranging from 91 to 98% for the calibration and 86 to 97% for the validation process. Moreover, applications of the model for climate change assessments indicated that corn yield in northern Philippines would be reduced by up to 44% in 2020 and 35% in 2050 due to changes in rainfall amount and rise in temperature which are indicators of climate change. The model was automated to provide a

O.F. Balderama (✉) · L.A. Alejo
Agricultural Engineering Dept., Isabela State University,
Echague, Isabela, Philippines
e-mail: orly_isu@yahoo.com

L.A. Alejo
e-mail: lhan_1023@yahoo.com

E.E. Tongson
Abuan Watershed Project, World Wide Fund for Nature, Quezon, Philippines
e-mail: etongson@wwf.org.ph

R.T. Pantola
IBM, Smarter Agriculture, Manila, Philippines
e-mail: rhia.trogo@ph.ibm.com

W. Leal Filho (ed.), *Climate Change Research at Universities*,
DOI 10.1007/978-3-319-58214-6_24

quick answers to farmer's operational decision making and crop and weather advisories for strategic and policy decision support by government agencies.

Keywords Corn model · Climate variability · Philippine farmers

1 Introduction

Corn is the second most important crop in the Philippines, next to rice, in terms of total cultivated area and amount of production. Isabela, the second largest province in the Philippines is the top corn producer of the Philippines, producing 22.57% of national production of corn and with a share of 20.26% of the total national area used for corn production (DA-BAS 2013).

Climate Change poses serious threats to agriculture and consequently to livelihoods of farmers. The increasing frequency of inter-annual anomalies from both typhoons and drought in recent years foretell the impacts from global warming. For instance, in 2005, dry spells followed by severe flooding caused extensive farm damage estimated at PHP838 million (USD 19.5 million) and in the early half of 2010, prolonged dry spells caused wilting of rice and corn crops (Tongson 2013).

In the long term, Climate Change poses serious threats to the livelihoods of farmers and to national food security. According to DOST-PAGASA (2011), the 2050 scenario in Isabela will result in a 1.9–2.1 °C increase in temperature; a 29% decrease in mean rainfall for dry months and a 1.7–25.1% increase in rainfall for wet months. This will translate to frequent and more intense flooding and dry spell events resulting in recurring crop damage and worsening poverty. With Climate Change, anomalies occurring the same year are becoming more common and are likely to get worse.

As the highest corn and second-highest rice producer in the country, the Isabela Province is a net exporter of rice and corn. The region is also the most visited by typhoons and has the highest temperatures ever recorded. Climate change impacts food production. Higher temperatures hasten phenology development processes resulting to shorter crop cycles and lower yields. While increases in evapotranspiration rates lead to higher demand for water and increased competition with other water uses. Increase in rainfall variability translates to more intense flooding during wet months and prolonged dry spells during summer. The timely provision of rainwater affects rain fed farms, where most of poor farmers depend. The PAGASA agency forecasts the present El Nino to be severe, similar to 1997–1998 El Nino, and is likely to persist until mid-2016.

This paper aims to present some results of a research and extension initiatives to help farmers cope with climate change and still increase crop productivity using information and communication technology (ICT) for corn farming in the Philippines. It utilized combinations of simulation and climate modeling, integration of automated weather station for real-time weather data inputs and SMS as decision support to government workers and farmers. Specifically it aims to develop a localized corn model using DSSAT-CEREZ Maize; assess future corn production under climate change scenarios and; and develop decision support system for corn production.

The next phase was model's automation to provide a quick answers to farmer's operational decision making and crop and weather advisories for strategic and policy decision support by government agencies in the Philippines and for future upscaling.

2 Use of Crop Models for Climate Change Adaptation

It is well accepted in the scientific community that our climate is changing due to rising anthropogenic greenhouse gas emission. According to Intergovernmental Panel on Climate Change, the future degradation of soil and water resources will be a major challenge facing global agriculture (IPCC 2001).

In recent years, several dynamic crop simulation models have been developed as information technology tools to support strategic decision-making in research, crop production and land use planning (Hoogenboom et al. 1992; Penning de Vries et al. 1991, 1993; Hoogenboom et al. 2004). These crop models can be used to evaluate agricultural production risk as a function of climatic variability, to assess regional crop yield potential across a wide range of environmental conditions and to determine suitable planting dates and other management factors for increasing crop yield (Egli and Bruening 1992; Meinke et al. 1995; Aggarwal and Kalra 1994; Hunt et al. 1993; Chapman et al. 2000).

A process-oriented Cropping System Model (CSM)—Ceres-Maize model has been developed to simulate vegetative and reproductive development, growth and yield as a function of crop characteristics, climatic factors, soil characteristics and crop managements. These models are part of suite of crop growth models that encompass the Decision Support System for Agrotechnology Transfer (DSSAT) (Hoogenboom et al. 2004).

Simulation models are thought to be an indispensable tool for supporting scientific research, crop management, and policy analysis (Fischer et al. 1999; Hammer et al. 2002; Hansen 2002), however, validation of such models needs to be conducted because these are developed under very specific edaphic and climatic conditions that do not necessarily prevail in other regions of the world. Therefore, before the model can be adopted or used in other locality, evaluation or validation study must be undertaken to establish its credibility (Balderama and Bareng 2009).

Advance crop production management support systems then needs appropriate process-based crop simulation model which utilizes site specific parameters such as weather, physical and chemical properties of soil, water management, and agronomic practices (Penning de Vries et. al. 1989). O'Toole and Stockle (1987) described the potential of simulation models in assessing trait benefits of winter cereals and their capacity to survive and reproduce in stress-prone environment. Crop growth models have been used in plant breeding to simulate the effects of changes in the morphological and physiological characteristics of crops which aid in identification of ideotypes for different environments (Hunt 1993; Kropff et al. 1995).

The policy management is one very useful application of crop simulation models. The issues range from global (impacts of climate change on crops) to field level (effect of crop rotation on soil quality) issues. Thornton et al. (1997) showed that in Burkina Faso, crop simulation modeling using satellite and ground-based data could be used to estimate millet production for famine early warning which can allow policy makers the time they need to take appropriate steps to ameliorate the effects of global food shortages on vulnerable urban and rural populations.

In the Philippines, Lansigan et al. (2004) used CERES-Maize, a module in the DSSAT software, in the analysis of corn yield gaps in major corn growing areas where crop genetic coefficients specific for the local corn variety IPB 911 was derived.

The DSSAT CERES Maize model is mainly a function of the genetic coefficients which are fundamental parameters to establish the potential performance of the crop given the limitations on environmental factors (soil, weather and crop management).

3 Methodology

The research study conducted is limited to corn production in a watershed ecosystem. It uses crop modelling, automated weather services and SMS platforms to generate data and send customize information to farmers. Below are the detailed methods in the implementation of the research undertakings.

3.1 Site Selection

The study was conducted in Abuan Watershed, Ilagan Isabela, Philippines. To develop parameters of the crop model, field experiment were conducted for two cropping seasons under various agro-ecological zones namely: floodplains, rolling and hilly croplands. Figure 1 shows the locations of these field sites as follows:

(a) Villa Imelda—rolling terrain, farmers field, rainfed
(b) Sindon Bayabo—hilly terrain, farmers field, rainfed
(c) Cabisera 10—flood plain, farmers field, rainfed
(d) Cagayan Valley Research Center, flood plain, rainfed and irrigated.

Fig. 1 Location of field sites

Table 1 Physical and soil characteristic of the study areas

Location	CVRC	Cabisera 10	Villa Imelda	Sindon Bayabo
Corn farmer	Research center	Charlito Servilla	Eulalio Paredes	Amado Bolda
Agro-zone	Flood plain	Flood plain	Rolling	Hilly
Slope (%)	≥ 1	≥ 1	8–18	18–30
Soil depth (cm)	>100	>100	>100	65
Texture	Silty clay loam	Loam	Clay	Sandy clay loam over clay
Soil reaction (pH)	5.6 (medium acid)	4.84 (very strongly acidic)	4.66 (very strongly acidic)	5.0 (very strongly acidic)
Organic matter (%)	0.41 (very low)	0.16 (very low)	0.59 (very low)	0.05 (very low)
Phosphorus (mg/kg)	8.3 (low)	6.99 (low)	15.96 (medium)	0.4 (very low)
Potassium (cmol/kg)	0.18 (very low)	0.24 (low)	0.3 (low)	0.36 (low)

Table 1 shows some physical and soil characterization of the study areas. Agro-zone is classified based on percent slope. Information on soil characteristics includes soil depth, texture and some chemical properties. As shown, typical corn soil in the watershed are acidic with low contents of organic matter, phosphorus and potassium.

3.2 Development of Local Crop Model for Corn

The CERES-Maize model as included in the Decision Support System for Agrotechnology Transfer (DSSAT) was used to simulate the phenology and yield of corn cultivar under different scenarios of climate change. The CERES-Maize model is designed to predict phenology and yield of different varieties for different climate and soil conditions and various management practices. The model also provides for validation of crop model outputs; thus allowing users to compare simulated outcomes with observed results. Crop model validation is accomplished by inputting the user's minimum data, running the model, and comparing outputs. By simulating probable outcomes of crop management strategies, DSSAT offers users information with which to rapidly appraise new crops, products, and practices for adoption. The DSSAT package incorporates models of 27 different crops with new tools that facilitate the creation and management of experimental, soil, and weather data files. DSSAT includes improved application programs for seasonal and sequence analyses that assess the economic risks and environmental impacts associated with irrigation, fertilizer and nutrient management, climate change, soil carbon sequestration, climate variability and precision management.

Model calibration was conducted to determine the cultivar coefficients for cultivar of corm that is normally grown in the Abuan watershed.

For model validation, thirty (30) farmers in the watershed participated in the validation of the model in the wet season period of 2015. There were eleven (15) farmer respondents in floodplain, five (8) from rolling areas and six (7) in hilly areas. Date of planting ranges from April 28 to June 3 but mostly in the second to third month of May.

3.3 Assessment of Impact of Climate Change

Projections are important as basis for adaptation interventions. The study used the 2020 and 2050 projected weather data from the Philippine's weather agency as input to the crop model to capture the effect of rise in temperature and change in rainfall amounts in rainfed corn yield.

3.4 Development of SMS-Based Decision Support System

A Farmers Decision Support System (FDSS), is an SMS and internet-based platform to deliver weather and crop advisories to farmers and extension agents to support farming decisions via cell phones or smart phones was developed and pilot tested in the area.

4 Results and Discussion

4.1 Project Setting: The Abuan Watershed

In the western buffer zone of the Northern Sierra Madre Natural Park lies the Abuan watershed, a 63,754 ha ecosystem that supports the livelihoods of some 2900 farming households. The upper catchment comprises 44,000 ha (or 69%) of the Northern Sierra Madre Natural Park, which is host to last remaining old-growth dipterocarp forests in the country. The lower sub-catchments consist of farm lands, residual forests and brush lands with an area of 19,000 ha (or 31%) of the watershed area. The watershed is named after the Abuan river, which, together with the smaller Bintacan River, drains into the Pinacanauan de Ilagan River, before merging with the Cagayan River. Figure 1 is a map of the Abuan watershed. The lower Abuan catchments, in yellow, consists of tenured lands with an area of 4057 ha. This area is the focus of project activities including the flood plain (Fig. 2).

4.2 Weather Characteristics

A 25-year weather data from the Cagayan Valley Research Center (CVRC) was used for the study area. Average annual rainfall is 2176 mm and rainy season starts in the month of May and ends in December. Highest rainfall occurs in the month of November with 319.16 mm and lowest during the month of March with 56.99 mm. The hottest month was observed in June with an extreme value of 35.64 °C while

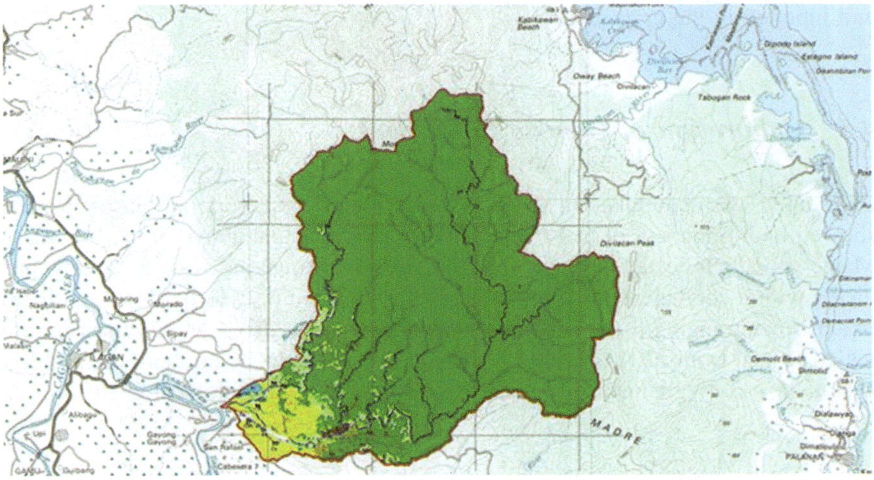

Fig. 2 Map of project area with inset map

coldest month is January with a minimum value of 19.88 °C. Sunshine duration is highest in the months of April, May and June with 6.9, 7.0 and 6.7 h, respectively.

4.3 Soil Characteristics

The soil at CVRC site is characterize as alluvial plain with slope of more or less 1%, soil depth is 100 cm with a silty clay loam texture of surface and subsoil. Cabisera 10 is also located in alluvial plain with slope of more or less 1%. Soil depth is more than 100 cm and soil texture of surface soil and subsoil is loam soil. At Villa Imelda site, the area is situated in rolling terrain with slopes ranging from 8 to 18%. Soil depth is more than 100 cm and soil texture of surface soil and subsoil is clay. Sindon Bayabo is a hilly farm with slope ranges from 18 to 30%. Soil depth is 65 cm and soil texture of surface soil is sandy clay loam, while subsoil is clay soil (Table 1).

4.4 Crop Management

In the CVRC site, planting was done in December with a spacing of 0.75 m by 0.20 m and fertilizer application followed the package of technology for corn. In Cabisera 10, Villa Imelda and Sindon Bayabo, farmers waits when soil moisture is wet enough to commence land preparation and planting. As a practice, planting usually takes place during the months of October and November. Plant population is approximately 60,000 per hectare and fertilizer application rates largely depends on the financial capability of farmers. All information on farm inputs were recorded and inputted in the crop model.

4.5 Calibration of Genetic Coefficients

Calibration was undertaken in all five sites during the first cropping season. Results showed high goodness of fit between simulated and actual above ground biomass. Maize growth reflected through the biomass accumulations was accurately predicted with a coefficient of determination ranging from 0.91 to 0.99 with an index of agreement ranging from 0.7 to 0.97 for the five sites. Also, RMSE values are low which ranged from 1201 to 2950 kg/ha of biomass (Table 2). Furthermore, Table 3 shows the close agreement of observed and simulated phenological events.

Table 2 Statistical analysis result between observed and simulated dry above ground biomass in kg

Site	Standard deviation		R^2	RMSE	d-Stat.
	Observed	Simulated			
CABISERA 10	3543.864	3419.42	0.929	1201.067	0.970
DA CVRC IRRIGATED	3899.12	5159.72	0.928	2950.313	0.902
DA CVRC RAINFED	2537.703	1937.85	0.988	2223.02	0.684
VILLA IMELDA	3033.281	1627.94	0.914	1846.862	0.847
SINDON BAYABO	3981.625	2592.47	0.985	2386.388	0.879

Table 3 Observed and predicted phenological events in days after planting

Site	Emergence		Silking		Beginning of grain filling		Physiological maturity	
	Actual	Simulated	Actual	Simulated	Actual	Simulated	Actual	Simulated
CABISERA 10	3	5	61	63	72	74	115	122
DA CVRC (IRRIGATED)	5	6	66	68	74	78	110	111
DA CVRC (RAINFED)	5	6	66	68	74	78	110	111
VILLA IMELDA	3	5	61	66	80	80	115	111
SINDON BAYABO	3	5	61	65	74	79	115	91

Table 4 Simulated and observed grain yield during the first and second cropping in kg

Site	Planting date	Observed	Simulated
CVRC IRRIGATED	June 12	8800	9163
CVRC RAINFED	June 12	8213	8777
CABISERA10	June 2	9867	10,202
VILLA IMELDA	July 3	5190	5024
SINDON BAYABO	June 27	7167	7566

4.6 Harvest Yield

The model predicted the second cropping season harvest yield very close with 3.2% to 6.87% difference from the actual yield (Table 4).

4.7 Model Validation from Farmer's Field

Thirty (30) farmers in the watershed participated in the validation of the model in the wet season period of 2015. There were eleven (15) farmer respondents in floodplain, five (8) from rolling areas and six (7) in hilly areas. Date of planting ranges from April 28 to June 3 but mostly in the second to third month of May. It was also noted that other hybrid varieties were used as planting materials. Simulated and actual yield shows high correlation with a coefficient of 0.8124 as shown in Fig. 3.

4.8 Assessment of Impact of Climate Change

Projections are important as basis for adaptation interventions. Table 5 shows the 2020 and 2050 projected weather data from DOST-PAGASA which was also used to capture the effect of rise in temperature and change in rainfall amounts in rainfed corn yield. Assumed planting dates is May for wet season and November for dry season and using recommended fertilizer application rate. Simulation results showed that corn yield would be sharply reduced by 44% in 2020 and by 35% in 2050 (Fig. 4).

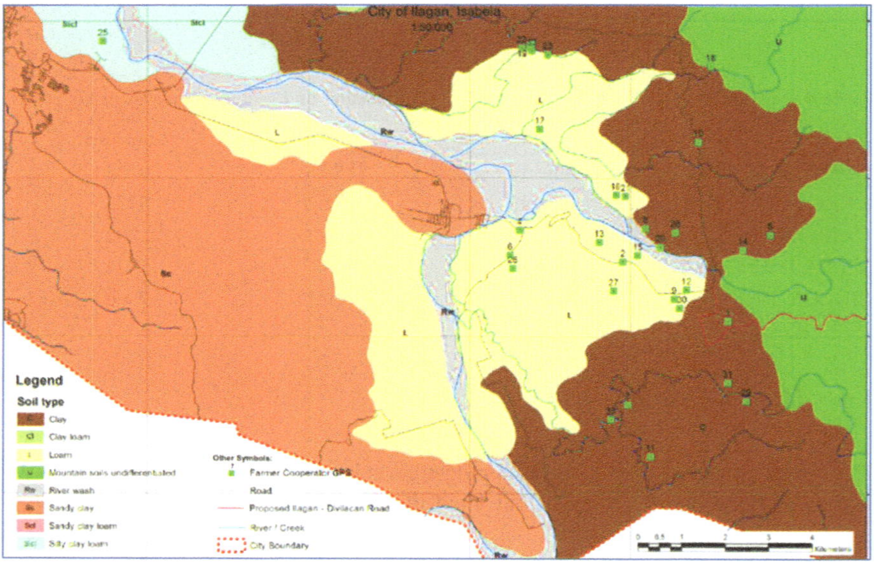

Fig. 3 Location of farms for model validation under various ecosystem and soil type

Table 5 Projected temperature and rainfall changes in 2020 and 2050 (DOST-PAGASA 2011)

Simulated year	% change in Rainfall, (PAGASA)				Increase in Temperature, (PAGASA)			
	Dec–Feb	Mar–May	Jun–Aug	Sep–Nov	Dec–Feb	Mar–May	Jun–Aug	Sep–Nov
2020 projections	3.9	−8.6	5.1	13.5	0.8	0.9	0.9	0.8
2050 projections	25.1	−29.2	8.7	1.7	2	2.1	2.1	1.9

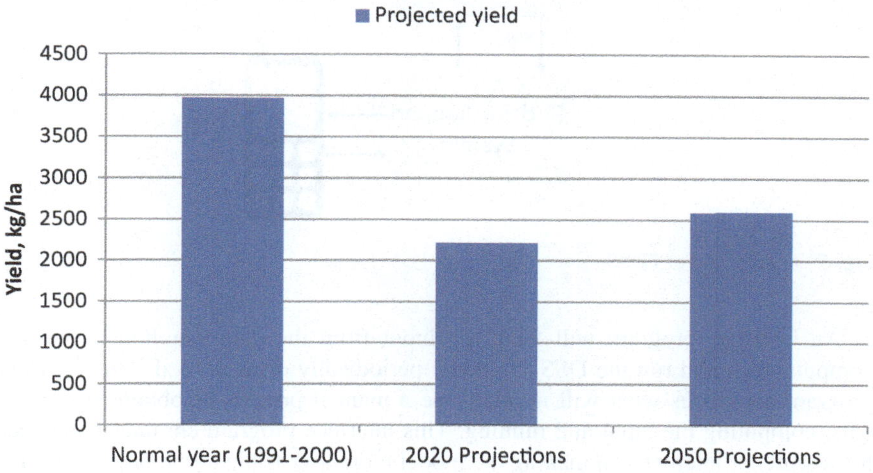

Fig. 4 Yield projections for the year 2020 and 2050

4.9 Development of Farmer Decision Support System(FDSS)

Figure 5 shows the architecture for the FDSS. The data warehouse will house the seasonal climate forecast and weather forecast from NOAH-WISE project. However, before these data can be stored, the data will have to go through extraction, transformation and loading (ETL) to ensure that the data will be clean before they get to the data warehouse. The same process will be undergone by the sensor data from AWS stored in the CSV files that are housed in Advanced Science and technology Insitute of Department of Science and Technology.

The data warehouse will be used as the data source for producing analytics reports and visualizations for analysis and reporting purposes. The data warehouse will also be used for producing the weather and crop bulletin that are meant to help the Extension Workers in the farms to assist the farmers.

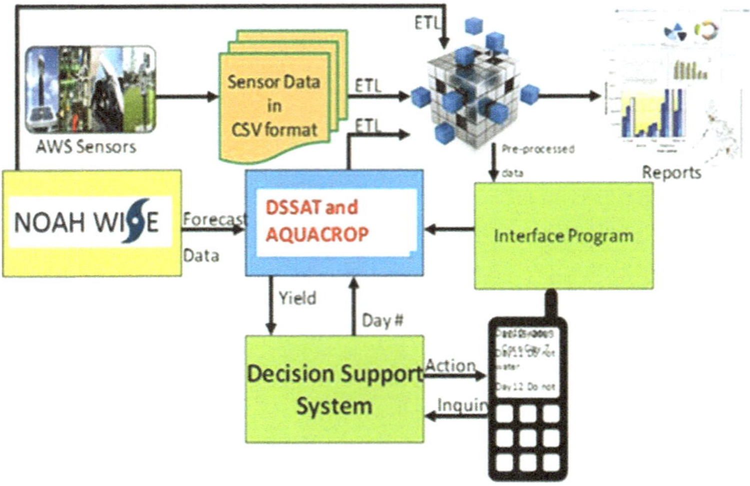

Fig. 5 Architecture of FDSS

An interface program will fetch the input from the data warehouse, perform computations, and run the DSSAT model periodically or as needed. This interface program automates what will normally be a manual process of obtaining the raw data, computing the input and running. This interface program eliminates the need for researcher intervention making the model more accessible to the farmers at their convenience.

The FDSS issues farm advisories including localized seasonal and 7-day daily weather forecasts to cell phones or via internet. Farmers can query and print crop calendars to help them monitor plant phenology and, together with rainfall forecasts.

In the dry season of 2015, the FDSS was piloted in five (5) barangays within 10 kms radius from the AWS in Vila Imelda. Majority of text queries from farmer cooperators were rainfall forecasts indicating their dependence of weather. Thirty farms were sampled for biomass and yields and compared with simulated values. Validation results showed good agreement between model and observed yields for both rain-fed and irrigated farms. The results prove the applicability and scalability of the model to other corn farms and to other weather stations.

Farmers may be able to inquire about the following using their SMS:

- Is it going to rain in the next 7 days and by how much?
- When is the best date to start planning?
- How much will I harvest?
- When will the growth stages occur?
- When do I apply fertilizer and by how much?
- When do I water the plants and by how much?
- When do I harvest?

Using the solution, the government may also get feedback from the farmers via SMS as farmers are allowed to report the following information:

- Report when I started planting
- Report how much I harvested at the end of the season
- Report how much I sold my harvest for
- Report initial capital
- Planting intentions for next season.

5 Conclusion

This paper presented results of research and extension initiative to help corn farmers in coping with climate variability. The undertaking utilized science tools such as simulation and climate modelling, automated weather station services and short messaging system (SMS) as decision support to government workers and farmers.

The field study provided an estimate of the genetic coefficient of hybrid corn cultivar in 2013–2015 using Ceres-Maize model of DSSAT software. Calibration was done to estimate the five (5) genetic coefficients for corn with input data taken from the dry season cropping. Using the genetic coefficient estimated from the data observed during the first cropping, validation was conducted using the data set observed from the five plots during the second cropping. Results showed that these genetic coefficients accurately predicted the actual corn biomass production and phenological stages. Yield data was also gathered from 30 farmers for further validation. The model yields high efficiency of 0.81 in predicting the yield of participating farmers in the watershed.

Application of the maize model was also use to assess the impact of climate change in corn production. Results indicated that corn yield would be reduced by up to 44% in 2020 and 35% in 2050 due to changes in rainfall amount and rise in temperature.

In developing the Farmer Decision Support System, the corn model was automated to provide a quick answers to farmer's operational decision making using SMS and crop and weather advisories for strategic and policy decision support by government agencies. In the future, the FDSS can be rolled out to other farming communities throughout the country and include other crops.

Acknowledgements The authors deeply acknowledge the financial support provided by USAID, World Wide Fund Philippines, Oscar M. Lopez Center for Climate Change Adaptation and Disaster Risk Mitigation and the Philippine Bureau of Agricultural Research.

Literature cited

Aggarwal, P K and N Kalra (1994). Simulating the Effect of Climatic Factors, Genotype and Management on Productivity of Wheat in India. New Delhi: Indian Agricultural Research Institute Publication. Pp. 156.

Balderama, O. F. and Bareng JL. R., 2009). Estimation of Soil Erosion and Water Yield in Community Watershed using the WEPP Model under Cagayan Valley Condition. Journal of researches of the CVPED 6th International Conference on Environment and Development, June 1-5, 2009.

Chapman, P., Clinton, J., Kerber, R., Khabaza, T., Reinartz, T., Shearer, C. & Wirth, R. (2000). CRISPDM 1.0. step-by-step data mining guide. Technical report, CRISP-DM.

DA-BAS (Department of Agriculture -Bureau of Agricultural Statistics). 2013. http://www.bas.gov.ph/.

DOST-PAGASA. 2011. Climate projection for Isabela Province. Quezon City, Manila.

Egli, D. and W. Bruening. 1992. Planting date and soybean yield: Evaluation of environmental effects with a crop simulation model: SOYGRO. Agric. for Meteorol. 62:19–29.

Fischer D, Barret C, Bryson K, Elofsson A, Godzik A, Jones D, Karplus KJ, Kelley LA, MacCallum RM, Pawowski K, Rost B, Rychlewski L, Sternberg MJE (1999): CAFASP1: Critical assessment of fully automated structure prediction methods. Proteins (Suppl. 3): 209–17.

Hammer GL, Kropff MJ, Sinclair TR, Porter JR (2002) Future contributions of crop modeling: from heuristics and supporting decision making to understanding genetic regulation and aiding crop improvement. Eur J Agron 18: 15–31.

Hansen, J.W., 2002. Realizing the potential benefits of the climate prediction to agriculture: issues, approaches, challenges. Agricultural Systems 74, 309–330.

Hoogenboom G, Jones JW, Wilkens PW, Porter CH, Batchelor WD, Hunt L.A, Boote KJ, Singh U, Uryasev O, Bowen WT, Gijsman AJ, du Toit A, White JW and Tsuji GY. 2004. Decision Support System for Agrotechnology Transfer Version 4.0 [CD-ROM]. University of Hawaii, Honolulu, Hawaii, USA.

Hoogenboom G, Jones JW and Boote KT. 1992. Modeling growth, development and yield of grain legumes using SOYGRO, PNUTGRO and BEANGRO: A review. Transactions of the ASAE 35, 2043–2056.

Hunt, L.A. 1993. Designing improved plant types: a breeders view point. p. 3–17. In F.W.T. Penning de Vries et al. (ed.) Systems approaches for sustainable agricultural development. Kluwer Academic Publishers, Dordrecht, The Netherlands.

IPCC (2001) In: McCarthy JJ, Canziani OF, Leary NA, Dokken DJ, White KS (eds) Climate change: impacts, adaptation and vulnerability contribution of working group II to the third assessment report of the intergovernmental panel on climate change. Cambridge University Press, Cambridge.

Kropff, M.J., A.J. Haverkort, P.K. Aggarwal and P.L. Kooman. 1995. Using systems approaches to design and evaluate ideotypes for specific environments. p. 417–435. In J. Bouma, A. Kuyvenhoven, B.A.M. Bouman, J.C. Luyten and H.G. Zandstra (eds.) Eco-regional approaches for sustainable land use and food production. Kluwer Academic Publishers, Dordrecht, The Netherlands.

Lansigan, F.P. W. L. de los Santos, R.E. V. Santos, M.T. Fabellar, J.K.S. Aunario, and B.S. Organo. 2004. Yield Gap Analysis in Corn Producing Areas in the Philippines: Project Terminal Report 2004. Institute of Statistics, University of the Philippines, Los Banos Laguna.

Meinke H and Hammer GL. 1995. A peanut simulation model: II. Assessing regional production potential. Agron. J. 87, 1093–1099.

O'Toole, J.C. and C.O. Stockle. 1987. The role of conceptual and simulation modelling in plant breeding. Presented at the Int. Symp. on Improving Winter Cereals under Temperature and Soil Salinity Stresses, 26–29 October 1987, Cordoba, Spain.

Penning de Vries, F.W.T., and Spitters, C.J.T. 1991. The potential for in crop yield simulation. p. 123–140. In R.C. Muchow and J.A. Bellamy (ed.) Climatic risk in crop production: Models and management for the semi-arid tropics and subtropics. Proc. Int. Symp. Climatic Risk. CAB. Int., Wallingford, UK.

Penning de Vries, F.W.T., and van Laar, H.H. 1989. Simulation of plant growth and crop production. Pudoc, Wageningen, The Netherlands.

Penning de Vries F.W.T, Teng P and Metselaar K. (Eds.). 1993. Systems approaches for agricultural development. Kluwer Academic Publishers, Dordrecht, The Netherlands.

Thornton, P.K., W.T. Bowen, A.C. Ravelo, P.W. Wilkens, G. Farmer, J. Brock and J.E. Brink. 1997. Estimating millet production for famine early warning: an application of crop simulation modelling using satellite and ground-based data in Burkina Faso. Agricultural and Forest Meteorology 83: 95–112.

Tongson, 2013. Corn Farming in the Abuan watershed. World Wide Fund for Nature. Quezon, City, Philippines.

Applied and Planning-Oriented Urban Climatology at the University of Kaiserslautern

Sascha M. Henninger and Arne Leitte

Abstract Urban authorities and stakeholders demand for answers on climatological and air quality related problems in the urban area. Applied and planning-oriented urban climatology provides the link between climatology and urban planning. Therefore, one aim of the Department of Physical Geography, Faculty of Regional and Environmental Planning, University of Kaiserslautern, Germany, is to understand how newly gained free space in shrinking cities can be integrated useful—in terms of urban climatology—in new utilization structures. For example, how can fresh and cold air production zones be incorporated in urban planning? Furthermore, the potential of the hardly noticed green roofs in order to enhance the urban micro climate is investigated, especially in arid and semi-arid zones. In addition, a database of roadside greenery and canopy is developed to characterize the influences on urban air quality. Parameters that are taken into account are amongst others canopy density and emission of volatile organic compounds. A dense canopy can reduce air movement leading to the accumulation of air pollution like within an urban canyon. Several tree species emit volatile organic compounds that contribute as a precursor to the formation of ground level ozone and are therefore not suitable for planting as urban green. Through the applied and planning-oriented urban climatology we provide recommendations of action for urban authorities.

Keywords Urban climatology · Urban planning · Applied sciences · Urban heating · Urban green · Urban air quality

S.M. Henninger (✉) · A. Leitte
Department of Physical Geography, Faculty of Regional and Environmental Planning, University of Kaiserslautern, Pfaffenbergstr. 95, 67663 Kaiserslautern, Germany
e-mail: sascha.henninger@ru.uni-kl.de

W. Leal Filho (ed.), *Climate Change Research at Universities*,
DOI 10.1007/978-3-319-58214-6_25

1 Introduction

In the future, applied urban climatology will play a bigger role due to the all present climate change. Urban areas can be defined as a climatological hot spot as many people live within these. Influences on the natural ecosystem by urban settlements are severe resulting in different problems. Thus urban planning must counteract these problems. One scope is to identify and categorize these urban areas with air quality and climatological related problems and based upon this knowledge to present urban climate maps and urban recommendation maps as countermeasures. This is the field of work of the Department of Physical Geography of the University of Kaiserslautern.

Primarily, we deal with questions on how to integrate newly gained free space in shrinking cities and how can they be integrated useful—in terms of urban climatology—in new utilization structures. For example, how can fresh and cold air production zones be incorporated in urban planning? Furthermore, the potential of the hardly noticed green roofs in order to enhance the urban micro climate is investigated, especially in arid and semi-arid zones. In addition, a database of roadside greenery and canopy is developed to characterize the influences on urban air quality. Parameters that are taken into account are amongst others canopy density and emission of volatile organic compounds. A dense canopy can reduce air movement leading to the accumulation of air pollution like within an urban canyon. Several tree species emit biogenic volatile organic compounds that contribute as a precursor to the formation of ground level ozone and are therefore not suitable for planting as urban green.

Essential aspects of environmental precaution and urban planning are the knowledge about the local climate, the air quality, and its connections. The natural resources climate and air are important aspects in the process of consideration in urban development, environmental impact tests, and location assessment. The aim of our applied and planning-oriented urban climatology is to identify and countermeasure urban climatological and air quality related problems by an applied and planning-oriented urban climatology (Henninger 2013a).

2 Research and Work Area of Applied Urban Climatology

The anthropogenic influences on the natural ecosystem causes the local phenomena urban climate. Compared to rural areas, densely built-up areas and areas with sealed surfaces create different microclimates and atmospheric trace elements. In general, there are the following reasons for the development of the urban climate:

- Large-scale sealing of former natural surfaces
- Reduction of vegetation and water bodies
- Specific 3-dimensional geometry by construction
- Anthropogenic input of heat and trace elements

These influencing factors lead to a reduction of evapotranspiration, increased surface roughness, and, in connection with the anthropogenic modified urban atmosphere, to changes in the radiation and energy balance (Henninger 2011a; Kuttler 2013).

At present, a main focus in applied urban climatology is on the work on climatological and air quality issues. In addition, the analysis of the negative impacts on the biotic and abiotic urban ecosystem became more important in recent years (Henninger 2011b). The applied and planning-oriented urban climatology is the link between climatology and urban planning. Thus, urban climatology is capable to provide a board spectrum of solutions for climatological and air quality related problem areas. Urban climatological analyses and from that produced synthetic climate function maps and planning recommendation maps allow authorities on the community level to respect climatic aspects within their planning (Overbeck et al. 2008; VDI 2015; Henninger 2013a).

3 Fields of Interest of the Applied and Planning-Oriented Urban Climatology

The primary goal of a planning-oriented urban climatology should be the improvement of climatological and air quality related urban problems. There exists a wide spectrum of fields of activity that can be deduced from urban climatological characteristics and problem areas (Henninger 2013a). The Department of Physical Geography seizes some of these fields of action and transfer them into the applied and planning-oriented research approach.

1. Planning-oriented urban climatology faces the disadvantage that most cities have been formed over a long time period without regarding climatological and air quality interests and now cannot be easily remodeled. Nevertheless, when looking at other climate zones, developments that have positive influences on urban development in terms of air quality and climatology can be identified. For example, narrow streets lead to a shading of the walls with passive cooling and thus can reduce the amount of energy of conventional air cooling systems. Thermal insulation might be part of a climate friendly built environment besides shading elements both reducing heat stress in hot summer times. A better insulation, resulting in a reduced amount of energy for cooling in summer, also has a positive effect in winter by reducing heat loss (Henninger 2010).

An "ideal urban climate" for urban dwellers has been defined by Mayer (1989) as a thermal bio-climate with lot of spatial variation and preferably no anthropogenic air pollution. This ideal urban climate only can be achieved when establishing new settlements and taking climatic and air quality aspects into account in urban planning from the beginning. It is not possible to achieve this ideal for already existing cities, but the goal of urban planning should be to approximate it with specific measures as near as possible. In this manner a "tolerable urban climate" can be implemented (Mayer 1989; Hartz 2011).

Most probably, air temperature in urban areas will increase due to climate change. The resulting thermal stress needs to be reduced by lowering the heat input, ensuring a good ventilation and improving air mass exchanges within the urban boundary layer. The use of certain building material can reduce the effect of urban heat islands. For example, bright materials with a high albedo reduce the heating of walls by reflecting more short wave sun radiation. A rigorous use of these materials in combination with considering planning aspects, could minimize the urban heat island effect (Henninger 2011a).

2. It is due to the demographic change and the urban exodus in many German cities that urban climatological knowledge can be incorporated in present and future urban planning. Newly gained space in shrinking cities can be integrated urban climatological useful in new utilization structures (Oswalt and Rieniets 2006). In this context, different fields of action of the applied and planning-oriented urban climatology can be highlighted. Especially, the creation of open, green and blue space within urban areas. In particular, existing open and green spaces are part of the research of the applied and planning-oriented Department of Physical Geography of the University of Kaiserslautern. All green spaces have a cooling effect on ambient air. During the day air temperature is reduced by energy consuming evapotranspiration and shading intercepting solar radiation by leaves and branches. During the night time cold air can be formed over green spaces and can initiate a so called "park breeze" resulting in air movement out of the park and thus exchanging and cooling surrounding air masses. Depending on their size, shape, and surrounding factors urban green spaces can have a tremendous impact on their surroundings. But not every green space has the same impact on its surroundings. The area of green spaces must be at least 50 ha to have a significant impact (Horbert 2000). On the other hand Bongardt (2006) has reported that smaller green spaces also have a positive effect on the thermal situation of directly neighboring areas. From the planning perspective, it has to be defined in advance which effect should be accomplished by a newly formed urban cool island for the users and the adjacent areas. In a cooperation with the University of Tel Aviv we could demonstrate, that the park cool island effect of a green urban area close to the sea changes over time depending on water supply and vegetation (Henninger and Saaroni 2014).
3. Natural open and green spaces can improve air mass exchange acting as ventilation paths and thus can transport cold air in built-up areas. In shrinking cities, new open spaces can underlie a change of use and are ideal for becoming green, blue, and open spaces. Potential air ventilation path areas like industrial and commercial wastelands, urban railways, and building gaps might be considered in a climate-oriented urban planning. In this case, a thoroughly examination of the situation is required to guarantee an improved urban ventilation system or if not already existing to develop one. If possible, such areas should not be

reassigned to development areas. Potential ventilation paths are also very sensible areas, from an urban climatological perspective, as already small obstacles can stop air movements. To ensure an appropriate urban ventilation via urban green spaces and ventilation paths these areas need to be preserved by limitations of building development (Henninger 2011a).

4. Green roofs, walls and facades are hardly recognized in order to enhance the urban micro climate. In particular, an intensive and extensive use of urban green roofs has the potential to not only have a local but also extensive positive climatic effect. One positive thermal effect of green roofs is the reduction of heat extremes. For example, the surface temperature of gravel roofs and bituminized papers heat up to 80 °C whereas green roofs only heat up to 20–25 °C (Henninger 2010). This positive thermal effect is not only present during summer, but also during winter times. Furthermore, vegetation lowers the heat transfer during winter resulting in a better insulation. Another positive consequence of green roofs is a higher evaporation. Also, flood risk resulting from heavy rainfall is reduced. Roof greening is capable to store a certain amount of surface water, comparable to open vegetation, depending on the kind of vegetation, and runs of delayed, minus the loss by evaporation and transpiration. The discharge coefficient is improved substantially by roof greening. The direct runoff to the canalization of a roof with a 15° roof inclination is 80–100% and of a green roof with the same inclination only 30% (Henninger 2010). The remaining part is stored and emitted via evapotranspiration into the urban atmosphere. The buffered water leads to an improved humidity of the air. The delay relieves the urban drainage system and lowers the risk of floods (Chifflard 2011). The knowledge from moderate climate zones can be transferred to other climate zones, in order to contribute to an enhancement of the urban thermal climatic situation especially in arid and semi-arid climate zone.

This knowledge is the basis for two research projects of the Department of Physical Geography in arid and semi-arid climate zones in Karaj (Iran) and the city of Bahrain (Bahrain). The basic idea of the projects is to apply exemplary a sustainable strategy in these areas. Above all, local decision-makers should be made aware of the sustainability concept. The triangle of sustainability (economy, ecology, and society) presents different goals that reveal that advantages in the mentioned three fields arise for the residential environment and the living quality in urban areas.

Economic advantages: Of course, investments are required in the initial phase of a green roof project for preparations, but with the duration of the project these costs can be saved. The advantages of green roofs are that they

- lead to a reduction of surface temperature, which in turn can have a positive effect on the efficiency of the heat sensitive photovoltaic systems,
- lead to a reduction of surface temperature, which in turn results in a lower warming of the underlying premises and thus limiting the amount of energy necessary for cooling, and

- can reduce, in dependence of the intensive and extensive utilization, the urban noise pollution by increasing the capacity to absorb sound.

Ecological advantages: Green roofs improve the local urban climate and also have a positive influence on urban air quality. Particulate matter and other air pollutants can be reduced with roof greening by providing additionally created deposition and filter areas. In addition, as part of a purposeful urban water management, roof greening can act as water storage and thus delays rainwater-runoff supporting the sewerage, especially for heavy rainfall (Henninger 2013b). Last but not least, green roofs provide a habitat for flora and fauna, in particular for extensive used areas.

Social advantages: Besides the economical and ecological benefits green roofs also can have positive impacts on social issues. Urban greening can improve the esthetic impression of an urban area and enhance quality of living by providing green oases for urban dwellers.

In Bahrain and Iran interdisciplinary projects like "Bahrain Regains Greenery—Green Roofs in Private & Public Buildings" were initiated in order to provide guidance by lighthouse projects for authorities. Public buildings provide large potential roof areas and the positive impact of green roofs is analyzed, adjusted, and promoted intensively (Henninger et al. 2015). Furthermore, besides the intensive and extensive use of roof greening and use of solar energy with photovoltaic, urban gardening/farming for sun exposed experimental roof areas is an important issue. Such areas can act as urban agricultural reserve areas and provide additional harvest yield in particular in climatic favored regions (Henninger 2012a).

5. Existing urban green has of course also a cooling effect resulting from shading and evapotranspiration. Thus, the Department of Physical Geography investigates among other things how urban green can mitigate the negative local impacts of urban heating. For example, silver linden (*Tilia tomentosa*) are investigated local climatic. The aim of the project is to quantify the cooling effect of silver linden on local surroundings and if there exists a positive mitigating effect on urban heating during summer. The focus is not on the already known effects of evapotranspiration and shading, but on the specific albedo. *Tilia tomentosa* is used and initial investigations have shown that especially young leaves of the treetop are able to guard themselves from intensive solar radiation. If temperature within the treetop is high during daytime the leaves turn over resulting in that the short wave reflecting silver bottom leave side faces up and the short wave radiation absorbing dark green upper leave side faces down (Wundsam and Henninger 2015). When the leave turns over, also the albedo changes, resulting in a reduction of the surface temperature at the treetop and a reduction of air temperature within the tree trunk and crown space and of the surrounding air. Analyses with trees of different age groups are conducted (age groups: 0–10 years, 11–25 years, >25 years) to analyze if the effects of the above described leave turning changes over time and to be able to better

compare different tree species. Afterwards, the results are transferred to the local climate and the responsible positions for urban planning including city administrations and urban planners are sensitized for the effect of cooling of silver linden.

In situ measurements are done within and outside the tree population. Stress points of leaves will be analyzed with a thermal imager to visualize changes and quantify when leaves react on air and surface temperature and turn their silver side up. Changes of meteorological parameters are measured close to the tree trunk and nearby the tree to detect the impacts of the leave rotation on the surroundings of the tree. Other tree species are analyzed as well to compare the thermal effects of silver linden with others to better illustrate the findings of *Tilia tomentosa*. Preliminary results are promising. If leave changing trees have a positive impact on the local urban climate these findings will be prepared and presented for different levels of urban planning in order to counteract the trend of urban heat islands (Wundsam and Henninger 2015).

6. Road side greenery leads, in general, to better local urban climatic conditions and acts as an air filter (Litschke and Kuttler 2008). From an air quality perspective the choice of tree species should take the type of treetop into account. If the treetop is to dense, air ventilation is reduced and concentration of air pollution can reach levels comparable to an urban canyon (Errel et al. 2011; Henninger 2015). This is only valid for roads with high traffic. Within the project "Endangerments on Schools" air quality in and nearby primary schools is analyzed. Thereby, the main focus is on areas where on first sight an air quality assessment would expect a low exposure like in outskirts. It could be shown, that air quality especially during morning rush hours in suburban primary schools and surroundings should be regarded as potentially polluted (Henninger 2013c).

In many cases, the combination of high traffic and bad air exchange within the urban boundary layer are the dominating negative influencing factors. It is hardly surprising for schools located in the city center, because of its narrow urban canyons. This effect of unsatisfactory air quality can be verified too for locations at the outskirt or suburban areas in the country. In order to do so, meteorological and air pollution parameters are measured simultaneously at urban, suburban, and rural school locations. First investigations show, that some locations suffer in the morning hours from bad air quality and that the situation is improving only slowly during the day. This effect was particularly apparent at suburban school locations where beside access road trees with a pronounced crown structure as roadside greenery were present. Such trees form an urban canyon like character. It could be shown that during summer the deterioration of the local air exchange was more distinct in the described suburban areas compared to typical urban canyons classified as air quality hot spots (Henninger 2013c).

7. The local climatic benefits from urban green areas are well known, nevertheless, some characteristics remain unclear and need to be investigated (Henninger 2015). Some tree species are capable to emit a certain amount of biogenic volatile organic compounds (Henninger 2012b, 2013a, 2014a, b), one being terpenoids. Some of the organic substances produced by vegetation serve as defense against herbivores and pathogens or attract pollinators. In addition, it is assumed that isoprene protects plants from heat or oxidative stress (Henninger 2013a, 2015). The emission rate of biogenic hydrocarbons depends on the meteorological conditions (air temperature and solar radiation or photosynthetic active radiation) and how it affects vegetation (Henninger 2015). During clear and calm weather conditions, that are characterized by high solar radiation and high air temperatures, the emission of isoprene from the leaves is high (Henninger 2014b, 2015). This implicates that during these days with already high ozone concentration the emission of biogenic isoprene results in an additional near surface ozone production, resulting in extra stress for urban retreat and recreation areas.

Relevant tree species emitting isoprene are in particular: *Quercus* (oak), *Populus* (poplar), *Robinia* (robinia) and *Salix* (willow). Conifers like the *Picea abies* (Norway spruce) play only a minor role in this context. It is a problem that many of the above mentioned tree species are characteristic urban vegetation. Thus, few tree species are responsible for the entire urban isoprene emissions. At this point, it can be noted that a great amount of biogenic isoprene can be emitted despite of a relative low vegetation density. This is due to the fact, that the composition of species of urban green is not necessarily comparable to the composition of species of a natural habitat (Henninger 2015). A popular example for an isoprene emitting tree is *Platanus acerifolia* (London plane). Originally, it is not a native tree species. Still, the London plane is planted often as roadside green or in parks because of its resistance against air pollution and insensitivity against dense soil (Wagner 2014; Henninger 2015).

Biogenic isoprene has a relatively high ozone formation potential compared to anthropogenic hydrocarbons, because of its high reaction rate constant for reactions with OH (hydroxyl) radicals. This means that even at low levels, isoprene may be regarded as a significant precursor for the production of ozone (Henninger 2015). Since the 1990s the terms "low emitter" and "high emitter" plants are used in urban ecology and urban climatology. The consideration of these high emitter plants in the planning of urban green can have a high impact on the formation rate of biogenic hydrocarbon and thus on the ozone creation potential (Henninger 2013a, 2015). The reason why this topic has been considered so little so far is that there exists no general method or approach for the investigation. The transferability to other places and the comparability is difficult. Of course there are plenty of publications concerning this issue (analyses about the ozone creation potential of urban green, urban trees, roadside greenery and urban park vegetation). However, the published emission potentials of the tree species are very diverse. The comparability is not given due to the many different location factors like genetic properties, age of the tree, tree species, leave size, soil properties, water availability, and atmospheric CO_2 concentration (Henninger 2015).

8. The human health impacts of ozone are diverse (Henninger 2012b, 2014b). Some harmful health effects of ozone could be:

   - penetrating the inferior respiratory tract,
   - irritation of the respiratory tract, dyspnea, tussis, headaches,
   - decrease of the physical working capacity,
   - increased frequency of asthma attacks, and
   - increased frequency of allergic reactions (delicate reactions by 10–20% of the people).

These indicate why the support of ozone formation by vegetation is an important and sensible topic (Henninger 2013a). Characteristic risk groups are outdoor workers, athletes, asthmatics, small children and babies as well as elderly people.

Our study basing on data from 2011 to 2014 shows a significant relationship between locations with high near surface ozone concentration and the number of respiratory illness, especially within urban park and recreation areas. This is because airborne allergens like pollens become more aggressive or allergenic by ozone resulting in possible strong and unexpected reactions of susceptible groups like asthmatics and allergy sufferers. This relationship is critical with respect to climate change, as stated earlier, the formation of near surface ozone is elevated by higher air temperature and higher solar radiation (Henninger 2012b, 2014b).

9. In order to present an adequate economic and ecological statement about the influence of changes of urban structures it is advisable to apply mathematic simulations. Urban problems and fields of action can be better assessed with the assistance of microclimatic models. The interactions between different meteorological parameters and the heterogenic urban building structure are complex making it difficult to predict precise results for the future planned situation, in particular for megacities in Asia (Ringhof and Henninger 2010; Henninger 2011c). In microclimatic simulation models different elements like urban vegetation and development can be incorporated and how they interact with the atmosphere. Through this proactive planning it is possible to avoid thermic and air quality related problems. As well, an optimization of already existing building structures is possible (Bruse 2000). These measures are often expensive and time consuming and thus the Department of Physical Geography has developed the so called "matrix method" (Fabisch and Henninger 2014). Urban climatic modifications are not only present in big cities, but also apparent on smaller scales. Small changes like an increase of air temperature can lead to an impairment of life quality and cause health problems in susceptible groups like children and elderly. Big cities with a high financial power can verify and investigate urban climatic effects by empirical surveys or complex models

whereas small communities cannot. An ecological urban development is only possible with the consideration of a reliable survey of the ecologic determining factors. The costs of empirical surveys and numerical models can be reduced by reducing the spatial resolution, but this also reduces the power of the results (Fabisch and Henninger 2015). The idea of the "matrix method" is to examine urban ecological relevant indicators like soil sealing and percentage of urban green. These indicators are used in combination with others to identify interdependencies that cause a certain urban climatological phenomenon. For example, a green area can be classified as a cool air production zone, but only in combination with a low surface roughness and inclination towards settlement the phenomenon of a cool air production zone is present in reality. The survey of the indicators is carried out using a grid with a resolution of 50–100 m (edge length). For the survey a questionnaire is developed and a user can retrieve, collect, and store information about indicators for every grid on-site in a database on a smartphone or tablet (Fabisch and Henninger 2014, 2015). This easy handling enables non-experts to conduct the survey and thus reduces the costs. Information from the database can be analyzed in geographic information systems (Allbach et al. 2014; Fabisch and Henninger 2014, 2015). The applied matrices allow for an estimation of the probability of the occurrence of an urban climatic phenomenon. Depending on the impact on humans, the probability can be rated as positive or negative. In this way, maps showing urban ecological problem areas and areas with positive potential within the settlement can be created for or by communities to promote an urban ecological-oriented urban planning and to develop recommendations of action (Fabisch and Henninger 2014, 2015).

4 Conclusions and Outlook

In future, applied urban climatology will play an increasingly important role due to climate change. The city, residence for many people must be described as a local climatic hot spot. The influence on the natural ecosystem by urban settlements and the resulting problem areas require action of urban planning. Climatological and air quality related problems need to be localized and countermeasures based on the findings of climate function maps and urban planning recommendation maps have to be presented. The awareness and knowledge about the local climate and the resulting influences on air quality and their climatic functional interactions are important aspects of urban planning and environmental precaution. Climate and air are important parts of urban planning, environmental impact assessment, and site evaluations. Thus, the aim of an applied and planning-oriented urban climatology is to identify and solve climatological and air quality related problems and to secure, restore, and develop urban climatic relevant land use structures.

References

Allbach, B., Henninger, S. & Deitche, E. (2014) *An urban sensing system as backbone of smart cities*, [Schrenk, M., Popovich, V., Zeile, P. & P. Elisei], REAL CORP 2014, Plan it Smart, pp. 55–64.

Bruse, M. (2000) *Anwendung von mikroskaligen Simulationsmodellen in der Stadtplanung*, [Bernhard, L. & Krüger, T.], Simulation raumbezogener Prozesse Methoden und Anwendung, Münster (in German).

Bongardt, B. (2006) *Stadtklimatische Bedeutung kleiner Parkanlagen – dargestellt am Beispiel des Dortmunder Westparks*, Essener Ökologische Schriften, Bd. 24, Hohenwarsleben (in German).

Chifflard, P. (2011) *Urbaner Wasserhaushalt*, [Henninger, S.], Stadtökologie, Paderborn, Ferdinand Schöningh Verlag (in German).

Erell, E., Pearlmutter, D. & Williamson, T. (2011) *Urban Microclimate – Designing the spaces between buildings*, London, Earthscan.

Fabisch, M. & Henninger, S. (2014) *Urban-ecological survey for small settlements*, Journal of Ecology - Biomedical & Life Science, Vol. 4, pp. 591–600.

Fabisch, M. & Henninger, S. (2015) *Smartphonegestützte Bestandsaufnahme zur ökologischen Bewertung von Siedlungsräumen*, [Schrenk, M., Popovich, V., Zeile, Elisei, P. & Beyer, C.], REAL CORP 2015, Plan Togehter - Right Now- Overall, pp. 561–570 (in German).

Hartz, A. (2011) *Neue Herausforderungen für die Stadtentwicklung – dargestellt am Beispiel des Klimawandels*, [Henninger, S.], Stadtökologie, Paderborn, Ferdinand Schöningh Verlag (in German).

Henninger, S. (2010) *Energieeffizientes Bauen*, Koblenzer Geographisches Kolloquium, Heft 32, pp. 53–65 (in German).

Henninger, S. (2011a) *Das Klima der Stadt*, [Henninger, S.], Stadtökologie, Bausteine des Ökosystems Stadt, Paderborn, Schöningh Verlag, pp. 63–96 (in German).

Henninger, S. (2011b) *Ökosystemkomplex Stadt*, [Henninger, S.], Stadtökologie, Bausteine des Ökosystems Stadt, Paderborn, Schöningh Verlag, pp. 11–33 (in German).

Henninger, S. (2011c) *Improvement of the thermal comfort within dense housing complexes*, [Herbert, M., Jankovic, V. & B. Webb], City Weathers - meteorology and urban design 1950-2010, Manchester Architecture Research Centre, pp. 81–84.

Henninger, S. (2012a) *Urban gardening and urban climate*, IGC 2012, Down to Earth, 32nd International Geographical Congress, Cologne.

Henninger, S. (2012b) *Biogenic isoprene and its impact on human health in dependence on meteorological conditions*, Journal of Environmental Protection, Vol. 3, No. 29A, pp. 1206–1212.

Henninger, S. (2013a) *Notwendiger Wandel im Umgang mit innerstädtischen Grünflächen?*, [Junkernheinrich, M. & K. Ziegler], Räume im Wandel - Empirie und Politik, pp. 73–92 (in German).

Henninger, S. (2013b) *Der urbane Wasserkreislauf - Veränderungen des lokalen Wasserkreislaufs durch Landnutzungsänderungen*, Praxis Geographie, Alles im Fluss - Ökosystemare Kreisläufe, 01/2013, pp. 14–16 (in German).

Henninger, S. (2013c) *Endangerments on Schools*, [Schrenk, M., Popovich, V., Zeile, P. & P. Elisei], REAL CORP 2013, Planning Times, pp. 285–292.

Henninger, S. (2014a) *Urban green and health*, [Pfaffenbach, C. & Schneider, C.], Global Demographic and Climate Challenges in the City - An interdisciplinary assessment of impacts, needs and strategies, Aachener Geographische Arbeiten, Heft 50, pp. 115–130.

Henninger, S. (2014b) *The Impact of Biogenic Isoprene in Dependence on Meteorological Conditions within Urban Green*, [Rauch, S., Morrison, G., Norra, S. und Schleicher, N.], Urban Environment - Proceedings of the 11th Urban Environment Symposium, Springer Verlag, pp. 153–162.

Henninger, S. (2015) *Kann innerstädtisches Grün die Luftqualität beeinträchtigen?*, Neue Landschaft - Fachzeitschrift für Garten- und Landschaftsbau, 2, pp. 31–35 (in German).

Henninger, S. & Saaroni, H. (2014) *Numeric simulation of microclimate modifications in a Mediterranean coastal urban park*, Changes – Challenges – Responsibility, International Geographical Union, Krakow.

Henninger, S., Elmarsafawy, H. & Tobias, K. (2015) *Bahrain Regains Greenery*, Journal of Environmental Protection, Vol. 6, pp. 929–934.

Horbert, M. (2000) *Klimatologische Aspekte der Stadt- und Landschaftsplanung*, Landschaftsentwicklung und Umweltforschung - Schriftenreihe im Fachbereich Umwelt und Gesellschaft, Berlin, Bd. 113 (in German).

Kuttler, W. (2013) *Klimatologie*, Paderborn, München, Wien, Zürich Ferdinand Schönigh Verlag (in German).

Litschke, T. & Kuttler, W. (2008) *On the reduction of urban particle concentration by vegetation - a review*, Meteorologische Zeitschrift, 17 (3), S. 229–240.

Mayer, H. (1989) *Workshop „Ideales Stadtklima"*, Mitteilungen der Deutschen Meteorologischen Gesellschaft, H. 3, S. 52–54 (in German).

Oswalt, P. & Rieniets, T. (2006) *Atlas of shrinking cities*, Ostfildern, Hatje Cantz Verlag.

Overbeck, G., Hartz, A. & Fleischhauer, M. (2008) *Ein 10-Punkte-Plan Klimaanpassung. Raumentwicklungsstrategien zum Klimawandel im Überblick*, Informationen zur Raumentwicklung, Heft 6, S. 1–18 (in German).

Ringhof, E. & Henninger, S. (2010) *Verbesserung der thermischen Behaglichkeit innerhalb dichter Wohnkomplexe am Beispiel einer südkoreanischen Stadt*, [Schrenk, V. Popovich, D. & Zeile, P.], REAL CORP 2010, CITIES FOR EVERYONE Liveable, prosper, healthy, pp. 889–897 (in German).

VDI - Verein Deutscher Ingenieure (2015) *Umweltmeteorologie – Klima und Lufthygienekarten für Städte und Regionen*, Düsseldorf, VDI-Richtlinie 3787, Bl. 1 (in English).

Wagner, P. (2014) *Analyse von biogenem und anthropogenem Isopren und seiner Bedeutung als Ozonvorläufersubstanz in der Stadtatmosphäre*, Essener Ökologische Schriften, Bd. 34, S. 142 (in German).

Wundsam, T. & Henninger, S. (2015) *Leaf-Turning tree species and their local climatic influence on the city*, ICUC `09, Toulouse, Frankreich, UCP-13.

Developing Indices for Adaptation and Adaptive Capacity in Indian Marine Fishing

Krishna Malakar, Trupti Mishra and Anand Patwardhan

Abstract The Indian marine fishing community faces multiple climate and non-climate related stresses, such as unpredictable and extreme weather events, declining fish stocks and pollution. Prompted by these changes, some community members have adopted strategies such as motorization and mechanization of their boats and using GPS (Global Positioning System) for navigation, to ensure a greater fish catch as well as safety in the sea. Capacity to adapt is crucial for retaining livelihoods. This study attempts to measure and develop indices denoting the strength of adaptation and adaptive capacity of the community, using suitable indicators. Percentage of the community implementing the adaptation strategies are used to develop the adaptation index. According to the Sustainable Livelihoods Approach, capacity to adapt can be determined by access to different capitals. Hence, adaptive capacity of the community is measured through indicators for human, physical, economic and social capitals. These community-level indices are developed for seventy Indian coastal districts. The results suggest that Srikakulam in Andhra Pradesh and Rajkot in Gujarat have the lowest adaptation and adaptive

K. Malakar (✉) · T. Mishra · A. Patwardhan
Interdisciplinary Program (IDP) in Climate Studies, Indian Institute of Technology Bombay, Mumbai 400076, Maharashtra, India
e-mail: krishnamalakar26@gmail.com; krishna.malakar@iitb.ac.in

T. Mishra
Shailesh J. Mehta School of Management, Indian Institute of Technology Bombay, Mumbai 400076, Maharashtra, India
e-mail: truptimishra@iitb.ac.in

A. Patwardhan
School of Public Policy, University of Maryland, College Park 20742, MD, USA
e-mail: apat@umd.edu

W. Leal Filho (ed.), *Climate Change Research at Universities*,
DOI 10.1007/978-3-319-58214-6_26

capacity index respectively. Further, the indices are validated by evaluating the relation between adaptation and adaptive capacity through regression and Monte Carlo simulation. Results indicate that adaptive capacity has low but significant influence on adaptation levels in the community, concluding that the indices are quite adept. The results of the study facilitate identification of the coastal districts in urgent need of policies and actions to develop adaptation and adaptive capacity of the community.

Keywords India · Marine fishing · Adaptation · Adaptive capacity · Index · Sustainable Livelihoods Approach (SLA) · Validation · Capital

1 Introduction

Climate change can severely impact the marine ecosystem by altering ocean temperatures, circulation, stratification, nutrients, oxygen and acidification levels (Doney et al. 2012). These transformations can lead to changes in migration and distributional patterns of marine species, population dynamics, ocean primary productivity and the marine food web (Roessig et al. 2004; Harley et al. 2006; Hoegh-Guldberg and Bruno 2010). Thus, climate change along with other human-induced stressors, like pollution and over fishing, pose severe threats to the marine fishing sector (Giulani et al. 2004; Daw et al. 2009). These multi-stressors have led to dwindling availability of fish and subsequently, a lower fish catch (Pauly and Zeller 2016), impacting marine livelihoods and national and global economics of fishing (Sumaila et al. 2011). Adapting to various stressors altering the ocean ecosystem is important for sustaining marine fishing livelihoods. Adequate capacity to adapt can lead to adequate adaptation (IPCC 2014). Thus, measures and indices for adaptation and adaptive capacity are crucial for planning adaptation and designing capacity-building programmes.

Developing indices for adaptation is still relatively unexplored in the literature. Adaptation indices help institutions and organizations, involved in executing policies that enhance adaptation, by giving a measure of success of their interventions. Indicators measuring adaptation can denote implementation as well as benefits of adaptation, that is, they can be based on the process and outcome of adaptation respectively (Harley et al. 2008; Hinkel 2011). In the past, both process and outcome based indicators have been proposed for assessing coastal management and governance (Ehler 2003). This study focuses specifically on marine fishing livelihoods and attempts to develop indices of adaptation and adaptive capacity in the community. Adaptation is measured through process-based indicators, that is, percentage implementation of different adaptation strategies in the community. Unlike outcome-based indicators, district-level data on a number of these process-based indicators are available for the community. Outcome-based indicators, such as increase in yield or fish catch over the years, can help in denoting adaptation in the community, but district-level data on these indicators are rarely available for India. Thus, usage of data-available process-based indicators can enable capturing a comprehensive index of adaptation in the community.

Possessing adaptive capacity can diminish vulnerability (IPCC 2007). The concept of adaptive capacity has been substantially used in the literature to define vulnerability of communities (Hahn et al. 2009; Nelson et al. 2010). Vulnerability of agriculture (Brien et al. 2004; Gbetibouo 2009) as well as of fisheries and coastal communities (Allison et al. 2009; Cinner et al. 2012; Mamauag et al. 2013) have also been widely studied through indices of adaptive capacity. Indicators of adaptive capacity of communities are generally socio-economic characteristics which can help in retaliating adverse impacts of stressors. These indicators can be segregated into human, social, natural, physical and economic capital according to the Sustainable Livelihoods Approach (SLA) (Scoones 1998; Serrat 2010). Table 1 provides a description of the various capitals. These capitals can furnish adaptive capacity of a community or a household (Hahn et al. 2009). Consequently, access to these capitals can reduce vulnerability to multi-stressors. Hence, the SLA framework can also be helpful in studying adaptive capacity of vulnerable marine fishing livelihoods (Allison and Horemans 2006; Iwasaki et al. 2009). This study has, thus further, proposed an SLA-based adaptive capacity index for the community.

India's marine fishing sector is one of the most vulnerable to climate change in Asia (Allison et al. 2005). Climate change can especially endanger marine fishing in South Asian countries, like India, because of their economic and dietary dependence on the sector (Barange et al. 2014). Future climate and marine population projections indicate that there might be extinction of numerous local specie in the tropics (Cheung et al. 2009). There can also be 40% decline in fish catch in the tropics by the mid 21st century (Cheung et al. 2010). India's marine resources are also threatened by pollution, destruction of mangrove habitats and overfishing (Bhatt and Venkataraman 2013; Bassi et al. 2014). In 2014, India saw a 5% decline in its fish catch (CMFRI 2015). India has a long coastline of around 7500 km on its east and west. Also, few union territories of India are islands located in the Bay of Bengal and Arabian Sea. In total, there are 13 states and union territories having 73

Table 1 Types of capital

Capital	Description	Example
Human	"A stock of capabilities" (Goodwin 2003, p. 5)	Skills, education
Social	"Stock of trust, mutual understanding, shared values, and socially held knowledge that facilitates the social coordination of economic activity" (Goodwin 2003, p. 6)	Networks, friends
Natural	Natural resources or ecosystem services	Land, fish stocks
Physical	"Assets that are generated by applying human productive activities" (Goodwin 2003, p. 4)	House, fishing gear
Economic	Direct source of money without having to trade any physical or natural asset. Though human and social assets may be used to settle a loan or credit	Availability of loans, savings

Adopted from Allison and Ellis (2001), Valdivia and Gilles (2001) and Goodwin (2003) coastal districts. Again, there are around 864,550 marine fishing households in India (CMFRI 2010a). Thus, the marine fishing population is spread across a large and diverse area.

In view of the current vulnerability of ocean resources to climate change and other stressors, this study attempts to measure adaptation and adaptive capacity of the marine fishing community in India. Indices are often deemed to be good measures of complex phenomena for understanding and initiating policy-making (Hinkel 2011). Thus, indices for adaptation and adaptive capacity of the marine fishing community have been developed by aggregating various proxies for the same. In the past, indices have been developed to assess the adaptive capacity (Cinner et al. 2015) as well as vulnerability (Islam et al. 2014) of fishing communities. These studies have provided in-depth understanding of the constructs through primary surveys of the targeted communities. In contrast to these studies which focussed on small regional scales, the present study attempts to utilize secondary data, representing the entire fishing population in India, to understand adaptation and adaptive capacity at a national level. Again, validating indices is a crucial part of index-development but there have been very rare attempts to do the same. Since capacity to adapt determines adaptation responses and its execution (IPCC 2014), this study has validated the indices by assessing whether availability of adaptive capacity, as defined in the study, is related to adaptation in the community.

2 Conceptual Framework: Measuring adaptation and adaptive capacity

The oceans are open-resources, and hence both public/cooperative and private responses are required for adaptation and sustainability. Private responses are often dictated by public decisions. The current study focuses on adaptation and adaptive capacity of individual fishermen, hence private adaptation responses are the subject of scrutinization in the analysis. Various types of private adaptation responses have been undertaken by marine fishing communities across the globe. Using variety and advanced fishing gear is one of the most important adaptation strategies (MacNeil et al. 2010; Johnson 2012; Blythe et al. 2014; Islam et al. 2014; Zamroni 2015). Thus, quantitative data on usage of fishing gear can be important measures of adaptation. In India, data on gear used by the community has been documented for the all the coastal states and union territories (Fishery Survey of India 2010; CMFRI 2010b). District-level data on percentage of motorized and mechanized boats, and percentage of fishermen having life saving equipments and GPS (Global Positioning System) are available. This data may serve to be useful indicators of adaptation in the community (Table 2). These are process-based indicators intended particularly for assessing strength of adaptation in marine fishing. Although using

Table 2 Indicators used in the study

Index for	Indicators	
Adaptation	Percentage of motorized boats	
	Percentage of mechanized boats	
	Percentage of fishermen using life saving equipments	
	Percentage of fishermen having GPS	
Index for	**Indicators**	**Proxy for**
Adaptive capacity	Percentage of schooled fisherfolk population	Human capital
	Number of landing centres	Physical capital
	Number of banks	Physical and economic capital
	Percentage of families below poverty line	Economic capital
	Number of community centres	Physical and social capital
	Percentage of fisherfolk population having membership in fishing cooperative	Economic and social capital

Indicators/proxies have partly been adopted from 'Malakar, K., Mishra, T. and Patwardhan, A. *Adaptation in the Indian marine fishing community*', currently under review and awaiting publication

life saving equipments is imperative to escape any life-risks in the sea and is rather a safety measure, it has been considered as an indicator of adaptation in the study, as usage of such equipments in many sections of the community is limited. Apart from adopting gear diversity, decisions to increase working hours and area are other responses to supplement fish catch (Daw et al. 2009; McIlgorm et al. 2010; Sievanen 2014). But such data could not be availed from secondary sources at country-level.

Adaptive capacity can be measured through availability of the capitals namely human, social, natural, physical and economic (Elasha et al. 2005; Wright et al. 2012). Oceans, which are the only natural resource/capital relevant for sustaining fishing livelihoods, are common and shared by the entire community. Hence, natural capital can be excluded while studying the private or individual adaptive capacity of marine fishermen. Indicators for the various other capitals are used to develop the adaptive capacity index of the community. Data on the indicators for the capitals (Table 2) are available in Marine Fisheries Census 2010 (Fishery Survey of India 2010; CMFRI 2010b). It is important to note that some indicators cannot be categorized into a single type of capital. A clear segregation is difficult as some entities can play multiple/dual roles. For example, the presence of banks in a district is a physical capital as well as a source of loans and finances, which makes it an economic capital. Similarly, community centres are physical capitals which help the community to congregate, thus reinforcing social capital. Membership in fishing cooperative can help in availing subsidies on fishing gear and other requirements, but can also help improve networks among the community and contribute to social capital.

3 Methodology

3.1 Indicators Used

The indicators of adaptation are the percentages of implementation of various adaptation strategies (Table 2). Following the Sustainable Livelihoods Approach (SLA), adaptive capacity is represented by various proxies of physical, human, economic and social capital (Table 2). SLA is a competent framework that can enable understanding of adaptive capacity in vulnerable livelihoods. It has been widely used to understand vulnerability and adaptation in various communities (Brocklesby and Fisher 2003; Hahn et al. 2009).

3.2 Data

India's coastline is shared by 13 states and union territories including the islands of Lakshadweep, and Andaman and Nicobar. The required data on the indicators (Table 2) for 70 coastal districts of India have been collected from the respective Marine Fisheries Census 2010 (Fishery Survey of India 2010; CMFRI 2010b) of the states and union territories. Out of the total 73 coastal districts of India, three districts, namely North 24 Parganas, Howrah and Yanam, have not been considered in the study because of missing data on a few indicators.

3.3 Constructing the Indices

Two composite indices of adaptation and adaptive capacity are constructed. Such indices can help in reducing a host of information into a single quantifiable measure, which can consequently facilitate identification of required strategies and designing of relevant policies (Hinkel 2011; Malakar and Mishra 2016). Thus, composite indices are advantageous as they can provide an outlook of the measured constructs. However, analyzing the contributing indicators can further assist in interpreting the indexed constructs.

Before computing the composite index, the indicators need to be normalized to enable relative comparison. All the indicators of adaptation and adaptive capacity have been normalized using Eq. 1 (Hahn et al. 2009; Malakar and Mishra 2016). Normalization facilitates a yardstick for comparison among the districts. Although this method of normalization may result in distortion of the index in presence of outlying maximum and minimum values, it is adopted as it allows comparison with the minimum. Hence, a greater value of the resulting index indicates greater strength of the indicator for a district. This enables easier comparison amongst the districts as well as easier aggregation of the indicators to form the composite indices. The districts having the highest and lowest values of the indicator are scored as 1 and 0 respectively.

$$Index_{A_d} = \frac{A_d - A_{\min}}{A_{\max} - A_{\min}} \tag{1}$$

where A_d is the value of the indicator for district d, and A_{min} and A_{max} are the minimum and maximum values of the indicator among all the districts. $Index_{A_d}$ represents the standardized value of an indicator of district d.

The normalized indices of adaptation and adaptive capacity obtained from Eq. 1 are averaged to obtain the overall indices for the same (Eq. 2). It is to be noted that the values obtained after normalizing the indicator 'percentage of the community below poverty line' are subtracted from 1, so that the final indicator value contributes positively to the overall adaptive capacity index. This is necessary, as the original indicator would have otherwise contributed negatively to the adaptive capacity index. Also, factor analysis and other similar techniques, which help in determining weights and reducing number of indicators, are not utilized in the present study as each index consists of a very few indicators (Malakar and Mishra 2016). The indicators have been selected using the SLA framework and from available data sources; and such data reduction techniques would diminish the dimensionality of the developed indices. Additionally, there is no available literature that can help judge the differential importance of the indicators. Thus, all indicators of the composite indices are considered to be equivalently important and are averaged.

$$A_d = \frac{\sum_{i=1}^{n} Index_{A_{di}}}{n} \tag{2}$$

where A_d is the adaptation or adaptive capacity index of district d; and n is the number of indicators. A higher value of the indices represents greater adaptation and adaptive capacity.

3.4 *Validation*

Indices are often validated by assessing their relation with the outcome of the indexed phenomenon, for instance, vulnerability index can be validated by assessing its relation with mortality rate, which is an outcome of being vulnerable (Adger et al. 2004; Fekete 2009). In the present case, the outcome of having adaptive capacity is implementation of adaptation strategies. Hence, the contribution of adaptive capacity to adaptation in the community has been assessed using Ordinary Least Square (OLS) regression. In other words, the relationship between adaptive capacity and adaptation, as developed and indexed in the study, is examined. This is an approach of ensuring construct validity, which is often defined as "the extent to which an operationalization measures the concept it is supposed to measure" (Bagozzi et al. 1991, p. 421). Regression offers an efficient and simple way to estimate the intensity of relationship between two variables (Gujarati et al. 2009). In this study, it is hypothesized that there is a significant relationship between the adaptive capacity and the adaptation index of the community. Thus, acceptance of the proposed hypothesis ensures the validity of the indices developed in the study. Equation 3 represents the regression model.

$$I_d = \alpha + \beta C_d + \varepsilon \tag{3}$$

where I_d and C_d are the indices for adaptation and adaptive capacity respectively of district d, α and β are the intercept and slope; and ε is the error.

The unbiasedness of the parameter estimates of the regression model (i.e. β_d and α_d) is, further, examined using Monte Carlo simulation (Briand and Hill 2013). This simulation is an effective way to check the unbiasedness of regression estimates (Meaney and Moineddin 2014). Monte Carlo analysis helps to simulate random data points iteratively based on the original data. The regression parameters are again estimated for each of these iterations. Averages of these parameters are taken to make comparison amongst the iterations as well as the true estimates from the regression on the original data. This technique is based on approximation, but greater number of iterations leads to higher precision. In the current study, the Monte Carlo simulation has had 10, 100 and 1000 runs. It is expected that increasing number of iterations would lead to improved parameter estimates nearing the true values. This would indicate that the regression parameters are unbiased, representative of entire population and would not differ with change in sample.

4 Results and Discussion

4.1 Adaptation and Adaptive Capacity Index

The indices for adaptation and adaptive capacity have been mapped in Fig. 1a, b respectively. The values of both the indices have been divided into five equal intervals and categorized into lowest, low, medium, high and highest in the maps. The adaptation and adaptive capacity indices range from 0.04 to 0.53 and 0.06 to 0.58 respectively. Many districts of Tamil Nadu on the east coast seem to have high levels of implementation of adaptation strategies (Fig. 1a and Appendix). High levels of motorization (46–84%) as well as usage of life saving equipments (42–97%) have resulted in a higher adaptation index in these districts. Few districts of Gujarat and Karnataka on the west also have high adaptation. Greater levels of mechanization in Gujarat (47–94%) contribute to higher adaptation indices, whereas the districts in Karnataka have higher motorization levels (61–62%).

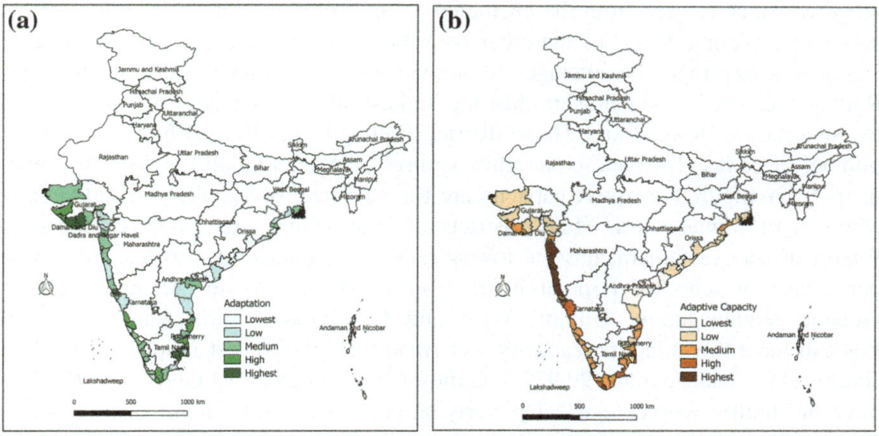

Fig. 1 Mapping of **a** adaptation index (Lowest = 0.04–0.13; Low = 0.13–0.23; Medium = 0.23–0.33; High = 0.33–0.43; Highest = 0.43–0.53) and **b** adaptive capacity index (Lowest = 0.06–0.17; Low = 0.17–0.27; Medium = 0.27–0.38; High = 0.38–0.48; Highest = 0.48–0.58) in the coastal districts of India

The percentage usage of life saving equipment (21–48%) and GPS (3–21%)[1] are also high in these districts of Gujarat and Karnataka. Overall, North Goa district of Goa, Amreli in Gujarat, Thanjavur in Tamil Nadu followed by Junagadh in Gujarat, Pudukkottai and Tirunelveli in Tamil Nadu have the highest adaptation indices. North Goa has the highest usage of GPS (33.5%) among all. Amreli and Tirunelveli have very high levels of mechanization (93.6%) and motorization (84.4%) respectively. Thanjavur and Pudukkottai have high usage of life saving equipment, 90 and 97% respectively. Thus, higher levels of adaptation of these strategies have led to a higher adaptation index. Many districts of Maharashtra and Kerala have medium levels of adaptation. Most districts of Andhra Pradesh, located on the east coast, have low adaptation index. Srikakulam in Andhra Pradesh has the lowest adaptation index, which is a resultant of low percentage of adaptation of all strategies. Other districts having low adaptation indices are Kendrapara and Ganjam in Odisha, Rajkot in Gujarat, West Godavari, Krishna, Vizianagaram and Visakhapatnam in Andhra Pradesh. It appears that Gujarat has districts having high as well as low adaptation, which shows that there is disparity within the state. Figure 1a suggests that largely, the west coast has adapted well as compared to the east coast. High levels of mechanization on the west have resulted in greater adaptation indices for the west.

In case of adaptive capacity as well, the west coast appears to have fared better (Fig. 1b and Appendix). All the districts of Maharashtra appear to have high

[1]The average percentage use of GPS in India is 2.7% (Appendix).

adaptive capacity. This may be attributed to high presence of community centres and above average levels of the other contributors of the index. Levels of poverty are also lower than the average in these districts and range from 15 to 37%. Ratnagiri district of Maharashtra has the highest adaptive capacity index followed by Sindhudurg, Raigad and Thane district. Ratnagiri and Raigad have the highest number of banks (190) and community centres (285) among all the districts. Other districts having high adaptive capacity are Kanyakumari of Tamil Nadu and Greater Mumbai of Maharashtra. Many districts of Odisha have low adaptive capacity. Rajkot district of Gujarat has the lowest adaptive capacity. It also has the lowest percentage of schooled population (0.19%) among all. Again, Anand district of Gujarat, Vizianagaram, Guntur, West and East Godavari districts of Andhra Pradesh have low adaptive capacity. As Anand has the lowest number of landing centres (1), Vizianagaram (99.92%), Guntur (98.97%) and East Godavari (96.73%) have the highest percentages of poverty-stricken population, these districts score poorly on adaptive capacity. Similarly, West Godavari does not have any bank and many of the other indicators have near minimum values, which contribute to its poor capacity to adapt. Rajkot in Gujarat, Vizianagaram and West Godavari in Andhra Pradesh also have low adaptation indices.

4.2 Assessing Validity

A comparison between the indices reveals that majority of the districts have a higher adaptive capacity index than adaptation (Fig. 2 and Appendix). However, many districts of Gujarat and Tamil Nadu, as well as a few in Goa, Karnataka, Kerala, West Bengal, Odisha, Andhra Pradesh and Andaman and Nicobar, have higher indices for adaptation than adaptive capacity, indicating these districts to be highly capable of utilizing their limited resources. Social capital through networks and bonds of trust, and other intangible sources of capabilities, like good physical and mental health (human capital), explicable through the ‘capability approach’ (Sen 1980, 1984, 1985, 1987, 1990, 1992, 1993, 1995, 1999 as in Robeyns 2005) can also improve capacity to adapt. Inclusion of such factors to further understand adaptive capacity can help decipher the higher adaptation than adaptive capacity in the districts.

Regression between the indices results in a coefficient (slope) of 0.286 for the dependent variable (adaptive capacity), which is significant at 99% confidence level. This indicates that higher adaptive capacity can indeed induce higher levels of adaptation. Results of Monte Carlo analysis (Table 3) also depict that the regression estimates are presumably unbiased. The average values of the parameters from a higher number of iterations result in values very close to those generated from the original data.

The regression model has an adjusted R^2 value of 8.2% which is again significant at 99% confidence level. Consideration of limited number of proxies for

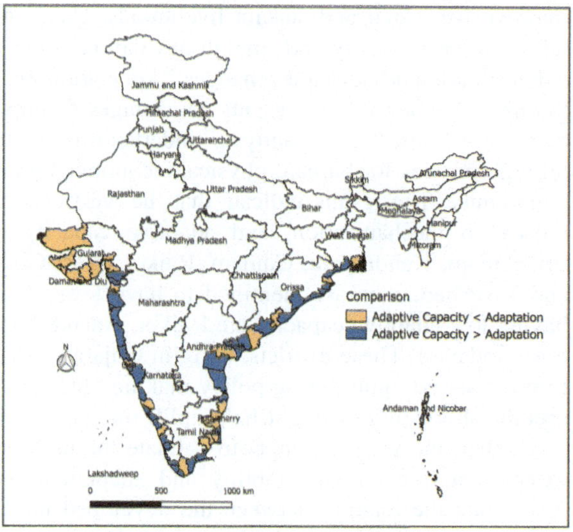

Fig. 2 Map comparing indices for adaptation and adaptive capacity in the coastal districts of India

Table 3 Results of Monte Carlo analysis and comparison with population regression

Number of iterations	10	100	1000	Parameter estimates from regression on population (Original data)
Average slope, β_d	0.251	0.278	0.288	0.286
Average intercept, α_d	0.190	0.180	0.176	0.177

constructing the adaptive capacity index must have led to low association values. Nevertheless, adaptive capacity has a low but strong influence on adaptation, indicating that having capacity to adapt might lead to adaptation. These sets of results suggest that the indices developed in the study have moderate construct validity and are quite competent in explaining adaptation and adaptive capacity.

5 Conclusion

Coastal fishing livelihoods are threatened by changing marine biodiversity, pollution, climate change, extreme weather events among others (Daw et al. 2009). Measuring adaptation and adaptive capacity are crucial for identifying potential vulnerable sections of the community and for adaptation management. Thus, this study has attempted to develop indices measuring adaptation and adaptive capacity in the community in India. Community-level indices for seventy coastal districts are developed. Adopting motorization and mechanization of fishing crafts can help improve fish catch and sustain livelihoods. Usage of life saving equipments and GPS can ensure safety and smooth navigation in the sea. Thus, these strategies can aid adaptation to multiple changes. Adaptation in the community is quantified through an index that represents percentages of implementation of different adaptation strategies. The capacity to undertake these strategies have been quantified through proxies for human, physical, economic and social capitals available in the community. The results indicate that the west coast has higher adaptation, owing mostly to mechanization, and adaptive capacity compared to the east coast. Srikakulam, Kendrapara, Ganjam, Rajkot, West Godavari, Krishna, Vizianagaram and Visakhapatnam are identified to have low adaptation. Similarly, the districts having low adaptive capacity are Rajkot, Anand, Vizianagaram, Guntur, West and East Godavari. These districts, lying in Gujarat, Odisha and Andhra Pradesh, need to be focussed upon during policy-making; and special interventions targeting the specific shortcomings might help uplift the community in these districts.

Further, the study attempts to validate the indices, which is very limited in the extant literature on vulnerability and adaptation indices. Regression is used to investigate the relation between the developed indices. One of the ways of validating indices is by correlating it with its outcome. Hence, regressing the two indices is a way of assessing construct validity. Communities adapt, which is an outcome, when there is capacity to adapt. Results show that adaptive capacity has a low but strong influence on adaptation, indicating moderate levels of validity and proficiency of the indices in explaining the phenomena.

Overall, the important contributions and inferences of the study are as follows:

(a) The study provides a useful way of quantifying adaptation and adaptive capacity in the Indian marine fishing community, which may also be applied for communities located in other regions of the world.
(b) There is large variability within states as many include districts having both low and high adaptation indices. There are a few states which show variability with respect to adaptive capacity in its districts.
(c) Many districts in the states of Goa, Gujarat, Karnataka and Tamil Nadu appear to have adapted (in comparison with the average adaptation index). Districts lying in Maharashtra, Kerala, Odisha and Andhra Pradesh seem to have poor adaptation.
(d) All districts of Maharashtra have high adaptive capacity. On the other hand, many districts of Gujarat, Odisha and Andhra Pradesh have poor adaptive capacity (in comparison with the average adaptive capacity index). Thus, adequate interventions to improve the capitals of the community in these states need to be planned.
(e) In the current literature on assessing vulnerability to climate change, indices have rarely been validated. This study has attempted validating the indices through regression, and a further check of the parameter estimates through Monte Carlo simulation.

The study has attempted to be comprehensive in developing the indices by exploiting secondary sources of data on the coastal fishing community in India. But data on other sources of capacity to adapt, like social capital which includes presence of networks and trust, can help in better estimation of the index. Limited number of indicators could be used for developing the indices because of availability of limited data on the community. This might have also led to lower association between adaptation and adaptive capacity. Greater and other relevant proxies for the capitals might result in different conclusions. Again, mechanization, which is considered as a form of adaptation in the study, is often deemed as an unsustainable technique of fishing as it can result in overfishing and can adversely impact the entire marine ecosystem (Giulani et al. 2004; Fernandes and Gopal 2012). Nevertheless, at the same time, high levels of mechanization indicates presence of well-off fishermen who have adapted and are sustaining their individual livelihoods.

This study has put a step forward in analyzing adaptation and adaptive capacity of the marine fishing community, whose livelihoods are one of the most vulnerable to climate change and other stressors. Approach and observations from the study may help in designing state and district-specific capacity-building policies for improving the livelihoods of the community.

Appendix

Descriptive statistics of indicators (inclusive of 70 districts used in the study)

S. no.	Indicator	Average	Minimum	Maximum
1	Motorized (in %)	39.1	0.0	88.8
2	Mechanized (in %)	33.2	0.0	100.0
3	Life saving equipment (in %)	18.6	0.0	97.1
4	GPS (in %)	2.7	0.0	33.5
5	Schooled fisherfolk population (in %)	51.96	0.19	89.1
6	Families below poverty line (in %)	53.8	1.0	99.9
7	No. of landing centres	21.9	1	100
8	No. of banks	22.2	0	190
9	No. of community centres	32.0	0	285
10	Membership in fishing cooperative (in %)	15.0	0.0	53.9

Adaptation and adaptive capacity indices of the districts

S. no.	State	Coast	District	Adaptation index	Adaptation grade	Adaptive capacity index	Adaptive capacity grade	Is adaptive capacity index higher than adaptation index?
1	Maharashtra	West	Ratnagiri	0.209	Low	0.583	Highest	Yes
2	Maharashtra	West	Sindhudurg	0.261	Medium	0.549	Highest	Yes
3	Maharashtra	West	Raigad	0.253	Medium	0.523	Highest	Yes
4	Maharashtra	West	Thane	0.246	Medium	0.501	Highest	Yes
5	Tamil Nadu	East	Kanyakumari	0.387	High	0.498	Highest	Yes
6	Maharashtra	West	Greater Mumbai	0.267	Medium	0.461	High	Yes
7	Tamil Nadu	East	Ramanathapuram	0.204	Low	0.431	High	Yes
8	Kerala	West	Alappuzha	0.134	Lowest	0.401	High	Yes
9	Karnataka	West	Uttara Kannada	0.225	Low	0.399	High	Yes
10	Goa	West	SouthGoa	0.284	Medium	0.398	High	Yes
11	Tamil Nadu	East	Nagapattinam	0.347	High	0.388	High	Yes
12	Kerala	West	Thiruvananthapuram	0.210	Low	0.384	High	Yes
13	Kerala	West	Kozhikode	0.260	Medium	0.376	Medium	Yes
14	Kerala	West	Ernakulam	0.264	Medium	0.376	Medium	Yes
15	Goa	West	NorthGoa	0.529	Highest	0.371	Medium	No
16	Puducherry	East	Puducherry	0.229	Low	0.362	Medium	Yes
17	Tamil Nadu	East	Tuticorin	0.321	Medium	0.357	Medium	Yes
18	Andaman and Nicobar	East	South Andaman	0.293	Medium	0.355	Medium	Yes
19	Tamil Nadu	East	Pudukkottai	0.431	Highest	0.354	Medium	No
20	Tamil Nadu	East	Thiruvallur	0.271	Medium	0.350	Medium	Yes
21	Kerala	West	Kannur	0.273	Medium	0.347	Medium	Yes

(continued)

(continued)

S. no.	State	Coast	District	Adaptation index	Adaptation grade	Adaptive capacity index	Adaptive capacity grade	Is adaptive capacity index higher than adaptation index?
22	Karnataka	West	Udupi	0.339	High	0.346	Medium	Yes
23	Puducherry	East	Mahe	0.288	Medium	0.338	Medium	Yes
24	Daman & Diu	West	Daman	0.283	Medium	0.337	Medium	Yes
25	Kerala	West	Thrissur	0.300	Medium	0.333	Medium	Yes
26	Tamil Nadu	East	Cuddalore	0.392	High	0.323	Medium	No
27	Karnataka	West	Dakshina Kannada	0.397	High	0.322	Medium	No
28	Tamil Nadu	East	Kanchipuram	0.373	High	0.322	Medium	No
29	Puducherry	East	Karaikal	0.279	Medium	0.322	Medium	Yes
30	Odisha	East	Balasore	0.210	Low	0.314	Medium	Yes
31	Tamil Nadu	East	Tirunelveli	0.409	High	0.312	Medium	No
32	Gujarat	West	Valsad	0.330	Medium	0.300	Medium	No
33	Kerala	West	Kasaragod	0.213	Low	0.298	Medium	Yes
34	Gujarat	West	Porbander	0.380	High	0.298	Medium	No
35	Kerala	West	Kollam	0.284	Medium	0.295	Medium	Yes
36	Gujarat	West	Junagadh	0.432	Highest	0.295	Medium	No
37	Andaman and Nicobar	East	N&M Andaman	0.274	Medium	0.291	Medium	Yes
38	Daman and Diu	West	Diu	0.288	Medium	0.289	Medium	Yes
39	Tamil Nadu	East	Thanjavur	0.455	Highest	0.285	Medium	No
40	Lakshadweep	West	Lakshadweep	0.220	Low	0.278	Low	Yes
41	Odisha	East	Kendrapara	0.087	Lowest	0.265	Low	Yes
42	Odisha	East	Jagatsinghpur	0.209	Low	0.263	Low	Yes
43	West Bengal	East	Purba Medinipur	0.175	Low	0.254	Low	Yes

(continued)

(continued)

S. no.	State	Coast	District	Adaptation index	Adaptation grade	Adaptive capacity index	Adaptive capacity grade	Is adaptive capacity index higher than adaptation index?
44	Gujarat	West	Navsari	0.192	Low	0.248	Low	Yes
45	Kerala	West	Malappuram	0.254	Medium	0.247	Low	No
46	Andhra Pradesh	East	Nellore	0.188	Low	0.244	Low	Yes
47	Gujarat	West	Surat	0.213	Low	0.240	Low	Yes
48	Gujarat	West	Kutch	0.249	Medium	0.239	Low	No
49	Gujarat	West	Bhavnagar	0.243	Medium	0.236	Low	No
50	Andhra Pradesh	East	Visakhapatnam	0.121	Lowest	0.231	Low	Yes
51	Tamil Nadu	East	Villupuram	0.291	Medium	0.230	Low	No
52	Gujarat	West	Jamnagar	0.339	High	0.227	Low	No
53	West Bengal	East	South 24 Parganas	0.330	Medium	0.223	Low	No
54	Gujarat	West	Amreli	0.521	Highest	0.216	Low	No
55	Andhra Pradesh	East	Srikakulam	0.035	Lowest	0.207	Low	Yes
56	Odisha	East	Puri	0.185	Low	0.205	Low	Yes
57	Tamil Nadu	East	Thiruvarur	0.218	Low	0.204	Low	No
58	Odisha	East	Ganjam	0.115	Lowest	0.201	Low	Yes
59	Odisha	East	Bhadrak	0.252	Medium	0.194	Low	No
60	Andaman and Nicobar	East	Nicobar	0.261	Medium	0.190	Low	No
61	Gujarat	West	Bharuch	0.228	Low	0.188	Low	No
62	Andhra Pradesh	East	Krishna	0.116	Lowest	0.172	Lowest	Yes

(continued)

(continued)

S. no.	State	Coast	District	Adaptation index	Adaptation grade	Adaptive capacity index	Adaptive capacity grade	Is adaptive capacity index higher than adaptation index?
63	Andhra Pradesh	East	Prakasam	0.149	Low	0.171	Lowest	Yes
64	Tamil Nadu	East	Chennai	0.272	Medium	0.167	Lowest	No
65	Andhra Pradesh	East	East Godavari	0.180	Low	0.165	Lowest	No
66	Andhra Pradesh	East	West Godavari	0.099	Lowest	0.150	Lowest	Yes
67	Gujarat	West	Anand	0.255	Medium	0.125	Lowest	No
68	Andhra Pradesh	East	Guntur	0.245	Medium	0.123	Lowest	No
69	Andhra Pradesh	East	Vizianagaram	0.119	Lowest	0.077	Lowest	No
70	Gujarat	West	Rajkot	0.114	Lowest	0.059	Lowest	No

Source Calculated by authors based on data in Marine Fisheries Census of the states and union territories (Fishery Survey of India 2010; CMFRI 2010b)

References

Adger WN, Brooks N, Bentham G, et al (2004) New indicators of vulnerability and adaptive capacity. Tyndall Centre Technical Report 7. Available at: http://www.tyndall.ac.uk/sites/default/files/it1_11.pdf. Accessed 3 August 2016.

Allison E, Adger W, Badjeck M-C, et al (2005) Effects of climate change on the sustainability of capture and enhancement fisheries important to the poor: analysis of the vulnerability and adaptability of fisherfolk living in poverty. Department for International Development. London, UK. Available at: https://assets.publishing.service.gov.uk/media/57a08ca340f0b652dd00145a/R4778Ja.pdf. Accessed 4 August 2016.

Allison EH, Ellis F (2001) The livelihoods approach and management of small-scale fisheries. Mar policy 25:377–388. doi: 10.1016/S0308-597X(01)00023-9.

Allison EH, Horemans B (2006) Putting the principles of the Sustainable Livelihoods Approach into fisheries development policy and practice. Mar Policy 30:757–766. doi: 10.1016/j.marpol.2006.02.001.

Allison EH, Perry AL, Badjeck MC, et al (2009) Vulnerability of national economies to the impacts of climate change on fisheries. Fish Fish 10:173–196. doi: 10.1111/j.1467-2979.2008.00310.x.

Bagozzi RP, Yi Y, Phillips LW (1991) Assessing Construct Validity in Organizational Research Organizational Research. Source Adm Sci Q 36:421–458. doi: 10.2307/2393203.

Barange M, Merino G, Blanchard JL, et al (2014) Impacts of climate change on marine ecosystem production in societies dependent on fisheries. Nat Clim Chang 4:211–216. doi: 10.1038/NCLIMATE2119.

Bassi N, Kumar MD, Sharma A, Pardha-Saradhi P (2014) Status of wetlands in India: A review of extent, ecosystem benefits, threats and management strategies. J Hydrol Reg Stud 2:1–19. doi: 10.1016/j.ejrh.2014.07.001.

Bhatt JR, Venkataraman E (2013) Coastal and marine biodiversity conservation in India. In: Regional Symposium on Ecosystem Approaches to Marine Fisheries & Biodiversity. CMFRI, Kochi, pp 15–18. Available at: http://eprints.cmfri.org.in/9683/1/Vivekanandan_Ecosystem_Approaches_to_the_Management_and_Conservation_of_Fisheries_and_Marine_Biodiversity_in_the_Asia_Region.pdf. Accessed 5 August 2016.

Blythe JL, Murray G, Flaherty M (2014) Strengthening threatened communities through adaptation: Insights from coastal Mozambique. Ecol Soc 19:1–6. doi: 10.5751/ES-06408-190206.

Briand G, Hill RC (2013) Teaching basic econometric concepts using Monte Carlo simulations in Excel. Int Rev Econ Educ 12:60–79. doi: 10.1016/j.iree.2013.04.001.

Brien KO, Leichenko R, Kelkar U, et al (2004) Mapping vulnerability to multiple stressors : climate change and globalization in India. Glob Environ Chang 14:303–313. doi: 10.1016/j.gloenvcha.2004.01.001.

Brocklesby MA, Fisher E (2003) Community development in sustainable livelihoods approaches – an introduction. Community Dev J 38:185–198. doi:10.1093/cdj/38.3.185.

Cheung WWL, Lam VWY, Sarmiento JL, et al (2009) Projecting global marine biodiversity impacts under climate change scenarios. Fish Fish 10:235–251. doi: 10.1111/j.1467-2979.2008.00315.x.

Cheung WWL, Lam VWY, Sarmiento JL, et al (2010) Large-scale redistribution of maximum fisheries catch potential in the global ocean under climate change. Glob Chang Biol 16:24–35. doi: 10.1111/j.1365-2486.2009.01995.x.

Cinner JE, Huchery C, Hicks CC, et al (2015) Changes in adaptive capacity of Kenyan fishing communities. Nat Clim Chang 5:872–876. doi: 10.1038/nclimate2690.

Cinner JE, Mcclanahan TR, Graham NAJ, et al (2012) Vulnerability of coastal communities to key impacts of climate change on coral reef fisheries. Glob Environ Chang 22:12–20. doi: 10.1016/j.gloenvcha.2011.09.018.

CMFRI (2015) Annual Report 2014 -15. Technical Report. CMFRI, Kochi. Available at: http://eprints.cmfri.org.in/10461/1/CMFRI%20Annual%20Report%202014-15.pdf. Accessed 1 May 2016.

CMFRI (2010a) India Marine Fisheries Census 2010. CMFRI, Kochi. Available at: http://eprints.cmfri.org.in/8998/1/India_report_full.pdf. Accessed 30 June 2016.

CMFRI (2010b) State Marine Fisheries Census (for Gujarat, Daman & Diu, Maharashtra, Goa, Karnataka, Kerala, Tamil Nadu, Puducherry, Andhra Pradesh, Odisha and West Bengal). CMFRI, Kochi.

Daw T, Adger WN, Brown K, et al (2009) Climate change and capture fisheries : potential impacts, adaptation and mitigation. In: Cochrane K, Young C De, Soto D, et al (eds). Climate change implications for fisheries and aquaculture: overview of current scientific knowledge. FAO Fisheries and Aquaculture Technical Paper. No. 530. FAO, Rome. pp. 107–150. Available at: http://www.fao.org/docrep/012/i0994e/i0994e03.pdf. Accessed 4 June 2016.

Doney SC, Ruckelshaus M, Emmett Duffy J, et al (2012) Climate change impacts on marine ecosystems. Ann Rev Mar Sci 4:11–37. doi: 10.1146/annurev-marine-041911-111611.

Ehler CN (2003) Indicators to measure governance performance in integrated coastal management. Ocean Coast Manag 46:335–345. doi: 10.1016/S0964-5691(03)00020-6.

Elasha BO, Elhassan NG, Ahmed H, Zakieldin S (2005) Sustainable livelihood approach for assessing community resilience to climate change: case studies from Sudan. AIACC Working Paper No.17. Available at: http://www.start.org/Projects/AIACC_Project/working_papers/Working Papers/AIACC_WP_No017.pdf. Accessed 26 June 2016.

Fekete A (2009) Validation of a social vulnerability index in context to river-floods in Germany. Nat Hazards Earth Syst Sci 9:393–403. doi: 10.5194/nhess-9-393-2009.

Fernandes A, Gopal S (2012) Safeguard or Squander? Deciding the future of India's fisheries. Greenpeace, Bengaluru. Available at: http://www.greenpeace.org/india/Global/india/report/Safeguard-or-squander-deciding-the-future-of-india's-fisheries.pdf. Accessed 30 June 2016.

Fishery Survey of India (2010) Marine Fisheries Census 2010: Union Territories of Andaman & Nicobar and Lakshadweep Islands. FSI, Mumbai.

Gbetibouo GA (2009) Understanding Farmers' Perceptions and Adaptations to Climate Change and Variability: The Case of the Limpopo Basin, South Africa. IFPRI Discuss Pap 00849 52. doi: 10.1068/a312017.

Giulani G, De Bono A, Kluser S, Peduzzi P (2004) Overfishing, a major threat to the global marine ecology. UNEP GRID-Europe, Geneva. Available at: http://www.grid.unep.ch/products/3_Reports/ew_overfishing.en.pdf. Accessed 20 June 2016.

Goodwin NR (2003) Five Kinds of Capital: Useful Concepts for Sustainable Development. G-DAE Working Paper No. 03-07. Medford, USA. Available at http://www.ase.tufts.edu/gdae/publications/working_papers/03-07sustainabledevelopment.PDF. Accessed 28 October 2015.

Gujarati DN, Porter DC, Gunasekar S (2009) Basic Econometrics, 5th edn. Tata McGraw-Hill Education, New York.

Hahn MB, Riederer AM, Foster SO (2009) The Livelihood Vulnerability Index : A pragmatic approach to assessing risks from climate variability and change — A case study in Mozambique. Glob Environ Chang 19:74–88. doi: 10.1016/j.gloenvcha.2008.11.002.

Harley CDG, Hughes AR, Hultgren KM, et al (2006) The impacts of climate change in coastal marine systems. Ecol Lett 9:228–241. doi: 10.1111/j.1461-0248.2005.00871.x.

Harley M, Horrocks L, Hodgson N, Van Minnen J (2008) Climate change vulnerability and adaptation indicators. ETC/ACC Technical Paper 2008/9. Available at: http://acm.eionet.europa.eu/docs/ETCACC_TP_2008_9_CCvuln_adapt_indicators.pdf. Accessed 7 June 2016.

Hinkel J (2011) " Indicators of vulnerability and adaptive capacity": Towards a clarification of the science-policy interface. Glob Environ Chang 21:198–208. doi: 10.1016/j.gloenvcha.2010.08.002.

Hoegh-Guldberg O, Bruno JF (2010) The impact of climate change on the world's marine ecosystems. Science 328:1523–1528. doi: 10.1126/science.1189930.

IPCC (2014) Climate Change 2014: Synthesis Report. Contribution of Working Groups I, II and III to the Fifth Assessment Report of the Intergovernmental Panel on Climate Change [Core Writing Team, R.K. Pachauri and L.A. Meyer (eds.)]. IPCC, Geneva, Switzerland.

IPCC (2007) IPCC Fourth Assessment Report (AR4). IPCC, Geneva, Switzerland.

Islam MM, Sallu S, Hubacek K, Paavola J (2014) Vulnerability of fishery-based livelihoods to the impacts of climate variability and change: Insights from coastal Bangladesh. Reg Environ Chang 14:281–294. doi: 10.1007/s10113-013-0487-6.

Iwasaki S, Razafindrabe BHN, Shaw R (2009) Fishery livelihoods and adaptation to climate change: A case study of Chilika lagoon, India. Mitig Adapt Strateg Glob Chang 14:339–355. doi: 10.1007/s11027-009-9167-8.

Johnson T (2012) Fisheries Adaptations to Climate Change. Alaska Sea Grant Marine Advisory Program 1-8. doi: 10.4027/facc.2012.

MacNeil MA, Graham NAJ, Cinner JE, et al (2010) Transitional states in marine fisheries: adapting to predicted global change. Philos Trans R Soc Lond B Biol Sci 365:3753–63. doi: 10.1098/rstb.2010.0289.

Malakar K, Mishra T (2016) Assessing socio-economic vulnerability to climate change: a city-level index-based approach. Clim Dev 1–14. doi: 10.1080/17565529.2016.1154449.

Mamauag SS, Al No PM, Martinez JS, et al (2013) A framework for vulnerability assessment of coastal fisheries ecosystems to climate change—Tool for understanding resilience of fisheries (VA–TURF). Fish Res 147:381–393. doi: 10.1016/j.fishres.2013.07.007.

McIlgorm A, Hanna S, Knapp G, et al (2010) How will climate change alter fishery governance? Insights from seven international case studies. Mar Policy 34:170–177. doi: 10.1016/j.marpol.2009.06.004.

Meaney C, Moineddin R (2014) A Monte Carlo simulation study comparing linear regression, beta regression, variable-dispersion beta regression and fractional logit regression at recovering average difference measures in a two sample design. BMC Med Res Methodol 14:1–22. doi: 10.1186/1471-2288-14-14.

Nelson R, Kokic P, Crimp S, et al (2010) The vulnerability of Australian rural communities to climate variability and change: Part II—Integrating impacts with adaptive capacity. Environ Sci Policy 13:18–27. doi: 10.1016/j.envsci.2009.09.007.

Pauly D, Zeller D (2016) Catch reconstructions reveal that global marine fisheries catches are higher than reported and declining. Nat Commun 7:1–9. doi: 10.1038/ncomms10244.

Robeyns I (2005) The Capability Approach: a theoretical survey. J Hum Dev 6:93–117. doi: 10.1080/146498805200034266.

Roessig JM, Woodley CM, Cech JJ, Hansen LJ (2004) Effects of global climate change on marine and estuarine fishes and fisheries. Rev Fish Biol Fish 14:251–275. doi: 10.1007/s11160-004-6749-0.

Scoones I (1998) Sustainable rural livelihoods: a framework for analysis. IDS working paper 72. Available at: https://www.staff.ncl.ac.uk/david.harvey/AEF806/Sconnes1998.pdf. Accessed 5 March 2015.

Serrat O (2010) The Sustainable Livelihoods Approach. Asian Development Bank, Washington DC. Available at: http://digitalcommons.ilr.cornell.edu/cgi/viewcontent.cgi?article=1207&context=intl. Accessed 6 June 2016.

Sievanen L (2014) How do small-scale fishers adapt to environmental variability? Lessons from Baja California, Sur, Mexico. Marit Stud 13:1–19. doi: 10.1186/s40152-014-0009-2.

Sumaila UR, Cheung WWL, Lam VWY, et al (2011) Climate change impacts on the biophysics and economics of world fisheries. Nat Clim Chang 1:449–456. doi: 10.1038/nclimate1301.

Valdivia C, Gilles J (2001) Gender and resource management: households and groups, strategies and transitions. Agric Human Values 18:5–9. doi: 10.1023/A:1007608717996.

Wright H, Kristjanson P, Bhatta G (2012) Understanding Adaptive Capacity: Sustainable Livelihoods and Food Security in Coastal Bangladesh. CCAFS Working Paper No. 32. CGIAR Research Program on Climate Change, Agriculture and Food Security (CCAFS), Copenhagen, Denmark. Available at: https://cgspace.cgiar.org/bitstream/handle/10568/24794/CCAFS_WP_32.pdf?sequence=1&isAllowed=y. Accessed 26 June 2016.

Zamroni A (2015) Socio-economics status and adaptations of purse seine fishermen in Bali coastal village, Indonesia. Int J Mar Sci 5:1–16. doi: 10.5376/ijms.2015.05.0020.

Building Capacity for Climate Change Mitigation and Adaptation Through Mainstreaming Climate Change in Curricula of Tertiary Training Institutions in Africa

Jackson Efitre, Vianny Natugonza, Laban Musinguzi, Mark Olokotum, Shamim Naigaga and Richard Ogutu-Ohwayo

Abstract Climate variability and change, which intensified since 1970s, are threatening natural resources and livelihoods in Sub-Saharan Africa where people depend on climate sensitive natural resources, such as agriculture and fisheries, but have limited capacity to adapt. Increasing human and institutional capacity to address threats posed by climate change to natural resources and livelihoods requires building capacity to generate and disseminate information and knowledge on climate change, its impacts, adaptation and mitigation through research, education and raising awareness by tertiary training institutions. Most tertiary training institutions in Africa have curricula covering basic and applied natural resources management but most of them do not include climate change. This paper presents a training curriculum and manual that was developed to fill this gap. The purpose of

J. Efitre (✉) · M. Olokotum
Department of Zoology, Entomology and Fisheries Sciences,
College of Natural Sciences, Makerere University, 7062, Kampala, Uganda
e-mail: jefitre@cns.mak.ac.ug; jefitre@gmail.com

M. Olokotum
e-mail: markolokotum@yahoo.com

V. Natugonza · L. Musinguzi · S. Naigaga · R. Ogutu-Ohwayo
National Fisheries Resources Research Institute (NaFIRRI),
P.O. Box 343, Jinja, Uganda
e-mail: viannynatugonza@yahoo.com

L. Musinguzi
e-mail: musinguzilaban@ymail.com

S. Naigaga
e-mail: wapsham@gmail.com

R. Ogutu-Ohwayo
e-mail: ogutuohwayo@gmail.com

W. Leal Filho (ed.), *Climate Change Research at Universities*,
DOI 10.1007/978-3-319-58214-6_27

the paper is to provide in-depth information on how Climate Change can be integrated into the fisheries and aquaculture curricula of tertiary training institutions. It also provides students, scientists, practitioners, and policy makers with an understanding of key concepts and approaches to climate change mitigation measures, adaptation strategies, and policies. The aim is to mainstream climate change in fisheries training. The specific objectives are to facilitate introduction of climate change in fisheries training in Uganda that can be developed further and adopted by other countries in Africa and elsewhere; Equip students with scientific and technical capacity to anticipate and evaluate changes in climate and its influence, communicate information to stakeholders, design, and test adaptation strategies and mitigation measures; and Increase human resource capacity to address climate change issues through reviewing and strengthening of the national education system. The curriculum consists of seven modules covering: Major threats to natural resources; Introduction to climate change; Implications of climate change on aquatic productivity processes and fisheries; Implications of climate change on aquaculture; Livelihoods, impacts, adaptation and mitigation; Aquatic ecosystem modeling in relation to climate change; Principles, policies, regulations and institutions required to address impacts of climate change. The modules will be delivered through lectures, discussions, case studies and field visits. It is recommended that the curriculum and manual be incorporated into training programs of tertiary training institutions to build the capacity required to address climate change challenges particularly for fisheries in Africa.

Keywords Climate change · Adaptation · Mitigation · Capacity building · Fisheries resources · Livelihoods · Tertiary institutions · Uganda

1 Introduction

Climate variability and change are increasingly becoming a major environmental, social and economic challenge contributing to poverty and food insecurity through increased temperature and extreme events such as floods and drought. Africa is among the most vulnerable continents due to its high poverty levels and dependence on climate sensitive rain-fed agriculture and natural resources comprising of livestock, forestry and fisheries (IPCC 2007). About 240 million people in Sub-Saharan Africa are poor and food insecure and 70–80% of the population of the region depends on the vulnerable agriculture and natural resources (FAO 2010). Unless action is taken to curtail impacts of the increasing variability and change in climate, Africa is not likely to attain the Sustainable Development Goals, particularly those that aim at ending poverty; ending hunger, achieving food security and improved nutrition and promoting sustainable agriculture; combating climate change and its impacts; and conserving and sustainably using the oceans, seas and marine resources for sustainable development.

Addressing impacts of climate variability and change in developing countries has concentrated on crops, livestock and forestry and less on fisheries despite the high economic and nutritional value of fish. Fisheries provide about 15% per capita animal protein for 7.3 billion people globally, employ 10–12% of the world's population and generate about $125 billion in exports (FAO 2012). Fisheries are also important at national level in different countries. For instance, in Uganda, fisheries contribute 2.5% to national GDP and 12.5% to agricultural GDP, employ about 1.2 million people, provide 50% of animal protein to the population, and fish is the most important non-traditional export commodity (MAAIF 2012). In addition, the highest human and livestock populations live in proximity to aquatic systems.

There is growing evidence that climate variability and change are either acting independently or adding to the other stressors to alter temperature, rainfall, water levels, water balance, aquatic productivity processes, life history and biological characteristics of fishes, diversity of aquatic organisms, and fish yield, which ultimately affect livelihoods (Sumaila et al. 2011; Cheung et al. 2013; Ogutu-Ohwayo et al. 2016). Ogutu-Ohwayo et al. (2016) discuss extensively how climate variability and change is influencing fisheries resources and livelihoods using experiences from different aquatic systems across Africa. Based on evidence presented in the paper, it is clear that climate variability and change has consequences on fisheries such as shifts in fish composition, and fish yield. For example, the contribution of stocks of *Rastreneobola argentea*, locally known as Dagaa/Omena/Mukene, in lakes Kyoga and Victoria and related small pelagic cyprinids in Lake Albert, Uganda, have increased to 40–80% of the fish yield over the period when climate variability and change intensified. This is in line with the prediction by FAO (2010) that climate change will shift fisheries to smaller, faster growing, and pelagic opportunistic species. In Lake Wamala (Uganda), an environmental change hot spot due to climate related conditions (UNEP 2009), the composition and life history of fishes have changed with rainfall and water level (Natugonza et al. 2015; NaFIRRI 2013). Fishers around Lake Wamala, like others around African lakes, have adapted by diversifying to other livelihoods such as crop agriculture among others (Musinguzi et al. 2016; Ogutu-Ohwayo et al. 2016). There is therefore need to build capacity, generate knowledge, develop adaptation and mitigation measures, develop and implement policies and increase awareness to address the impacts of climate variability and change on inland aquatic ecosystems, fisheries and livelihoods.

Success in addressing the influence of climate variability and change on inland aquatic ecosystems and livelihoods will require local academic and research institutions with capacity to anticipate and evaluate changes in climate, communicate information to policy makers and stakeholders, and design, test and implement appropriate adaptation strategies and mitigation measures, which is however inadequate in most institutions in Africa.

1.1 The Role of Tertiary Training Institutions in Building Capacity for Climate Change Mitigation and Adaptation

A key goal of tertiary training institutions has been to provide solutions to global development challenges (Bloom et al. 2005). Tertiary training institutions play a critical role in helping society to adapt to impacts of climate variability and change by creating awareness through research and training focusing on climate change impacts, adaptation strategies and mitigation measures. They can therefore act as role models and serve as 'hubs' in their local communities for creating, testing, and disseminating knowledge about climate change impacts, mitigation and adaptation strategies. However, these roles can only be fulfilled if climate change is integrated in the curricula of the tertiary training institutions. There is therefore need for mainstreaming climate change into the curricula of tertiary education institutions as climate change is becoming a challenge to Africa's development efforts that depend on climate sensitive sectors for livelihoods (Allison et al. 2009). Indeed, the incorporation of climate change in all aspects of education and training is recommended by current national, regional and international policies because the challenge posed by the increasing variability and change in climate is now recognized at all the levels. Efforts should be put to ensure that climate change is comprehensively incorporated into curricula of tertiary institutions in Africa. Currently, most tertiary training institutions with fisheries curricula for training professionals to conduct basic and applied research in capture fisheries and aquaculture and manage fisheries resources have limited or no coverage of climate change. As a result, students and graduates of tertiary institutions will be well-versed with the challenges posed by climate change to be better prepared to advise the communities on appropriate climate change mitigation and adaptation.

It was against this background that the Department of Zoology, Entomology and Fisheries Sciences, College of Natural Sciences, Makerere University, in collaboration with the National Fisheries Resources Research Institute (NaFIRRI), Uganda, with support from The Rockefeller Foundation, developed this training curriculum and manual as a contribution to building capacity in addressing climate change issues in inland aquatic ecosystems, fisheries and aquaculture.

1.2 The Climate Change Project

The curriculum and training manual was developed as part of a climate change project titled "Equipping Small Scale Fishers and Riparian Communities with Adaptation Strategies to Cope with Impacts of Climate Variability and Change" that was implemented from 2011 to 2014. The purpose of the project was to mainstream climate change in fisheries research on aquatic ecosystems and develop tools that would be applied to address impacts of climate change in fisheries in Uganda, the African Great lakes and elsewhere. The overall objective of the project was to equip small scale fishers and riparian communities with knowledge, adaptation strategies, mitigation measures and policy guidelines to cope with impacts of climate variability and change to increase their resilient to sustain livelihoods. The specific objectives were to: (i) Improve knowledge systems; (ii) Provide policy guidance; (iii) Increase awareness; and (iv) Improve capacity. The objectives were achieved by literature review supplemented with empirical field observations on two satellite lakes (Wamala and Kawi in Uganda) which have manifested changes in the environment, ecological and biological characteristics of organisms with changes in climate variables. Given the multidisciplinary and multi-sectoral nature of climate change and the fact that climate change interacts with other environmental factors in affecting fisheries resources and livelihoods, holistic understanding of the impacts of climate change on fisheries resources and livelihoods could not be achieved within the scope and time frame of the project that lead to the development of this curriculum and training manual. Moreover, climate change analysis requires time-series data which was not always available or accessible. However, the empirical field observations and literature review upon which this training manual is based provide some start-up for addressing climate variability and change in inland aquatic and riparian ecosystems.

Evidence from literature indicated that changes in climate variables influence riparian and aquatic ecosystems, aquatic productivity processes, composition, distribution, life history and yield of fishes, aquatic weeds, parasites and disease interactions (NaFIRRI 2013; Namboowa 2015; Natugonza et al. 2015) and impacted livelihoods of small scale fishers and riparian communities who were forced to adapt to increase resilience to sustain their livelihoods (Musinguzi et al. 2016). This was supported by field observations which showed that temperatures increased by 0.02–0.03 °C annually since the 1980s and rainfall was above average since 1990s in line with what had been reported for the East African region (NaFIRRI 2013; Natugonza et al. 2015). The composition of fishes changed with the contribution of the African catfish and the lungfish increasing and that of the Nile tilapia decreasing (Natugonza et al. 2015). The changes in variety and quantity of fish affected the livelihoods of fishers with some fishers adapting by switching to the African catfish and the lungfish, diversifying to crops and livestock and irrigating crops during drought. More innovative fishers diversified to high value crops such as pineapples, oranges, tomatoes, cabbages, and livestock and poultry (Musinguzi et al. 2016). It was noted that adjustments to international, regional and national policies on management of environment, water, pollution, biodiversity, fisheries, wetlands, lake shores and river banks, land use, forests, and agriculture are needed to accommodate the impacts of climate change.

1.3 Purpose and Scope

The purpose of this training curriculum and manual is to contribute to building capacity to address climate change issues in inland aquatic systems, fisheries and aquaculture by initiating the process of introducing climate change in capture fisheries and aquaculture training in Uganda, although it can be developed further and adopted by other countries in Africa and elsewhere. This is because the capacity to address climate change challenges in fisheries is still weak or non-existent and there is therefore need to provide manpower that can generate knowledge to address climate change and its effects on resources and livelihoods. Existing efforts in addressing climate change on natural resources have concentrated on forestry, crops and livestock and less on fisheries despite the high economic importance in some countries like Uganda. In addition, fisheries research and training on inland aquatic ecosystems has concentrated on lake productivity processes and threats such as over-exploitation, invasive species, habitat loss and pollution with very little focus on climate change. As a result, there is limited capacity and knowledge on how the riparian and aquatic ecosystems, productivity processes, fisheries and livelihoods are affected by the increasing variability and change in climate to facilitate development of planned adaptation and mitigation strategies. This training curriculum and manual is intended to reverse this trend. The proposed training will complement existing fisheries Certificate, Diploma, Bachelor and Master of Science, and Doctor of Philosophy (PhD) programs in tertiary training institutions in Uganda, Africa and elsewhere.

1.4 Goal and Objectives

The overall goal of the training curriculum and manual is to provide guidelines for mainstreaming climate change issues in fisheries and aquaculture training in tertiary education institutions.

The specific objectives are to:

1. Facilitate introduction of climate change in fisheries and aquaculture training in tertiary education institutions;
2. Equip students with scientific and technical capacity to anticipate and evaluate changes in climate and its influence, design, test and implement adaptation strategies and mitigation measures and communicate information to stakeholders; and
3. Increase human resource capacity to address climate change issues through reviewing and strengthening of the national education system.

1.5 Learning Outcomes

By the end of the course, participants should be able to:

(i) Explain the fundamentals of climate change science;
(ii) Articulate the implications of climate change to natural resources and livelihoods of local communities;

(iii) Provide a rationale for climate change mitigation & adaptation measures and propose actions for the fisheries sector;
(iv) Present national, regional and international climate change legal, policy and institutional frameworks;
(v) Communicate climate change issues to stakeholders in the fisheries sector.

1.6 Target Audience

The target audience of this training curriculum and manual includes: Fisheries and aquaculture certificate, diploma, university undergraduate and graduate students. The training curriculum and manual is designed to complement the ongoing fisheries and aquaculture courses through incorporating climate change into the curricular and developing human resource capacity.

2 Course Structure and Content

The manual is organized into seven modules, each consisting of sub-sections covering: Major threats to fisheries resources; Introduction to climate change; Implications of climate change on aquatic productivity processes and fisheries; Implications of climate change on aquaculture; Livelihoods, impacts, adaptation and mitigation; Aquatic ecosystem modeling in relation to climate change;

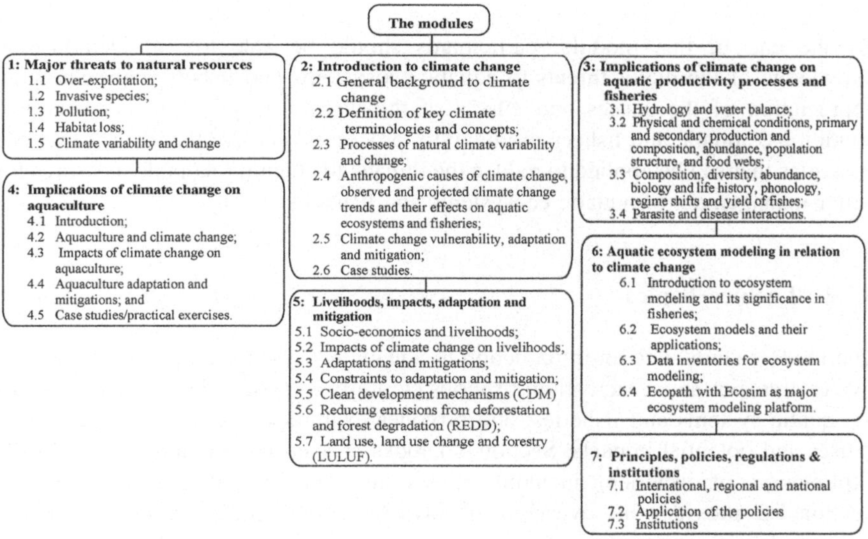

Fig. 1 Description of the modules and subsections of the training curriculum and manual

and Policies for addressing climate change on inland aquatic systems. Descriptions of each module are provided in Fig. 1 and Sect. 3. Each module consists of units to be delivered through lectures, and practical/field excursions in collaboration with research and training institutions in Uganda with mandates relevant to the course including: the Department of Zoology, Entomology and Fisheries Sciences, Makerere University; the National Fisheries Resources Research Institute (NaFIRRI); the Fisheries Training Institute (FTI); the Uganda National Meteorological Authority, the Climate Change Department of the Ministry of Water and Environment; the Directorate of Water Resources Development; the Wetland Division; and institutions dealing with climate change policy.

3 Description of the Modules

3.1 Module 1: Major Threats to Natural Resources

This module describes how different threats influence aquatic ecosystems and fisheries. It examines causes of the threats and how they influence physical and chemical conditions, aquatic and fishery productivity processes, and fish yield and how they may interact with climate change. It then discusses strategies that can be applied to mitigate the influence of the threats and thus help in better management of aquatic ecosystems and fisheries.

3.1.1 Learning Objectives

By the end of this module, participants should be able to: (a) Demonstrate knowledge of the major threats to aquatic ecosystems and fisheries; (b) Show an appreciation of the causes and effects of the threats on physical and chemical conditions, aquatic and fishery productivity processes and livelihoods; (c) Propose strategies that can be applied to reduce the threats; (d) Identify suitable measures for better management of aquatic ecosystems and fisheries.

3.1.2 Lecture Topics

The module starts with an introduction to aquatic ecosystems and their importance (water, transportation, recreation, fisheries etc.); gives an overview of major threats to aquatic systems and fisheries; and provides examples of different threats, their causes and potential impacts. Section 1.1 looks at over-exploitation (unsustainable exploitation of fisheries) including types and causes and control measures. Section 1.2 provides an overview of invasive alien species: definition; major sources of invasive species; their impacts on ecological and ecosystem functioning;

control of invasive species and prevention programmes using the Nile perch and water hyacinth as examples in Uganda. Section 1.3 describes the nature and sources of pollution (microbial, eutrophication, chemical) and management of pollution (legislation and environmental quality monitoring). Section 1.4 presents habitat loss including its causes; examples of habitat degradation such as land use, wetland loss; ecological effects; remediation and control. Section 1.5 introduces climate variability and change as a major threat to aquatic ecosystems and fisheries and its potential interaction with the other threats. The lectures will be complemented with field visits.

3.2 Module 2: Introduction to Climate Change

This module provides an introduction to climate science, its importance and historical development and basic concepts of climate change science such as weather, climate, the greenhouse effect, global warming and natural climate fluctuations. It then defines key climate change terminologies, discusses the main causes and elements of anthropogenic (human caused) climate change, including observed and projected changes in climate at global and regional scales. The module concludes with a discussion of potential impacts of climate change, adaptation and mitigation options for reducing and managing the effects of climate variability and change.

3.2.1 Learning Objectives

By the end of the module, participants should be able to: (a) Explain the basic concepts of climate science and climate change; (b) Define key climate change terminologies; (c) Identify the anthropogenic drivers of climate change (greenhouse effect); (d) Explain observed and projected trends in climate and their impacts on the aquatic ecosystems and fisheries; (e) Explain the concepts of vulnerability, exposure, sensitivity, resiliency, adaptation and mitigation and their importance in preparation for and coping with effects of climate change.

3.2.2 Lecture Topics

Section 2.1 introduces participants to basic climate change concepts such as climate system, climate change, climate variability, vulnerability, sensitivity, adaptation, adaptive capacity, resilience, cryosphere, greenhouse gas emission scenarios, extreme weather events, global warming, green house effect, mitigation, palaeoclimate, etc. Section 2.2 provides definitions of key climate change terminologies and components of the climate system: the atmosphere, ocean, cryosphere, and biosphere; interactions among the components (e.g. earth's energy balance, surface energy balance, hydrological cycle, atmospheric circulation, ocean circulation). Section 2.3 discusses processes of natural climate variability and change with emphasis on Africa. Section 2.4 presents observed and projected climate change trends and their impacts on the aquatic ecosystems and fisheries. Section 2.5 introduces the concepts of climate change vulnerability, adaptation and mitigation and highlights how to assess vulnerability to climate change as well as highlighting adaptation and coping strategies related to aquatic and riparian ecosystems and fisheries. The section also presents some of the key international, regional and national levels mitigation mechanisms including afforestation, reforestation, promotion of energy efficient systems, improvement of crop production systems, waste management, and promoting Clean Development Mechanisms (CDM). The lectures will be supplemented with practical exercises and case studies.

3.3 *Module 3: Implications of Climate Change on Aquatic Productivity Processes and Fisheries*

This module examines the influence of climate variability and change on hydrology, water balance, physical and chemical conditions such as transparency, nutrient loading, stratification, circulation dynamics and oxygen levels which drive plankton production, composition, population structure, food-webs, life history characteristics, physiology, composition, distribution, phenology, regime shifts, biological characteristics of fishes, fisheries, fish yield and ultimately livelihoods.

3.3.1 Learning Objectives

By the end of the module, participants should be able to:

(a) Describe the processes through which climate change may affect hydrology, water balance, physical properties of inland aquatic ecosystems, primary and secondary production and composition, population structure, and food-webs; (b) Articulate how abundance, biology and life history and yield of fishes are influenced by climate variability and change; (c) Explain the effects of climate variability and change on the physiology, phenological changes and regime shifts; and d) Explain the effect of climate change on species invasions and disease organisms (parasites and diseases interactions).

3.3.2 Lecture Topics

Section 3.1 examines the effect of climate change on hydrology and water balance (water levels and water budget). Examples will be drawn from small water bodies in Uganda (Lakes Wamala and Kawi) where changes in climate parameters (i.e., temperature and rainfall) have been reported to influence hydrology and water balance (Natugonza et al. 2016). The section also describes the effects of climate change on physical and chemical conditions such as circulation dynamics, production processes, stratification and mixing, recycling of nutrients, and oxygen circulation; primary production and phytoplankton composition; secondary production (zooplankton) and benthic organisms; composition, diversity, distribution, abundance, population structure and foodwebs. This will be illustrated with case studies from literature in various tropical lakes (e.g., Hecky et al. 2010; O'Reilly et al. 2003; Ogutu-Ohwayo et al. 2016). Evidence from literature will also be used to demonstrate that climate variability and change can operate independently or interact with other factors to modify circulation dynamics, stratification, loading and recycling of nutrients and oxygen in aquatic systems (Hecky et al. 1994, 2010; Lehman et al. 1998; Lorke et al. 2004; Verburg et al. 2003; Vollmer et al. 2005; Barange and Perry 2009; Sitoki et al. 2010; Ogutu-Ohwayo et al. 2016). Section 3.3 explores the influence of climate change on abundance, biology and life history and yield of fishes. Various examples from literature will be used to illustrate that climate factors contribute to changes in composition, diversity and yield of fishes in different African lakes (O'Reilly et al. 2003; Allison et al. 2007; Barange and Perry 2009; FAO 2010; Ndebele-Murisa et al. 2011, Marshall 2012; Marshall et al. 2013; Ogutu-Ohwayo et al. 2013, 2016; Natugonza et al. 2015). Section 3.4 examines the effects of climate variability and change on the physiology of fishes, phenological changes (timing of life history events) and regime shifts (how changes in climate might trigger unpredictable biological responses as ecosystems shift from one state to another). Section 3.5 looks at how climate variability and change influences species invasions and disease organisms (parasites and diseases interactions) including: life-cycle of parasites; relationship between temperature and susceptibility of fish species to parasite attacks and diseases; indirect effects of climate change on parasites and their hosts in aquatic systems through alteration in water levels, eutrophication and ultra violet radiation. Participants will learn from case studies which suggest that climate warming contributes to changes in host-parasite interactions and proliferation of parasites and diseases in both temperate and tropical regions (e.g., Marcogliese 2001; Cochrane et al. 2009; Macnab and Barber 2012; NaFIRRI 2013; Namboowa 2015).

3.4 Module 4: Implications of Climate Change on Aquaculture

This module will introduce participants to the potential impacts of climatic variability and change on aquaculture production. It will highlight potential impacts and the required adaptation strategies and mitigation measures.

3.4.1 Learning Objectives

By the end of the module, participants should be able to: (a) Discuss the relationship between climate change and aquaculture production; (b) Explain the direct and indirect influence of climate change on aquaculture; and (c) Discuss the adaptation strategies and mitigations measures needed to address the impacts of climate change on aquaculture.

3.4.2 Lecture Topics

Section 4.1 provides an overview of global, regional and national relevance of aquaculture to food, income and employment; trends in aquaculture production and consumption; aquaculture and climate change. Section 4.2 discusses impacts of climate change on aquaculture highlighting the potential positive and negative impacts of climate change on aquaculture; direct and indirect impacts on water, land, seed, feed and energy. Section 4.3 presents aquaculture adaptation and mitigations strategies including: aquaculture diversification; proper site selection and aquaculture zoning to reduce threats associated with climate change; assessment of the required adaptations of cultured species using lessons learnt from the expansion of farming species outside their natural range of distribution; identifying new disease interactions and measures to address them; identifying species that are better adapted to emerging conditions. Lectures will be supplemented with practical/field excursions to aquaculture facilities to observe and discuss possible effects of extreme climate events like drought and floods on aquaculture farms.

3.5 Module 5: Livelihoods, Impacts, Adaptations and Mitigation

This module introduces participants to the impact of climate variability and change on livelihoods of fishers, adaptations strategies and mitigations measures needed to address the impacts, constraints to adaptation and the required interventions. It will cover initiatives such as the Clean development mechanisms (CDM); Reducing emissions from deforestation and forest degradation (REDD) and Land use, land use change and forestry (LULUF) that can contribute to reduction of greenhouse gases.

3.5.1 Learning Objectives

By the end of the module, participants should be able to: (a) Discuss socio-economic characteristics and livelihoods of fishers; (b) Describe the impacts of climate variability and change on livelihoods of fishers; (c) Discuss adaptation strategies and mitigation measures that can be applied by fishers to mitigate impacts of climate variability and change; (d) Identify the constraints to adaption and mitigation and the required interventions; (e) Design appropriate climate adaptation and mitigation initiatives at local community level.

3.5.2 Lecture Topics

Section 5.1 will provide an overview of socio-economic characteristics and livelihoods of fisher and riparian communities. The livelihoods will include fishing, boat renting, gear renting, fish trading, fish processing, boat crew and others including those out of fisheries. Section 5.2 will examine impacts of climate variability and change on fisheries livelihood assets, activities and outcomes. Section 5.3 will address adaptation strategies such as migration, changing fishing gear, using more nets, changing target species, reverting to non-fishing activities, exiting the fishery, changing fishing grounds, and increasing time on fishing grounds. Section 5.4 will highlight mitigation measures used to address impacts of climate change including tree planting; protection of wetlands and riparian zones; mulching gardens; and fuel efficiency. Section 5.5 will identify constraints to adaption and mitigation and how they can be overcome. Section 5.6 will introduce participants to other climate adaptation and mitigation initiatives such as CDM, REDD and LULUF. The lectures will be illustrated with practicals, field visits and case studies on impact of climate change on fisheries and livelihoods of fishers and adaptation and mitigation strategies (e.g., Aiken et al. 1992; Broad et al. 1999; Jallow et al. 1999; Sarch and Allison 2000; Turner et al. 2007; Badjeck 2008; Badjeck et al. 2009; Iwakasi et al. 2009; Ogutu-Ohwayo et al. 2013; Musinguzi et al. 2016).

3.6 Module 6: Aquatic Ecosystem Modeling in relation to Climate Change

This module equips participants with skills for designing and implementing ecosystem models for research to inform fisheries policy and decision making. Specifically, it aims at creating awareness of uses and limitations of different modeling approaches and developing skills required to construct ecosystem models and to use them to study ecosystem dynamics and biogeochemical processes in aquatic systems to guide prediction of changes in the ecosystem. Participants will learn how to develop models using Ecopath with Ecosim (EwE), an ecosystem level model that has been widely used to address ecological questions, management policy options and effects of environmental changes such as those that may be brought about by climate change. The above model will be coupled with other models like the Multi-scale Integrated Model of Ecosystem Services (MIMES) to accommodate nonlinear and indirect effects to explore the effects of climate change on human and natural systems coupling under varied climate change scenarios.

3.6.1 Learning Objectives

By the end of the module, participants should be able to: (a) Identify the strengths and weaknesses of using ecosystem models as a means of studying changes in aquatic ecosystems; (b) Design different ecosystem models for developing relationships between different components of the ecosystem; (c) Evaluate the performance of the models applied to the ecosystems; (d) Assess different available models to describe ecosystem responses to stress and inform decision making and policy.

3.6.2 Lecture Topics

Section 6.1 introduces ecosystem modeling and its significance in fisheries including: the purpose of ecosystem modeling; model formulation and parameterization; analysis of model outputs and validation model inter-comparisons; and importance and limitations of ecosystem modeling. Section 6.2 presents ecosystem models and their applications focusing on components of fisheries ecosystems and their inter-comparisons, models and their application for the different ecosystems components, their strengths and weaknesses. The emphasis will be on when, where and how different systems can be modeled. It will also involve defining and scoping research problems addressed by each model, formulation of conceptual diagrams, their strengths and weaknesses and validation. Section 6.3 explores data inventories for ecosystem modeling including: data types and sources for different models; data availability and gaps in Uganda's aquatic ecosystems and fisheries; and limitations and challenges in data availability. Section 6.4 presents Ecopath with Ecosim (EwE) as a major ecosystem modeling tool and provides an introduction to the trophic modeling software, Ecopath and its ecological network analysis interface, development of models using EwE, modeling and dynamics of aquatic ecosystems. This will give the participants an understanding of the theory behind the software and the confidence to construct and interpret basic ecosystem models.

3.7 Module 7: Policies for Addressing Climate Change on Inland Aquatic Systems

This module introduces participants to available policies that are applicable to address impacts of climate variability and change and associated stressors.

It emphasizes how the policies can be improved or revised to accommodate changes due to climate variability and change, including policies specific to climate change and those targeting management of the resources impacted by climate change such as land use, forestry, wetlands, water and fisheries.

3.7.1 Learning Objectives

By the end of the module, participants should be able to: (a) Articulate the international, regional, and national policies that can be applied to address climate change and resources impacted by it; (b) Discuss the application of different policies and the possibilities for their improvement to address climate change and impacted resources; (c) Design strategies, action plans and institutional frameworks for implementing the policies.

3.7.2 Lecture Topics

Section 7.1 provides an overview of International policies and legal instruments including the United National Framework Convention on Climate Change (UNFCCC); the Kyoto Protocol; the Vienna Protocol for Protection of the Ozone layer; the Montreal Protocol on ozone layer; and the Convention to Combat Desertification; the World Heritage Convention; the Convention on Wetlands of International Importance, the Ramsar Convention; the Convention on Biological Diversity (CBD); the Convention on International Trade in Endangered Species of Wild Fauna and Flora (CITES); the Convention on the Conservation of Migratory Species of Wild Animals; the Code of Conduct for Responsible Fisheries; the International Plan of Action on Illegal, Unregulated and Unreported (IUU) fishing; and the International Plan of Action to Regulate Fishing Capacity.

Section 7.2 presents regional legal instruments that include: the African Convention on Conservation of Nature and Natural Resources under the AU; the African Union Strategy on Climate Change, the Treaty for the Establishment of the East African Community; the East African Community (EAC) Climate Change Policy; the Protocol on Environment and Natural Resources Management; The Protocol for Sustainable Development of Lake Victoria Basin; the Convention for Establishment of Lake Victoria Fisheries Organization; the Regional Plan of Action to prevent, deter and eliminate Illegal Un-regulated and Unreported fishing (IUU); and the Regional Plan of Action for Management of Fishing Capacity (RPOA-Capacity).

Section 7.3 presents national legal instruments including: the Constitution of the Republic of Uganda; the Uganda National Climate Change Policy; the National Environmental Policy and Statute; the Land Act; the National Forestry Policy and the National Forestry and Tree Planting Act; the National Policy for the Conservation and Management of Wetland Resources; the Water Policy; National Fisheries Policy and the Fish Act and associated regulations. Section 7.4 lists institutional frameworks to implement the above instruments including secretariats, national ministries and departments for implementing the policies at international, regional, national and local levels.

4 Conclusions

Incorporating climate change into training curricula of tertiary education institutions is critical to building capacity and knowledge required to understand implications of the increasing variability and change in climate for fisheries, and adaptation and mitigation measures required to reduce the impacts on resources and livelihoods. This paper demonstrated that climate change was little understood and appreciated and requires concerted effort to promote it. In addition, addressing impacts of climate change effectively needs a multi-disciplinary and multi-sectoral approach that encompasses the entire process from climate factors to physical and chemical processes, primary and secondary productivity, fishery productivity, livelihoods, adaptation and mitigation measures, policy, and awareness raising. Furthermore, evidence from the paper points to inadequate knowledge, training and institutional capacity to address impacts of climate change on fisheries in most African countries. There was therefore need to build human resources capacity that can generate knowledge and increase awareness on climate variability and change in inland aquatic systems and livelihoods through mainstreaming climate change in the curriculum of training institutions. To address the inadequacy in human resources capacity, the future plan is to test the curriculum and training manual in tertiary institutions in Uganda and improve it over time. A strategy for continued training of young scientists has also been developed to provide a critical mass of manpower to continue addressing impacts of climate change on fisheries. The tools developed for addressing climate variability and change in inland aquatic and riparian ecosystems will also be up-scaled to larger lakes within Uganda and other aquatic ecosystems in Africa.

Acknowledgements This curriculum and manual was developed as part of a climate change pilot project based at the National Fisheries Resources Research Institute (NaFIRRI) with funding from The Rockefeller Foundation to whom we are very grateful.

References

Aiken KA et al (1992). Recovery after Hurricane Gilbert: implications for disaster preparedness in the fishing industry in Jamaica. *Proceedings of Gulf and Caribbean Fisheries Institute* 41: 261–83.

Allison EH et al (2007). Enhancing the resilience of inland fisheries and aquaculture systems to climate change. *Journal of Semi-Arid Tropical Agricultural Research*; 4(1): 35.

Allison EH et al (2009). Vulnerability of national economies to the impacts of climate change on fisheries. *Fish and Fisheries*; 10(2): 173–196. DOI:10.1111/j.1467-2979.2008.00310.x.

Badjeck MC (2008). Vulnerability of coastal fishing communities to climate variability and change: implications for fisheries livelihoods and management in Peru. *University of Bremen, Bremen, Germany. (also available at* http://elib.suub.uni-bremen.de/diss/docs/00011064.pdf. *Ph.D. thesis.) Climate change and capture fisheries: potential impacts, adaptation and mitigation,* 145.

Badjeck MC et al (2009). Climate variability and the Peruvian scallop fishery: the role of formal institutions in resilience building. *Climatic Change* 94(1–2): 211–32.

Barange M, Perry RI (2009). Physical and ecological impacts of climate change relevant to marine and inland capture fisheries and aquaculture. In: K. Cochrane, C. De Young, D. Soto T. Bahri (eds) Climate change implications for fisheries and aquaculture: overview of current scientific knowledge. *FAO Fisheries and Aquaculture Technical paper 530*. Rome, FAO pp. 7–106 (eds).

Bloom D et al (2005). Higher Education and Economic Development in Africa. World Bank Report. 85 pp. Available at: http://siteresources.worldbank.org/EDUCATION/Resources/278200-1099079877269/547664-1099079956815/HigherEd_Econ_Growth_Africa.pdf.

Broad K et al (1999). Climate information and conflicting goals: El Nino 1997–98 and the Peruvian fishery. *Public philosophy, environment, and social justice, Carnegie Council on Ethics and International Affairs*, New York, 21–22.

Cheung WWL et al (2013). Shrinking of fishes exacerbates impacts of global ocean changes on marine ecosystems. *Nature Climate Change*, 3, 254–258.

Cochrane K et al (2009). Climate change implications for fisheries and aquaculture: overview of current scientific knowledge. *FAO Fisheries and Aquaculture Technical Paper*. No. 530. Rome FAO. 212p.

FAO (2010). *The state of food insecurity in the world. Addressing food insecurity in protracted areas*. FAO, Rome pp. 62.

FAO (2012). *The state of world fisheries and aquaculture, 2012*. FAO Fisheries and Aquaculture Department, Food and Agriculture Organization of the United Nations, Rome.

Hecky RE et al (1994). Deoxygenation of the deep water of Lake Victoria, East Africa. *Limnology and Oceanography* 39(6): 1481–1486.

Hecky RE et al (2010). Multiple stressors cause rapid ecosystem change in Lake Victoria. *Freshwater Biology* 55(1): 19–42.

IPCC (2007). Summary for Policy Makers. In Parry ML, Canziani OF, Palutikof JP, van Linden PJ & Hansen CE (eds), *Climate Change: Impacts, Adaptation and Vulnerability*. Contribution of Working Group II to the Fourth Report of the Intergovernmental Panel on Climate Change, Cambridge University Press, UK, pp 7–22.

Jallow BP et al (1999). Coastal zone of The Gambia and the Abidjan region in Côte d'Ivoire: sea level rise vulnerability, response strategies, and adaptation options. *Climate Research* 12(2–3): 129–36.

Lehman JT et al (1998). Lake Victoria plankton ecology: mixing depth and climate driven control of lake condition, pp. 99–116. In: Lehman J. T. (ed). Environmental Change and Response in East African Lakes. Kluwer Academic Publishers, NL.

Lorke A et al (2004). Response of Lake Kivu stratification to lava inflow and climate warming. *Limnology and Oceanography* 49, 778–783.

MAAIF (Ministry of Agriculture Animal Industry and Fisheries) 2012. Operationalisation of the Non-ATAAS Component of the Development Strategy and Investment Plan (DSIP), Situation Analysis Report, Fish Production, Entebbe, Uganda. 90 pp.

Macnab V and Barber I (2012). Some (worms) like it hot: fish parasites grow faster in warmer water, and alter host thermal preferences. *Global Change Biology* 18(5): 1540–1548.

Marcogliese D J (2001). Implications of climate change for parasitism of animals in the aquatic environment. *Canadian Journal of Zoology* 79: 13331–1352.

Marshall BE et al (2013). Has climate change disrupted stratification patterns in Lake Victoria, East Africa? *African Journal of Aquatic Science* 38: 249–253.

Marshall BE (2012). Does climate change really explain changes in the fisheries productivity of Lake Kariba (Zambia-Zimbabwe)? *Transactions of the Royal Society of South Africa* 67: 45–51.

Musinguzi L et al (2016). Fishers' perceptions of climate change impacts on their livelihoods and adaptation strategies in environmental change hotspots: a case of Lake Wamala, Uganda, *Environment, Development and Sustainability,* 17(4). DOI 10.1007/s10668-015-9690-6.

NaFIRRI, 2013. Vulnerability, impacts and adaptations of inland aquatic and riparian aquatic ecosystems and fisheries to climate variability and change: A case study of lakes Wamala and Kawi (Uganda). Technical Progress Report, NaFIRRI, Jinja, Uganda.

Namboowa, S. 2015. *The occurrence and effects of helminth parasites infesting the African Catfish (Clarias gariepinus, Burchell, 1822) in lake Wamala, Uganda.* MSc. dissertation, Makerere University, Kampala, 48 pp.

Natugonza V et al (2015). The responses of Nile tilapia *Oreochromis niloticus* (Linnaeus, 1758) in Lake Wamala (Uganda) to changing climatic conditions. *Lakes & Reservoirs: Research & Management* 20(2): 101–119.

Natugonza, V et al (2016). Implications of climate warming for hydrology and water balance of small shallow lakes: A case of Wamala and Kawi, Uganda. *Aquatic Ecosystem Health & Management* (just accepted), DOI:10.1080/14634988.2016.1142167.

Ndebele-Murisa M R et al (2011). The implications of a changing climate on the Kapenta fish stocks of Lake Kariba, Zimbabwe. *Transactions of the Royal Society of South Africa* 66(2): 105–119.

O'Reilly CM et al (2003). Climate change decreases aquatic ecosystem productivity of Lake Tanganyika, Africa. *Nature* 424: 766–768.

Ogutu-Ohwayo R et al (2013). Variations and changes in habitat, productivity, composition of aquatic biota and fisheries of the Kyoga lake system: lessons for management. *African Journal of Aquatic Sciences*, 38(1):1–14.

Ogutu-ohwayo et al (2016). Implications of climate variability and change for African lake ecosystems, fisheries productivity, and livelihoods *Journal of Great lakes Research*, doi:10. 1016/j.jglr.2016.03.004.

Sarch MT, Allison EH (2000). Fluctuating fisheries in Africa's Inland Waters: Well-Adapted livelihoods, maladapted management. Proceedings of the 10th International Conference of the Institute of Fisheries Economics and Trade. Corvallis, Oregon, July 9–14th 2000, 11 pp.

Sitoki L et al (2010). The Environment of Lake Victoria (East Africa): Current status and historical changes. *International Review of Hydrobiology* 95: 209–223. doi:10.1002/iroh.201011226.

Sumaila et al (2011). Climate change impacts on the biophysics and economics of world fisheries. *Nature Climate Change*, 1: 449–456.

Turner R et al (2007). Declining reliance on marine resources in remote South Pacific societies: ecological versus socio-economic drivers. *Coral Reefs* 26(4): 997–1008.

UNEP. (2009). Atlas of our changing environment. www.na.unep.net/atlas.google.php.

Verburg P et al (2003). Ecological consequences of a century of warming in Lake Tanganyika. *Science* 301: 505–507.

Vollmer MK et al (2005). Deep-water warming trend in Lake Malawi, East Africa. *Limnology and Oceanography* 50: 727–732.

Evaluation of a Transdisciplinary Research Project Aimed at the Development of Climate Change Adaptation Measures

Andrea Heilmann and Hardy Pundt

Abstract An important characteristic of transdisciplinary research is the integration of real-life, practical problems and scientific knowledge. In many cases problems related to climate change concern primarily scientists who analyze vast amounts of data aiming at forecasting potential effects of climate change. However, if climate change adaptation measures are developed, and implemented, this concerns not only science, but especially local administrations, as well as companies, organizations and the public. They look at the problems from their specific perspectives, and evaluate potential measures in close relation to their context. In order to achieve sustainable and widely accepted problem solutions, it is essential to integrate the different views and related contextual information. Within the framework of a pilot project, aimed at the development and implementation of climate change adaptation measures, this article envisages to present the underlying research process. The development and critical evaluation of the project will be described, proving that only the close cooperation of all relevant actors from concerned sectors, accompanied by scientists, leads to sustainable decisions and problem solutions aimed at the mitigation of risks and adaptation to potential threats. The pilot project, carried out in two phases between 2011 and 2016 in the administrative district of Mansfeld-Südharz, Germany, embraces the development of a local climate change adaptation strategy, and the implementation of specific adaptation measures. Success factors, as well as identified obstacles, are discussed critically.

Keywords Climate change adaptation · Adaptation measures · Participation · GIS · Vulnerability maps

---

A. Heilmann · H. Pundt (✉)
Harz University of Applied Sciences, Friedrichstr. 57-59, 38855 Wernigerode, Germany
e-mail: hpundt@hs-harz.de

A. Heilmann
e-mail: aheilmann@hs-harz.de

W. Leal Filho (ed.), *Climate Change Research at Universities*,
DOI 10.1007/978-3-319-58214-6_28

1 Introduction

Concerning the IPCC (International Panel of Climate Change), adaptation measures are suitable to reduce harmful and undesirable effects of climate change significantly (IPCC 2014, p. 19). It is emphasized that measures, carried out in different sectors and complementing each other, can support the process of adaptation. This is true concerning measures implemented by unique actors, as well as local and regional administrations governments: "Local government and the private sector are increasingly recognized as critical to progress in adaptation, given their roles in scaling up adaptation of communities, households and civil society and in managing risk information and financing" (IPCC 2014, p. 19).

This insight is reflected in the German adaptation strategy (DAS 2008, p. 60), as well as in the German Action Plan Adaptation (APA 2011): Local administrations are central actors regarding climate change adaptation, because many effects of climate change show local impacts (APA 2011, p. 27).

The administrations must understand climate change adaptation as cross-sectional task. Specific knowledge and expertise of all relevant sectors has to be integrated adequately. The actors are responsible for the identification of interdependencies that cross sectional borders imply and which must be taken into account *before* decisions are made.

Today it is undisputed that the handling of complex systems (e.g. ecosystems which are prone to climate change) require a holistic approach (GeoDZ 2016). Apart from well-known methods and strategies, new pathways toward problem solving have to be identified to enhance the adaptive capacity of regions.

"(…) Research can expand the options for climate change adaptation and—capacities through technical, economic, and social innovations. To prove these innovations practically, and to guarantee that they are applied immediately, a close cooperation between actors from science, administration and civil society is essential" (Fortschrittsbericht 2015, p. 64; translated).

Within the framework of a pilot project, taking place in the administrative district of Mansfeld-Südharz, Germany, a local climate change adaptation strategy has been developed and finalized 2012. An important outcome was the definition of several measures in different sectors, such as urban planning, forestry, and water management, taking into account the different contextual perspectives of involved actors. Strongly following a transdisciplinary approach, the process of finding consensus between the actors was permanently accompanied by a moderating team from science. Additionally to the moderating tasks a fundamental research question was defined:

> What are the critical success factors for a (transdisciplinary) research and development process that envisages local climate change adaptation?

2 Development of Local and Municipal Measures for Climate Change Adaptation Within a Model Region: The Use Case of "Mansfeld-Südharz"

Within the mentioned pilot project the development of a local climate change adaptation strategy, as well as the preparation and implementation of selected measures, has been carried out. The pilot region for this project was the administrative district of Mansfeld-Südharz, including the city of Sangerhausen, in Germany.

At the beginning of the project it has been identified that climate change was acknowledged by many employees as a future problem, but that it was not comprehensively included into their daily work yet. Therefore, a central goal of the project was the enablement of administrational employees to include adaptation to climate change into their common thinking and their daily workflows.

2.1 Project Process and -Progress

The administrational district of Mansfeld-Südharz, in which at least 150,000 inhabitants live, is situated in the south of the German federal state of Saxony-Anhalt. The district covers an area of nearly 1450 km^2. Land use in the rural areas is dominated by agriculture and forestry. The city of Sangerhausen with around 31,000 inhabitants is the largest agglomeration in the area.

The development of local adaptation measures was grouped into two phases. The goal of the first phase ("Klimpass", 2011–2012) took twelve months and included the compilation of a local adaptation strategy.

The strategy suggests four steps (Pundt et al. 2012). In order to determinate prioritized measures, it has been discovered that an indispensable step is the identification of vulnerable areas within the region. This has to be done prior to the definition of measures. It includes answers on the questions to what extent a region is exposed to future risks. The detection of vulnerable areas must consider of the exposition, the sensitivity, and currently executed adaptation measures, as well as the capacity to adapt in general (see Fig. 1).

In various cases vulnerabilities cannot be determined based on singular sectors or processes. This is due to the fact that vulnerabilities have to be seen from different sectoral perspectives, or contexts. The representatives of the unique sectors often evaluate the vulnerability of a certain area very differently. Measures, based on the assessment of vulnerabilities, must therefore take into account the different contexts carefully. In such a way, the outcoming measures are a result of a complex appreciation of varying views, thus compromises.

The following example should clarify the necessity to determine the vulnerability as basis for defining adaptation measures. The erosion of streams concerns the banks, and sole of the watercourse. From a water management perspective, the

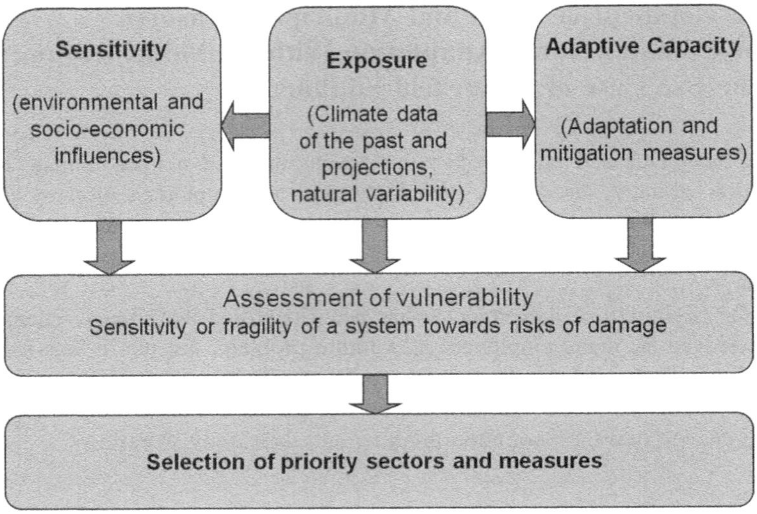

Fig. 1 Determination of vulnerable areas (*Source* Schmidt 2011, p. 10, modified)

erosion is often seen as a problem of high significance. From a forestry perspective, the erosion might play only a minor role. In agriculture one fear to lose land, and therefore erosion gets significant attention. Whereas nature protection might argue for a freer, natural development of watercourses, other sectors might vote for artificial fastenings of bank sides, the building of dams, or the reduction of water flow through barrages. This example shows that the vulnerability of a stream in terms of erosion, as well as measures to deal with the problem, is evaluated very differently by these sectors. Concerning the definition of measures, the ignorance of one or more sectors would lead necessarily to decisions prone to criticism after implementation. In the worst case, the measures are useless due to the fact that the problem was not observed in its full extent.

This simple finding is an important aspect of the stepwise approach suggested in the climate change adaptation strategy: the identification of vulnerabilities must take place in a transdisciplinary process, taking into account all relevant contexts, practice, and science. The aim of this approach is to make qualitatively better decisions about measures, aiming at satisfying as many actors as possible.

As a result some key aspects were identified. They were chosen as starting point to develop concrete adaptation measures in Mansfeld-Südharz:

1. Reduction of soil erosion in agriculturally used areas.
2. Change of the management of water courses.
3. Concept to dealing with neophytes.
4. Adaptation of forest management.
5. Improvement of disaster management (especially in conjunction with extreme weather events).

6. Change of administrational methods to include climate change adaptation in urban planning.
7. Adaptation of water provision and wastewater removal.

To develop complete, coordinated, and balanced adaptation measures, the coordination and integration of relevant sectors was one of the most important goals of the project. This aim has been achieved through the following activities:

- Identification of 'interfaces' between relevant sectors
- Definition of measures for project-management
- Definition of measures for project-controlling
- Commitment to selected measures for public relations.

Within the second project phase ("Klimpass-Aktiv", 2013–2016), the advancement, implementation, and evaluation of specific measures were envisaged. During this phase, the implementation of adaptation measures was intended as a transdisciplinary, multi-contextual process.

Several interfaces exist between the different concerned sectors (Fig. 2). To guarantee the access to information for all actors, and to enable them to discuss the proposed solutions, a web-based information and participation platform has been implemented as an integrative element. An important part of this platform is an online, interactive mapping system, based on GIS technologies and acknowledged technical standards. Applying this system, users are enabled to integrate spatial data from different sectors. They can look at unique, sector-specific data, but furthermore they are able to combine different data layers. They are empowered to investigate if there exist interdependencies between different sectors. In such a way, the system supports the identification of vulnerabilities under consideration of a multi-contextual approach. As an additional idea an online forum was implemented, thus complementing the mapping system. Citizens, for instance, often possess knowledge about specific locations which could be important within the framework of assessing vulnerabilities or developing measures. They can use the platform after registration.

The platform has been reviewed through a usability study carried out by external specialists for websites. Even there were some remarks for improvement concerning

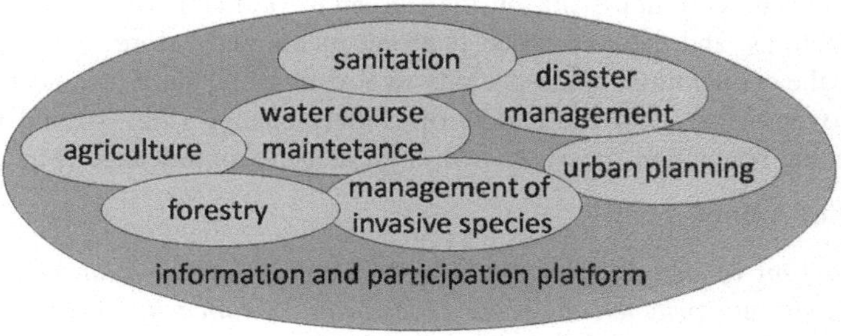

Fig. 2 Cross-linking of sectoral core aspects within the project Klimpass-Aktiv

the user interface, the overall usefulness of the platform has been confirmed during this study. The development, usage, and evaluation of the platform are characteristics of the integrative research approach within the pilot project Klimpass-Aktiv (www.klimpass.de).

2.2 *Implementation of the Transdisciplinary, Multi-contextual Research Approach*

The measures defined in the adaptation strategy, are aiming at an improvement of the adaptation capacity of the pilot region Mansfeld-Südharz. The capability to adapt to climate change includes the competence of individuals, as well as societal systems, to minimize the ongoing and expected changes, but also to profit from potentially occurring advantages (IPCC 2014).

2.2.1 Analysis of Important Characteristics of a Transdisciplinary Process

A transdisciplinary research process includes four essential requirements. In the following section it is discussed, whether the Klimpass-Aktiv project complies with these requirements.

Problem orientation and transformation:

Transdisciplinary research picks up real-life questions and transforms them into research questions. The formulation and development of the research questions is carried out by the affected project partners who originate from different specific sectors. The structure of the project is arranged in such a way that a common goal is achievable. (Bergmann et al. 2005, p. 16, translated).

During Klimpass-Aktiv, an overall research task was concerned with "Networking and Cooperation of the Actors". The aim was to identify relevant contexts and to connect them adequately. For instance, actors came from water management, forestry, urban planning and nature protection. The networking activities and the common work on the (research) problem were realized through regular meetings, and the web-based platform mentioned before.

During the cooperation, and especially as a result of an audit (see Sect. 3), it has been discovered that both the effects of climate change (Fig. 3), and the fixed, sector-specific adaptation measures own high real-life significance (Fig. 5).

Inclusion of Actors:

The interaction of actors from practice is seen as an essential characteristic. Project results are oriented at target groups and included in societal spheres of activity (Bergmann et al. 2005, p. 16, translated).

Fig. 3 Erosion of the banksides of a small stream in the administrative district of Mansfeld-Südharz

Apart from the concrete inclusion of all relevant actors during workshops, as well as the usage of the web-based platform, context-specific and regional information was provided, especially findings concerning adaptation measures. A direct participation of all actors was possible during the workshops, and by using the platform. Knowledge and experiences, as well as good-practice-examples, were exchanged and discussed. Finally, they were harmonized between the parties and included in the website www.klimpass.de.

Transdisciplinary integration concept

Such a concept includes a summary of results on the one hand. On the other hand it provides stimuli for changed procedures and behaviors in practice and research, resulting from new findings during the project (Bergmann et al. 2005, p. 16).

A considerable result from the scientific point of view is the web-based platform developed in Klimpass-Aktiv. It provides the GIS-based, online mapping system which enables to access and integrate context-specific, spatial information in various sectors, and to develop (vulnerability-) maps. The online platform will be presented more comprehensively in Sect. 2.3. In such a way, especially the online platform contributed to convince actors from practice concerning the necessity to include scientific results into their daily work, and to overcome one-dimensional thinking.

Context-sensitivity

Local conditions and options for action, as well as interdependencies between potential sectoral solutions, are perceived as common framework (Bergmann et al. 2005, p. 16, translated).

The consideration of regional, and local, environmental and societal conditions has been a substantial aspect of the project Klimpass-Aktiv. The reciprocal influences between project-specific and other variables of sustainable development have been incorporated by using a multi-criteria assessment method, called PRIMATE.

This method has been adapted to the purposes of Klimpass-Aktiv and is now usable for other decision making processes as well.

The brief discussion of the four essential elements shows that the project Klimpass-Aktiv owns various indicators of a transdisciplinary research process. However, practical requirements concerning the target group-specific processing of information in selected key areas (e.g. urban planning, disaster management) complement the research themes in other sectors (e.g. water management, dealing with neophytes). The development, implementation and usage of the online system www.klimpass.de is a considerable element for both an effective adaptation process and a successful transdisciplinary project.

2.2.2 The Sequence of the Transdisciplinary Research Process in Klimpass-Aktiv

A transdisciplinary research process includes three phases (Pohl et al. 2008, p. 9):

1. Identification of the problem and deduction of an overall research subject
2. Analysis of the problem and generation of specific knowledge
3. Implementation of results/integration

In Klimpass-Aktiv, the overall research goal was to implement both, the transdisciplinary approach, and, as a supporting tool, the web-based platform. The platform is an essential prerequisite to represent transdisciplinary cooperation (see Sect. 2.3). Additionally, a special expert's group with around 30 members was built to accompany the project critically. The members were chosen from different sectors, as well as diverse administrational units and levels. Besides, four smaller teams dedicated to work on specific themes, were formed (urban planning, water management, forestry, public relations in climate change adaptation). They were also composed of persons from practice and science.

For selected key aspects (see Fig. 2), specific research questions were defined. To demonstrate the methodological approach, only the sector of water management is presented in the following part as an example. Concerning this sector it has been discovered that currently processed measures to maintain small rivers are not sufficient anymore due to expected heavier rainfalls that result in larger discharge, as, among others, Trenberth states: "(…) it is evident (…) that climate change has large direct impacts on several aspects of the hydrological cycle and, in particular, increases extremes, making managing and using water resources more challenging." (Trenberth 2011, p. 136). Apart from the increasing danger of flooding, especially the enlarged erosion is seen as a serious threat (Fig. 3).

To deal with this problem, a new method has been developed which suggests a stepwise approach to model soil and riverbed erosion under changing climatic conditions. The results led to different options of how to solve the problem. During the next step it was assessed, how a systematic comparison under explicit consideration of sustainability criteria can be performed. To support this, a multi-criteria assessment method was applied (Hansjürgens 2015; Wittmer and Gundimeda 2012). This method was adapted to the regional circumstances in close cooperation with the relevant actors who provided immensely important local knowledge and facts. The method was tested afterwards. The results, presented at a workshop, were quite promising and convinced all participating actors from practice, as well as science. In doing so, a new methodological approach has been established concerning the cross-sectoral development of adaptation measures.

The potential solutions concerning this specific case resulted in a prioritization of options. For the implementation of the prioritized solution, the cooperation between water management, agriculture, and nature protection was without alternative. The online platform was applied constantly as an important supportive tool during the derivation of measures.

2.3 Development and Usage of the Interactive, Web-Based Platform "www.klimpass.de"

Climate change adaptation strategies require the cooperation of actors from politics, administration, economy, science, and citizens ("multi-level-governance", EEA 2012). The development of adaptation measures should not end at the borders of unique sectors, but overcome them by applying a cross-sectoral approach (Klimaschutz and Partizipation 2015). Apart from diverse print media, the development of a web-based platform for interactive participation of all actors has been envisaged from the beginning of the project Klimpass-Aktiv. The functionalities of the website www.klimpass.de should contribute to

- The provision of relevant knowledge about climate change facts, climate data (including projections).
- The linkage to other relevant websites concerned with climate change.
- The provision of GIS-functionalities, especially relevant datasets from concerned sectors.
- The cross-sectoral discussion of problems and eventually occurring conflicts during the planning phase of adaptation measures.
- In general an improved acceptance of adaptation measures by the actors, as well as the public.

2.3.1 The Interactive Mapping System

The interactive mapping system enables internal (project partners) and external (e.g. citizens) users to select data from a great variety of spatial data layers. They can integrate different datasets based on their specific goal-settings. The system facilitates users to produce vulnerability maps based on transdisciplinary considerations

(Fig. 4) which supports decision making in general. The value of such online, participatory methods was already discovered many years ago: "As the complexity of resource policy decisions has increased, the managers have found a need for new approaches, information and new analytical tools to integrate the multiple interests and viewpoints of stakeholders in official decision making (Walker 2002)". Integrating multiple interests and viewpoints is an evident aspect of transdisciplinary cooperation. Within this framework, Kyem and Saku ask critical questions about the role of participatory GIS. However, they come to the conclusion that, apart from the value for scientists, the practitioners have advantages in using participatory GIS tools: "(…) the internet provides a means for sharing, maintaining and disseminating data besides the possibility of conducting interactive analysis over the web to expand participatory opportunities to interested parties" (Kyem and Saku 2009, p. 13).

Whereas making spatial data visible is one aim of the mapping system, the analysis of different data layers aiming at discovering new spatial patterns is the other. Due to its interactivity the system does provide the opportunity to select different layers and combine them in dependence of the users' careful considerations. This way, the system supports the development of cross-sectoral maps, explicitly taking into account varying contexts, and furthermore considering climate data and projections. The maps are an important means to look on specific locations from different perspectives. In such a way, the maps support a critical discussion of potential future climate change effects. Adaptation measures can be debated among all relevant actors, and alternatives can be appreciated in a scientifically sound

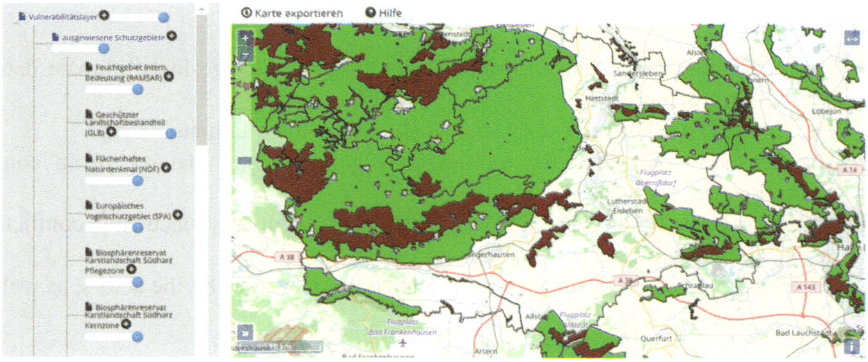

Fig. 4 The interactive online mapping system as part of the Klimpass-Aktiv platform www.klimpass.de

manner. The online platform became therefore a very acknowledged tool that strengthens an effective cooperation between practitioners and scientists, thus a transdisciplinary approach. The mentioned properties (mapping system, forum, see Sect. 2.3.2) are close to typical characteristics of public participation GIS (PPGIS) which are aimed at concrete inclusion of citizens into administrative planning processes (Brown and Weber 2011).

Figure 5 shows an example of a vulnerability map produced using the online mapping system. It contains, apart from topographic information, climate data of a future projection, as well as vulnerable areas threatened with flooding. Furthermore, areas that are prone to erosion are shown, especially those areas where easily erodible soils exist on steep slopes. Hazards can occur here especially after heavy rainfall. Heavy rainfall events are forecasted to increase due to climate change in the future (Trenberth 2012). As a consequence, adaptation measures aimed at the minimization of soil erosion, or more severe events like landslide, can be discussed and possibly implemented.

Based on such maps, the discussion and evaluation processes concerning measures can happen on a fundament that includes different contexts, practically and scientifically.

2.3.2 The Interactive Forum

The interactive forum is an additional feature of the platform aimed at strengthening the discussion between administrational employees, scientists, and citizens (Brennan et al. 2012). All groups should participate in the development of adaptation measures, and an online forum can support the cooperation between different parties if it is used comprehensively (Kyem and Saku 2009). However, extensive usage can fail due to low acceptance of such systems on the one hand, or lacking familiarity on the other. The latter problem has been discovered during Klimpass-Aktiv. Several actors assessed the forum as a good idea, but did not use the forum so far. This is not a question of the age of users (Madden 2010), but more of how daily work is done. Many actors did not use social media or similar systems at all, and therefore they were somehow inhibited to apply the forum. A conclusion resulting in Klimpass-Aktiv is that online forums must be advertised among project participants more effectively, or that they are simply suggested as an obligatory means of communication. Online participation is important because the inclusion of citizens is vital to achieve acceptance for measures implemented by administrations. This has been shown in other studies as well (Granzow and Stephan 2015; Tang 2006). Table 1 presents a balance of the forum, listing the usage of different thematic items provided by the platform. A special hope is that especially local knowledge of citizens about local peculiarities is taken into account during decision making. Such knowledge is often only available for local people, often elderly citizens, and prone to be ignored during administrative planning processes.

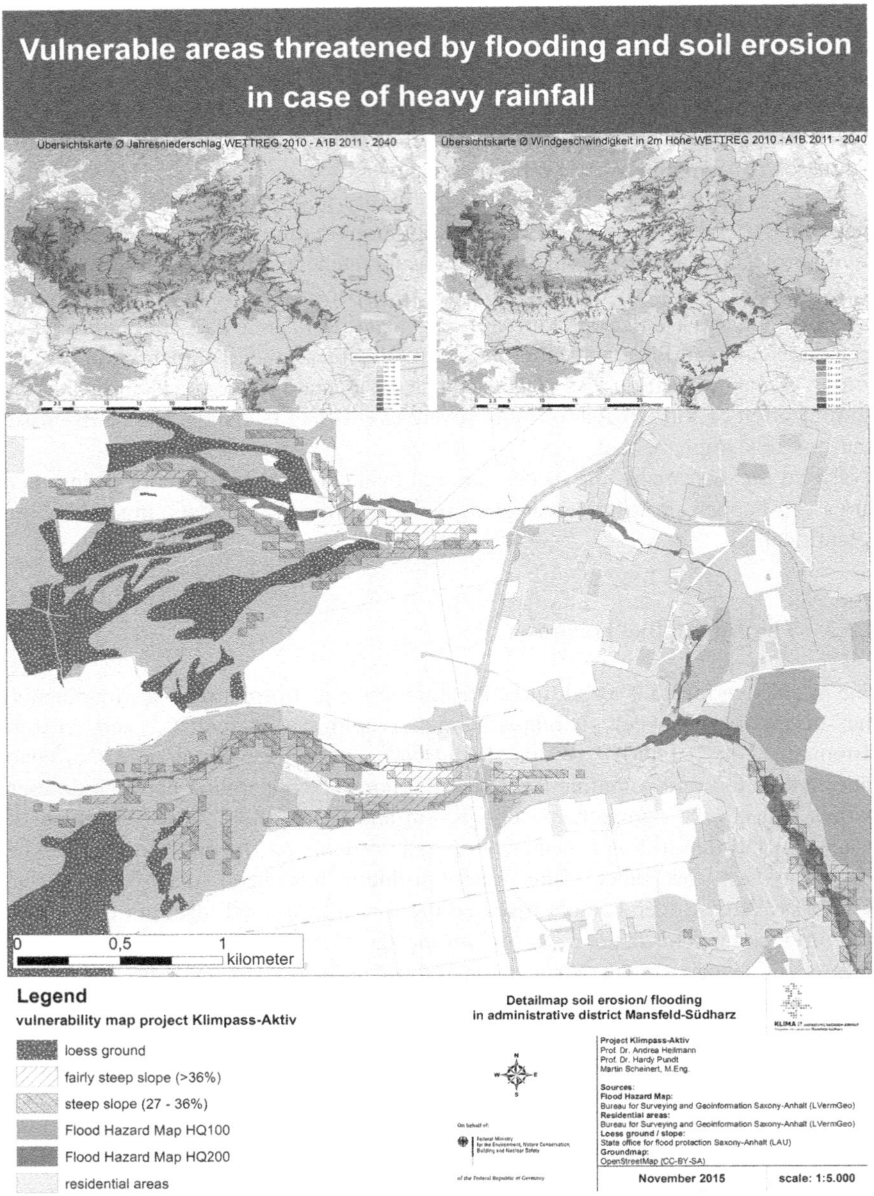

Fig. 5 Vulnerable areas threatened by flooding and soil erosion in case of heavy rainfall

Table 1 Topics, provided via www.klimpass.de, and number of user calls at a specific timestamp

Topic	User calls (total)
Inclusion of the public/enterprises	81
Role of local policy	78
Governmental support opportunities	57
Obstacles during the implementation of climate change measures	64
Already implemented measures	61
Inclusion of new proposed measures	71

3 Evaluations Carried Out During the Project Klimpass-Aktiv

The evaluation of climate change adaptation projects involves both the transdisciplinary research approach and the results, thus considering the newly implemented structural changes and processes.

3.1 Formative Evaluation of the Research Approach

As proved in Sect. 2.2.1, Klimpass-Aktiv complies with the requirements of a transdisciplinary process. In the final phase of the project eight selected actors, two of them coming from practice, were asked to evaluate the whole project. The goal of the evaluation was to find out how the structures and processes, underlying the project, can be improved in future.

The evaluation relied on a questionnaire which considered comprehensively the document "Quality criteria of transdisciplinary research: A guide for formative evaluations of research projects" (Bergmann et al. 2005). However, the guideline has been completed in terms of some project-specific details. The authors of the guideline recommend a discursive-formative evaluation. Differently to this proposal, the participants in Klimpass-Aktiv executed the questionnaires separately. The reason for this variation was that they should discuss and synchronize their answers with other involved employees within their sectors.

The questionnaire includes 20 questions, separated into the following themes:

- Complex I—Description of the problem (3 questions)
- Complex II—Appraisal of the project management, the selection of and cooperation between the actors from science and practice (6 questions)
- Complex III—Assessment of the project process and methodology (3 questions)
- Complex IV—Results (8 questions)

The partly relatively open questions enabled the participants to assess specific aspects on the basis of a school score system (from 1 to 5, meaning 1 = "very good", and "5" = "poor").

The evaluation resulted in some notable findings. For the actors from practice, the scientific problems were not the focus of their goal-settings. Following them, an overall research problem concerning the integration of knowledge has not been identified. Instead, the actors from science saw the central research problem in the new methodological approach concerning the participative development and evaluation of local climate change adaptation measures.

These different perspectives of scientists and practitioners were also reflected. Scientists saw clearly the identification of innovative research questions during the processing of specific tasks. This aspect was not considered by the practitioners at all. The cooperation between science and practice, however, is assessed very positively by both sides. This point has been repeated several times, especially in conjunction with the interactive mapping system. Furthermore, the workshops finalizing project milestones were appreciated as important parts of a transdisciplinary process to develop adaptation measures commonly.

All participants of the evaluation assessed Klimpass-Aktiv as "very good" or at least "good" in terms of contributing to solve climate change adaptation problems. This was especially emphasized by the practitioners who evaluated the cooperation with scientists as being very valuable. They did not criticize the achieved results as such, but more in terms of practical constraints, foremost lacking financial resources to implement the suggested measures. Besides, they stressed the particular importance that the solution of an everyday life problem should be as much focused as possible. This is close to Pohl et al. (2011). They state that the specific purposes of a project have to be clearly identified.

3.2 Audit for the Analysis of the Adaptation Process

To assess objectively the adaptation process carried out in Mansfeld-Südharz, and to identify success factors as well as obstacles, an audit has been carried out in May 2015. This audit was also aimed at observing if the adaptation process can be continued even after the project will be finished. Once implemented, an adaptation strategy results in a perpetual process and requires continuation. It demands a clear idea of how to proceed with the transdisciplinary cooperation, thus guaranteeing sustainable progress.

The audit involved actors from the relevant sectors, it included employees of all administrational levels, and scientists. 25 persons participated in the audit. The audit involved five thematic complexes:

Thematic complex I—Self-assessment of the participant
Thematic complex II—Effects of climate change
Thematic complex III—Usage of relevant information
Thematic complex IV—Status quo concerning the implementation of the adaptation strategy

Thematic complex V—Ideas and hints for the continuation of the adaptation strategy.

A majority of the participants, especially from practice, sees effects of climate change not only in future, but happening even now. Many believe that these effects will increase in future and that this will have consequences for their specific fields of work (Fig. 6).

Special attention is dedicated to the increase of heavy rainfall events, thus playing an important role for actors from different sectors. Nearly 80% of the participants emphasize the necessity to implement additional adaptation measures in this area.

For all participants of the audit, the target group-specific provision of relevant information is a major issue. Therefore, within the second thematic complex, questions about the current status quo of available knowledge were asked. From a scientific perspective, it was interesting to learn which sources of information are used in daily work of administrational employees.

Figure 7 shows that most participants assess their status concerning their personal level of information relatively positive. An information deficit was detected in conjunction with socio-economic effects that could occur in context with local adaptation measures. The respondents underlined that these effects should not be ignored and that the information deficits must be removed. This result corresponds with findings of (Mahammadzadeh 2013, S. 90), who states that at least 40% of employees of local administrations indicated to be insufficiently informed about social and economic effects of climate change.

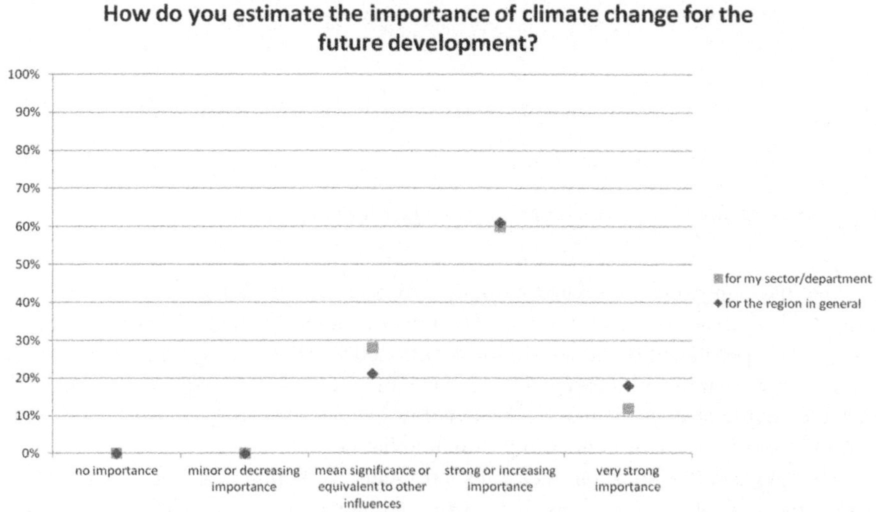

Fig. 6 Overall importance of "Climate change" for the participants of the audit

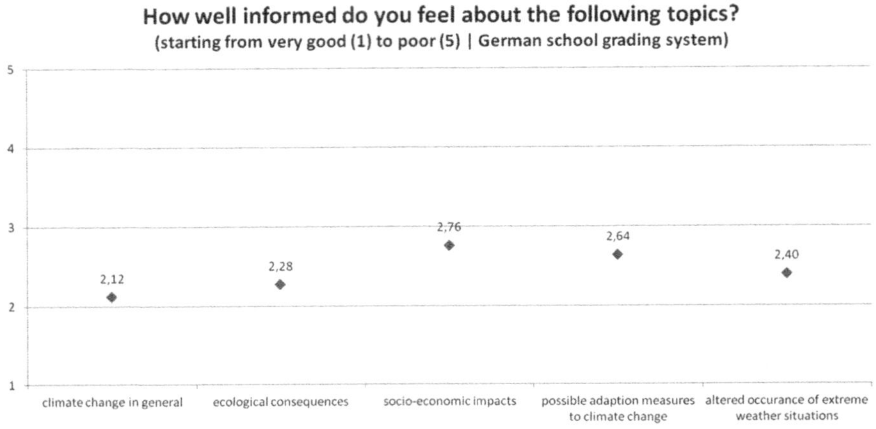

Fig. 7 Status of information concerning climate change adaptation

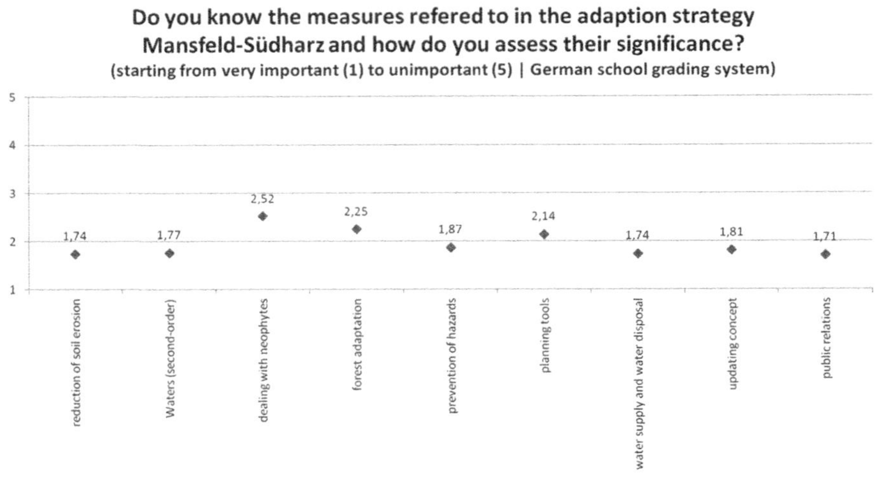

Fig. 8 Significance of the measures proposed in the climate change adaptation concept

The measures that are known in the climate change adaptation strategy for Mansfeld-Südharz are acknowledged by most of the participants. This is especially due to their permanent inclusion in the mentioned working groups, and therefore in the decision making processes. This is an indicator that the project structure was suitable to increase exchange of knowledge and experiences, as well as a more continuous engagement concerning climate change.

All suggested adaptation measures are seen as being relevant. The specific measures dedicated to soil erosion, water management, and to deal with water provision and waste water removal, were rated higher in comparison with others.

All these measures are influenced or caused by heavy rainfall in particular. They show also dependencies to other contexts. This confirms that the necessity to take a cross-sectoral approach and to include multi-contextual information is the only way in order to achieve sustainable problem solutions.

Additionally, the communication of measures is quite important, especially in terms of informing the public. All measures suggested dealing with soil erosion and water management affect somehow the citizens, and should be communicated transparently. The online-platform, as well as analogue materials, can support this. *Both* forms of communicating project results to the public have been realized in Klimpass-Aktiv and were assessed as necessary (Fig. 8). However, all participants agreed that information of the public had to be intensified in future to achieve a larger degree of acceptance of measures.

4 Success Factors and Obstacles of a Transdisciplinary Climate Change Adaptation Process

Based on the results of the formative evaluation and the audit success factors as well as obstacles, of climate change adaptation processes can be identified. They should be considered for future activities in this domain, especially if results from the pilot region are transferred to other regions. Figure 9 includes answers concerning obstacles that are seen within the framework of a successful local climate change adaptation process. All items have a large, at least medium significance for the participating actors.

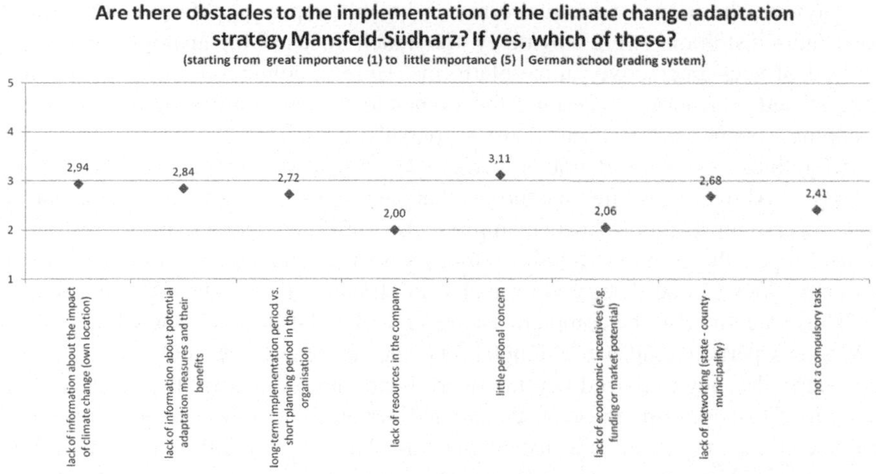

Fig. 9 Obstacles concerning the implementation of a local climate change adaptation process

The connection of results of the formative evaluation and the audit leads to other remarkable findings. For instance, a division in external (obstacles), and internal influences (success factors) is possible. External influences cannot be avoided from the perspective of a local administration, whereas internal ones can be steered actively.

External influences are, among others:

- The consternation of actors and/or the public, and the long-term temporal horizon of measures
- Legal conditions which have to be taken into account
- Lacking financial incentives.

Internal influences are, among others:

- Information/Education of participants
- Personal and material resources
- Networking of actors
- Financial or other incentives.

The internal influences can be transferred into success factors through goal-driven activities of the local administrations. This has been proved during seven other projects, related thematically to Klimpass-Aktiv, under the umbrella of the framework program KLIMZUG (see also EEA 2012).

The provision and combination of information (including informal information, experiences) coming from different sectors, as well as the exchange of experiences must be increased in local administrations significantly. The online platform www. klimpass.de has been appreciated widely as a tool to support both the region-specific information about spatial conditions and interdependencies, as well as the enablement of transdisciplinary networking between all relevant actors.

Anyway, the usage of the platform can be still improved. Many employees of local administrations were obviously too much focused on analogue materials, instead of such interactive online-platforms. Here, a continuous process of information and education concerning the tremendous opportunities delivered by the mapping system and the forum must be pursued in future.

Resources comprehend not only financial, but also personal aspects. During Klimpass-Aktiv, it was often mentioned that participating in such a project means to invest personal resources, which implies also time and money. Both are crucial issues, especially in public administrations with low budgets. The outcome is that no employees can be delegated to deal with climate change adaptation problems.

This concerns also the comprehensive usage of a platform like www.klimpass.de . Whereas participating actors emphasized the importance of applying such supporting tools, they criticized that this work is not appreciated if other tasks are left undone. This criticism must be considered seriously in future to guarantee that climate change is treated in a sustainable way. This might require investments into personnel and/or a re-organisation of administrative departments (EEA 2012).

The demand for supporting existing and new networks is an important aspect of future climate change adaptation projects (vgl. Biebeler et al. 2014). This concerns horizontal (cross-sectoral) and vertical networking (levels in administration).

At last, it has to be mentioned that climate change adaptation processes may require long-term decisions and measures (often more than five years). To hold transdisciplinary networks active over such long periods can be a problem. Therefore, a certain institutionalization of the necessary horizontal and vertical networking activities is desirable. Current administrational regulations and structures hinder such processes and require more openness and flexibility in administrations. Increased usage of platforms like www.klimpass de, but also further information about financial sources, as well as new software technologies like data mining, context-aware systems and the increased usage of automatic sensor networks will also contribute to a better understanding of climate change, as well as supporting climate change adaptation strategies and their implementation.

Adaptation projects which produce added value in various forms, enhance the readiness to participate (EEA 2012). It has to be checked which (financial) incentives, given by a local administration, can help to motivate local actors (e g. citizens, enterprises, and other organizations) to provide contributions to deal with climate change adaptation. There are examples for this, for instance the reduction of fees (e.g. for the discharge of rainwater or the removal of wastewater) or taxes (e.g. for the waste avoidance or the removal of paved areas to improve infiltration). Administrations could also reward good ideas in conjunction with climate change adaptation within the framework of challenges or public competitions.

However, there are opportunities for administrations to influence the success factors. To achieve this, one of the most pressing barriers, the fact that employees in local administrations still feel not or only few affected by climate change problems, must be overcome. In other words: the willingness to support the adaptation process proactively and in the long run must grow in future.

5 Conclusions and Outlook

Based on a pilot project, during which a local climate change adaptation strategy has been developed and specific measures have been defined and implemented, an evaluation and an audit prove the necessity of a transdisciplinary approach to achieve sustainable climate change adaptation measures.

Climate change adaptation requires very often decisions concerning measures that do not concern only one sector, or field of knowledge. Most problems are multi-dimensional, and require the contextual knowledge, as well as practical skills of different actors. The actors, however, come from science and from practice, e.g. administrational departments, organizations, or companies. Even the citizens should be included if measures are about to be implemented. Therefore, a transdisciplinary approach aiming at finding solutions that are harmonized between actors is indispensable.

During the evaluation and audit of the pilot project, it has been discovered that the methodological process which has been applied during the project Klimpass-Aktiv was successful. Apart from meetings and workshops, the development of a web-based information platform, including an interactive, GIS-supported mapping system and a discussion forum, played an essential role in implementing a transdisciplinary project. Such technologies can support the participation and teamwork of different actors significantly. Another aspect concerning which new and innovative strategies are required is the provision of sufficient personnel and financial resources. They often lack, especially in regions where public budgets are limited. For an effective adaptation to climate change, they are vital. No-regret measures are a first step towards climate change adaptation, but there will be measures that are not achievable without spending money.

An overall conclusion is that the evaluation and the audit of the pilot project indicate the operational reliability of the proposed methods and processes. Other regions and administrations can use the guidelines and suggestions resulting from the project Klimpass-Aktiv, thus indicating that the project has led to innovative and transferable procedures and methods which are potentially eligible to become good practice. The success factors that were identified and which are controllable to a certain extent by the (local) actors, especially the administrative units, play an important role within the overall process.

However the willingness to cooperate and thereby crossing administrational borders, vertically and horizontally, on the one hand, and to remove barriers between science and practice on the other hand, are irreplaceable prerequisites for the implementation of a sustainable, transdisciplinary process.

References

Aktionsplan Anpassung APA (2011) http://www.bmub.bund.de/fileadmin/bmu-import/files/pdfs/allgemein/application/pdf/aktionsplan_anpassung_klimawandel_bf.pdf.

Bergmann et al (2005) Qualitätskriterien transdiziplinärer Forschung; Ein Leitfaden für die formative Evaluation von Forschungsprojekten, ISOE Studientexte 13, Frankfurt/Main.

Brennan, J., Heilmann, A. & Pundt, H. (2012) An Information Systems Approach to Developing Adaptation Strategies. Proc. of the European, Mediterranean & Middle Eastern Conference on Information Systems 2012. EMCIS2012, Munich, Germany, pp 231–241.

Brown, G., Weber, D. (2011) Public Participation GIS: A new method for national park planning. Landscape and Urban Research. 102(1), pp. 1–15.

DAS (2008) http://www.bmub.bund.de/fileadmin/bmu-import/files/pdfs/allgemein/application/pdf/das_gesamt_bf.pdf (accessed 26.04.2016).

EEA - European Environment Agency (Hg.) (2012) Urban adaptation to climate change in Europe, EEA Report 2/2012. Verfügbar unter: http://www.eea.europa.eu/publications/urban-adaptation-to-climate-change (accessed 09.02.2016).

Fortschrittsbericht (2015) http://www.bmub.bund.de/fileadmin/Daten_BMU/Download_PDF/Klimaschutz/klimawandel_das_fortschrittsbericht_bf.pdf. (accessed 24.03.2016).

GeoDZ (2016): GeoDataZone. http://www.geodz.com/deu/d/holistischer_Ansatz (accessed 21.01.2016).

Granzow, P., Stephan, H. (2015) Viernheim auf dem Weg zu einer „klimafreundlichen Bürgerkommune“. In: Klimaschutz & Partizipation, Service- und Kompetenzzentrum kommunaler Klimaschutz, Köln. S. 22–33.

Hansjürgens, B. (2015) Ökonomische und soziale Bewertungsansätze. In: Kaltschmitt, M., Schebeck, L. (eds.): Umweltbewertung für Ingenieure. Methoden und Verfahren, Berlin, Heidelberg: Springer, S. 127–146.

IPCC 2014 Climate Change 2014 Synthesis Report, Summary for Policymakers http://www.ipcc.ch/pdf/assessment-report/ar5/syr/AR5_SYR_FINAL_SPM.pdf (accessed 19.03.2016).

Klimaschutz & Partizipation (2015) Akteure in der Kommune informieren und beteiligen. Service & Kompetenzzentrum kommunaler Klimaschutz beim Dt. Institut für Urbanistik, Köln.

Kyem, P.A.K., Saku, J.C. (2009) Web-based GIS and the future of participatory GIS applications with local and indigenous communities. The Electronic Journal on Information Systems in Developing Countries 38(7), pp. 1–16. Available under: http://www.ejisdc.org/ojs2/index.php/ejisdc/article/view/584/292 (accessed 08.03.2016).

Madden, M. (2010) Older Adults and Social Media. http://www.pewinternet.org/2010/08/27/older-adults-and-social-media/ (accessed 27.03.2016).

Mahammadzedeh, M., Chrischilles, E. & Biebeler, H. Institut der deutschen Wirtschaft (Hg.) (2013) Klimaanpassung in Unternehmen und Kommunen; Forschungsberichte aus dem Institut der deutschen Wirtschaft Köln 83. Köln: Institut der dt. Wirtschaft Medien GmbH.

Pohl, C., von Kerkhoff, L., Hirsch Hadorn, G., Bammer G., (2008) “Core Terms in Transdisciplinary Research”. In Handbook of Transdisciplinary Research, ed. by G. Hirsch Hadorn, H. Hoffmann-Riem, S. Biber-Klemm, W. Grossenbacher-Mansuy, D. Joye, C. Pohl, U. Wiesmann, and E. Zemp, pp 427–432. Dordrecht: Springer. http://www.transdisciplinarity.ch/d/Transdisciplinarity/documents/HB_Core_terms.pdf.

Pohl C. et al. (2011) “Questions to evaluate inter- and transdisciplinary research proposals”, working paper, td-net for transdisciplinary research, Berne.

Pundt, H., Heilmann, A., Kerwel, E. (2012) Development of a guideline for regional and local authorities to adapt to climate change-results of the project “KLIMPASS”. In: R. Seppelt, A., A. Voinov, S. Lange, D. Bankamp (Eds.): Int. Congress on Environmental Modelling and Software-Managing Resources of a Limited Planet, Leipzig, Germany, http://www.iemss.org/society/index.php/iemss-2012-proceedings, pp 876–883.

Schmidt, C. (2011) Klimaanpassung auf Regionaler Ebene, am Beispiel der Region Westsachsen und Oberlausitz-Niederschlesien. In: Technische Universität Dresden (Hg.), Dresdner Planergespräche; Klimaanpassungsstrategien in der Landschafts- und Raumplanung (Band 2), Dresden.

Tang, M. Y. (2006) Design and Implementation of a GIS-enabled online discussion forum for participatory planning. Department of Geodesy and Geomatics Engineering University of New Brunswick, Technical Report 244. Available under: http://www2.unb.ca/gge/Pubs/TR244.pdf (accessed 14.03.2016).

Trenberth, K. E. (2011) Changes in precipitation with Climate Change. In: Climate Research, Vol. 47, pp 123–138.

Trenberth, K. E., 2012: Framing the way to relate climate extremes to climate change. Climatic Change, 115, Issue 2, 283–290.

Walker, D.H. (2002) Decision Support, Learning and Rural Resource Management. Agricultural Systems, 73, 1, pp 113–127.

Wittmer, H., Gundimeda, H. (eds., 2012) The Economics of Ecosystems and Biodiversity in Local and Regional Policy and Management. Earthscan (Taylor and Francis), London, Washington.

The Role of Risk and Vulnerability Science Centres in Enhancing the Resilience of Local Government and Local Communities to Climate Change in South Africa

Julia Mambo and Miriam Murambadoro

Abstract Climate change and variability negatively affect Africa the most, this is particularly because of their dependence on sectors and livelihoods that are largely dependent on climatic and environmental factors (such as agriculture), high levels of poverty, low levels of human, social and physical capital. South Africa in particular is susceptible to changes in climate due to its location at the southern-most point of Africa and the rich terrestrial and marine biodiversity that support many livelihoods in the country, make responding to climate change a priority for the government. Academia in South Africa has played a key role in climate change science developing down scaled climate projections used in planning and responding to climate change impacts. However, the interpretation, utilisation as well as the integration of climate science in planning at all levels of government has lagged behind, especially at local municipal level. This paper assesses the role that has been played by Risk and Vulnerability Science Centres in promoting resilience of local government and local communities to climate change. The Risk and Vulnerability Science Centres are one mechanism that has been developed to assist in bridging this gap, by providing a link between scientists and local governments and communities. Three Risk and Vulnerability Science Centres are hosted in three previously disadvantaged universities i.e., the University of Limpopo, University of Fort Hare and Walter Sisulu University. The centres present a great opportunity for the interpretation, packaging and dissemination of climate and other scientific data to local government and communities. They are also well placed to co-produce knowledge as they can

J. Mambo (✉)
Council for Scientific and Industrial Research Natural Resources
and the Environment, Climate Studies Modelling and Environmental Health,
Carlow Road, 91230, Auckland Park, Johannesburg 2006, South Africa
e-mail: JMambo@csir.co.za

M. Murambadoro
Council of Scientific and Industrial Research—Natural Resources
and the Environment, Sustainability Science and Resource Economics,
Carlow Road, 91230, Auckland Park, Johannesburg 2006, South Africa
e-mail: MMurambadoro@csir.co.za

W. Leal Filho (ed.), *Climate Change Research at Universities*,
DOI 10.1007/978-3-319-58214-6_29

conduct research to fill in the knowledge gaps identified by local government and local communities, however they have struggled to perform as expected especially addressing the knowledge gaps. The paper explores the challenges and the opportunities faced by Risk and Science Centres in understanding local level climate change risk and vulnerability and the complex systems in which local governments and community operate. An understanding of this would help design appropriate response options in order to build the resilience of local government and local communities to climate change. The centres have provided an opportunity for the capacity building of students and could play a key role in facilitating participatory platforms for social learning and co-production of knowledge at various levels of government that integrate both indigenous and experiential knowledge. Challenges experienced include lack of commitment by government officials to integrate the research recommendations into policy practice and lack of opportunities to present research in local government forums to inform development planning.

Keywords Climate change · Local government · Universities · Risk and vulnerability science centres · South Africa

1 Introduction

Climate change has emerged as a "wicked problem" that requires a transdisciplinary approach to enhance understanding on the comprehensive impacts of weather and climate related extremes. The climate change phenomenon can be classified as a planetary-level difficulty, being produced by both nature and society with long term far reaching consequences for the environment and society. Multi-disciplinary approaches have often failed to address climate change because they lack a theoretical and methodological base (or thread) that can facilitate the integration of knowledge produced by the different disciplines (Mebratu 2001; Audouin 2011). The result is that the multi-disciplinary approaches become a combination of disciplinary principles, methods and tools and the limitations of these are merely transferred to this new setting (Audouin 2011:3). The production of new transdisciplinary knowledge is hypothesised as being capable of producing practical and useful knowledge for the solving of real-world difficulties by applying all forms of knowledge such as scientific, traditional and experiental knowledge for improved understanding of complex problems such as climate and global change.

A transdisciplinary process is one that incorporates and goes beyond, disciplinary and multi-disciplinary approaches and seeks to bridge the gap between knowledge producers (e.g. universities, science councils) and users (e.g. government, business, local communities) (Burns et al. 2006). This gap is one of the reasons why knowledge produced through research on the environment and society, is failing to effectively inform policy and decision-making by government among others. Tackling "wicked problems" such as climate change requires co-production of new knowledge given that the new problems cannot be approached or solved by mono-disciplinary methods as has been done in the past (Swilling 2014). This means that the science and social disciplines can no longer be considered as supposedly two fundamentally diverse and independent realities that produce partial information as that does not address the complex problems such as climate change. Instead, integrated solutions premised in integrated knowledge co-production are pushed forward (Dhansay 2015; Swilling 2014; Morin 1999).

While climate change occurs concurrently at both planetary and local level scales, its impacts will be felt more at the local level. In order to begin to address the issues of climate change, there is need for a range of actors to be involved in the initial design of the problem and its solution space (Gibbons et al. 1994; Lang et al. 2012; Leach 2014; Swilling 2014; Leck and Roberts 2015; Patel et al. 2015; Reyers et al. 2015; Murambadoro 2016). The South African government has been pro-active in responding to climate change and this is shown by initiatives at national level aimed at enhancing the country's understanding of climate change risk, vulnerability of sectors and response. This includes initiatives by the Department of Environmental Affairs through the Long Term Mitigation Scenarios (DEA 2007), National Climate Change Response Policy (DEA 2011), Long Term Adaptation Scenarios (DEA 2013). The Department of Science and Technology (DST) in particular adopted a Ten-Year Innovation Plan in 2008 aimed at propelling the country towards a knowledge economy. The "Innovation Towards a Knowledge Economy" plan identified four key grand challenges, one of which is Science and Technology in Responding To Global Change. Projects under this Global Change Grand Programme include the South African Risk and Vulnerability Atlas and the Risk and Vulnerability Science Centres (previously known as the Risk and Vulnerability Assessment Centres).

The South Africa Risk and Vulnerability Atlas (SARVA) provides information on the vulnerabilities and risks associated with global change, including climate change, for various sectors. The first two phases of SARVA developed the architecture and infrastructure of an online platform (Phase I). This online platform contains an electronic spatial database of risk and vulnerability datasets aimed at providing local government decision-makers with climate change information for planning purposes. Phase II included the development, population and outreach of the Atlas. From these two phases of SARVA, it has connected to a wide range of stakeholders through thousands of freely available global change data resources. The Atlas has earned a place in informing adaptation efforts and information needs at various levels, including, parliament, the National Climate Change Response Policy, Long-Term Adaptation Scenarios, and individual municipal risk and vulnerability assessments. However, this success has been overshadowed by the poor uptake by local government and business sector who are the target users of the Atlas. Science based institutions that produce knowledge have not fully exploited the opportunity to co-produce and integrate scientific information with experiential knowledge held by local government officials and other user groups in order to enhance uptake of information.

Currently there are three science centres at the University of Limpopo (Limpopo Province), the University of Fort Hare and Walter Sisulu University (both in the

Eastern Cape Province). It was anticipated that the centres would assist municipalities and communities to effectively use information on the South Africa Risk and Vulnerability Atlas (SARVA) while also providing support for the collection and brokering of data, knowledge and information. While the Global Grand Challenge is funded by DST, the funding and management of the Risk and Vulnerability Science Centres is managed by the National Research Foundation. The science centres are located in different municipalities that have different information and support needs and as such climate change information provided to them needs to be useful and relevant at local level. This information can be scientific and indigenous knowledge. Climate change response in these municipalities is guided by the context in which the municipality operates. To meet these information needs, institutional systems such as the science centre's can play a key role in connecting the knowledge providers and the users (Srinivasan et al. 2011; Murambadoro 2016). The right information enables local governments and business to plan in the right context whereby they have an understanding of the local risks and vulnerabilities; the sources of greenhouse gases so that they can develop the most effective response strategies which are aligned to other government sectors and institutions.

The aim of the paper is to investigate the effectiveness of Risk and Vulnerability Science Centres in promoting resilience to climate change for local government and local communities. This has been done by highlighting the mandate of the centres, their strengths and challenges as well as recommendations for the way forward so as to enhance climate change resilience in their respective locations.

2 Methodology

The research technique used for this study was a mix of methods which included a detailed review of literature from two key government institutions, the Department of Science and Technology and the National Research Foundation that played a key role in the establishment and administration of the centres. Other information on operational, research and administration issues on the science centres was collected through informal discussions and a structured questionnaire that was sent through email to all the centre managers and the post graduate students who assist in the mentoring of undergraduate students in the centres. Four key questions that were enquired in the questionnaire were:

1. What is the mandate of the Risk and Vulnerability Science centres and what are the key deliverables?
2. What type of collaborations both local and international have the centres established to advance climate change research and knowledge sharing for the benefit of students enrolled for post graduate study through the centre.

3. Based on the key deliverables expected from the centre what have you been able to achieve to date? Are there any additional things you have accomplished although not highlighted as a key deliverable (bonus points).
4. What challenges have the centre faced in the past two phases of the programme that may have constrained performance? What should be done to address this?

The responses from the centre managers and post graduates were assessed for common issues and variances among the three centres, especially on the challenges reported. These responses were used in the write up of this paper and included some recommendations on how to address some of the challenges encountered.

2.1 Limitations and Constraints

While an extensive literature review was conducted, key government documentation from the Department of Science and Technology and the National Research Foundation which would have explained the rationale behind the establishment of the centres were not found. There was no document that clearly stated the mandate of the centres. The documents that were reviewed mentioned the Risk and Vulnerability Science Centres in different capacities, as well as the expected role of the centres under the Global Grand Challenge, however, the appointed centre managers did not receive a written mandate or terms of reference from both the Department of Science and Technology and the NRF. All centre managers agreed that the "*mandate*" that they are using was based on verbal agreements and discussions that were held at the inception meeting. Further, only the information from the centres as well as the opinions of the managers and observations of the authors who have worked closely with the centres, have been used as a basis of this paper.

3 Results and Analysis

3.1 Complexity of Risk and Vulnerability Science Centres and Their Role in Climate Change Response

Higher education systems are an integral part of both society and the economy. The education system plays a key role in knowledge production and dissemination. To remain functional it needs to have an impact on society. It is influenced by its environment and adapts over time. At inception, three Risk and Vulnerability Science Centres where established at previously disadvantaged rural-based universities, with the aim of strengthening the capacity of these institutions in conducting research in climate, global and environmental change. The three science centres (and universities) have been used in this study specifically to analyse the role that these higher institutions play in enhancing climate change resilience because unlike other higher institutions they are part of the Global Change Grand Challenge and are mandated and funded to play this role through the science centres. Risk and Vulnerability Science Centres are a component of a complex university system and hence there is need to understand how this university system operates and how they can facilitate climate change resilience at local level as well as the challenges and the opportunities presented. The three universities often face challenges such as limited funding, lack of skilled professionals to lecture and supervise students and student protests. These challenges are part of the complex system in which the science centres have to function in and often constrain their ability to meet their objectives.

Box 1: What are previously disadvantaged universities?
The universities referred to as previously disadvantaged, are the dominantly black student tertiary institutions, most of which were located in or near former homeland areas created under the apartheid era. These include institutions such as the Universities of Fort Hare, Zululand, Venda, Limpopo, Western Cape and Walter Sisulu. Despite the ending of apartheid and its policies of inequality, these institutions are still characterised by a backlog of underdevelopment and face a myriad of challenges which include the lack of funding for infrastructure development, for both the institution as well as the student residences. The lack funding has also affected the development of systems to ensure the sustainability of such institutions, to strengthen its academic focus areas and as such they still lag far behind other tertiary institutions in terms of academic output. To address some of these challenges in these universities the Department of Higher Education and Training has invested R2.5 billion to establish financially sustainable institutions that are academically sound and contribute to the knowledge economy through research.

Box 1: The box provides a summary of what are previously disadvantaged universities in South Africa.

The three established Risk and Vulnerability Science Centres are expected to play a key role in the promotion of resilient rural communities, based on scientific knowledge and research. They also assist municipalities identify the best decision support tools and development options through research to effectively and efficiently respond to the risks posed by climate change and other environmental stressors. The Risk and Vulnerability Science Centres are multi-disciplinary, conducting research that provides information on the vulnerability to climate and global changes as well as the risks that are associated with these changes and other social-cultural stressors in South Africa. The provinces in which the centres are located have a significant proportion of the population that is mostly rural and hence the centres have been designed to also empower local communities to respond to climate change by identifying risks and vulnerabilities as well as the adaptive capacity through studies that identify local risk and vulnerability. The information gathered would ideally feed into the South Africa Risk and Vulnerability Atlas, with the centres also playing a knowledge brokering support role for the Atlas.

The focus areas for the Risk and Vulnerability Science Centre as determined by the National Research Foundation are centred on;

- Building capacity of students to conduct local level research and production of transdisciplinary information for the key sectors that would be disseminated on the SARVA portal.
- To provide local government support by playing a key role in the science – policy interface as well as assist in the interpretation of data in the Risk Atlas to the local government and communities;
- Local and international collaborations with other universities and research institutions to build the capacity of students, local government and communities in climate change research.

The following section evaluates how the sciences centres have fared in these focus areas. It should be noted that science centres are only but one of the initiatives by the universities to enhance climate change response in their respective areas. These universities have several other initiatives that are done and are funded separately.

3.2 *Achievements and Challenges in the Science Centres*

The Science centres have been operational since 2010 and this assessment was done from March 2015 to June 2016 and is based on interviews with key stakeholders from the science centres. The achievements by the science centres are not uniform as some centres have done well on some aspects and poorly on others. The following is a discussion on what the science centres have done to date.

(I) Building capacity of students to conduct local level research and production of transdisciplinary information for the key sectors that would be disseminated on the SARVA portal

The ultimate objective of the three science centres is to empower rural based universities to contribute to the generation of global change knowledge and its application at local level, as well as develop the capacity for global change at local government level, academic institutions and communities. All three centres are functional within their respective academic institutions providing the much needed scientific guidance for climate and global change research. Most centres are well equipped with state of the art geographic information systems (GIS) laboratories and software and have successfully recruited academic staff from various academic departments in the university to provide multi-disciplinary research expertise as well as post graduate student supervision. Apart from the Centre managers, the centres also have post-doctoral research fellows who give guidance to post graduate students as well as directing the research undertaken by the centre. The centres also have administrative staff to assist in the day to day running of the centre.

Another crucial role that the science centre's play is capacity building of post-graduate students to conduct global change research that will contribute to the knowledge economy. The science centre at the University of Limpopo has focussed primarily on agricultural production and climate change as well as issues of food security in the rural areas of Limpopo Province. Extensive research is being conducted on new crop varieties such as canola and moringa for livestock fodder. Other research areas include waste minimisation, rehabilitation of mining sites, human health-focussing on malaria, diarrhoea and pneumonia.

The science centre at the University of Fort Hare focusses its research on crop and livestock production, studied in the context of global and climate change at the local scale in the Eastern Cape Province. The placing of the science centre at the university of Fort Hare in the Faculty of Science and Agriculture might have influenced this outcome for the centre while being based in a rural area where agriculture is the main livelihood is also a likely explanation for its placement in this faculty. The centre at the Walter Sisulu University's research spans into the coastal and marine ecosystems, agriculture, food security, human health and waste management.

Although all centres have been conducting research a closer analysis on this indicates that the research focus by some of the centres has been very narrow. This has been attributed to challenges in getting buy in from other disciplines (e.g. social anthropology, geography, built environment) with in the university to collaborate on identifying suitable research topics, students and supervisors. Some of the initial research topics by some of the post graduate students did not fit under the global change theme and had to be re-directed to fit into the required theme to enable the students to access the funding through the National Research Foundation. Although a lot of research has been conducted by post graduate students from the three centres this research has not been uploaded on the SARVA portal for dissemination.

(II) Provide local government support by playing a key role in the science–policy interface as well as assist in the interpretation of data in the Risk Atlas to local government and communities

The centres have also been mandated to facilitate access to and understanding of scientific data by policy makers and decisions makers at different levels of government i.e. provincial, district and local municipality. Building the capacity at local government level will assist in contribute to effective planning and provision of services for the local communities as well as building the resilience of the communities to changes in climate and its impacts. The centres aim to create networks of expertise to support structures cross the various environmental aspects as well as facilitate the coordination of communities, municipalities, government departments and academic institutions towards sustainable economic development.

Walter Sisulu University centre has been involved in research collaboration with the OR Tambo District Municipality and Alfred Nzo District Municipality and this collaboration extends to the capacity building of local government officials. The University of Limpopo has been actively collaborating with the Limpopo provincial government, especially the Limpopo Department of Economic Development, Environment & Tourism (LEDET) that is leading climate change response in the province. This has included joint workshops hosted by the centre and facilitated by SARVA as well as the inclusion of the centre to various provincial committees on climate change to guide climate change into planning at provincial level. The piloting of SARVA was a key event for both the provincial government and the university. The University of Fort Hare has been working with the Amatole District Municipality which is the local municipality where the university is located.

The outreach programmes for local government include the provision of technical assistance in risk and vulnerability assessment and other related fields such as disaster risk reduction and management as well as the interpretation of scientific information for policy makers using evidence based knowledge. The centres are also meant to play a key role in rolling out of the South Africa Risk and Vulnerability Atlas (Personal communication with key informants at science centre, April 2016).

(III) Local and international collaborations with other universities and research institutions to build the capacity of students, local government and communities in climate change research

The universities hosting the science centres do not always have all the facilities required for research and as such collaboration with other research institutions provides access to scientific facilities that are not available locally. Collaborations with other researchers especially those from more scientifically advanced countries and institutions provide a good way of enhancing local scientific competences through the exchange of knowledge and experience. It also allows for co-production of knowledge to address some of the key pressing issues such as climate change. The Centres are involved in local level collaborative research, mainly on projects around food security, water and agriculture. The University of Fort Hare science centre has substantial local collaborations for research with the University Cape Town (UCT), University of South Africa (UNISA), Water Research Commission (WRC), University of Venda and University of KwaZulu Natal. The centre is working with the traditional leadership through collaboration with the Nkonkobe traditional communities research group based at the university (Personal Communication with key informants at science centre 2016). The local level research includes the centres outreach and partnerships in projects with the local municipalities as outlined in Box 2.

Box 2: Local collaborations established by science centres
Local Collaborations - University of Fort Hare (UFH) and Walter Sisulu University (WSU)

- South Africa Risk Assessment Committee (SAVAC)
- South African Institute for Aquatic Biodiversity (SAIAB)—WAS Marine Projects

- Eastern Cape Anti-Poverty Food Security Advisory Committee
- Eastern Cape Integrated Food Security and Nutrition Programme
- Eastern Cape Provincial Disaster Management Forum
- Eastern Cape Rural Development and Agrarian Reform
- Eastern Cape Provincial Stats Advisory Forum

Collaborations with Local Municipalities

- UFH—Amatole District Municipality—Waste Management Advisory Forum
- WSU—OR Tambo District Municipality—Water Management, Food Security, and Disaster Risk
- WSU—Alfred Nzo District Municipality—Umzimvubu Catchment Partnership Programme for ecological Integrity

Box 2: Local collaboration initiated by the science centres at University of Fort Hare and Walter Sisulu University.

In terms of building international networks to advance research data production, the centre based at the University of Limpopo has made great strides in establishing international research networks compared to the other centres (see Box 3). The networks built were established in conjunction with the University's research department and these include collaboration on food security, crop production and the human health as well as technological innovations such as early warning systems for small holder farmers and malaria. The other two centres have limited international and regional collaborations with academic or research institutions and there is limited information on the details of these collaborations.

Box 3: International collaborations established by science centres University of Limpopo

- **Limpopo Living Landscapes**—Understanding the dynamics of ecological and cultural landscapes, in the face of global change: collaboration with University of Gottingen, University of Bonn, University of Cologne and University of Frankfurt
- **Infectious diseases early warning systems (IDEWS)**: in partnership with University of Nagasaki
- **Building large evergreen agricultural network in Southern Africa (BLEANSA)**: partnership with the International centre for research in agroforestry (ICRAF)/World Agroforestry Centre

Walter Sisulu University

- Developmental Studies in partnership with University of Addis Ababa
- National Food technology Research Centre (Botswana)
- Association of Africa Universities—Food Security

University of Fort Hare

- **Climate Impact Research Capacity Leadership Enhancement (CIRCLE)** funded by UK-AID

Box 3: The international collaboration projects by the Risk and Vulnerability Science Centres.

3.3 Challenges with the Institutionalisation of the Risk and Vulnerability Science Centres Within the Academic Institutions

It is pertinent to clarify at the beginning of this section that the challenges discussed here are the author's perceptions[1] and also includes feedback from the science centre managers. The challenges discussed are not comprehensive but highlight some of the factors that may or may not have affected the ability of the centres to effectively meet all their objectives.

Administrative and management challenges

- At the inception of the Risk and Vulnerability Science Centres, the centres awarded to the respective universities through a tender process, where by a call was made by the Department of Science and Technology and the National Research Foundation, for universities to host the centres. The call was followed by site visits and more deliberations before the Universities were finally awarded the contract to host the science centre for three years. While the centre managers are aware of their key focus areas and what is expected of them, there is no formal documentation on this or any other information on the setting up of and the administration of the science centres. This absence of specific guidelines has resulted in several cumulative effects for some centres such as,
  - The centres being placed in different departments that have different structures and processes within the respective universities. The Department of Science and Technology and the NRF did not give guidance or instructions to the host institutions on how to they can best facilitate the hosting of the science centre to augment its impact. At Walter Sisulu University, the science centre was established in the Research, Development and Innovation department, where the centre manager reports to the director of research. This set up was meant to be a temporary arrangement but has not changed and the post of manager of the science centre is still recognised as an acting position.

[1]Authors have been working closely with the science centre managers since 2010.

The centre at the University of Limpopo was established as a Centre of Excellence under the office of the Deputy Vice Chancellor, where the centre manager reports directly to the vice chancellor. The centre at the University of Fort Hare was established under the Faculty of Science and Agriculture and most of the research done by the centre is related to agriculture.

- A narrow focus on the research themes that have been selected by the science centres as there is uncertainty on what should/should not be included. As a result the centres focus more in the research areas in their host department or where the expertise of the manager lies and there is very little diversity. This situation has been exacerbated by the difficulty encountered by the centre managers to recruit additional academic staff from other departments in the university, to widen the areas of research as well as provide academic supervision for the post graduate students. Although the situation has improved in some science centres in recent years there is still room for improvement and more academic staff especially from the social sciences are needed to widen the research focus areas and make the research by the science centre's truly transdisciplinary.
- Uncertainties on how each centre has recruited the staff as well as the post-graduate students eligible for funding under the Global Change theme. This has resulted in the centres managers operating without the guidance of the NRF. The staff working in the centres, the managers, the post-graduates and the administration staff have also cited the lack of competitive remuneration, with the salaries being static over the last three years (Personal Communication with key informants at science centre 2016).

• There has been poor communication and consultation between the key role players i.e. the DST, the NRF and the science centres. Secondly, communication between the science centres, the NRF, DST and the CSIR which manages the SARVA has been lacking thus the collaboration between the centres and SARVA, has not been as effective as it should be. This has resulted in the centres not being able to feed local level data onto SARVA or assist local government in understanding the data on SARVA, which is one of their key deliverables. Currently, these two initiatives despite the numerous synergies between them are managed by two different units within DST.
• The lack of progress on this deliverable has been exacerbated by the lack of staff in the science centres that are knowledgeable on SARVA data and how to effectively disseminate it to users such as local governments. The incentives to take on these tasks that are considered to be additional to what the science centres are already doing are also lacking.

Integration of science centres into university processes and structures

• The integration of the science centres into the university processes and structures plays key role helping the centres achieve their objectives such as building capacity of students in knowledge production. Problems have emerged in some science centres as a result of where they are placed within the university structures which include a lack of control over the administration of the budgets and the procurement systems for the university which often results in bottlenecks, especially where the university systems are slow. Internal university systems and structures at times also cause frustration as they constrain the use of allocated resources in areas where the university considers are not a priority. This results in under spending of financial resources allocated to meet some of the key objectives.

4 Recommendations

While the three centres were meant to be pilot sites, more centres are earmarked for the rest of the historically black universities. To ensure the success of the current and new centres it is essential to note what has worked and what has not worked to date as a point of departure. The recommendations suggested are based on literature as well the gaps and opportunities identified in this document. These findings are preliminary and are intended to highlight areas that need further investigation and analysis.

- Develop a business model where the science centre managers have more autonomy over the day to day running and the budgets administered to ensure all objectives are met while also ensuring accountability for these budgets (public funds) are in line with treasury regulations.
- To improve the knowledge base and expertise of the academics involved in the centres, there needs to be a formalised mandate and clear articulation of the objectives and how these can be achieved. While there is appreciation that every institution is different, there should be common traits that need to the part of every institution, and consistency across the science centres.
- There is also a need to consider research incentives to academic staff especially in the social sciences to give more support to the science centre through transdisciplinary research collaboration and supervision of students.
- The programmes being run by the science centres need to align with the current research focus in other departments and the research needs/gaps in their communities for it to be relevant and useful to users such as local governments.
- The network of Risk and Vulnerability Science Centres will also benefit through knowledge sharing platforms between staff and students in the different institutions. This can be done either online or through contact sessions (workshops or conference). Other stakeholders involved in global change research (e.g. SARVA and ACCESS) can also be invited to attend to share knowledge on global change research in South Africa and beyond.
- In terms of the administration of the centres by the National Research Foundation, there is a need for more communication with the centre managers as well consultation, where changes are proposed to the programmes and

administration. To date, the management of the programmes has been very much of a top—down approach without much consideration of the university's administrative processes and governance structures which often constrain the effective running of the centre.
- There is a need to improve communication between the DST, NRF and the CSIR, which manages the SARVA, to ensure that all parties are on the same level of understanding on what is expected from the collaboration as well as the roles and responsibility of each actor.
- While collaboration currently exists between SARVA and the science centres, this needs to be streamlined to be effective. It might also be a good idea to have both SARVA and the Risk and Vulnerability Centres under one department within the Department of Science and Technology. Having the two initiatives managed by one unit in DST would enhance the effectiveness and efficiency of the centres to meets its mandate.

5 Conclusion

The two initiatives under the DST 's Global Grand Challenge, the Risk and Vulnerability Science centres and the SARVA have the potential to take transdisciplinary research and co-production of local knowledge to new levels. SARVA currently struggles to provide local governments and other users with local level information which can inform decisions and planning at local level. Science Centres can play a key role in bridging this gap and use a transdisciplinary approach to produce local level information that is informed by the knowledge gaps in those areas. The positioning of the centres, in the vast rural areas of South Africa present an opportunity for the centres to build the resilience of these areas where most of the country's poor population groups reside, and where the impacts of climate change will be severely felt. The universities where the centres are located have benefited from hosting the science centres as this profiles their research which is presented at several conferences nationally and internationally and highlights their importance as a contender in the co-production of knowledge. While more work and coordination is needed between the science centres and their key partners, the centres have been making their mark on the rural academic landscape. There is also need for the responsible national departments to facilitate and officiate the engagements between the science centres and lower levels of government (provincial, district and local municipality) so as the impacts of climate change continue to manifest and intensify affecting local populations the Risk and Vulnerability Science Centres, despite the challenges faced to date, are essential in building the resilience of local government and local communities to climate change. The appropriate institutionalisation of the centres within the academic setting, will determine the success or failure of the centres in the delivery of their mandate, given that the issue of institutionalisation brings with it a host of other problems facing the centres affecting the day to day running of the centres. Further, there is a need to streamline the role of the centres in the co-production of local level knowledge, while the preparation of terms of reference for the deliverables, setting up of timelines and targets for the centres will improve the efficiency of the centres and will allow the monitoring and evaluation on the performance and its ability to impact on local, district and provincial governments to use this information for planning and decision making. There is need for continued support of the centres so they can conduct research and support local governments in building the resilience of poor rural communities in South Africa against the adverse impacts of climate change.

Acknowledgements We would like to thank the substantial information that we received from the key informants at the science centres which was required to write this paper.

References

Audouin, M., 2011. Transdisciplinary Research for Sustainability: User Guide. Council for Scientific and Industrial Research (CSIR), Stellenbosch South Africa.

Burns, M.E.R., Audouin, M.A. and Weaver, A, 2006. Advancing Sustainability Science in South Africa. *South Africa Journal of Science* 102: 379–384.

Department of Environmental Affairs, 2007.Long Term Mitigation Scenarios. Pretoria South Africa.

Department of Environmental Affairs, 2011. National Climate Change Response Policy. Republic of South Africa.

Department of Environmental Affairs, 2013. Long Term Adaptation Scenarios Pretoria South Africa.

Dhansay T, Serper A, Linol B, Ndluvo S, Perumal L, De Wit M. Transdisciplinarity within South Africa's global change research: How (well?) are we doing? S Afr J Sci. 2015;111(5/6), Art. #a0107, 4 pages. http://dx.doi. org/10.17159/sajs.2015/a0107.

Gibbons, M., Limoges, C., Nowotny, H., Schwartzman, S., Scott, P., Trow, M., 1994. The New Production of Knowledge: The Dynamics of Science and Research in Contemporary Societies. SAGE, London.

Lang, D. J., Wiek, A., Bergmann, M., Stauffacher, M., Martens, P., Moll, P., Swilling, M., and Thomas, C.J., 2012. Transdisciplinary research in sustainability science: practice, principles, and challenges. Sustainability Science 7 (Supplement 1):25–43.

Leach, M., 2014. "Co-design for relevance and usefulness"- Q&A with Melissa Leach. Future Earth. Accessed online 21/09/2015 http://www.futureearth.org/blog/2014-jul-23/co-design-relevance-and-usefulness-qa-melissa-leach.

Leck, H. and Roberts, D., 2015. What lies beneath: understanding the invisible aspects of municipal climate change governance. Current Opinion in Environmental Sustainability 2015, 13: 61–67.

Mebratu, D., 2001. The Knowledge Dimension of the Sustainability Science Challenge. In: International Journal of Economic Development 3(1). Online: http://www.spaef.com/IJED_PUB/v3n1_mebratu.html.

Morin, M., 1999. Seven Complex Lessons in education in the future. United Nations Educational and Scientific Organisation. Paris, France. Online: http://unesdoc.unesco.org/images/0011/001177/117740eo.pdf.

Murambadoro, M., 2016. Assessing the role of learning systems in supporting local government climate change adaptation. Wits University PhD proposal.

Patel, Z., Greyling, S., Parnell, S. and Pirie, G., 2015. Co-producing urban knowledge: experimenting with alternatives to 'best practice' for Cape Town, South Africa. IDPR, 37 (2) pp 187–203.

Personal Communication with key informants at science centres, 2016.

Reyers, B., Nel, J.L., O'Farrell, P.J., Sitasa, N., and Nel D.C., 2015. Navigating complexity through knowledge coproduction: Mainstreaming ecosystem services into disaster risk reduction. PNAS June 16, 2015 Vol. 112:24 pp 7362–7368.

Swilling, M., 2014. Rethinking the science–policy interface in South Africa: Experiments in knowledge co-production. South African Journal of Science 2014; 110(5/6).

Srinivasan, G., Rafuisura, K.M, and Subbiah, A.R., 2011. Climate information requirements for community-level risk management and adaptation. Climate Research Vol. 47: 5–12, 2011.

University of Fort Hare. http://ufh.ac.za/centres/rsvc.

Walter Sisulu University. http://www.wsu.ac.za/waltersisulu/index.php/ravac.

Opportunities and Barriers for Research and Actions in Climate Change Adaptation in Tanzania

Nathalie Jean-Baptiste and Wilbard Kombe

Abstract Research in climate change adaptation in Tanzania has expended considerably over the last ten years. In Dar es Salaam, such research efforts include assessing the vulnerability of exposed communities in informal settlements, developing risks maps, investigating landscapes based storm water management for increasing climate resilience and supporting participatory decision-making and prioritization of actions on climate change adaptation. Often, scholars face high expectations and must generate knowledge and provide solutions in a short amount of time regardless of the challenges they face. This paper provides an overview of the opportunities and barriers for research and implementation of climate change adaptation in Tanzania. It pinpoints the necessity of overcoming traditional assessments that largely focus on the deficits and/or failures of urban systems in Africa. Instead, university scholars aim to provide context-based solutions through techniques such as urban morphology types, participative design charretes or multi criteria analysis to steer adaptation efforts. Yet, the available scholarly work comes with many challenges. The results show a need to manoeuver around rigid bureaucracies and reluctance to change, technocratic urban ideals, short time spent for project implementation and lack of reliable information. This suggests that high education institutions require other ways of advocating and communicating climate change adaptation and planning in rapidly urbarnising African cities.

Keywords Climate change adaptation · Tanzania · Universities · Opportunities · Barriers

N. Jean-Baptiste (✉) · W. Kombe
Institute of Human Settlements Studies, Ardhi University,
Off University Road, Kinondoni Municipality, Dar es Salaam, Tanzania
e-mail: Jean-Baptiste@Stadtforschung.Tu-Darmstadt.De

W. Kombe
e-mail: kombewilbard@yahoo.com

W. Leal Filho (ed.), *Climate Change Research at Universities*,
DOI 10.1007/978-3-319-58214-6_30

1 Introduction

Universities are generally regarded as lead institutions and agents of change in the process of social transformation and development. They are responsible for building a highly skilled workforce and a knowledge-based society. It is through higher education and research that knowledge creation and transmission can be meaningfully achieved towards advancing careers and improving quality of life. Universities are also responsible for meeting labour market requirements for highly skilled human capital. It is in Universities that theories are built and innovative ideas are developed as they have mandate to move societies towards a more enlighten path.

Manuel Castells (2001) argues that universities have historically played four major functions in societies. First traditionally, universities have been centres of training and producing highly trained and skilled human resources. Second, they are seen as pinnacles or platforms for the ideological agenda of societies, third, they are constituted for a socialization of dominant elites. And finally, universities have been generators of knowledge and innovation. This latter function however has been increasingly played by specialized non-university institutions (Brennan et al. 2004).

Upon independence, most African countries established universities mainly to help fight poverty, low production and unemployment, hunger, illiteracy as well as diseases (Mosha 1986). African universities then largely concentrated on training and in recent years, increasingly on research both basic and applied. In terms of championing social change and development, universities in most African countries have had limited impact. Nonetheless, they have somehow contributed in building the capacities of non-state actors including NGOs and civil societies. In turn, these have become particularly critical actors concerning citizenry, governance and democracy (Kombe 2012).

Challenges that have generally undermined the functions of African universities include poor dissemination, restrained intellectual autonomy, limited financial resources and ill-conceived research design. The latter includes the tendency to under-estimate the role of end-users in research design as well as lack of engagements with end-users in research execution. This in turn perpetuates a situation in which research results from universities are not effectively used; and thus remain marginal to contemporary societal challenges.

These problems not only remain dominant in African countries. There is evidence of limited communication of research results across the globe. This suggests that many universities are still today facing challenges with regards to fulfilling their role in supporting social transformation and development.

2 Selected Projects and Actions Addressing Climate Change Adaptation in Tanzania

Climate change (CC) research in Tanzania has expended over the past 10 years. Several Universities have built knowledge base through Master and PhD programs, research projects within large international consortiums and commissioned work from Tanzanian local government agencies. As a result data on mitigation and adaptation has increased significantly.

The six projects presented below address different problems. However, what these projects have in common is a focus on climate related impacts and adaptation actions. The choice of these was greatly influenced by the possibilities of generalizing not only the key findings but also the lessons learned from implementing them. A set of criteria was considered to redraw extreme of deviant cases that do not necessarily represent the typicality of research efforts in climate change adaptation in Tanzania:

- **Criteria I**. The selected projects needed to be information rich in relation to their policy implication. Indeed they needed to offer evidence to inform polices in both short and long term and provide data that could support decision making in a context of uncertainty and unclear parameters;
- **Criteria II**. The projects needed to be designed, implemented and completed largely by universities. This was important to determine how involved university scholars are in the provision of knowledge and how useful the science coming from the scientific community is in Tanzania;
- **Criteria III**. The cross-sector research collaboration and multidisciplinary approach were important factors to be considered. The projects selected need to flag more visibly the value of partnership across disciplines, but also with non-academic organizations; and
- **Criteria IV**. The projects needed to be relevant and applicable on the ground. Meaning that it needed to include details of arrangement of data sharing and exploitation of results by the communities themselves.

2.1 Project 1. Climate Change and Vulnerability in Africa (CLUVA)

The CLUVA project (www.cluva.eu) was an integrated effort between seven European institutions and six African universities, which included close to 60 experts in climate hazard, risk management, urban planning and governance as well as social vulnerability. One main aspect of CLUVA was to improve the capacity of African scientific institutions to produce knowledge on climate related impacts. The main goal was to develop methods and products—such as prototype software and applications—that can be readily used by the scientific community, policy makers and city administrators.

CC is likely to rapidly exacerbate development challenges in Africa, it threatens cities and fast urbanizing coasts where about 38% of Africa's population lives (Pauleit et al. 2015). Unemployment, inadequate standards of housing and impacts on human health are seen in many cities. Weather-related events are increasingly damaging to the existing critical infrastructure leaving water supply, energy, transport and telecommunication networks more vulnerable.

A central task in CLUVA was to assess the social vulnerability of exposed settlements in five cities (Saint Louis, Senegal; Ouagadougou, Burkina Faso; Douala, Cameroon; Addis Ababa, Ethiopia; and Dar es Salaam, Tanzania). The assessment was based on a framework that combines four main dimensions shown in Fig. 1.

- Asset vulnerability refers to the livelihood and resources of individuals or a community. Asset based assessment explores what resources are available to exposed groups;
- Institutional vulnerability addresses the state of local authorities and civil action groups that operate to prevent, adapt or mitigate the effect of extreme weather events;

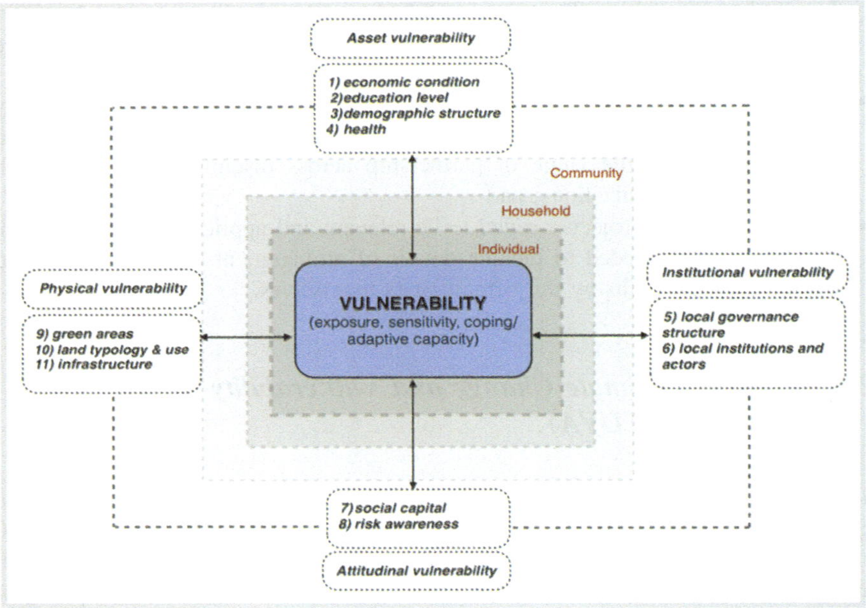

Fig. 1 CLUVA framework for assessing social vulnerabilities (Jean-Baptiste et al. 2013)

- Attitudinal vulnerability is related to the perception of risk, prevention attitude as well as previous hazard experiences from individuals or groups, assuming that such behaviour can foster different adaptation modes; and
- Physical vulnerability is closely related to anthropogenic causes and covers components such as the state of urban ecosystems, green infrastructure systems, land use and building conditions (Jean-Baptiste et al. 2013).

CLUVA findings show that vulnerability and exposure may increase significantly due to factors such as uncontrolled land development growth and the reduction of urban green areas. This implies that the overall climate change related risks will increase in the next decades. The project showed a way to approach these challenges in a scientifically sound and easily interpreted manner through maps, social vulnerability survey and multi risks analysis.

2.2 *Adapting to Climate Change in Coastal Dar es Salaam (ACC Dar)*

ACC Dar (http://www.planning4adaptation.eu) aimed to contribute to the operationalization of the National Adaptation Programme of Action (NAPA) in Tanzania. This was done through enhancing the capacity of Dar es Salaam municipal authorities in understanding CC issues specific to coastal areas, and in assessing their impacts on the livelihoods of coastal dwellers depending on natural resources. In addition, the project also intended to uncover and enhance knowledge on autonomous adaptive capacity. Such knowledge could potentially reveal weaknesses, gaps and possibilities for dealing with CC at municipal and community levels.

The coast of Dar es Salaam is increasingly urbanized. The city's coastal front features residential settlements, tourism infrastructure and commerce largely established in unplanned and un-serviced land and affecting natural resources and fishermen living from these. The concern surrounding this increased urban character is how to balance development in ways that are climate proof as well as socially and environmentally just.

With this in mind, three case study areas (Kawe, Kunduchi and Bunju wards) were studied. These are located along the most extended urbanized sea front of Dar es Salaam. ACC Dar found there is a strong central control and inadequate financing of local governments in matters related to CC. Indeed, Tanzanian local government authorities have little autonomy and limited resources to tackle adaptation, despite various reforms undertaken over the last 20 years (Kombe and Namangaya 2016; Shemdoe et al. 2015). In light of this, the project's activities included short courses, on-the-job exercises, joint identification and planning of adaptation activities and mainstreaming of climate change adaptation into municipal services. The knowledge and skills gained were used to initiate and support local communities to identify design and implement adaptation context relevant activities. Principal outcomes were: (1) Monitoring the quality and quantity of ground water sources in coastal areas. (2) Shared database and information on the challenges associated with increasing salinity of ground water sources.

ACC Dar was funded by the European Union and was jointly executed by Ardhi University in Tanzania and Sapienza University in Italy within a three years timeframe. The project targeted scientists, technical experts as well as federal and municipal government officers.

2.3 Climate Change Impacts, Adaptation and Mitigation in Tanzania (CCIAM)

CCIAM (http://www.suanet.ac.tz/cciam/) addressed the question of better management of natural resources through appropriate adaptation and mitigation strategies. The project focused on promoting natural forest conservation, afforestation, reforestation and better agricultural practices for improved livelihoods related to the Reduced Emissions from Deforestations and Forest Degradation (REDD) initiative.

Deforestation in Tanzania results in carbon emissions of approximately 100 million tons per year (CCIAM 2016). The causes of deforestation include over-exploitation, poor agricultural practices and illegal use of resources. CCIAM was conceived as a five-year program that regrouped a bundle of research projects, capacity building and student exchange between Tanzanian and Norwegian partners. Funded by the Norwegian Embassy in Tanzania, the projects had four specific objectives:

- To determine and develop appropriate climate change mitigation and adaptation strategies in forestry, other land uses, ecosystems and biodiversity management;
- To assess climate change impacts and the vulnerability of ecosystem services and livelihoods under REDD initiatives;
- To conduct policy and legal framework analysis of climate adaptation and mitigation with emphasis on economic efficiency, ecological effectiveness and wider political legitimacy; and
- Finally to develop and undertake capacity building, dissemination and strategic interventions for adaptation and mitigation to CC (CCIAM website 2016).

One aspect of CCIAM was to strengthen and disseminate information relevant to climate change impacts, adaptation and mitigation in Tanzania. For that effect, the program supported the establishment of a number of platforms such as (1) an open access repository (OAR), (2) an information resource centre, (3) Q&A services, (4) the CCIAM website and (5) a document delivery service to communicate the most updated information in Tanzania.

2.4 Partners Enhancing Resilience for People Exposed to Risks (Periperi U)

Periperi U (http://www.riskreductionafrica.org) initiated in 2006 and was conceptualized as a progressive and collaborative interdisciplinary African-led project. It focused on cross-sectorial and multi-level capacity strengthening and knowledge generation on disaster risks for the next generation of African professionals and academics (Malele 2009). Periperi U addresses disaster risks and climate variability, rapid population growth and urbanization, environmental and natural resources degradation as well as economic under-development. What makes this initiative unique is that it was embedded in higher learning institutions primarily as a measure to ensure sustained actions. Since 2006, Periperi U has grown to include 11 universities across Africa simultaneously developing and piloting postgraduate training programs, preparing and running short courses, and conducting strategic engagement with policy makers and practitioners.

During its last phase (2011–2015), Periperi U put particular emphasis on research and policy advocacy on urban and hydro-meteorological risks. Support was given for at least one to two short courses annually in disaster risk management, community based disaster risk management and food/livelihood security. In addition, Periperi U universities established units or programmes of either undergraduate and/or graduate modules related to managing risks and developing vulnerability profiles. Several research outputs were disseminated and are readily available online.

Periperi U's more recent input was the identification of weaknesses and challenges in existing institutional structures and policies in Tanzania. It was found that some prior policies excluded private sector actors, whilst other actors were more inclined towards reactive responses rather than being pro-actives. Thus review by and largely informed Ardhi University's research agenda as well as helped revised curricula for undergraduate and postgraduate studies in environmental science and management.

Other achievements were the development of short courses targeting practitioners and experts outside the University. In total 12 courses which trained over 120 professionals were conducted during the last 4 years. Periperi U also facilitated the increase of publications on risk management and CC, a total of 88 papers were produced over the 2011–2015 period. This further reflects the growing expertise in disaster risks domain that is growing in the continent.

2.5 Multi-criteria Analysis Application to Respond to Flooding at a Community Level Case of Dar es Salaam (MCA-Dar)

MCA-Dar (https://mcadaressalaam.wordpress.com) aimed at investigating the capacity of local authorities to consider flood risk reduction measures and climate adaptation in their local plans and budgets. A particular focus was given to decision-making processes and prioritization of action. To that effect, MCA-Dar took an action research approach where scientists from Ardhi and key government officials investigated the feasibility of identified adaptation and flood reduction action in Dar es Salaam.

Most planning and budgeting activities across Tanzania are consolidated at municipal levels. The challenge is that precisely at that level, CC knowledge and information is limited. This in turn makes yearly budget allocations that incorporate CC adaptation and removing the bottlenecks for the mainstreaming of adaptation action in municipal plans difficult.

With this in mind, MCA-Dar aimed at empowering government officials to integrate adaptation measures in their development activities. The project used a multicriteria analytical tool to identify a wide range of decision criteria and set objectives to assess and prioritize climate actions. The tool is known as Climact Prio,[1] which stands for CLImate ACTion Prioritization; it is a capacity building and decision support tool developed by the Institute for Housing and Development Studies (IHSS) in Rotterdam.

Through a series of meetings involving different steps, scientists and government officers were able to: (1) identify more clearly a set of feasible adaptation actions (e.g. improve roads with storm water drainage, tree planting, setting up early warning systems) (2) select a set of criteria through which these actions can be evaluated (e.g. public acceptance, flood reduction, cost), (3) score such actions using expert judgments and (4) weigh them based on priorities and level of importance. Figure 2 illustrates on the left an initial adaptation actions list before using Climact Prio and on the right the final results based on the systematic steps taken using the multicriteria analytical tool.

2.6 *Water Resilient Green Cities for Africa (WGA)*

WGA[1] is an ongoing water resilience project jointly implemented by the University of Copenhagen and two African universities: The Ethiopian Institute of Architecture, Building Construction and City Development (EiABC) and Ardhi University. WGA focuses on conservation of urban green and blue structures as a climate change adaptation strategy in Addis Ababa and Dar es Salaam. The project's objective is mainly to leap-frog towards water resilience using landscape-based green infrastructure approach to solve urban flooding downstream and water scarcity upstream. WGA's premise is that using landscape may

---

[1]Climact Prio is a climate awareness decision support and capacity building tool for assessment, screening and prioritization of climate adaptation actions. Developed at the Institute for Housing and Development Studies in Rotterdam the tools has been applied in Tanzania to support decision making and identification of flood reduction measures in Dar es Salaam. For more information about the tool see (REF. Stelios).

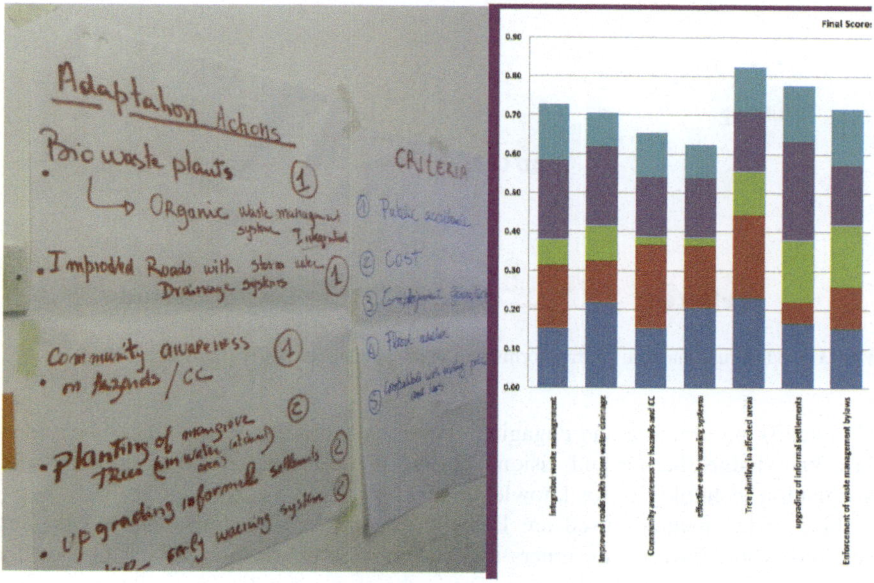

Fig. 2 Shifting from a random list of adaptation actions towards a set of best ranking actions through Climact Prio

'leap-frog' development of urban water services by means of eco-technologies which seems to be more affordable, more flexible and adaptable than conventional 'grey' technologies that African cities are currently using and/or rectifying. Funded by the Denmark Development Cooperation (DANIDA), African and European scholars have jointly prepared a framework, defined several case sites and engaged with important stakeholders in each city. A case catchment with three specific local sites has been selected. These sites feature different water stresses from water shortage, erosion and flooding and also exemplify different urban residential typologies and green spaces. The work done in WGA focused on improving water resilience by establishing a strong green infrastructure that can reduce the impact of floods and simultaneously provide day-to-day services to local residents who occupy the land (Mguni et al. 2015).

In the case of Dar es Salaam, water supply is a major concern upstream. The key concept for the Mbezi River catchment is a storm-water management train that spans across spatial scales from the building level to the river catchment level while targeting new green field developments in an increasingly dense suburban landscape (Fig. 3).

WGA addresses water resilience using stakeholder workshops, interviews, orthophotos and GIS analysis to generate knowledge on green space and its use, coping with water stress and current institutional structures. The project team has developed a vision, hypothesis and an integrated approach to urban water management in Addis Ababa and Dar es Salaam through: (1) drafting catchment plans,

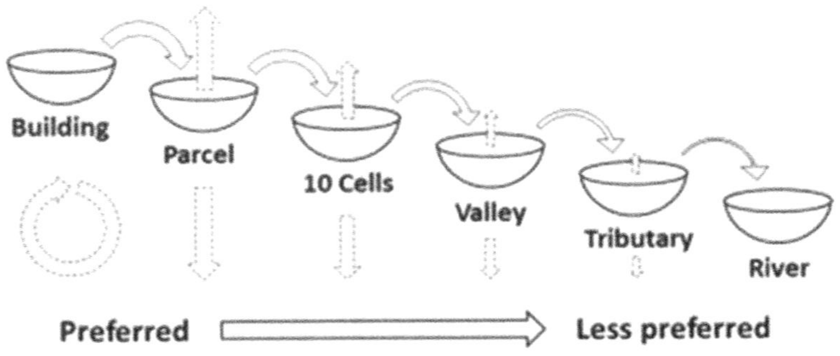

Fig. 3 Conceptual diagram of the stormwater management for MCR (Fryd et al. 2016)

(2) identifying, training and engaging champions among relevant city stakeholders in a knowledge sharing and vision building process, and (3) adopting smart dissemination technologies for knowledge sharing.

One of the methods used are Design Charettes. These are intensive planning sessions where citizens, designers and stakeholders work together on a vision for the development of the landscapes. These charettes were 3 days exchanges, which involve immediate feedback and critics to designers. The focus is to analyze critically how and when to substitute drainage for landscape green options and how to ensure the sustainability of such technologies. The project idea and initial results have also been disseminated to a broader audience by the Project Newsletter and News in the public media, as well as papers published and presented at the international conference "Global Challenges: Achieving Sustainability" (Table 1).

3 Opportunities and Barriers for Research and Implementation of Climate Change Adaptation in Tanzania

If we look at the body of research that has been undertaken by universities and other research institutions, many of which have addressed a wide range of development challenges, one can identify quite a number of innovative ideas that can make a difference in human livelihoods if implemented. Yet, reflecting upon the amount of research which has been undertaken on urban land management or informal/squatter housing since the 1960s, no one could have imagined that almost 60 years later, the bulk of urban inhabitants would still be living in informal settlements and informality would still continue to defy all norms.

In order to understand this paradox—which concerns not only the urban land or the housing sector but cuts across different sectors and issues including community service, environment, governance, and climate change—it is central to recognize

Table 1 Overview of research projects conducted on climate change adaptation at Ardhi University

No.	Project title	Time frame	Main topic	Outstanding methods	Multidisciplinary component
1	Climate Change and Urban Vulnerability in Africa (CLUVA)	2011–2013	Vulnerability and risk assessment, as well as CC adaptation in 5 cities in West and East Africa. Development of methods to manage climate, to reduce vulnerabilities and to improve the coping capacity at multi-scale (city, neighbourhood, household)	• Urban morphology types (UMT) • Social vulnerability survey • Building material tests	Integration of 2 opposite vulnerability perspectives coming from climatologists and hydrologists (CC community) and planners and sociologists (disaster risk and development community)
2	Adapting to Climate Change in Coastal Dar es Salaam (ACC DAR)	2011–2013	Operationalization of the National Adaptation Programme of Action (NAPA) in Tanzania. The project addressed the question of autonomous adaptive capacity and its integration into larger strategies for the development of un-serviced coastal settlements	–	
3	CCIAM (Climate Change Impacts, Adaptation and Mitigation in Tanzania)	2009–2014	Better management of natural resources through appropriate adaptation and mitigation Strategies. Dealing in particular with promoting natural forest conservation, afforestation and reforestation	• Forest carbon assessment tools, monitoring and verification	Integration of land use and farming systems for adaptation to and mitigation of climate change to support REDD initiatives
4	Partnership Enhancing Resilience for People Exposed to Risks (Periperi U)	2005 to date	Reduction of disaster risks in Africa through improved national and local disaster risk management, due to enhanced strategic human capacity to integrate risk reduction into critical developmental sectors and programs.	–	Pan African effort on disaster risk knowledge production stretching across 11 universities with a strong applied focus initiated by international organizations involved in Periperi U

(continued)

Table 1 (continued)

No.	Project title	Time frame	Main topic	Outstanding methods	Multidisciplinary component
5	Multi-criteria Analysis Application to Respond to Flooding at a Community Level Case of Dar es Salaam (MCA-DAR)	2014–2015	Enhancing the capacity of local authorities to consider flood risk reduction measures and climate adaptation in their local plans. Focus on prioritization and informed decision making	• Mental Mapping Using Ketso • MCA using Climact Prio	MCA-Dar's approach lies in the integration of economic tools to support decision-making efforts in flood risk reduction. In practice, the tools fostered multi-sector collaboration at municipality levels
6	Water Resilient Green Cities for Africa (WGA)	2013–2017	The project investigates the options for using landscape based storm water management (LSM) as a core element for increasing the climate resilience of larger African cities and for providing inclusive urban green spaces, and better local planning and governance. Vulnerable human settlements can be made more resilient by means of adaptive urban landscapes that can cope with changing precipitation patterns and the increased risk of floods and droughts	• Charettes • Orthophotos and GIS analysis	

Ketso is a participatory tool in a handbag that can be used for generating/exchanging ideas in group discussions. Originated in Lesotho and South Africa, the tool was developed in the mid 90s by Dr. Joanne Tippet. The word "ketso" means "action" in Lesotho and the tool itself was inspired by the work of scholars such as Freire (1968), Chambers (1994) and others. For more information see: Ketso http://www.ketso.com/

what have been key barriers to the utilization of research results as well as what opportunities exist to best harness climate change research.

The projects undertaken have revealed barriers and opportunities, which can be explored to enhance the utilization of research results.

Barriers

- **Weak or non-involvement of end-users in the research activities**: Participatory research engagement is not a very common practice in research engagement undertaken by universities in Tanzania. Where end-users participate, they are largely engaged to provide data, information or to solicit comments/inputs on the findings or recommendations. They therefore do not participate as partners. In the case of WGA, CCIAM, CLUVA end-users have extensively participated but not as partners.
- **Unknown specific targets of end-users**: Research results that concern issues such as climate change have many potential end-users. In case where research results are not targeted to specific end-users, the utilization of such research is restrained. This is particularly the case where implementation requires collective intervention of cross-sectoral actors working at different levels/scales. This was the case in CLUVA and WGA projects. However, attempts were made to share the findings with specific expected end-users. In WGA, design charrettes proved particularly useful tool for engagement and co-production of catchment strategy with end-users/local communities and professionals.
- **Weak or lack of skills to communicate research results to end-users (policy makers and practitioners)**: Too often research findings are put in technical writings (e.g. publication), which are either not easily understandable or rarely translated into implementable user-friendly forms. The situation is further complicated by the fact that funding agencies hardly cover costs beyond the production of research reports or journal articles. This means that most research results may at times not even be known or disseminated to end users, let alone be productive in a suitable form. In the case of the CLUVA project, much was done to involve potential end-users especially Local Government Authorities (LGA), however there were almost no hands-on output dedicated to the end-users or for practical use.

As highlighted above, research activities on CC are undermined by a number of impediments. These include structural and non-structural barriers that weaken research efforts undertaken by universities and other tertiary institutions. Structural barriers are mostly related to research designs and implementations, which feature little or non-involvement of end-users as well as untargeted research results with ill-defined beneficiaries. Non-structural barriers are rather related to weak capacities among researchers to facilitate user-friendly communication and the lack of skills to translate research results into practice and key messages for different target audiences. The barriers related with translating research into practice are becoming more sharply into focus as evidence based guidelines are being produced to support wider dissemination. There are also a number of opportunities for actions in climate change adaptation research.

Opportunities

- **Increasing appreciation from policy makers and practitioners:** Non-state actors have shown keen interest to adopt ideas generated in WGA and the Periperi U projects. Interests in the WGA project increased following the co-production of water management strategies using charrettes.
- **Advances in visual communication and ICT-based tools:** Technical advancements used to prepare 3D imageries and to simulate real conditions offer opportunities to enhance communication between researcher and end-users. They have proven to be effective in the co-production of strategies during the design charrettes under the WGA project.
- **Increasing awareness of impact and use of research results:** Increasingly calls for proposal are requiring applicants to clearly show how the research result will be utilized to inform policy or in real practical terms. This suggests that there is a shift towards more action orientated and applied research activities with a clear significance to transformation.

Funding agencies including local and international organizations are increasingly calling for research proposals and activities that can make a difference in the lives of local communities. There is also an interest in promoting the production of a more focused and engaging literature that can better inform policy makers. There are a few signs of change and improvement in the traditional one–way data sharing processes we often observed in workshop settings, thus other forms of learning are being experienced in Tanzania. In addition, practitioners and local communities alike are showing increasing interests to co-produce knowledge and strategies intended to address local problems. This opens up new opportunities for creating better links between research and practice in particular in ICT-based tools. As seen above, advances in 3-D visual communication skills are likely to be further applied in research.

4 Conclusion

One of the greatest successes of African Universities over the last ten years, arguably, is the increase of research activities on climate change and related disciplines. This paper has shown that in Tanzania and Ardhi University in particular, several research projects have been jointly undertaken, most of them with partners from the Global North. Yet, the role and impact of universities in championing social change and development remain limited despite efforts made in recent years. In climate change research, challenges which have generally undermined the functions of universities in many African countries include weak relationships between universities and local communities; poor dissemination and communication of research result to end–users; restrained institutional/intellectual autonomy; limited financial resources and ill-conceived research designs.

Table 2 Overview of partners involved in the six selected research projects

Projects	Institutions	Countries
Climate change and Urban Vulnerability in Africa (CLUVA)	• Analysis and Monitoring of Environmental Risks Scarl • The University of Manchester • Kobenhavns Universitet • Technische Universitat München • Ecole nationale superieure Polytechnique Universite de Yaounde I • Universite Gaston Berger de Saint Louis • University of Ouagadougou Burkina Faso • Norwegian Institute of Urban and Regional Research • Council for Scientific and Industrial Research • Euro-mediterranean Center of Climate Change • Helmholtz Centre for Environmental Research Leipzig • Ardhi University • Ethiopian Institute of Architecture	• Italy • UK • Denmark • Germany • Cameroon • Senegal • Burkina Faso • Norway • South Africa • Italy • Germany • Tanzania • Ethiopia
Adapting to Climatic Change in Dar es Salaam	• Ardhi University • Sapienza University	• Tanzania • Italy
Climate Change Impact, Adaptation and mitigation in Tanzania (CCIAM)	• Sokoine University of Agriculture (SUA) • University of Dar es Salaam • Ardhi University • Tanzania Meteorological Agency • Norwegian University of Life Sciences	• Tanzania • Tanzania • Tanzania • Tanzania • Norwegian
Partners Enhancing Resilience for People Exposed to Risks (Periperi U)	• University of Bab Ezzouar • Bahir Dar University • University of Ghana • Mio University • University of Antananarivo • Ahmadu Bello University • Gaston Berger University • Stellenbosch University • Ardhi University • Makerere University • Universidade Tecnica de Mozambique	• Algeria • Ethiopia • Ghana • Kenya • Madagascar • Nigeria • Senegal • South Africa • Tanzania • Uganda • Mozambique

(continued)

Table 2 (continued)

Projects	Institutions	Countries
Multi-criteria Analysis (MCA-Dar)	• Helmholtz Centre for Environmental Research—UFZ • Ardhi University • Institute for Housing and Urban Development Studies • Institute for Environment and Development Sustainability	• Germany • Tanzania • Netherlands • Tanzania
Water Resilient Green Cities for Africa (WGA)	• Institute of Human Settlements Studies (IHSS) at Ardhi University • Ethiopian Institute of Architecture, Building Construction and City Development in Addis Ababa University • The Department of Geosciences and Natural Resource Management (IGN) at the University of Copenhagen	• Tanzania • Ethiopia • Denmark

Tanzanian institutions are becoming more aware of these barriers and taking steps to rectify or overcome their limitations partly by taking advantages of the momentum of Climate Change research in Africa and the increasing appreciation from policy makers and practitioners to adopt ideas coming from research projects. In the last 10 years, Ardhi University has taken a pioneer role in training critical advocates for change in environmental governance. The opportunities coming in visual communication and ICT-based tools have fostered a new kind of relationship between researchers and end-users, new mapping and 3D technologies are enabling better communication between different groups and ultimately these allow more sustainable adaptation actions. The task at hand is to encourage a new generation of knowledge brokers that will have to navigate between advocacy and science to deliver design and ideas for a more equitable and climate-proof development in the years to come. The aforementioned opportunities offer an important window for African universities to tear down walls, which have tended to isolate them from real problems local communities are facing (Table 2).

References

Brennan J, King R, and Labeau Y, (2004) The role of Universities in the transformation of societies. An International Research Synthesis Report, Association of Commonwealth University, London.

Castells M, (2001) Universities as dynamic systems of contradictory functions. In: Muller J et al (eds) Challenges of globalisation. South African debates with Manuel Castells. Maskew Miller Longman, Cape Town, pp 206–223.

Chambers, R. (1994). Participatory Rural Appraisal (PRA): Challenges, Potentials and Paradigm. World Development, 22(10): 1437-1454.

Freire, P. (1968). Pedagogy of the oppressed. Seabury: New York.

Fryd, O., Backhaus, A., Adugna, D., Assefa, A., Jensen, M.B., Justin, J., Mguni, P., Mpyanga, S., Workneh, A., Yeshitela, K. and Herslund, L. (2016). Draft landscape-based stormwater management strategies for the Jemo and Mbezi River catchments. Water Resilient Green Cities for Africa, Report 2, Work package 2. Retrieved on 17 October 2017 from http://ign.ku.dk/english/research/landscape-architecture-planning/landscape-technology/water-green-africa/project-publications/filer/WGA_Report_2_WP2_2015_160601.pdf

Jean-Baptiste N, Kabisch S, and Kuhlicke C (2013) Urban Vulnerability Assessment in Flood-Prone Areas in West and East Africa. In: Rauch S et al (eds) Urban Environment. Springer, Heidelberg, New York, pp 203–218.

Kombe W J, (2012) The Role of Research in the National Social-Economic Development. A Paper presented at the Annual National Conference Organized by the Tanzanian Commission of Universities, Arusha, Tanzania.

Kombe W, Namangaya A, (2016) Decentralisation and Urban Governance: Trends and Lessons from Cities in Tanzania. In: Decentralisation and Regional Development. Springer International Publishing, Heidelberg, pp 71–88.

Malele B F, (2009) The contribution of ineffective urban planning practices to disaster and disaster risks accumulation in urban areas: the case of former Kunduchi quarry site in Dar es Salaam, Tanzania. In: Jàmbá: Journal of Disaster Risk Studies 2(1):28–53.

Mguni P, Herslund L, and Jensen M B, (2015) Green infrastructure for flood-risk management in Dar es Salaam and Copenhagen: exploring the potential for transitions towards sustainable urban water management. In: Water Policy 17(1):126–142.

Mosha H P, (1986) The role of African universities in national development. In: Higher Education 15(1):113–134.
Pauleit S, Coly A; Fohlmeister S; Gasparini P; Jørgensen G, Kabisch S; Kombe W J; Lindley S; Simonis I; Yashitela K; (eds) (2015) Urban Vulnerability and Climate Change in Africa—A Multidisciplinary Approach. Springer International Publishing, Heidelberg.
Shemdoe R, Kassenga G, and Mbuligwe S, (2015) Implementing climate change adaptation and mitigation interventions at the local government levels in Tanzania: where do we start?. Current Opinion in Environmental Sustainability. Elsevier Science, 13:32–41.

Adoption of Product and Process Eco-innovation for Developing Low Carbon Economy: A Rough Set Theory Based Analysis

Chiranjit Das and Sanjay Jharkharia

Abstract Adoption of different environmental policies and regimes to reduce the economic externalities has been mainly studied in institutional economics. Effectiveness of environmental management system has witnessed a positive relationship with economic performance in the literature. Therefore, institutions adopted several eco-innovation practices to gain long term sustainability. This paper addresses the relative importance's of different eco-innovation practices such as product and process eco-innovation; incremental and radical eco-innovations for improving eco-efficiency. Conditional attributes and decisional attributes criteria have been identified through the review of extant literature. Rough set theory based analytical model is applied for finding the most important innovation strategies and their relative importance. Industry experts' opinions have been collected through telephonic interview. Product stewardship, Life Cycle Analysis (LCA) of product innovation, cleaner technology, and radical product eco-innovation are the most important strategies for developing low carbon economy. These results will be useful for organizations to adopt the effective eco-innovation strategies for improving eco-efficiency.

Keywords Eco-innovations · Eco-efficiency · Rough set theory

1 Introduction

Global warming has been a serious issue for the world. Serious consequences of global warming have already been witnessed by human beings. Increasing sea level, increasing surface temperature, shrinking volume of iceberg in Antarctic region, frequent draught and cyclone, changing eco-system, and acidification of water are some serious implications of global warming (IPCC 2014). Intergovernmental

C. Das (✉) · S. Jharkharia
Department of Operations Management, Indian Institute of Management,
Rohtak, M. D. University Campus, Rohtak 124001, Haryana, India
e-mail: fpm02.003@iimrohtak.ac.in

W. Leal Filho (ed.), *Climate Change Research at Universities*,
DOI 10.1007/978-3-319-58214-6_31

Panel on Climate Change (IPCC) has assessed that the surface temperature has already been increased by 0.5 °C in comparison to the 1950 level and it will be increased further up-to 2 °C by the end of the 21st century. IPCC has also assessed that the increasing concentration of anthropogenic greenhouse gases is the main reason for global warming and human influence on climate change is clear (IPCC 2013). To combat these daunting challenges of climate change several nations, non governmental bodies, and intuitions such as United Nation Federation on Climate Change (UNFCC) and IPCC were signed an agreement, called Kyoto protocol in 1997, to restrict the emissions level at least 5% by the end of 2012 in comparison to the 1990 level.

Furthermore, IPCC has adopted several adaptation strategies to reduce these human induced environmental impacts such as circular economy, cleaner development program (CDP), and joint implementation etc. But, only these strategies will not be effective until environmental technological innovations or eco-innovations are not being promoted. Jänicke (2008) defined eco-innovation as an ecological modernization and also it needs marketable technological solution while at the same time it will reduce environmental impact. EU has already invested 185 eco-innovation projects through their EU's Entrepreneurship and Innovation Programme (EIP) to make eco-innovation marketable (EACI/ECO/2013/001 2016). These projects include greening business, recycling, water, building and construction, and food and drinks (EACI/ECO/2013/001 2016). European Commission (EC) has assessed that climate change mitigation strategies need co-operation among the technology, economics and social system. Hence, organizations have adopted different eco-innovation practices and strategies to address degraded environmental impact.

On the other hand, it is also found that the different product innovations perform well when the product is in growth stage of its life cycle and also improve organization's financial performances by increasing market share and sales volume; whereas, the process innovation strategies perform well when the product is in mature stage of its life cycle (Damanpour and Gopalakrishnan 2001). Furthermore, radical innovation strategies performed well in long-term (Dewar and Dutton 1986), while, incremental innovation strategies performed well in short-term (Dewar and Dutton 1986). Therefore, organization's innovations decisions have relationships with firm's economic performance as well as it require firm's financial and technical capabilities. But, little efforts have been given to analyze different eco-innovation strategies and their impact on the firm environmental performances. Moreover, organizations will adopt more number of eco-innovation strategies if they found substantial environmental and economic benefits out of adoption of eco-innovation strategies. Therefore, in this paper, an effort has given to analyze firm's adoption of different eco-innovation strategies such as product and process; radical and incremental for improving environmental performances, more specifically, eco- efficiency. Furthermore, The Organisation for Economic Co-operation and Development (OECD) (2015) defined low carbon economy as "Establishing a resource-efficient economy that improve resource productivity, sustainable natural resource, help to improve the environment, sustain economic growth by securing adequate supplies of materials, investing in new technologies and innovation, and improving competitiveness" (OECD/IEA/NEA/ITF 2015). Therefore, eco-efficiency is used as a proxy for low carbon economy in this paper. Eco-efficiency means reduction in resources requirement, energy requirement, wastes generation, and improvement in material recyclability (Worrell et al. 2009).

Therefore, contributions of this paper are twofold: firstly, this paper find an analytical empirical relationships between different eco innovation strategies such as product, process, incremental, and radical eco-innovation to improve eco efficiency; secondly, the relative importance of these strategies are identified through an analytical rough set based methodological model. To find the major important eco-innovation strategies and their relative importance for improving eco-efficiency, industry experts' responses from Indian manufacturing firms have been collected through telephonic interview and rough set based analytical model is applied for the analysis. Rough set theory is a mathematical analytical model for finding inter-relationships, data reduction, and identification of hidden patterns. By analysing collected data through rough set theory, finally, a set of rules are developed. These rules provide relative importance of eco-innovation. Managerial implications of this study are discussed in discussion section. Detailed methodological steps are provided in subsequent sections. Rest of paper is structured as follows: Sect. 2 provides detailed literature review on environmental economics, innovation economics, and eco-innovation; Sect. 3 provides detailed theoretical relationship between eco-innovation and eco- efficiency; and Sect. 4 provides detailed methodological analysis of data through rough set theory. Detailed discussions and conclusions are also provided in the Sects. 5 and 6.

2 Theoretical Background

This section describes the underline theories behind the environmental economics, innovation economics and eco-innovation.

2.1 Environmental Economics

Environmental economics has been studied since long years ago. Early in 1960's, these studies have mostly focused on the regulations, policy regimes, development of environmental instruments, and environmental impact assessment (Cropper and Oates 1992). In particular, these instruments regulate the natural outcome of the economic transaction on the air, water, soil, and other natural resources. These environmental externalities are the natural outcome of economic transaction. A firm uses natural inputs of the environment and generates valuable outputs. In the process of doing this at the same time it spills over into the water, air, and soil. Surroundings of the firm are victimized by these spills over. An organization has a little interest to neutralize externalities until institutions will impose some strict rules and regimes as it incurs cost and hamper the economic benefits of the firm (Elbasha and Roe 1996; Bithas 2011). In addition to that, there has been a debate on the effectiveness of free market economy and regulated market economy; whether the free market will perform well in comparison to the regulated market economy. Later on a number of analytical tools have been developed such as Cost benefit analysis, Life Cycle Analysis (LCA), life cycle costing, and Input- Output analysis (Curran 1996; Hendrickson et al. 1998; Atkinson and Mourato 2008). Generally, cost benefit analysis is carried out to assess the economic aspects of the environmental regulations (Atkinson and Mourato 2008), whereas the lifecycle analysis is carried out to assess the impact of the products and services throughout its useful life (Curran 1996; Hendrickson et al. 1998). Hence, these analytical tools are very important assessment techniques for the development of low carbon economy.

2.2 *Innovation Economics*

Oslo manual defines innovation as "The implementation of a new or significantly improved product (good or service), or process, a new marketing method, or a new organizational method in business practices, workplace organization or external relations" (Manual O 2005). Therefore, innovation economics is primarily economic analyses of different factors regarding innovations. These factors include market pull, technological characteristics, structural characteristics of organization, and product characteristics etc. (Cooper 1998). It is not always true that all innovation will achieve economic success or every organization will be succeeded by adopting same innovation. An innovation must fulfilled the present needs of the consumer through providing innovative product or services, only then it will be economically successful and adopted by others in its diffusion stage (Abernathy and Utterback 1978). Technological characteristics push the innovation in large extend, for example an organization can gain high productivity, reduction in wastes, and high quality of products etc. by adopting the highly technological efficient equipment and changing process technology (Cooper 1998). Organizational characteristics are also having an important role while adopting innovation strategy. Highly decentralize organization can be very efficient for innovation whereas highly centralize organization may be less effective for the same (Ettlie et al. 1984). Innovation needs firm's research and development capability and also needs investment. Therefore, a large organization has enough R&D and investment capability to deploy in innovation in comparison to a small organization (Ettlie et al. 1984). Another important factor of innovation is product characteristics. This means type of product whether it is an innovative product or functional product. If it is an innovative product, more radical changes will be effective; whereas, for the functional product incremental innovation will be more effective because efficiency is the main objective for functional product (Dewar and Dutton 1986; Damanpour and Gopalakrishnan 2001).

2.3 Eco-innovation

European commission defines eco-innovation in MEI project as "The production, assimilation or exploitation of a product, production process, service or management or business method that is novel to the organization (developing or adopting it) and which results, throughout its life cycle, in a reduction of environmental risk, pollution and other negative impacts of resources use (including energy use) compared to relevant alternatives (Kemp and Pearson 2007)". Therefore, eco-innovation is an innovation focusing on environmental aspects. Early studies on eco- innovation focused on end-of-pipe technologies. Moreover, it was primarily pollution prevention system or processes at each and every process to reduce pollution discharges either continuously by changing in the equipment or changing in the process (Frondel et al. 2004). It is also noted that by changing in product or process design, environmental impact can be reduced. Similarly, product stewardship is an environmental impact analysis of product throughout its lifecycle, from material extortion stage to end of life of a product (Machiba 2010; Hellstrom 2007). Besides these organizational aspects of eco-innovation, there are also other aspects of eco-innovation such as social system, co-evolutionary environmental and institutional aspects, but, these are out of scope of this study (Ekins 2010; Foxon 2011; Demirel and Kesidou 2011). Several eco-innovation strategies are elaborately discussed below.

2.3.1 Adoption of Product and Process Eco-innovation

The relationships of product and process innovation with the overall organizational competitiveness i.e. market share, sales volume, cost efficiency, product and process quality, have been studied so far in innovation economics (Utterback and Abernathy 1975; Dewar and Dutton 1986; Damanpour and Gopalakrishnan 2001). Product innovation is relatively more important in comparison to the process innovation for enhancing firm's competitiveness (Damanpour and Gopalakrishnan 2001). But, product and process eco-innovation has not received substantial attention of academicians. Product eco-innovation means capability of product design that reduces the environmental impact including product stewardship, life cycle analysis of product, and product design for reduce, recycle and recover (Rennings 2000). Product stewardship is the product design capability that reduces environmental impact by minimizing the use of less toxic material, redesigning the existing product with less material consumption, minimizing use of non-renewable material (Hart 1995). Therefore, it can reduce material consumptions and process time that leads to improvement in eco-efficiency. On the other hand, Life Cycle Assessment (LCA) is an environmental impact analysis of a product from its raw material extraction stage to end of use lifetime of the product. Hence, an organization can make necessary changes in product design, substitute alternative material that leads to reduce the environmental impact as well as less material consumptions

(Curran 1996). Therefore, eco-efficiency can be improved by adopting LCA methodologies during product design stage. Similarly, eco-efficiency can also be substantially improved by adopting product design for reduce, recycle and recover (Herring 2006). Reduction in material consumptions, minimization of wastes, and improvement in process can be possible by using material that recyclable, changing in product designs that consumes less material, and reduces wastes during production. Thus, above mentioned product design capabilities can reduce material and energy consumption and also improve product and process quality.

Similarly, environmental impact can be reduced by changing in the process technologies such as cleaner technology, end-of-pipe control, and pollution prevention (Klassen and Whybark 1999; Klassen 2000; Frondel et al. 2004; Ekins 2010). Cleaner technology means the process that uses eco-efficient equipment which requires less energy, produces more output, reduces environmental impact, and management practices for reducing environmental impact (Frondel et al. 2004; Unnikrishnan and Hegde 2007). Therefore, an organization can reduces material and energy consumption, enhance product and process quality by adopting these cleaner technologies. On the other hand, end-of-pipe control means strict control over pollution, controlling toxic water discharge of each and every process, for example, installing air filtration equipment and water treatment plant that generates less pollution (Frondel et al. 2004). Hence, organizations can control over pollution and emissions by adopting these end-of-pipe technologies. Besides these, pollution prevention is the more proactive environmental management strategies of firm. This defines as changing in fundamental process technologies, controlling over the wastes generation, and some housekeeping strategies that reduces the pollution (Klassen 2000; Ekins 2010). Thus, organization can improve their eco-efficiency, product and process quality by adopting eco-innovation strategies. Details of these eco-innovations strategies are summarised in Table 1.

2.3.2 Adoption of Incremental and Radical Eco-innovation

Based on the depth of knowledge, innovation can be classified into two broad categories: radical and incremental innovation. Radical innovation defines as remarkable changes in product and process that needs high depth of knowledge, higher organizational capability and resources (Hellstrom 2007). Whereas, incremental innovation is a continuous process through which gradual changes in product and process can be possible and it needs low depth of knowledge, and lesser organizational capability. Empirical studies posit that radical product innovations outperform over incremental product innovation and incremental process innovation outperform over radical process innovation (Ettlie et al. 1984). This paper considers four innovation strategies regarding radical and incremental innovation, such as radical and incremental product eco- innovation and radical and incremental process eco-innovation. Resource efficient radical product innovation can significantly improve eco-efficiency while pollution prevention incremental process innovation can effectively reduce environmental impact (Hellstrom 2007; Ekins 2010).

Table 1 Description of questionnaire items

Questionnaire items	Descriptions	References
Product stewardship (P1)	It is the product design capability that reduces environmental impact	Hart (1995)
Life Cycle Analysis of product (P2)	It is an environmental impact analysis of product life cycle	Curran (1996)
Product design for Reduce Recycle and Recovery (P3)	Reduction of material consumption and use of recyclable material	Rennings (2000)
Cleaner technologies (M1)	Use of resource efficient equipment	Frondel et al. (2004), Unnikrishnan and Hegde (2007)
End of pipe control (M2)	Pollution control on every process output	Klassen (2000), Ekins (2010)
Pollution prevention (M3)	Proactive pollution control mechanism	Hellstrom (2007)
Radical changes in resource efficient process (R1)	Significant changes in production process	Hellstrom (2007)
Radical changes in resource efficient product design (R2)	Significant changes in product design for less environmental impact	Hellstrom (2007)
Incremental changes in pollution prevention process (I1)	Continuous but small changes in process	Ekins (2010)
Incremental changes in less environmental impactful product design (I2)	Continuous but small changes in product design	Ekins (2010)
Improvement in eco-efficiency (D1)	Measurement for resources conservation	Patterson (1996), Herring (2006)
Improvement in product and process quality (D2)		Abernathy and Utterback (1978)

3 Methodologies

In this section, we discuss the formal methodology that is being followed for the analysis of this study. Empirical studies, case study based research, methodological descriptions, and analytical theoretical based studies have been dominating methodologies in the Eco-innovation studies (Rennings 2000; Hellstrom 2007; Ekins 2010; Machiba 2010). For example, Cleff and Rennings (1999) empirically investigated determinants and policy instruments of eco-innovation; Hellstrom (2007) theoretically described several dimensions of eco-innovation; and Machiba (2010) analytically described eco-innovation and used cases to describe eco-innovation practices. Formal methodology based analytical model has not sufficiently carried out in this domain so far. In addition to that, the objective of this study is to find relative importance of product and process eco-innovation for improving eco-efficiency. Therefore, analytical model based methodology is adopted for addressing the issues of this study. A set of analytical tools have found in literature such as Analytical Hierarchy Process (AHP), Analytical Network Process (ANP), and TOPSIS etc. But, these tools have some limitations inherent in nature. For example, inter-relationships among the criteria cannot be possible through AHP besides the advantage of its ease of interpretation. While, ANP considers the interrelationships but identification of relative importance of the criteria will not possible through ANP. These formal multi-criteria- decision making (MCDM) models will give only the crisp results for the decision making process, formal rules cannot be possible to develop through these methodologies. A rough set theory based analytical method is best suited to find inter relative rules that related to the objectives of our study. Hence, relative importance of product and process eco-innovation for improving eco- efficiency can be developed through rough set theory. Another approach is also suitable for finding the rules from the data that is fuzzy inference system (FIS) but this need a membership function and interpretation required complex computational expertise. Advantages of rough set theory for data analysis are as follows: it does not required any preliminary and additional information regarding data such as probability distribution and membership function, it is easy to understand, it generates set of decision rules from data, and it does not required any computational complexities for analysis (Pawlak 1982). Due to these advantages, rough set theory based analytical model excessively used in data analysis, data reduction, data mining, and multi criteria decision making model in recent times (Bai and Sarkis 2010; Shyng et al. 2007; Pawlak 2002; Pawlak 1997, 1998). Besides these advantages some inherent limitation of rough set theory is there such as decisions rules are formed based on logical approximation of rough set but rules or propositions will be more robust if it will be developed through mathematical topology. However, this limitation can be avoided as logical approximation is also valid interpretation for the managerial decision making. Hence, a rough set theoretical model is adopted in this study. Bai and Sarkis (2010) have also applied the similar method in their analysis. For finding relative importance, industry experts' opinions have been collected through telephonic interviews of the manufacturing and production manager of Indian manufacturing firms. Detailed questionnaire items are provided in Table 1. Interviews were taking over an hour approximately and responses were collected after carefully explaining all the items of the questionnaire so that respondents' biasness can be avoided. Respondents are from wide range of industries including textile, steel, thermal power, oil and gas, paper, cement, and heavy engineering etc. Details of all respondents are provided in the Appendix. Before going to the detailed analytical model of rough set theory through illustration of collected data, some important notes on rough set theory are explained below.

3.1 Introduction to Rough Set Theory

Rough Set Theory (Pawlak 1982, 1998), is a mathematical approach for data analysis and data mining when data are ambiguous and vague in nature. Vagueness of data is found in many real life cases where data are not precisely defined. For example, in this study, relative importance of adopting product and process innovation for improving eco-efficiency are not precisely defined, it can be defined as very low to very high in relative terms. Vagueness can be defined in rough set theory through lower and upper approximations of rough set. Lower approximation defines as the set of objects which definitely belong to the subset of interest and upper approximation defines as the set of objects that can possibly belong to the subset of interest. The difference between these two, upper approximation and lower approximation, is called boundary region. This boundary region is implies to the vagueness of data. If the boundary of region is null, the data will be crisp data, otherwise the date will be rough. By analyzing the relative dependency of rough set, Reduct can be developed, that means we can eliminate some data from the main universal set without changing the consistency of information system. Meaningful rules can be developed from these Reduct that are useful for management decision making process. In this study, these final meaningful rules are implied to the relative importance of adoption of product and process innovation. Some important definitions of rough set theory are provided below that will be useful for further analysis.

Definition 1 Let us suppose U be a universal set of non-empty object and X be a subset of U having R equivalence relation of set of objects if and only if the all attribute values are equal. The main issues of rough set theory are the R approximation of set for the subset X ∈ U; the lower and upper approximations are define as

$$\underline{R}X = \{X \in U | [X]_R \subseteq X\} \quad (1)$$

$$\bar{R}X = \{X \in U | [X]_R \cap X \neq \Phi\} \quad (2)$$

$$BND_R(X) = \bar{R}X - \underline{R}X \quad (3)$$

$\underline{R}X$ defines as lower approximation of X, means the set of objects that can be certainly defined by the set of interest, whereas $\bar{R}X$ defines as upper approximation of set X, means the set of objects that can be possibly defined by the set of interest. $BND_R(X)$ is the boundary region of set X, if $BND_R(X) = \Phi$, the set will be crisp, and if $BND_R(X) > \Phi$, the set X will be rough. $\underline{R}X$ also defines as positive R relation of set X ($POS_R(X)$) and ($U - \bar{R}X$) defines as $NEG_R(X)$, negative R relation of X.

Definition 2 Let S = (U, C) is a complete information system. Information content or entropy of attributes set C is defines as (Liang and Shi 2004)

$$I(C) = 1 - \frac{1}{|U|^2} \sum_{i=1}^{|U|} |Xi| \tag{4}$$

|U| is the cardinality of universal set, |Xi| is the cardinality of indiscernible set of attributes C. Indiscernible set means the set of Xi that are not discernible in terms of conditional attributes. Expression 5 will shows the indiscernible binary relation of set.

$$\text{SIM}\ (C) = \{(i,j) \in U \times U | \forall\ c \in C, c(i) = c(j)\} \tag{5}$$

This means two sample i and j are the member of the same set if and only if c (i) = c (j), for every elements of c ∈ C, means each and every elements of conditional attributes set is equal. Xi is defined as the set of $\{i \in U|\ (i,j) \in \text{SIM}\ (C)\}$.

Definition 3 The significance of information of conditional set c defines as

$$\text{Sig}\ C \backslash \{c\}(C) = I(C) - I(C \backslash \{c\}) \tag{6}$$

$I(C \backslash \{c\})$ is the information entropy without considering the conditional attribute c. Only positive Sig $C \backslash \{c\}(C)$ of conditional attributes can be considered as a core conditional attribute set. Thus, we can form a set of Reduct and further find the relational rules from the reduced conditional attributes set.

3.2 *Illustration of Rough Set Theory Through Industry Experts' Responses*

In this section detailed illustration of collected data through rough set theory to indentify the relative importance of product and process eco-innovation for improving eco-efficiency is provided. A structured methodology has been carried out for the analysis. Firstly, information system table is developed considering the industry experts' responses on several eco-innovation strategies. Then, elementary set of indiscernible relation, significance of each criterion, Reducts, meaningful rules, and final relational rules are discussed in details. Each methodological step is elaborately illustrated below.

3.2.1 Introduction to Information System Table

Firstly, information system table is formulated (Table 2). In our study, U = {S_i, i = 1, 2 … 20} is the universal set i.e. 20 expert responses on the relative importance of total ten conditional attributes C = {C_j, j = 1, 2 … 10}, for each conditional attributes there are the two decisional attributes D1 and D2. Conditional attributes set consist of three product based eco-innovation attributes i.e. product stewardship (P1), Life Cycle Analysis (LCA) of product (P2), and product design for reduce recycle and recovery (P2); three process based eco-innovation attributes set i.e. cleaner technologies (M1), end of pipe control(M2), and pollution prevention(M3); two radical eco-innovation i.e. radical changes in resource efficient process (R1) and radical changes in resource efficient product design(R2); and two incremental eco-innovation attributes set i.e. incremental changes in pollution prevention process (I1) and incremental changes in less environmental impactful product design(I2). Two decisional attributes are improvement in eco-efficiency (D1) and improvement in product and process quality (D2). Details of these conditional attributes set are provided in Table 1.

Table 2 shows the complete information system data. In the Table 2 twenty respondents responded to the 10 conditional attributes and two decisional attributes that have explained above. Expert responses have been collected for conditional

Table 2 Information system table of industry expert's responses

Responses	Conditional attributes (C)										Decisional attributes	
	P1	P2	P3	M1	M2	M3	R1	R2	I1	I2	D1	D2
1	M	H	VH	L	VL	M	H	VH	VH	VH	H	M
2	H	M	M	H	H	M	M	M	H	H	M	H
3	M	M	H	H	H	H	H	M	M	M	H	M
4	M	H	H	H	H	H	H	M	M	M	H	M
5	H	H	VH	VH	M	VH	VH	H	VH	M	H	H
6	VH	M	M	M	M	VH	M	M	M	M	H	H
7	H	H	H	H	H	H	H	M	M	M	H	M
8	VH	H	VH	VH	H	VH	VH	VH	VH	VH	H	H
9	M	M	H	M	H	H	H	M	M	M	H	M
10	H	VH	M	H	H	M	M	M	H	H	M	H
11	H	VH	H	M	M	H	VH	VH	H	H	H	H
12	L	M	M	M	M	L	M	M	M	L	L	M
13	M	H	L	M	H	H	H	H	VH	VH	H	H
14	H	H	H	H	H	H	H	H	M	M	H	M
15	VH	M	VH	H	M	H	H	H	H	VH	H	H
16	VH	VH	VH	VH	VH	VH	H	H	M	M	H	H
17	M	H	H	H	H	H	H	M	M	M	H	M
18	M	H	M	VH	VH	VH	VH	M	H	H	H	H
19	H	H	M	M	M	M	H	M	H	L	H	M
20	H	H	M	H	M	H	H	H	H	H	H	H

Note Representation of notation are as follows: *VL* Very low, *L* Low, *M* Moderate, *H* High, *VH* Very high attributes on a five point LIKERT scale, i.e. VL = Very low, L = Low, M = Moderate, H = High, VH = Very high and for decision attributes on a three point scale i.e. L = low, M = Moderate, and H = High. For example, f(5, M1) = VH, which means the respondent 5 responded very high to the relative importance of cleaner technologies (M1) for improving eco-efficiency and product and process quality. For decisional attributes, respondents were asked to respond on relative impact on these two decisional attributes by adopting above mentioned conditional attributes. For example, f (1, D1) is H, which means as per respondent 1 opinion impact of these conditional attributes on improvement of eco-efficiency is high. Similarly, all data point of Table 2 can be explained.

3.2.2 Determination of Elementary Sets and Overall Information Content

Rough set theory eliminates superfluous data that are not important to make conclusion from the data without changing consistency of the data. In this step, we find the indiscernible set by using Expression 5 and build an elementary set of indiscernible set which is shown in Table 3. In the data set (Table 2), responses 4 and responses 17 have indiscernible binary relation that means each and every element of these two responses is same. Rest all other responses are unique in nature. Hence, responses 4 and 17 have indiscernible binary relation that forming one elementary set i.e. Type 4 (Table 3) and cardinality of $|X_4|$ and $|X_{17}|$ are 2. Cardinality values of all elementary set and the responses included on the particular set are provided in Table 3. Thus, the overall information content or entropy of the information system ($I(C)$) is calculated by using the Expression 4 and the dataset of Table 3, explained below:

$$I(C) = 1 - \left(\frac{1}{|20|^2}(1+1+1+2+1+\cdots+2)\right) = \left(1 - \frac{22}{400}\right) = 0.945$$

3.2.3 Determination of Significance of Each Conditional Attributes, Core Attributes, and Reducts

In this section, significance of each and every conditional attributes is calculated by using the Expressions 4 and 6. For example, in case of attributes P1, $I(C \setminus \{P1\})$ will be calculated by using Expression 4 and data of information system Table 2 excluding the data of column P1. If we exclude data set of the column P1, we will find another response that is equivalent to response 4 and 17 is response 7.

Table 3 Elementary set of indiscernible relation

Elementary set type	Responses included	Number of responses in set |X|	Elementary set type	Responses included	Number of responses in set |X|
Type1	Response 1	1	Type11	Response 11	1
Type2	Response 2	1	Type12	Response 12	1
Type3	Response 3	1	Type13	Response 13	1
Type4	Response 4, Response 17	2	Type14	Response 14	1
Type5	Response 5	1	Type15	Response 15	1
Type6	Response 6	1	Type16	Response 16	1
Type7	Response 7	1	Type17	Response 18	1
Type8	Response 8	1	Type18	Response 19	1
Type9	Response 9	1	Type19	Response 20	1
Type10	Response 10	1			

Hence cardinality of response 4, 7, and 17 are respectively $|X_4| = 3$, $|X_7| = 3$, and $|X_{17}| = 3$ and calculation of $I(C\backslash\{P1\})$ is explained below

$$I(C\backslash\{P1\}) = 1 - \left(\frac{1}{|20|^2}(1+1+1+3+1+\cdots+3+3)\right) = \left(1 - \frac{26}{400}\right)$$
$$= 0.935$$

$I(C)$ has already been calculated and the value is 0.945. Hence, significance of conditional attributes P1 is $\text{Sig } C\backslash\{P1\}(C) = I(C) - I(C\backslash\{P1\}) = 0.945 - 0.935 = 0.01$. Similarly, we calculate significance value of every attributes set. The other positive significance values of conditional attributes are P2, M1, and R2; and the values are respectively, 0.005, 0.005, and 0.005.

The attributes are considered to be core attributes set if and only if the $\text{Sig } C\backslash\{c\}(C) > 0$ i.e. significance of attributes must be positive for considering core attributes set. Therefore attributes P1, P2, M1, and R2 will be considered as core conditional attributes set. The core set will be considered as Reduct of information system if and only if I(Core(c)) = I(C). By using the Expression 4 and

Table 4 The reduced core attribute matrix

Responses	Conditional attributes (C)				Decisional attributes	
	P1	P2	M1	R2	D1	D2
1	M	H	L	VH	H	M
2	H	M	H	M	M	H
3	M	M	H	M	H	M
4	M	H	H	M	H	M
5	H	H	VH	H	H	H
6	VH	M	M	M	H	H
7	H	H	H	M	H	M
8	VH	H	VH	VH	H	H
9	M	M	M	M	H	M
10	H	VH	H	M	M	H
11	H	VH	M	VH	H	H
12	L	M	M	M	L	M
13	M	H	M	H	H	H
14	H	H	H	H	H	M
15	VH	M	H	H	H	H
16	VH	VH	VH	H	H	H
17	M	H	H	M	H	M
18	M	H	VH	M	H	H
19	H	H	M	M	H	M
20	H	H	H	H	H	H

Note Representation of notation are as follows: *VL* Very low, *L* Low, *M* Moderate, *H* High, *VH* Very high

data of information system Table 2, considering only core attributes data i.e. data of P1, P2, M1, and R2, we calculated I(Core(c)), which is also 0.945. Hence, P1, P2, M1, and R2 are reduced core attribute set or Reduct set and shown in Table 4.

3.2.4 Determination of Rules Between Core Conditional Attributes and Decisional Attributes

In this section we explore the reduced core attributes matrix and find the relationship rules from the conditional attributes to the decisions attributes. For finding the rules, Bai and Sarkis (2010) used RSES software for exploration, but we did our analysis manually as the sample size is less. Starting of our analysis, we will find the decisions rule only for high performance outcomes i.e. performances of D1 and D2 both will be "High". Initial rule set of reduced conditional attributes that have both D1 and D2 are "High" is shown in Table 5 after considering the feasibility conditions of the rules. The feasibility condition is explained by the following example. The response 11 having D1 and D2 both are "High" and having conditional attributes value for P1 and P2 are "High" and "Very High" respectively, in another case, the response 10, conditional attributes P1 and P2 having the same value "High" and "Very High" but the value of D1 is "moderate" (Table 2). Hence, this rule is contradictory in nature. We eliminate such rules that have infeasible relations. For checking feasibility, we prioritize only performance of D1 as the main objective of this study is to improve eco-efficiency.

In Table 5 we provide all initial rule set considering only high decisional attributes set. For example, from Table 4, 'response 5' have both D1 and D2 high, and corresponding P1 and P2 values are high and high, and 'response 20' also have the same combination, both D1 and D2 are high and both P1 and P2 are high. Hence, first rule set P1 and P2 is both high $\{(P1 = H) \wedge (P2 = H)\}$ and the number of occurrences is 2. Similarly, we calculate all the other initial rule set by analyzing the data of Reduct (Table 4).

After that, we refine our initial rule set provided in Table 5 by imposing one more priority criterion. We consider only those rules set for final rules set which have more than two numbers of occurrences. As the higher the number of occurrences means the rules set are more dominating. Threshold number of occurrence value for this study is assumed as two. Reduced dominant rules set are provided in Table 6. Finally, the final rule set of conditional attributes for getting higher environmental and business performance are provided in Table 7. The final rules set are found after analyzing the relative dominance of reduced dominant rules set provided in Table 6. The relative dominance can be explained by following example. For example, in the case of P1, 'VH' is dominated over 'H' and 'M', as it is come in "P1 and M1" as well as in "P1 and R2". Hence, "VH" is selected for the

Table 5 Initial feasible rule set of conditional attributes for D1 and D1 = "High"

D1 and D2 = "High"	Number of occurrences	D1 and D2 = "High"	Number of occurrences
$(P1 = H) \wedge (P2 = H)$	2	$(P2 = H) \wedge (M1 = VH)$	2
$(P1 = VH) \wedge (P2 = M)$	2	$(P2 = M) \wedge (M1 = VH)$	1
$(P1 = M) \wedge (P2 = H)$	2	$(P2 = H) \wedge (M1 = M)$	1
$(P1 = VH) \wedge (P2 = H)$	1	$(P2 = VH) \wedge (M1 = VH)$	1
$(P1 = VH) \wedge (P2 = VH)$	1	$(P2 = H) \wedge (M1 = H)$	1
$(P1 = VH) \wedge (M1 = VH)$	2	$(P2 = H) \wedge (R2 = H)$	3
$(P1 = H) \wedge (M1 = VH)$	1	$(P2 = H) \wedge (R2 = VH)$	1
$(P1 = VH) \wedge (M1 = M)$	1	$(P2 = VH) \wedge (R2 = VH)$	1
$(P1 = H) \wedge (M1 = M)$	1	$(P2 = M) \wedge (R2 = H)$	1
$(P1 = M) \wedge (M1 = M)$	1	$(P2 = VH) \wedge (R2 = H)$	1
$(P1 = VH) \wedge (M1 = H)$	1	$(P2 = H) \wedge (R2 = M)$	1
$(P1 = M) \wedge (M1 = VH)$	1		
$(P1 = H) \wedge (R2 = H)$	2	$(M1 = VH) \wedge (R2 = H)$	2
$(P1 = VH) \wedge (R2 = H)$	2	$(M1 = H) \wedge (R2 = H)$	2
$(P1 = VH) \wedge (R2 = M)$	1	$(M1 = VH) \wedge (R = VH)$	1
$(P1 = VH) \wedge (R2 = VH)$	1	$(M1 = M) \wedge (R2 = VH)$	1
$(P1 = H) \wedge (R2 = VH)$	1	$(M1 = M) \wedge (R2 = H)$	1
$(P1 = M) \wedge (R2 = H)$	1	$(M1 = VH) \wedge (R2 = M)$	1
$(P1 = M) \wedge (R2 = M)$	1	$(P1 = H) \wedge (P2 = H) \wedge (R2 = H)$	2

Table 6 Reduced feasible rule set after imposing dominance criteria

D1 and D2 = "High"	D1 and D2 = "High"
(P1 = H) ∧ (P2 = H)	(P1 = VH) ∧ (R2 = H)
(P1 = VH) ∧ (P2 = M)	(P2 = H) ∧ (M1 = VH)
(P1 = M) ∧ (P2 = H)	(P2 = H) ∧ (R2 = H)
(P1 = VH) ∧ (M1 = VH)	(M1 = VH) ∧ (R2 = H)
(P1 = H) ∧ (R2 = H)	(M1 = H) ∧ (R2 = H)
	(P1 = H) ∧ (P2 = H) ∧ (R2 = H)

Table 7 Final rule set for getting high environmental and organizational performances

(P1 = VH) ∧ (P2 = H)
(P1 = VH) ∧ (M1 = VH)
(P1 = VH) ∧ (R2 = H)
(P2 = H) ∧ (*M*1 = VH)
(P2 = H) ∧ (r2 = H)
(M1 = VH) ∧ (R2 = H)
(P1 = H) ∧ (P2 = H) ∧ (R2 = H)

relative importance of P1. Similarly we select relative importance of P2, M1, and R2 as "H", "VH", and "H" respectively. Detailed explanations of rules are provided in the discussion section.

4 Discussions

In this section, the detailed interpretation of results is discussed. Firstly, none of the incremental innovation strategy, incremental product eco-innovation and incremental process eco-innovation, has been considered in the core attribute set. The reason may be the effectiveness of these incremental innovation strategies for improving the eco-efficiency and as well as for improving product and process innovation is relatively low in comparison to the radical innovation strategies. Whereas only radical product eco-innovation is found in the core attribute set. Reason for not finding the radical process eco-innovation in the core attribute set, may be the investment requirement for radical process eco-innovation is very high; hence, it is less adopted by industry. For process eco-innovation, only adoption of cleaner technology is included in the core attribution set. This implies that industries can significantly achieve high level of eco-efficiency by adopting cleaner technology. From the product eco-innovation strategies, only product stewardship and life cycle analysis of product are significant. Hence, these two strategies are very important for improving eco-efficiency. By adopting product stewardship it can reduce environmental impact as well as improve product quality. Therefore, organization should invest more on product stewardship to reduce environmental impact. Thus, from a practical point of view eco-efficiency can be significantly improved by adopting product stewardship, life cycle analysis of product, cleaner technologies, and radical product eco-innovation. Product design for reduce, recycle, and recovery is not significant in our analysis although it is an important strategy for improving eco-efficiency. Moreover, this study is based on analytical model of less sample size; further statistical empirical studies are required to posit more robust relationships.

The objective of this paper is to find the relative importance of eco innovation for improving eco-efficiency. Hence, rough set theoretical analysis methodology is adopted for analyzing industry experts' responses. Finally, a set of rules are derived through analysis that is shown in Table 7. From Table 7 it can infer that relative importance of product stewardship is very high (P1 = VH) and radical product eco-innovation is high (R2 = H) that means organization can invest more on radical product eco-innovation at product design stage. It is also inferred that organizations having very high cleaner technology and relatively high life cycle analysis strategy will relatively more efficient to combat environmental impact $\{(P2 = H) \wedge (M1 = VH)\}$. Overall relative importance of Life Cycle Analysis (LCA) is high (P2 = H). Furthermore, organization can identify emissions hot spot area and important areas for improvement through LCA analysis of product carbon emissions; hence, emissions can also be reduced. In addition to that, organizations having high product stewardship strategy, high LCA analysis capability, and high radical product innovation capability will be more efficient with less investment. Combine effect of these three conditional criteria may be the reason for that. Furthermore, overall relative importance of adoption of cleaner technologies, such as energy efficient equipment, waste water treatment plant etc. is very high (M1 = VH). But these cleaner technologies need very high initial investments; hence, organizations need to trade off between economic and environmental objectives. Lastly, these results are highly effective for decisions maker of heavy pollution intensive organization, to adopt effective strategies for improving eco-efficiency that indirectly related to the development of low carbon economy.

5 Conclusions

Global climate change is a serious issue nowadays. To combat against climate change, several governments and institutions have already taken several initiatives. But, these will not sufficient enough against climate change. Industries have to accept the need for climate change mitigation. Industries will not accept adoption of climate change strategies until they will find substantial environmental and economic benefits out of it. Furthermore, improvement in eco-efficiency is the one of the most important strategies for developing long term sustainability. Therefore, in this study we analyse several eco-innovation strategies for improving eco-efficiency and operational performance. Rough set theoretical model has adopted for analysis. Major findings of this study are, firstly product stewardship, life cycle analysis of product emission, cleaner technology, and radical product innovation strategies are found to be significant for improving eco-efficiency. Secondly, radical product eco innovation strategy is the most important strategy for improving environmental performance for managerial perspective. Investment in cleaner technology based process innovation is also important strategies for reducing environmental impact. Contrastingly, product innovation for reduce, reuse, and recycling is not found to be significant in this study but it is an important strategy for improving resource efficiency in the literature. Reason may be the consciousness of recycling and remanufacturing has not got substantial attention in Indian manufacturing firm. Moreover, relative importance of radical eco-innovation during product design stage is very high and cleaner technology based process innovation is also very high for improving eco-efficiency. The inherent limitation of rough set theory is its analysis will be based on approximation. The accurate proposition will be found by using mathematical topology. But, for managerial implication approximation will be sufficient enough. This study is based on broad eco-innovation concept. So it is expected that in the near future there will be more research on exhaustive eco-innovation construct. Large sample size based elaborative statistical empirical investigation is needed for more robust propositions. Role of other factors for adoption of eco- innovations such as circular economy, industrial symbiosis and co-operative society are needed to be researched further. Lastly, this study is an initial theoretical investigation on eco-innovation and eco-efficiency. Empirical investigations still need to be carried out in the near future.

Appendix: Respondents' Profile

Respondents designation	**Respondents years of experience**
Bottom level Manager (6 respondents) Middle level manager (6 respondents) Corporate head (1 respondents) Senior engineer (5 respondents) Consultants (2 respondents)	More than 10 years (2 respondents) 5–10 years (3 respondents) 3–5 years (5 respondents) 1–3 years (9 respondents) Less than 1 year (1 respondents)
Industry type Textile (5 respondents) Cement (2 respondents) Power (3 respondents) Steel (5 respondents) Oil (1 respondents) Heavy engineering (1 respondents) Consumer product (1 respondents) Paper (2 respondents)	

References

Abernathy W J, Utterback JM (1978) Patterns of industrial innovation. Technology review 64(7): 254–228. Available at: http://teaching.up.edu/bus580/bps/Abernathy%20and%20Utterback,%201978.pdf.

Atkinson G, Mourato S (2008) Environmental cost-benefit analysis. Annual review of environment and resources 33: 317–344. Available at: http://www.annualreviews.org/doi/abs/10.1146/annurev.environ.33.020107.112927. Accessed 30 June 2016.

Bai C, Sarkis J (2010) Green supplier development: analytical evaluation using rough set theory. Journal of Cleaner Production 18(12): 1200–1210. Available at: http://dx.doi.org/10.1016/j.jclepro.2010.01.016. Accessed 04 September 2016.

Bithas K (2011) Sustainability and externalities: Is the internalization of externalities a sufficient condition for sustainability?. Ecological Economics 7(10):1703–1706. Available at: http://dx.doi.org/10.1016/j.ecolecon.2011.05.014. Accessed 30 June 2016.

Cleff T, Rennings K (1999) Determinants of environmental product and process innovation. European environment 19(5):191–201. Available at: http://onlinelibrary.wiley.com/doi/10.1002/(SICI)1099-0976(199909/10)9:5%3C191::AID-EET201%3E3.0.CO;2-M/abstract. Accessed 23 July 2016.

Cropper M L, Oates W E (1992) Environmental economics: a survey. Journal of economic literature 30 (2): 675–740. Available at: http://www.jstor.org/stable/2727701?seq=1#page_scan_tab_contents. Accessed 23 July 2016.

Curran MA (1996) Environmental life-cycle assessment. The International Journal of Life Cycle Assessment 1(3):179–179. Available at: http://link.springer.com/article/10.1007%2FBF02978949?LI=true. Accessed 30 June 2016.

Damanpour F, Gopalakrishnan S (2001) The dynamics of the adoption of product and process innovations in organizations. Journal of management studies, 38(1): 45–65. Available at: http://onlinelibrary.wiley.com/doi/10.1111/1467-6486.00227/abstract. Accessed 02 July 2016.

Demirel P, Kesidou E (2011) Stimulating different types of eco-innovation in the UK: Government policies and firm motivations. Ecological Economics 70(8):1546–1557. Available at: http://dx.doi.org/10.1016/j.ecolecon.2011.03.019. Accessed 04 August 2016.

Dewar R D, Dutton J E (1986) The adoption of radical and incremental innovations: An empirical analysis. Management science 32 (11): 1422–1433. Available at: http://dx.doi.org/10.1287/mnsc.32.11.1422. Accessed 30 June 2016.

EACI/ECO/2013/001 (2016) Analysis of results achieved by CIP Eco-innovation market replication projects EXECUTIVE SUMMARY. Executive Agency for Competitiveness & Innovation (EACI), Brussels. Available at: https://ec.europa.eu/easme/sites/easme-site/files/1_executive_summary.pdf.

Ekins P (2010) Eco-innovation for environmental sustainability: concepts, progress and policies. International Economics and Economic Policy 7(2–3):267–290. Available at: http://link.springer.com/article/10.1007/s10368-010-0162-z. Accessed 13 August 2016.

Elbasha EH, Roe T L (1996) On endogenous growth: the implications of environmental externalities. Journal of Environmental Economics and Management 3(2): 240–268. Available at: doi:10.1006/jeem.1996.0043. Accessed 13 August 2016.

Ettlie J E, Bridges W P, O'keefe R D (1984) Organization strategy and structural differences for radical versus incremental innovation. Management science 30(6): 682–695. Available at: http://dx.doi.org/10.1287/mnsc.30.6.682. Accessed 10 August 2016.

Foxon T J (2011) A co-evolutionary framework for analysing a transition to a sustainable low carbon economy. Ecological Economics 70 (12):2258–2267. Available at: http://dx.doi.org/10.1016/j.ecolecon.2011.07.014. Accessed 30 June 2016.

Frondel M, Horbach, Rennings K (2004) End-of-pipe or cleaner production? An empirical comparison of environmental innovation decisions across OECD countries. An empirical comparison of environmental innovation decisions across OECD countries: 04–082. Available at: ftp://193.196.11.222/pub/zew-docs/dp/dp0482.pdf.

Hart S L (1995) A natural-resource-based view of the firm. Academy of management review 20(4): 986–1014. Available at: http://amr.aom.org/content/20/4/986.short. Accessed 02 July 2016.

Hellstrom T (2007) Dimensions of environmentally sustainable innovation: the structure of eco-innovation concepts. Sustainable Development-Bradford 15 (3): 148. Available at: http://onlinelibrary.wiley.com/doi/10.1002/sd.309/abstract. Accessed 23 July 2016.

Hendrickson C, Horvath A, Joshi S, Lave L (1998) Peer reviewed: economic input–output models for environmental life-cycle assessment. Environmental science & technology 32(7): 184A–191A. Available at: http://pubs.acs.org/doi/abs/10.1021/es983471i. Accessed 30 June 2016.

Herring H (2006) Energy efficiency—a critical view. Energy 31(1):10–20. Available at: http://dx.doi.org/10.1016/j.energy.2004.04.055. Accessed 28 July 2016.

IPCC (2013) Summary for Policymakers. In: Climate Change 2013: The Physical Science Basis. Contribution of Working Group I to the Fifth Assessment Report of the Intergovernmental Panel on Climate Change Cambridge University Press, Cambridge, United Kingdom and New York, NY, USA. Available at: https://www.ipcc.ch/pdf/assessment-report/ar5/wg1/WG1AR5_SPM_FINAL.pdf.

IPCC (2014) Climate Change 2014: Synthesis Report. Contribution of Working Groups I, II and III to the Fifth Assessment Report of the Intergovernmental Panel on Climate Change. IPCC, Geneva, Switzerland, 151 pp. Available at: https://www.ipcc.ch/pdf/assessment-report/ar5/syr/SYR_AR5_FINAL_full_wcover.pdf.

Jänicke M (2008) Ecological modernisation: new perspectives. Journal of cleaner production 16(5): 557–565. Available at: http://dx.doi.org/10.1016/j.jclepro.2007.02.011. Accessed 25 September 2016.

Juett R Cooper (1998) A multidimensional approach to the adoption of innovation. Management Decision 36 (8): 493–502. Available at: http://dx.doi.org/10.1108/00251749810232565. Accessed 23 July 2016.

Kemp R, Pearson P (2007) Final report MEI project about measuring eco-innovation. UM Merit, Maastricht, 32(3):121–124. Available at: http://www.oecd.org/env/consumption-innovation/43960830.pdf.

Klassen R D (2000) Exploring the linkage between investment in manufacturing and environmental technologies. International Journal of Operations & Production Management, 20(2): 127–147. Available at: http://dx.doi.org/10.1108/01443570010304224. Accessed 09 January 2016.

Klassen R D, Whybark D C (1999) The impact of environmental technologies on manufacturing performance. Academy of Management Journal 42(6):599–615. Available at: http://www.jstor.org/stable/256982?seq=1#page_scan_tab_contents. Accessed 09 January 2016.

Liang J, Shi Z (2004) The information entropy, rough entropy and knowledge granulation in rough set theory. International Journal of Uncertainty, Fuzziness and Knowledge-Based Systems 12(01): 37–46. Available at: http://dx.doi.org/10.1142/S0218488504002631. Accessed 25 September 2016.

Machiba T (2010) Eco-innovation for enabling resource efficiency and green growth: development of an analytical framework and preliminary analysis of industry and policy practices. International Economics and Economic Policy 7(2–3): 357–370. Available at: http://link.springer.com/article/10.1007/s10368-010-0171-y. Accessed 28 July 2016.

Manual O (2005) The measurement of scientific and technological activities. Proposed guidelines for collecting and interpreting innovation data. Organisation for Economic Cooperation and Development, OECD Publishing. Paris. Available at: http://www.oecd-ilibrary.org/science-and-technology/oslo-manual_9789264013100-en.

OECD/IEA/NEA/ITF (2015), Aligning Policies for a Low-carbon Economy, OECD Publishing, Paris. http://dx.doi.org/10.1787/9789264233294-en.

Patterson M G (1996) What is energy efficiency?: Concepts, indicators and methodological issues. Energy policy 24(5): 377–390. Available at: doi:10.1016/0301-4215(96)00017-1. Accessed 28 July 2016.

Pawlak Z (1982) Rough sets. International Journal of Computer & Information Sciences 11(5): 341–356. Available at: http://link.springer.com/article/10.1007/BF01001956. Accessed 25 September 2016.

Pawlak Z (1997) Rough set approach to knowledge-based decision support. European journal of operational research 99(1):48–57. Available at: doi:10.1016/S0377-2217(96)00382-7. Accessed 25 September 2016.

Pawlak Z (1998) Rough set theory and its applications to data analysis. Cybernetics & Systems 29(7): 661–688. Available at: http://dx.doi.org/10.1080/019697298125470. Accessed 25 September 2016.

Pawlak Z (2002) Rough sets and intelligent data analysis. Information Sciences 147(1):1–12. Available at: http://doi.org/10.1016/S0020-0255(02)00197-4. Accessed 25 September 2016.

Rennings K (2000) Redefining innovation—eco-innovation research and the contribution from ecological economics. Ecological economics 32(2):319–332. Available at: http://dx.doi.org/10.1016/S0921-8009(99)00112-3. Accessed 23 July 2016.

Shyng J Y, Wang F K, Tzeng, G H, Wu K S (2007) Rough set theory in analyzing the attributes of combination values for the insurance market. Expert Systems with Applications 32(1):56–64. Available at: http://dx.doi.org/10.1016/j.eswa.2005.11.002. Accessed 25 September 2016.

Unnikrishnan S, Hegde D S (2007) Environmental training and cleaner production in Indian industry—A micro-level study. Resources, conservation and recycling 50 (4):427–441. Available at: http://dx.doi.org/10.1016/j.resconrec.2006.07.003. Accessed 13 August 2016.

Utterback J M, Abernathy W J (1975) A dynamic model of process and product innovation. Omega 3(6): 639–656. Available at: doi:10.1016/0305-0483(75)90068-7. Accessed 23 July 2016.

Worrell E, Bernstein L, Roy J, Price L, Harnisch J (2009) Industrial energy efficiency and climate change mitigation. Energy Efficiency 2(2):109–123. Available at: http://link.springer.com/article/10.1007/s12053-008-9032-8. Accessed 28 July 2016.

Case Study of Uttarakhand in Perspective of Extreme Climatic Events: Fire, Ecosystem and Livelihoods

Harshit Pant

Abstract The phenomenon of climate change is widely known as statistical change in climatic properties over a specified period of time. The forest sector had been mostly reactive towards it as some effects of climate change are already noticeable and there is a need to be better prepared for future change. Present study gives an idea about impact of extreme climatic events on forest ecosystem of Uttarakhand. The results showed that a long dry period without rains had lead to efficient conditions for fire initiation which is a regular phenomenon during summer season in this region. Eventually the extreme fire incidents peaked during April–May 2016 and almost every forest of 13 districts of this region was affected. These incidents ultimately affected the precious ecosystem of this biodiversity rich region, increased the temperature and affected the air quality and lead to decline in livelihood services depending upon native forests. Findings of the present study will be beneficial for the further assessment of the likely impacts of climate change events on forests, forest-dependent people and their vulnerabilities which are important for enhancing climate change adaptation as it forms the basis for sustainable development options to avoid harmful effects of climate change in near future.

Keywords Climate change · Himalayan region · Uttarakhand · Adaptation · Sustainable development

1 Introduction

Ongoing climate change over succeeding decades will likely to have additional negative impacts across the hill regions which will include significant cascading effects on river flows, groundwater recharge, natural hazards, and biodiversity; ecosystem composition, structure, and function; and human livelihoods (Xu et al. 2009).

H. Pant (✉)
GBPNIHESD (G.B. Pant National Institute of Himalayan Environment and Sustainable Development), Kosi-Katarmal, Almora, Uttarakhand 263643, India
e-mail: drharshitpant@gmail.com

W. Leal Filho (ed.), *Climate Change Research at Universities*,
DOI 10.1007/978-3-319-58214-6_32

India produces about 2 tons of CO_2 equivalents (CO_2e) per capita, versus 20 tons and 8 tons, respectively, in respect to United States and China. Being one of the fastest growing major economies in the world, India is the fourth largest green house gases (GHG) emitter, accounting for 5.8% of global cumulative energy-related emissions since 1850, compared to 16 and 15% for the United States and China. India's emissions have increased by 67.1% between 1990 and 2012, and are projected to grow 85% by 2030 under a business-as-usual scenario (CCES 2015).

India has the climate policy for climate agreement which includes following key points:

1. To reduce the emissions intensity of its gross domestic product (GDP) by 33–35% by 2030 from 2005 level.
2. To achieve about 40% cumulative electric power installed capacity from non-fossil fuel based energy resources by 2030, with the help of transfer of technology and low cost international finance including from Green Climate Fund (GCF).
3. To create an additional carbon sink of 2.5–3 billion tons of CO_2 equivalent through additional forest and tree cover by 2030.

India has pledged under the Copenhagen Accord to reduce its CO_2 intensity (emissions per gross domestic product) by 20–25% by 2020 compared to 2005 levels (GOI 2010). India appears on track to achieve its voluntary pledge, though emissions are not projected to peak until around 2050 or later. On October 1, 2015, India formally submitted its intended nationally determined contribution (INDC) to the climate agreement in Paris and on 2nd October 2016, India has ratified to Paris agreement with keeping in view of following low carbon path to progress.

India has a number of policies that contribute to climate mitigation by reducing or avoiding greenhouse gas emissions. One of the main contributing policies in this aspect is 'National Action Plan on Climate Change' which was initiated by the Prime minister in 2008. It incorporates eight core missions related to climate change adaptation and mitigation which will be running through 2017. Similarly India's current five year plan (2012–2017), which guides overall economic policies includes goals to achieve average 8% annual GDP growth along with reducing emissions intensity in line with India's Copenhagen pledge and adding 300,000 MW of renewable energy capacity.

The impact of climate change upon India has been reported in the area of upper Himalayas, northern and central parts of Western Ghats and parts of central India which are most vulnerable to projected impacts of climate change. Northeastern forests are more resilient to climate change impacts in comparison to other parts of India (Chaturvedi et al. 2011).

For the mitigation and reduction in GHG emissions, the initiative of curbing the emission also involves the control of climate and human induced emissions such as agriculture residue burning and forest fires. On one hand the black carbon produced from crop biomass burning has long lasting impacts on atmosphere as well as the nearby ecosystems (Kumar et al. 2015), while on the other hand forest fires have been predicted to increase under climate change (Cai and Yang 2016).

A fire regime refers to the character of fire occurring over long time periods and the immediate effects of fire that generally characterize ecosystems. Descriptions of fire regimes are general and broad because of the enormous variability of fire in time and space. The fire regime concept brings a degree of order to a complicated body of fire behaviour and fire ecology knowledge; and provides a simplifying means of communicating about the role of fire among technical and non-technical audiences (Brown and Smith 2000).

The role of fire can be understood and communicated through the use of fire regimes. Significant changes in the role of fire due to management actions or possible shifts in climate can be described by shifts in fire regimes. Thus, fire regimes can be used to understand the past role of fire, current changes in fire regimes due to management actions, and as indicators to guide future management through more sustainable practices and policies. Fires have negative impacts on native plant diversity, with varying effects on species and ecosystems, including the potential for localized extinction (Franklin et al. 2005).

The effects of fire on forests vary, depending upon weather, topography, and fuel type (Podur and Martell 2009). The observations in the past 20 years show that the increasing intensity and spread of forest fires in Asia were largely related to rises in temperature and declines in precipitation in combination with increasing intensity of land uses (IPCC 2007). The uncertainty lies in whether the expected increase in temperature would be enough to trigger more frequent and severe fires despite the projected increase in precipitation.

Climate change is expected to make impacts on boundaries of forest types and areas, primary productivity, species population and migration, occurrence of pest and diseases and forest regeneration. The interaction between climate change and forest fire is like a feedback system. As climate change has an impact on forest fire, the forest fire may also have an impact on climate change (Shijo et al. 2009).

Forest fire tends to change the microclimate of the area Climate-change influences on wildfire extent; severity and frequency depend on interactions among several factors including forest management history, drought frequency and severity, insect outbreaks and many others. There is evidence of both increase and decrease in fire activity at regional scales (Easterling et al. 2007). Forest thinning, controlled burning and other measures to reduce fuel loads and other aspects of wildfire hazard can be effective in reducing forest vulnerability to fire-mediated effects of climate change.

2 Indian Scenario

India has forest cover of 701,673 km^2 which occupies 21.34% of the total geographical area of the country. The Indian Himalayan region (IHR) has total forest cover of 221,652 km^2 covering 41.24% of the IHR geographical area and 6.74% of country's geographical area (ISFR 2015). These forests contain variety of different climate zones, including the tropical south, north-western deserts, Himalayan mountains, and the wet northeast. Forests are widely distributed in the country. India's forests are endowed with a variety of biomes and biological communities. The forest vegetation in the country varies from tropical evergreen forests in the West Coast and in the Northeast to alpine forests in the Himalayas in the North. In between the two extremes, there are semi-evergreen forests, deciduous forests, sub-tropical broad-leaved hill forests, sub-tropical pine forests, and subtropical montane temperate forests.

With increasing population pressure, the forest cover of the country is deteriorating at an alarming rate. Along with various factors, forest fires are a major cause of degradation of Indian forests. Such losses can be summed up as the loss of valuable timber, loss of bio-diversity, loss of wildlife habitat, global warming, soil erosion, loss of fuel-wood and fodder, damage to water and other natural resources along with loss of natural regeneration.

The vulnerability of the Indian forests to fire varies from place to place depending upon the type of vegetation and the climate. The coniferous forests in the Himalayan region comprising of fir (*Abies pindrow.*), spruce (*Picea smithiana*), *Cedrus deodara, Pinus roxburgii* and *Pinus wallichiana* etc. are highly prone to fire. Every year there are one or two major incidences of forest fire in this region. The other parts of the country dominated by deciduous forests are also damaged by fire (Bahuguna and Singh 2001a).

Historical incidents of severe forest fire incidents are reported during the 1990s, several forest fires occurred in the then hills of Uttar Pradesh (presently Uttarakhand) and Himachal Pradesh. From 1995 to 1999, fire hazards in these two states assumed dangerous dimensions. An area of 677,700 hectares was affected by these fires. The estimated timber loss from these hazards was US$43 million. Other losses due to these fires included loss of soil fertility, soil erosion, loss of employment, drying up of water resources, and loss of bio-diversity. These fires brought a major change in the microclimate of the region in the form of soil moisture balance and increased evaporation. The dense smoke from the fires affected visibility up to 14,000 ft (Bahuguna and Singh 2001b).

In the year 2016, severe forest fires were reported in Hills of Uttarakhand state (SFD 2016). As the Himalayan forests are reported to have multiple functions as they harbor biodiversity, anchor soil and water, provide carbon sinks, regulate climate, and temper stream flow. They also supply forest products for local livelihoods and economies. In aspect of this the community forest of Uttarakhand plays an important role as they are the longest institution of community forest management (Guha 1989) and the main source of the livelihood needs of the dependent communities.

Present study is focused on the impacts of long dry spell and temperature increase in 15 community managed forests (*Van Panchayats*) in the Garhwal region of Uttarakhand. The *Van Panchayats* (hereafter VPFs) have been the most prominent form of community based forest management in the Uttarakhand region of India and were came in limelight after the much talked '*Chipko Protest*'. Being the soul resource of community livelihood resources, there VPFs have been a harbour of rich forest diversity and various ecosystem services.

The aim of present paper is to ascertain the impact of temperature increase and long dry spells over the forest resources and its relation to the fire incidents and livelihood opportunities of the dependent communities of VPFs. The paper focuses upon the effect of extreme climatic events (i.e. forest fire) upon the livelihood and ecosystem of the local communities.

3 Methodology

The study focuses upon the VPFs of Uttarakhand state of Indian Himalayan region. For the present study, 15 VPFs were selected. These VPFs were located between 29.87°N latitude and 78.83°E longitudes to 30.48°N latitude and 79.06°E longitude across Pauri and Rudraprayag district of Garhwal region of Uttarakhand (Table 1). The studied VPFs were located at an elevation ranging between 800 and 1800 m, covering a wide variety of forest types.

The study was based upon the impact of extreme climatic events (i.e. forest fire, temperature increase and rainfall) upon the resource availability from the community forest. The study was conducted using open ended questionnaire (Table 2). The questions were basically designed to be understood easily by the community members and were asked from the members of VPF management committee and were structured following Lise (2000). The questions were analyzed for the response percentage obtained by community members of both study regions.

Table 1 Description of the study sites

Sr. No.	Name of the VPF	District	Altitude (m)	Forest type
1.	Gaihar	Pauri	1500	Pine-oak mixed
2.	Nisni	Pauri	1650	Pine-oak mixed
3.	Gangwara	Pauri	1600	Oak mixed
4.	Tamlag	Pauri	1500	Pine-oak mixed
5.	Balori Gaon	Pauri	800	Pine-oak mixed
6.	Chamkot	Pauri	1000	Pine-oak
7.	Ujyara	Pauri	1800	Oak mixed
8.	Baigwari	Rudraprayag	1400	Pine mixed
9.	Udaltyun	Pauri	1000	Pine mixed
10.	Bhanadhar	Rudraprayag	1450	Pine mixed
11.	Makkumath	Rudraprayag	1400	Mixed forest
12.	Tilwara	Rudraprayag	1300	Pine oak mixed
13.	Bhiri	Pauri	1300	Pine mixed
14.	Tilni	Rudraprayag	1340	Pine oak mixed
15.	Baswara	Pauri	1600	Pine oak mixed

Table 2 Sample set of the questions asked during questionnaire survey

Q.No	Questions	Answers
1.	Effect of climate change	Yes/No/Don't Know
2.	Extreme climatic events during last 10 years	Yes/No/Don't Know
3.	Fire events during last 10 years	Yes/No/Don't Know
4.	Rainfall variations during last 10 years	Yes/No/Don't Know
5.	temperature variation during last 10 years	Yes/No/Don't Know
6.	Impact of climatic events upon livelihood	Yes/No/Don't Know
7.	Impacts of climatic events upon forest ecosystem	Yes/No/Don't Know

The study was limited to the lower and mid elevation VPFs of the region as it was hypothesized that the dry spell and temperature anomalies were limited to mid hill elevations, thus the VPFs located in higher elevations (i.e. more than 1800 m) were not taken up for the study.

For the assessment of forest fire incident during past decade the data was obtained from the Forest survey of India (FSI 2016) and ENVIS (Environmental Information system) center on Himalayan Ecology of GBPNIHESD (G.B. Pant National Institute of Himalayan Environment and Sustainable Development), Kosi-Katarmal, Almora. The obtained data was compared for assessment of fire incidents in study region.

The study suggested the vulnerability assessment of the fire affected areas of Uttarakhand is necessary for development of future fire vulnerability indices as present study was limited to the assessment status and further research in the aspect of after fire scenario is highly required. Similarly the maximum severe fire incidents of 2016 have been reported in the Pauri region of Uttarakhand, in which some the high altitudinal areas were also affected by fire, thus present study recommends an analysis of high altitudinal forest is also necessary to assess the extent of damage after forest fire.

4 Results

The study provided an idea of extreme climatic events upon the livelihood activities of community. The VPFs of Pauri district were more affected from fire incidents in comparison to the VPFs of Rudraprayag region. Although both of the regions faced change in climatic incidents which signifies that even though the fire events can be region specific, extreme climatic events are not specified towards a specified region.

Most of the respondents in both regions agreed towards the effects of climate change (67% in Pauri region, 33% in Rudraprayag region) and impacts of forest fire (87% in Pauri region, 33% in Rudraprayag region). The Pauri region showed higher response in aspect of forest fire which was similar with the yearly fire incidents as reported by FSI (2016) (Fig. 5).

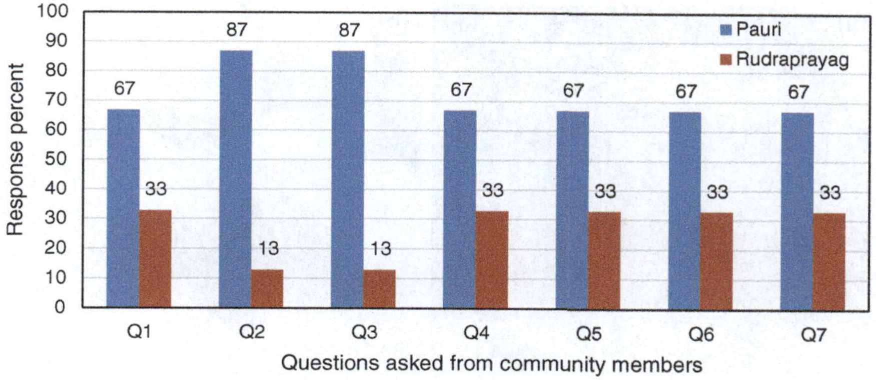

Fig. 1 Responses of community members across both regions

Rainfall and temperature variation was reported in both regions (100%) while both of the regions reported the effect of climatic events upon the livelihood and forest ecosystem with 67% in Pauri region and 33% in Rudraprayag region (Fig. 1).

The community responses indicated that the VPF members are well aware of the fact that climate change in happening and it is having various affect upon their forest resources such as crop drying and seedling mortality. They have mentioned that as there has been a long dry spell since September 2015, it is affecting the horticulture crop growth, as due to higher than usual temperature the crops are drying.

Although most of the respondents were skeptical about the aftermath of this dry spell, their fears come true as the fire incidents increased and took a heavy toll on the forest resources. The long duration of dry spell which continued from September 2015 to May 2016 has affected the fruit crop growth as well as drying of natural water sources and acted as a pathway for forest fire. As the fire season approached these adverse conditions increased the risk of fire incidents which were started to show from 26th to 27th April 2016.

Dry weather spell also led delayed ripening of fruits in the hill forests of the State (e.g., *Myrica esculenta* which contributes significantly to local economy, and *Berberis asiatica* which is an important medicinal plant of the region). Similarly the subsequent incidents of forest fire which have caused seedlings and saplings mortality in forests that will affect the regeneration in long run (Fig. 2a). The unripe fruits on trees indicated that the increasing temperature has affecting the growth vigor, eventually leading to either mortality or late ripening which was delaying the marketing and effecting the livelihood (e.g., *Myrica,* Fig. 2b). Moreover after the fire disaster there is a wide scale loss of forest crops and subsequent vegetation. A total loss of Rs. 4.6 million rupees was estimated by the forest fire of 2016 (SFD 2016) which equals to 68,328 USD.

This forest fire disaster has affected total of 4429.35 ha of forest in Uttarakhand, while highest fire damage was reported in District Pauri with 1037.75 ha. of affected

Fig. 2 Effect of forest fire on forest resources and livelihood options. **a** Showing the effect of fire on forest resources. **b** Showing the unripe fruit of *Myrica*

forest area. The Pauri region had highest forest fire incidents (402) while Rudraprayag region had 82 fire incidents during the tenure which continued till May 3rd 2016. These fires have consumed about 537.65 ha. area of reserve forests and 494.1 ha area of VPF and civil soyam forests in Pauri region. In the Rudraprayag region total forest area affected by fire was 159.7 ha which included 68.7 ha area of reserve forest and 91 ha. area of VPF and civil soyam forests (Fig. 3).

These forest fires caused expected loss of Rupees 89,89,75 in Pauri region while expected loss of rupees 221,050 was reported for Rudraprayag region which includes the forest resources such as NTFPs, fruits and *resin* extraction which is a big revenue resource for VPF committees as well as for the forest department.

As the severity of forest fire was very high, it became very difficult for forest department to extinguish fire in remote and rugged forest terrain (Fig. 4a). Thus for combating this MI-17 Helicopter of Indian Air force with a specified water bucket were deployed for firefighting which fetched the water from nearby sources such as from Bhimtal lake, for extinguishing fire in the forest area of Nainital district (Fig. 4b). These fires were eventually burned out after 3rd and 4th May 2016's rainfall, which helped in controlling the fire to some extent.

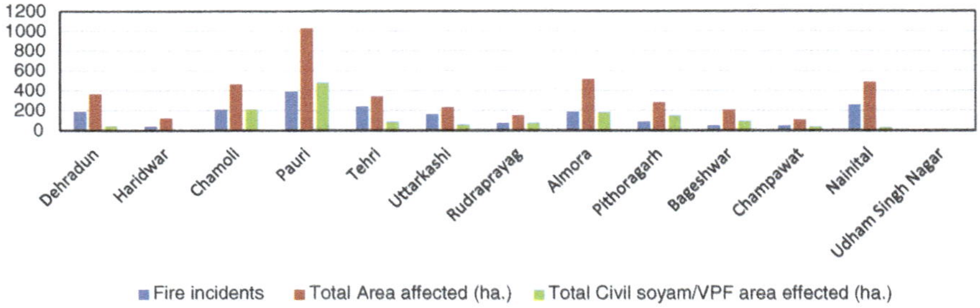

Fig. 3 Forest fire affected area (ha) across all districts in Uttarakhand. *Source* FSI (2016)

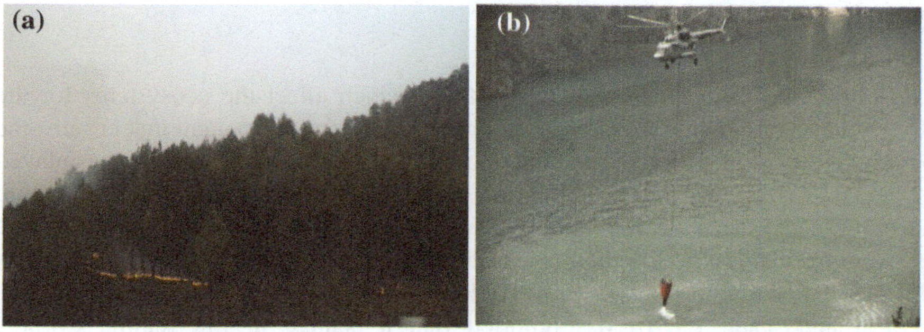

Fig. 4 Fire incidents and rescue operations in Uttarakhand. **a** Forest fire in Gangwara VPF (Pauri). **b** MIG 17 carrying water from Bhimtal lake (Nainital)

The results and fire aftermath showed that though there was community awareness about the climatic incidents and its subsequent after effects, there were lesser precautions to combat any emergency situation. Though the VPF committee members are active in handling the fire incidents in their own way, there is lack of sufficient resources to fight it back in proper way. The high fire incidents in Pauri region during the last ten years shows that due to the far flung areas combating forest fires could be difficult which eventually results in increased fire incidents (Fig. 5) and greater damage to forest ecosystems.

This kind of emergency situation eventually leads up to the resource provision capacity of forest department, which needed to be upgraded with the involvement of more people centered trainings and ease of resource availability. Similarly there is strong need of developing a linkage between the fire alert system for the common man as their daily livelihood is related with the forest resources, and any damage to the forest deprives them from the benefits they sustain from it.

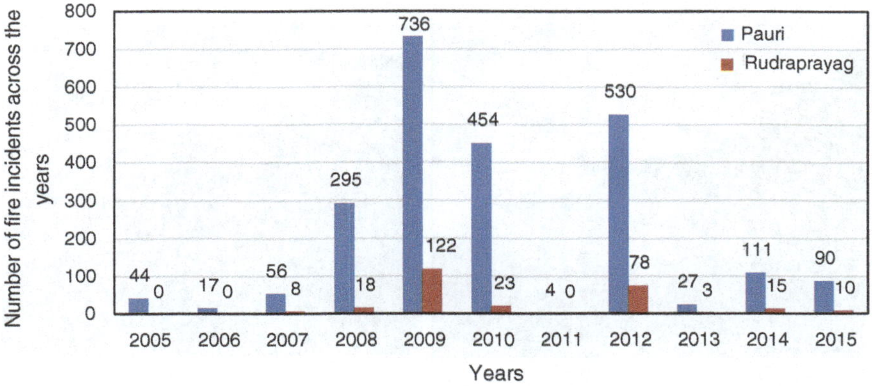

Fig. 5 Fire incidents during the past ten years across the study regions

5 Discussion

The forest fire has various cascading effects which affect the ecosystems for the long time. Fire in many cases, is an essential ecosystem component that ensures the sustainability of its processes and communities. Since its emergence, fire has played a key role as an environmental filter, selecting for species and their traits, and shaping ecosystems' communities. However, not all fires are the same and ecosystems exposed to fire regimes out of their historical range of variability might become vulnerable and suffer irreversible changes (Aponte et al. 2016).

Cai and Yang (2016) has reported that a change in fire regime towards high-severity fires could result in a shift in forest structure, which may influence future forest's capacity to sequester carbon.

In aspect of present study the forests suffered from high intensity fires during 27th April–3rd May 2016. The fires were severe which can be seen through the images obtained from the geoportal of National remote sensing centre of Indian Space Research Organization (NRSC ISRO 2016). It shows that the devastation of forest was very high (Fig. 6a, b) and was spread to a wider extent than any earlier reported forest fires.

Jha et al. (2016) has also reported the extent of forest area devastation by forest fire as 7.35% of the forest area of the state was burned in this event. As the burned area was largely in the moist deciduous (55.29%) and subtropical pine forests (29%), it has affected the forest dependent communities and their livelihoods as well.

Though in India there are reported measures (Fig. 7) to monitor, assess and combat forest fire incidents, it becomes hard to act it on ground at the time of chaos. For the development of adaptive capacity to combat climate change, the initial stage of local level monitoring, information sharing and timely response from higher authority is a necessary part. The proper training and capacity building for the VPF forest managers is essential as they know how to handle and protect their own forest, but due to the resource efficiency they are not able to achieve the success up to the mark.

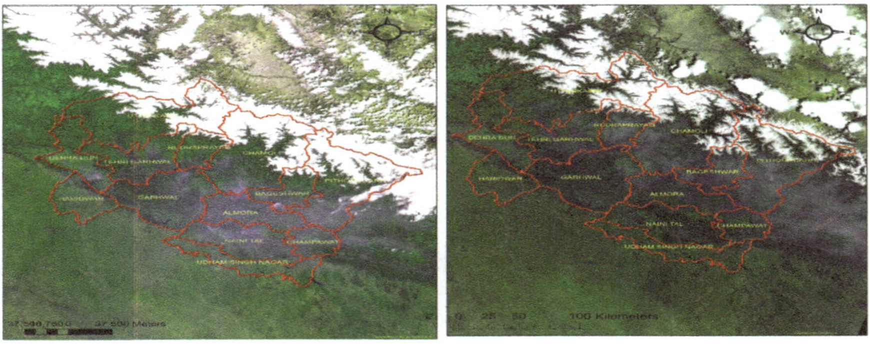

Fig. 6 Fire extent in Uttarakhand. **a** Fire extent on 30th April 2016. **b** Fire extent on 2nd May 2016

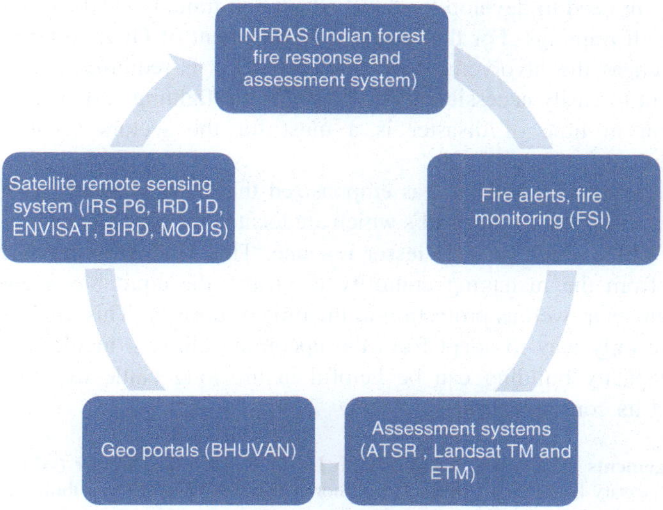

Fig. 7 Description of some fire assessment and reporting systems in India

There is a strong need to develop the people oriented trainings and operation modules which can be easily understood by the common man residing in village, so that they can eventually be successful in protecting their forest and its resources, which will eventually lead to their livelihood improvement and sustainable development of their precious resources.

6 Conclusion

The extreme climatic events affects the forest dependent communities in negative way, either it is excessive rainfall, increase or decrease in temperature or extensive forest fires; it eventually lead to the loss in the valuable forest resources. Present study gives an idea of extent of climatic anomaly as the long dry spell dry spell led to severe forest fires. These severe forest fires are either initiated by local causes (i.e. illicit felling, poaching or land mafia) or by the natural means (i.e. high temperature and absence of rainfall, dry weather conditions), eventually affected the forest resources of Uttarakhand. The high severity of these forest fires has impacted rural economy as the decline of valuable forest based livelihoods was affected by the vast damage from fires.

In India the measures to combat forest fire are sufficient, such as alert system and fire maps developed by FSI, but their outreach is needed to be ensured to common man so that this kind of emergency situations could be tackled with efficiency. To make this suitable for a common man residing in the interior village community, there is strong need to develop localised adaptation trainings and capacity building through small trainings. For this purpose involvement of Governmental organizations as well as the involvement of efficient NGOs is required. Furthermore the development of easily accessible technology for fire fighting and ensuring resource conservation in time of disaster is a must for the sectors involved in VPF management.

During the present study it was emphasized that there is lesser involvement of governmental support for the VPFs which are located far from the main road as well as for the VPFs which provide lesser revenue. This kind of situation needs consideration from the managing authority to ensure the equitable management of forest resources as well as protection in the time of disaster. This kind of combined effort is the only way to adapt from the upcoming climatic incidents as only the adaptive capacity building can be helpful in the long battle to combat climate change and its consequences.

Acknowledgements The author is thankful to Dr. P. P. Dhyani, Director (GBPNIHESD) for providing necessary facilities for work. The author is also thankful for Dr. Subrat Sharma and Dr. R. C. Sundriyal for their valuable suggestions

References

Aponte, C. De Groot, W.J. and Wotton, B.M. 2016. Forest fires and climate change: causes, consequences and management options. *International Journal of Wildland Fire*, *25*, i–ii http://dx.doi.org/10.1071/WFv25n8_FO.

Bahuguna, V.K. and Singh, S. 2001a. Fire situation in India. In Global forest resource assessment 1990–2000. Forest Resources Assessment Programme Working Paper 55. Pp172–177.

Bahuguna, V.K. & Singh, S.. 2001b. The forest fire situation in India. *Int. Forest Fire News* 26.

Brown, J.K. & Smith, J.K., eds. 2000. *Wildland fire in ecosystems: effects of fire on flora*. Gen. Tech. Rep. RMRS-GTR-42, Vol.2. Ogden, Utah: U.S. Department of Agriculture, Forest Service, Rocky Mountain Research Station. 257 pp.

Cai, W, Yang, J (2016) High severity fire reduces early successional boreal larch forest above ground productivity by shifting stand density in northeastern China. *International Journal of Wildland Fire* **27**, 861–875. doi:10.1071/WF15026.

CCES (2015). Centre for climate and energy solutions. India's climate and energy policies. 1–3. 2101 Wilson Blvd. Suite 550, Arlington, VA 22201 703-516-4146.

Chaturvedi, R.K., Gopalakrishnan, R., Jayaraman, M,.Bala, G., Joshi, N.V., Sukumar, R. and Ravindranath, N.H. 2011. Impact of climate change on Indian forests: a dynamic vegetation modeling approach. Mitig Adapt Strateg Glob Change 16(2): 119–142. doi:10.1007/s11027-010-9257-7.

Easterling, D.R., Meehl, G.A., Parmesan, C., Changnon, S.A., Karl, T.R. & Mearns, L.O. (2007) Climate extremes: observations, modeling, and impacts. *Science* 289, 2068–2074.

Franklin, J., Syphard, A.D., Hong, S.H., Mladenoff, D.J., 2005. Altered fire regimes affect landscape patterns of plant succession in the foothills and mountains of southern California. *Ecosystems*, 8: 885–898.

FSI (2016). Overlay analysis of Forest Fire incidences based on FSI Fire monitoring system. Archived from ENVIS centre of Himalayan ecology, GBPNIHESD. G.B. Pant National Institute of Himalayan Environment and Sustainable Development, Kosi-Katarmal, Almora. Uttarakhand. Date 27/6/2016.

GOI,2010. Government of India, 2010. Copenhagen pledge, available at http://unfccc.int/files/meetings/cop_15/copenhagen_accord/application/pdf/indiacphaccord_app2.pdf.

Guha. R. (1989). The Unquite Woods: Ecological Change and Peasant resistance in the Himalaya. Oxford University press, Delhi. 1–214.

India state of forest report (2015). Forest and tree resources in states and union territories. Forest survey of India. Pp 110–288.

IPCC (2007). Climate Change Synthesis Report. Contribution of Working Groups I, II and III to the Fourth Assessment Report of the Intergovernmental Panel on Climate Change [Core Writing Team, Pachauri, R.K and Reisinger, A. (eds.)]. IPCC, Geneva, Switzerland, 104 pp.

Jha, C.S., Gopalakrishnan, R., Thumaty, K.C., Singhal, J., Reddy, C.S., Singh, J., Pasha, S.V., Middinti, S., Praveen, M., Murugavel, A.R., Reddy, S.Y., Vedantam, M.K., Yadav, A., Rao, G.S., Parsi,. G.D., Dadhwal. V.K. (2016). Monitoring of forest fires from space – ISRO's initiative for near real-time monitoring of the recent forest fires in Uttarakhand, India. *Current Science*. 110 (11) 2057–2060.

Kumar, P., Kumar, S. and Joshi, L. 2015. Socioeconomic and Environmental Implications of Agricultural Residue Burning: A Case Study of Punjab, India. ISBN 9788132221463. DOI 10.1007/978-81-322-2014-5. Springer briefs in Environmental Sciences. 1–141.

Lise W (2000) Factors influencing people's participation in forest management in India. Ecol Econ 34(3):379–392.

NRSC ISRO, 2016. BHUVAN geoportal Resourcesat-2 AWiFS data images. http://nrsc.gov.in/Forest_Fire_Uttarakhand. archived on 27th May 2016/personal user.

Podur, J.J., and Martell, D.L. 2009. *The Influence of Weather and Fuel Type on the Fuel Composition of the Area Burned by Forest Fires in Ontario, 1996*–2006. Ecological Applications. 19; 1246–1252.

SFD, 2016. State forest department, Uttarakhand. Data acquired from http://www.vigilanceandlegalcell.co.in/fire/Office-Fire.html on 28/7/2016.

Shijo, J. Anitha, K. and Murthy, M.S.R. 2009. Forest fire in India: a review of the knowledge base. *J. For Res.*14:127–134DOI 10.1007/s10310-009-0116-x.

Xu, J., Grumbine, R.E., Shrestha, A. Eriksson, M., Yang, X., Wang, Y. and Wilkes, A. 2009. The Melting Himalayas: Cascading Effects of Climate Change on Water, Biodiversity, and Livelihoods. *Conservation Biology*, 23(3), 520–530, DOI:10.1111/j.1523-1739.2009.01237.x.

Financing Low-Emission and Climate-Resilient Infrastructure in the Arab Region: Potentials and Limitations of Public-Private Partnership Contracts

Damilola S. Olawuyi

Abstract The aim of this chapter is to discuss the application of the public private partnership (P3) model as a contractual arrangement for financing climate smart infrastructure development in the Arab region, emphasizing the preconditions for implementation and the barriers and difficulties for its application. The methodology approach is based in a doctrinal analysis and desk-based survey of the scope and status of P3 implementation in the Arab region. Due to the scope of the study and the nature of the methodological approach, the survey, by no means, can be regarded as representative. However, since it is combined with the review of the literature, it builds a rough profile of the salient law and institutional barriers to P3 implementation in the Arab region. Larger studies can be undertaken at a later stage, provided that the necessary staff time and funding are available. The results indicate that the absence of P3 laws; lack of standardized contractual framework; absence of P3 institutions; and lack of human capacity are the main legal barriers that must be addressed if the P3 model is to enhance climate resilient infrastructure development in the Arab region. This chapter will be useful to academics and researchers interested in the topic of climate change adaptation in the Arab region.

Keywords Climate infrastructure · Financing · Public private partnerships · Law · Policy · Adaptation strategy

1 Introduction

With increased manifestation of extreme weather events, such as Cyclone Pam in Vanuatu, devastating floods in Nigeria, droughts in South Sudan, hurricanes and fatal heat waves across the United States, there is now increased consensus that the

D.S. Olawuyi (✉)
College of Law and Public Policy, Hamad Bin Khalifa University, Doha, Qatar
e-mail: dolawuyi@hbku.edu.qa

W. Leal Filho (ed.), *Climate Change Research at Universities*,
DOI 10.1007/978-3-319-58214-6_33

effects of climate change are here; and that if left unchecked, climate change may threaten food security, energy supply and public infrastructure in nations of the world (Olawuyi 2015, 2016; IPCC 2013; Christensen et al. 2007). Climate change, however, poses even more serious economic, social, and environmental threats to Arab countries—arguably, more so than any other region in the world. Apart from climate-induced fatal heat waves and debilitating sea level rise, climate change could have wide-ranging effects on extant and future infrastructure in the Arab region (Assaf 2009; Black et al. 2011; Born et al. 2008). This could include potential failure of facilities, reduced life expectancy of buildings and increased operational and maintenance expenditure of electricity, water and aviation infrastructure, amongst others (European Commission 2013; Seiber 2013).

For example, even without climate change, Arab countries, such as Qatar, are currently subjected to tough arid conditions and extreme heat, which typically affect the structural integrity, operation and life span of water, energy, coastal and transportation infrastructure. A region is classified as arid if it is characterized by severe lack of available water. The higher intensity and frequency of hot days in many Arab countries already affect the reliability of agricultural and food systems, water supply and wastewater, transportation networks, and energy supply systems (Assaf 2009). Climate change would only escalate these pre-existing conditions (Zereini and Hötzl 2008; IPCC 2013).

The projected increase in intensity and prolongation of very hot days can result in increased pressure and consequent damage to water, transportation and energy infrastructure and buildings in Arab States (Assaf 2009; Schaeffer et al. 2012; Bollinger et al. 2014). For example, excessive heat could geometrically increase demand for cooling, placing additional demand and pressure on electricity infrastructure others (European Commission 2013; Seiber 2013). Similarly, higher temperature could reduce the efficiency of transportation and energy transmission infrastructure, thereby increasing the overall cost of constructing and maintaining critical infrastructure and buildings in the region. Furthermore, given that a significant percentage of greenhouse gases (GHGs) that cause dangerous climate change come from high emission buildings and structures, failure to build infrastructure that reduce GHG emissions, and that are resilient to the risks posed by climate change, could increase the vulnerability of Arab states to climate change (Zawahri 2010; Assaf 2009).

To effectively address the impacts of climate change on critical infrastructure in the Arab region, there is need for increased investment in climate-smart infrastructure, i.e. buildings, structures and systems that reduce GHG emissions, and improve society's ability to adapt to, and cope with, the risks posed by climate change (Carter et al. 2015; Resilient City Project 2016). Arab countries and cities must develop capacities to absorb future climate related shocks and stresses to social, economic, and technical systems and infrastructures.

Enhancing the adaptive capacity of extant and emerging infrastructure to the risks posed by climate change will require a strategic commitment by Arab countries to integrate and reflect climate objectives in the design, approval, finance and implementation of infrastructure projects. For example, Article 7 (5) of the Paris

Agreement specifically encourages countries to integrate climate adaptation into relevant socioeconomic and environmental policies and actions, where appropriate. This will include redesigning extant water, transportation and energy infrastructure and buildings to enhance their resilience and adaptive capacity; upgrading infrastructure operating and maintenance practices to enhance climate resilience; upgrading building technologies to incorporate low-carbon and energy efficient materials; and investing in new climate-smart infrastructure to replace ageing and dirty ones (Article 9 (e) Paris Agreement). Promoting and achieving climate smart infrastructural development in the Arab region will also require a robust national and regional adoption of key principles that reinforces the integration of climate risk and climate objectives in national infrastructure policies and plan. For example as many Arab countries continue the construction of legacy projects and infrastructure, for example World Cup stadia in Qatar; expansive road construction projects across Kuwait; and rail transportation projects in Saudi Arabia to mention a few; preparedness for extreme weather conditions and events must be robustly incorporated into project design, approval, financing and implementation plans.

However, shifting to climate smart infrastructure is a capital-intensive commitment. About US$90 trillion in infrastructure investment is needed globally by 2030 to achieve climate-smart infrastructure (Rydge et al. 2015; Bhattacharya et al. 2015; Meltzer 2016). Considering the high level of capital required to design, construct and upgrade extant and future infrastructures to become more resilient, government, with its many responsibilities in other sectors of the economy, cannot fund climate smart infrastructure projects alone. Mobilizing adequate finance to build and maintain climate smart infrastructure will require leveraging both public and private sector capital. There is therefore a need for governments to provide the right legal and policy incentives for private sector participation and investment in climate smart infrastructure development across the Arab region. According to analysis by Brooking Institute, about half of the additional $3 trillion per annum that will be needed for sustainable infrastructure over the next 15 years must come from private investment, while other financing will come from public sources such as governments, development agencies, and multilateral, regional, and national development banks (Meltzer 2016).

P3 contracts have been increasingly promoted globally as legal and contractual instruments that provide opportunities for governments to collaborate with private sector stakeholders to finance and share the typically high cost of developing climate adaptation infrastructure projects (Tolani 2013). Over 1000 P3 projects worth at least $1.5 trillion have been completed during the last 25 years (McNichol 2013). In recent years, annual volume of P3 transactions has averaged around $50 to $60 billion. P3s establish a long-term contract between a private party and a government entity for providing public infrastructure; allocate risks and management responsibility to the private sector party and link returns to performance (World Bank 2016). By reducing public sector involvement in the design, financing and maintenance of public infrastructure, P3s have significantly stimulated private sector investment and participation in the design of climate smart infrastructure in many nations of the world (QFC 2012).

Despite the potential and strengths of the P3 model in financing energy infrastructure development, its specific viability for financing climate resilient infrastructure in the Arab region has not been given detailed analysis and consideration in the literature. In order to attract the significant investment required to develop climate-smart infrastructure through P3 frameworks, innovative governance reforms are required in the way infrastructure projects are designed, approved, financed and implemented to remove limitations and barriers.

This paper examines the potentials and limitations of the P3 model as a contractual arrangement for financing climate adaptation infrastructure in the Arab region. First, it outlines the rationale and preconditions for implementing P3s in climate adaptation context. Following this, the paper assesses the main legal barriers that must be addressed if the P3 model is to enhance climate resilient infrastructure development in the Arab region. Finally, consideration is given to how law, policy and governance can be strengthened across the Arab region to enhance the application of P3 frameworks in climate smart infrastructure projects.

2 Public Private Partnerships Models and Climate Adaptation

2.1 Background

A P3 contract is an arrangement between a public agency (federal, state or local) and a private sector entity, in which the resources, expertise, skills and assets of each sector (public and private) are shared in delivering public infrastructure and/or services for the use of the general public (NCPPP 2016; World Bank 2016). Under this arrangement, the risks and responsibilities for designing, financing, including maintaining, enhancing or constructing the necessary infrastructure are typically allocated to the private sector firm, while the government provides policy, planning and regulation support for the project. For example, Article 2 of the Egypt PPP Law expressly entrusts the responsibility and risk of designing, financing, equipping, operating, constructing and rehabilitating public infrastructure to the project company during the duration of the P3 contract, while government is tasked with providing administrative licenses and clearances for the project company. This arrangement provides a real incentive for the project company to serve as a key stakeholder in project design and execution.

P3 contracts have been increasingly promoted globally as legal and contractual instruments that provide opportunities for governments to collaborate with private sector stakeholders to finance and share the typically high cost of developing climate adaptation infrastructure projects (Tolani 2013; McNichol 2013). Countries such as the United States, United Kingdom, Australia, Canada and New Zealand have achieved varying levels of success in adopting P3 models for large scale renewal of public infrastructure including roads, airports, hospitals, railroads, power projects, stadiums, prisons, schools, and waste management systems (QFC 2012). Similarly in India, over 880 PPP projects have been implemented or are under implementation in the last 11 years (Public Private Partnerships in India 2016).

The incentive model of P3s makes them particularly suited for climate smart infrastructure projects. Unlike traditional 'design-bid-build' procurement or service contracts in which the government invites private sector firms to build and operate a public infrastructure for a fee, rewards and remuneration to private sector firms under a P3 model is linked to long-term efficacy and sustainability of the project (Tolani 2013). The 'value for money' predicate of the P3 model can incentivize private sector participant to, as primary and long-term stakeholders, deliver projects that are sustainable and durable. By allocating project risks, governments can effectively leverage the resources, expertise and skills of the private sector to build and deliver legacy infrastructure.

Similarly, the P3 model is particularly very important in a low oil price world. For example, in many Arab States, the drop in the price of oil has weakened government income and earnings, making it very difficult for governments to continue or complete several ongoing infrastructure projects (Husain et al. 2015). Governments across the Arab region have also been forced to cut services and jobs, while several projects have had to be postponed. Similarly, in countries such as Kuwait and Qatar, governments have over the last decade already committed significant resources to building roads, energy and water infrastructure. Qatar for example has set out plans to invest over US$ 30 billion in several ongoing infrastructure projects; Dubai is completing the world's largest airport, Saudi Arabia is completing four mega cities with over $100 billion expenditure; while several road projects are ongoing in Kuwait (QFC 2012). Upgrading these infrastructures, or building new structures that are climate resilient, will be a significant endeavor, which governments may struggle to undertake and achieve given current shortfalls in projected and actual incomes, and the need to continue high priority projects to meet the demands of a growing population. Given the typically high costs and resource commitment required to finance climate smart infrastructure, the opportunity to leverage private sector participation and investment makes P3 not only desirable, but very essential for climate change adaptation.

Third, one of the key questions facing infrastructure projects in the Arab region is their long-term sustainability and maintenance. This is the question of how legacy infrastructure projects, such as the several road networks and energy infrastructure projects, will be maintained over the long term. Maintaining public infrastructure is a complex and demanding commitment that can be entrusted to a private sector entity under a P3 arrangement. A P3 model draws on the expertise of the private sector to maintain, enhance and preserve public infrastructure thereby reducing the responsibility of governments in this regard. Private sector supervision and oversight of adaptation infrastructure could significantly enhance the sustainability prospects of climate smart infrastructure projects.

Similarly, P3 projects provide a platform for countries to make fast and effective progress in meeting the urgent requirements of climate change adaptation. About 80% of P3 projects are delivered quicker than projects constructed under traditional procurement methodologies, which often take longer to complete (Conference Board of Canada 2015). Similarly, according to a 2007 study conducted by the Infrastructure Partnerships Australia, 18% of traditionally procured projects had timing overruns, while only 10% of P3 projects missed deadlines (Infrastructure Partnerships Australia 2007). In addition, traditional projects were delivered 26% later than originally expected, while P3 projects were delivered only 13% later than expected. Delivering climate resilient infrastructure on time is a specifically important consideration for limiting social, environmental and economic impacts of dangerous climate change. Article 7 of the Paris Climate Agreement specifically encourages countries to take into account the urgent and immediate need of climate adaptation while designing and planning responses to the adverse effects of climate change. Infrastructure required to address the impacts of climate change must be designed and completed in the fastest and most effective manner given the urgent nature of climate change. P3 projects could provide fast and reliable financing that would help fast track the completion of climate smart infrastructure projects.

Despite the potential and strengths of PPP contractual model in financing energy infrastructure development, and despite the adoption of P3 models across the world in energy, transportation and water sectors, P3, as a contractual and financial model, remains largely underutilized in the Arab region. For example, currently only four Arab countries—Egypt, Morocco, Kuwait and Dubai in the United Arab Emirates (UAE)—have enacted P3 laws. Furthermore, while Saudi Arabia, Kuwait and the UAE have achieved considerable success in applying P3 models for power and water sector projects, lack of coherent and comprehensive legal framework on P3 investments has stifled further adoption and growth of the P3 model across the region.

Given that P3s allocate and share a bundle of rights, obligations and risks between public and private sector stakeholders, there is need for a robust legal and institutional framework that will spell out and protect the respective obligations and expectations of both public and private participants. To attract private sector confidence and interests in P3 projects, there is need to address and remove institutional and regulatory hurdles that currently hinder private sector participation in P3 projects. Without a strong regulatory and institutional foundation that provides clarity and certainty for investors, it would be difficult to attract private sector financing for climate smart infrastructural development.

2.2 *Methodology*

Given the expository nature of its subject, the methodology approach of the paper is based in a doctrinal analysis and desk-based survey of the scope and status of P3 implementation in the Arab region. The paper relies primarily on published literature in P3 to survey the barriers and limitations to the adoption of P3 contractual arrangements in the Arab region. Desk-based analysis of published literature is adopted as the methodical basis of this paper because existing work in the field have satisfactorily compiled the scope and status of P3 implementation in the Arab region. What remains absent is a detailed examination of the legal barriers to the entrenched implementation of P3 arrangements in the region. This paper moves the debates forward by discussing how legal barriers and limitations to the adoption of P3 arrangements can be addressed.

Due to the scope of the study and the nature of the methodological approach, the survey, by no means, can be regarded as representative. However, since it is combined with the review of the literature, it builds a rough profile of the salient law and institutional barriers to P3 implementation in the Arab region. Larger studies can be undertaken at a later stage, provided that the necessary staff time and funding are available.

3 Results and Analysis

3.1 *P3 and Climate Smart Infrastructure Projects in the Arab Region: Survey of Barriers and Limitations*

Given the tendencies to mix up P3 arrangements with procurement, service or turn-key construction agreements, the development of P3 projects require clarity on the nature and model of relationship envisaged and created; and the how risks are allocated by the parties. P3, as a financial and contractual model, has therefore thrived in countries that have put in place adequate legal contractual and institutional frameworks to incentivize and govern the design, approval, financing and implementation of P3 projects. This section examines legal paradoxes and gaps that must be addressed if the P3 model is to enhance climate resilient infrastructure development in the Arab region.

(a) ***Lack of legal framework on P3s***: One of the most important barriers to the adoption of P3 models in the Arab region is the absence of clear and coherent legal framework on the design and implementation of P3 projects. Many Arab countries are yet to enact P3 laws that will ascertain the requirements and process for developing, financing and implementing P3 projects. As earlier noted, while P3 laws have been expansively espoused and effectively utilized in Europe, North America, Latin America and Asia, P3 laws have only been enacted in three Arab countries, Egypt, Kuwait and Dubai in the United Arab Emirates. In the United States alone, more than thirty (30) P3 legislation have been enacted across state and federal levels (McNichol 2013).

The absence of P3 laws in Arab States creates legal uncertainty for investors who would generally be wary of investing in a country where there is no legal protection for their investments. Also, investors will want to have a clear knowledge of a country's P3 framework and goals, the types of P3 arrangements in the country; approval process, project and sector eligibility; contract formation process;

responsibilities and risks of the parties; registration and certification of projects; baselines; verification, validation, reporting and monitoring. These are methodological questions that must be carefully laid out in a legal framework designed to clarify and govern P3 investments.

P3 laws can also be very helpful in addressing barriers and gaps in commercial or concession laws that may hinder the successful implementation of P3 projects. For example, many Arab countries already have commercial and procurement laws that grant national authorities higher participation and ownership rights in business ventures. This could range from local laws that stipulate that more than 50% of a business venture must be owned by a government entity or laws that stipulate that public infrastructure can only be managed and maintained by the national authority. While such laws may work in traditional procurement models, they do not provide adequate legal basis for the effective implementation of P3 projects. The adoption of clear and specific P3 laws could provide robust and tailored requirements that align with the P3 methodology. Enacting P3 laws will also send a positive signal to investors about a country's recognition, interest and commitment to accelerating infrastructure development and growth though P3.

P3 laws provide important statutory and legislative foundation for the implementation of P3 projects. Considering the importance of P3 projects to effective climate adaptation, it is important for governments across Arab States to provide legislative foundation for the recognition, financing and implementation of P3 projects. In Qatar for example, the adoption of a P3 law has been in the pipeline. Fast tracking the adoption of this law could open up P3 investment in the country and in other Arab countries that are yet to enact P3 legislation.

(b) ***Lack of Standardized Contractual Framework***: One of the key causes of risks and disputes in P3 investments is the lack of clear, detailed contractual framework that allocates risks. Given that a P3 investment is a contractual relationship between the public and private sectors for the execution of a project or service, clear and detailed information on how risks, returns and responsibilities are allocated must be comprehensively delineated in a P3 contract. Having a contract in place will reduce the possibility of conflict or risk at all stages of the project life cycle. Limiting risks in P3 projects will on the part of a prudent investor involve a careful investigation of the scope and nature of projected risks, especially how risks of governmental interference, project confiscation or nationalization and midterm changes in contractual terms have been allayed in the P3 contract.

Due to limited practice and adoption of P3 models in the Arab region, there has been very sparse information on the contractual and fiscal terms to expect when investing in P3 projects in the Arab region. Currently, many Arab countries are yet to evolve standardized P3 contracts, making it difficult for investors to clearly ascertain the underlying commercial terms, risks and responsibilities for designing, financing, building and maintaining infrastructure project. In countries with mature P3 markets, these information are available in model contracts that have been compiled by national authorities, after consultation with private sector stakeholder, to clearly establish the underlying commercial principles that PPP projects should adhere to (See for example New Zealand Treasury Department 2015). A standard form P3 agreement makes it possible for investors to perform long-term risk analyses and assessment while determining investment locations. It also enhances the transparency standardization and certainty of a P3 process.

Furthermore, considering the risk allocation components of P3 projects, banks and insurance companies are typically interested in scrutinizing the contractual terms in providing financing or insurance coverage for a private sector participant. Standardizing and setting out contractual terms also increases the possibilities of attracting financing and loans (Arab Financing Facility for Infrastructure 2015).

To catalyze access to private finance sources, insurance and loans for climate smart infrastructure projects, it is important for Arab countries to, in conjunction with private sector stakeholders, evolve clear, credible and market-tested precedent for P3 projects. Standardized P3 models developed in mature P3 markets in other regions can provide guidance for developing standardized P3 contracts across the Arab region. Such templates can be reworked to reflect local market realities, and to ensure consistency with existing legislation and government requirements, for example local content requirements, Sharia compliant contractual terms, responsible product sourcing and local employment practices.

(c) ***Absence of dedicated P3 institutional framework***: The effective implementation of P3 models in any country requires the strengthening of existing government institutions, creation of special implementation institutions, and the establishment of a direct synergy between all government institutions, both new and old, to fast track and simplify the implementation of P3 investments. The absence of legally recognized national authority for P3 projects, coupled with the lack of coordination amongst existing government institutions and ministries, remains one of the serious institutional challenges facing the successful implementation of P3 projects in the Arab region.

According to a recent study, at least 50 countries have established a PPP unit at the federal government level, however only 2 of those dedicated PPP units are in the Arab region, i.e. Kuwait and Egypt (QFC 2012). Qatar for example, like many other Arab states, is yet to establish a dedicated PPP Unit. Successful implementation of P3 projects in Australia, Canada, United Kingdom, and New Zealand have been attributed to strong and well-managed P3 units, while lack of an institutional framework for P3 projects has hampered the growth of P3 projects in Qatar (QFC 2012). In addition to promoting public awareness on P3 projects, a P3 unit can coordinate and spearhead the development and design of P3 projects in alignment with government priorities and enabling laws. The P3 Unit can also guide investors in proposing and developing clean and sustainable projects that are in line with a country's National Vision; and more specifically that can result in real, measurable and long-term benefits to climate change adaptation and mitigation.

For a P3 unit to function optimally, it must be created and backed by a national law. The law creating the unit is expected amongst other things to contain clear statements regarding the unit's authority, objectives, organizational structure, functions, priorities and mode of operation. It should also contain clear statements regarding the extent of the legal authority, which it is granted. The P3 unit will also serve as the national authority on P3 projects, with powers to enter into P3 arrangements for and on behalf of the government.

Establishing a P3 unit as a separate entity could also enable it to function as a one-stop shop that can coordinate with other relevant agencies and ministries to reduce the administrative and licensing requirements for P3 projects. This is particularly important considering that climate infrastructure project, for example a waste water management project, could require permits and approvals from different government ministries, organizations and departments. For example, the Ministry of Finance has roles to play in currency importation and approval of payment instruments, while the Ministry of Interior is responsible for granting approvals for project execution. Also, the Ministry of Economy and Commerce will be needed to grant approvals for foreign agencies to carry out investment activities while Ministries such as power, energy and transportation have prominent roles to play in approving all projects to be executed under their ministries. This underscores the importance of coordination among these ministries. With solid intergovernmental coordination, granting and getting approvals for P3 projects will be less cumbersome and straightforward. Such coordination will also simplify the process of passing information between government ministries, thereby removing inefficiencies and bureaucracies.

Establishing central P3 units can help accelerate the process of obtaining relevant approvals and licenses for climate smart infrastructure projects through the P3 model. To promote P3 investments for climate change adaptation infrastructure, adequate institutional frameworks must be established in line with national preferences to empower the P3 unit as a recognized national authority on P3 investments.

(d) ***Capacity gaps***: Flowing from the lack of comprehensive legislation and institutions on P3 across many parts of the Arab region is that climate change administrators and lawyers are yet to evolve sufficient technical capacity and skills that are directly relevant for structuring P3 investments and projects. For example, what are the required due diligence steps and contractual negotiation steps for allocating project risks and minimize disputes between private and public sector participants? These are questions that must be sufficiently understood by administrators, policy leaders, project evaluators, investment bankers, auditors and lawyers involved in formulating P3 policies and project requirements.

The required sophistication to successfully navigate the entire project chain of a P3 project can be improved in many of the surveyed Arab countries. To promote P3 projects, it is important to provide capacity development opportunities for administrators to acquire technical knowledge about the methods, requirements and challenges of a P3 project.

The foregoing gaps and barriers in the legal and institutional architecture for P3 investments in the Arab region do not call for intellectual surrender. Rather, they can be addressed holistically by governments across the Arab region by putting in place adequate law and policy frameworks that will enhance and support the application of P3 models in the design and implementation of climate adaptation infrastructure projects.

3.2 Pre-conditions for the Effective Implementation of P3 Projects in the Arab Region

National authorities across the Arab region must realize the importance of the P3 model as an alternative, and mutually supportive, platform through which the costs of financing climate change adaptation can be shared with private sector investors. With reduced oil income, revenue drop and increased demand on government budget, countries must look to private sector participation in developing the infrastructure required to achieve climate resilient cities. Success lies in the abilities and commitment of governments across the Arab States to address the above noted limitations and barriers in an efficient and pragmatic manner.

To improve the efforts for the implementation of P3 models in financing climate smart infrastructure development across the Arab region, some points should be considered:

3.2.1 Update and Reform Existing Investment Laws

The starting point is for national authorities across the region to reform and update current commercial and investment laws that hinder P3 investments. For example, existing laws and regulations on foreign investment that stipulate stringent requirements for participation for foreign nationals or investors may prohibit or impede a proposed P3 project (World Bank 2016). Furthermore, stringent local laws on repatriation of earnings, investment taxation, insurance and dispute resolution may also serve as disincentives for private sector participation and investments in P3 projects. It may therefore be necessary for countries to review and fundamentally reform extant investment laws and institutions that could hinder the robust implementation of P3 projects.

3.2.2 Establish Comprehensive P3 Laws and Policies

In addition to reforming extant investment laws to simplify private sector participation, it is also pertinent to establish comprehensive laws on P3 investments that can provide clear guidelines on P3 investments especially in the climate adaptation context. The law should amongst other things provide fiscal and tax incentives for investments in the renewable energy sectors and harmonize the conflicting laws and regimes on P3 investments. Most importantly, the P3 legislation should establish a designated institution or focal point as a one-stop shop for P3 investments. This would help simplify the processes and procedures for seeking and obtaining regulatory permits and investment approvals; it would also serve as a rallying point that could foster intergovernmental coordination and linkages amongst the many institutions that currently play important roles in the approval and execution of public infrastructure projects.

3.2.3 Establish Focal Institutions on P3 Projects

To promote P3 projects, it is important to establish a focal institution or administrative unit that will coordinate the design, approval and implementation of P3 projects across various sectors, including in the climate change adaptation context. Apart from serving as a one stop shop that will streamline the approval processes for projects, such an institution would also provide capacity development opportunities for administrators to acquire technical knowledge about the methods, requirements and challenges of a P3 project. By empowering and establishing a focal institution on P3 projects, investors across multiple sectors can obtain relevant information and develop an institutional understanding about the process and methodology for implementing P3 projects.

3.2.4 Promote Regional Cooperation and Knowledge Sharing

Regional interaction and knowledge sharing between countries with experience and practice on P3 can help promote expertise on how P3 can accelerate infrastructure development in the region, especially in the context of climate smart infrastructure. While countries such as Kuwait and Egypt have some experience with P3, several other countries within the Arab region have little to no experience at all. It is therefore important to promote the cooperation and knowledge sharing between regional networks and institutions, within and outside of the Arab region, on how P3 models can help facilitate climate smart infrastructure development. Regional centers and platforms that can also enhance the exchange of ideas, best practices and knowledge on existing P3 project opportunities, model contracts and practical steps for planning and implementing P3 projects in the climate adaptation context. A good example is the United Nations P3 platform which provides an international platform for sharing ideas, learning materials, opportunities and best practices on P3 projects (http://platform.un-ppp.org/).

A regional knowledge sharing platform could help capture the common challenges and approaches in the design and implementation of P3 climate infrastructure projects within the Arab region.

4 Conclusion

The P3 model provides viable opportunities for countries in the Arab region to attract additional financing, technical expertise, skills and commitment from the private sector in the development, financing, implementation and maintenance of climate smart infrastructure projects. Lack of widespread and sustained implementation of P3 initiatives, especially to stimulate required climate change adaptation infrastructure in the region, is exacerbated by the absence of robust legal, contractual and institutional frameworks on P3 projects. These barriers must be addressed if Arab countries are to leverage on the potentials of P3 investments to attract climate smart infrastructure projects and developments.

The adoption of P3 laws in Kuwait, Egypt and Dubai show how P3 models could be applied for financing large-scale infrastructure projects in the Arab region. With regard to the unique challenges facing the maintenance and sustainability of climate adaptation projects financed under the P3 model, future research will be necessary to develop a better understanding of the current situation of existing and future P3 climate infrastructure projects, and ways to better organize and promote sustainability in the design, approval, finance and implementation of such projects.

References

Arab Financing Facility for Infrastructure (2015) Arab Financing Facility for Infrastructure Developing infrastructure for growth and regional integration in Arab countries. 1–2. http://www.ifc.org/wps/wcm/connect/d75c10804aee4412a0f3fa888d4159f8/AFFI+Brochure.pdf?MOD=AJPERES (Last Accessed 21 September 2016).

Assaf H (2009) Impact of Climate Change: Vulnerability and Adaptation Infrastructure 1–5, http://www.afedonline.org/afedreport09/english/char9.pdf (Last Accessed 21 September 2016).

Bhattacharya A, Oppenheim J, Stern N, (2015) Driving sustainable development through better infrastructure: key elements of a transformation program pp 1–10. https://www.brookings.edu/research/driving-sustainable-development-through-better-infrastructure-key-elements-of-a-transformation-program/ (Last Accessed 12 October 2016).

Black E, Brayshaw D, Slingo J, Hoskins B (2011) Future climate of the Middle East, in S Mithen & E Black (eds), Water, Life and Civilisation: Climate, Environment and Society in the Jordan Valley. Cambridge University Press, Cambridge, pp. 51–62.

Bollinger LA, Huibregtse JN, · Maas N, Snelder M, Tavasszy L.A, Wit M.S. de · Bogmans, CWJ, Chappin EJL, Dijkema GPJ, Schenk T, Thienen P, van · Wols B, (2014) Climate adaptation of interconnected infrastructures, Regional environmental change 14 (3), 1–15.

Born K, Fink AH, Paeth H (2008) Moroccan climate in the present and future: combined view from observational data and regional climate scenarios", in Zereini, F, Hötzl H (2008), Climatic changes and water resources in the Middle East and North Africa. Springer, Heidelberg, pp. 29–45.

Carter J, Cavan G, Connelly A, Guy S., Handley J, Kazmierczak A, (2015). Climate change and the city: Building capacity for urban adaptation. Progress in Planning 95(1): 1–66.

Christensen JH, Hewitson B, Busuioc A, Chen A, Gao X, Held I, Jones R, Kolli RK, Kwon WT, Laprise R, Magaña Rueda V, Mearns L, Menéndez CG, Räisänen J, Rinke A, Sarr A, Whetton P, 2007: Regional Climate Projections. In: Climate Change 2007: The Physical

Science Basis. Contribution of Working Group I to the Fourth Assessment Report of the Intergovernmental Panel on Climate Change [Solomon, S., D. Qin, M. Manning, Z. Chen, M. Marquis, K.B. Averyt, M. Tignor and H.L. Miller (eds.). Cambridge University Press, Cambridge, United Kingdom and New York, NY, USA 847–940.

Conference Board of Canada (2015), Experience pays: the growing value of public-private partnerships in Canada 1–10. http://www.conferenceboard.ca/press/newsrelease/13-08-21/experience_pays_the_growing_value_of_public-private_partnerships_in_canada.aspx (Last Accessed 12 October 2016).

Conference of the Parties to the United Nations Framework Convention on Climate Change, Paris Agreement, Twenty-first session Paris, 30 November to 11 December 2015. https://unfccc.int/resource/docs/2015/cop21/eng/l09.pdf. (Last Accessed 12 October 2016).

European Commission (2013), Adapting infrastructure to climate change 1–5. https://ec.europa.eu/clima/policies/adaptation/what/docs/swd_2013_137_en.pdf (Last Accessed 12 October 2016).

Infrastructure Partnerships Australia (2007), Performance of PPPs and Traditional Procurement in Australia 1–5. http://infrastructureaustralia.gov.au/policy-publications/publications/Allen-Consulting-Performance-of-PPPs-and-Traditional-Procurement-in-Aust-2007.aspx (Last Accessed 12 October 2016).

Intergovernmental Panel on Climate Change (IPCC) (2013). Working Group I Contribution To The IPCC Fifth Assessment Report. Climate Change 2013: The Physical Science Basis. Summary For Policymakers (Intergovernmental Panel on Climate Change, Switzerland), Sept. 27, 2013.

Husain A, Arezki R, Breuer P, Haksar V, Helbling T, Medas P, Sommer M, IMF Staff Team. Global implications of lower oil prices. International Monetary Fund, July 2015 http://www.imf.org/external/pubs/ft/sdn/2015/sdn1515.pdf. (Last Accessed 12 October 2016).

McNichol, D., (2013). The United States: the world's largest emerging P3 market: rebuilding America's infrastructure pp 1–15. https://www.aig.com/content/dam/aig/america-canada/us/documents/insights/final-p3-aig-whitepaper-brochure.pdf. (Last Accessed 12 October 2016).

Meltzer, J., (2016) Financing sustainable infrastructure pp 1–5. https://www.brookings.edu/wp-content/uploads/2016/08/global_20160818_financing_sustainable_infrastructure.pdf (Last Accessed 12 October 2016).

New Zealand Treasury Department, (2015). Standard Form Public Private Partnership (PPP) Project Agreement. http://www.treasury.govt.nz/statesector/ppp/standard-form-ppp-project-agreement (Last Accessed 12 October 2016).

National Council for Public-Private Partnerships (2016), Public-Private Partnerships Defined. http://www.ncppp.org/ppp-basics/7-keys/ (Last Accessed 12 October 2016).

Nierop, Sam C.A., (2014). Envisioning resilient electrical infrastructure: a policy framework for incorporating future climate change into electricity sector planning Environmental Science & Policy 40: 78–84.

Public Private Partnerships in India (PPP India) (2016), Database of public private partnership projects. https://www.pppinindia.gov.in/ (Last Accessed 12 October 2016).

Resilient City Project (2016) Resilient Cities. http://resilientcities2016.iclei.org/ (Last Accessed 12 October 2016).

Olawuyi, D (2015) Proposal for a Climate Compensation Mechanism for Small Island States: Response to Maxine Burkett, Santa Clara Journal of Int'l L. 13 (1): 133.

Olawuyi, D (2016) The Human Rights Based Approach to Carbon Finance. Cambridge University Press, Cambridge, pp 1–15.

Rydge, J, Jacobs, M., Granoff, I., (2015) Ensuring new infrastructure is climate-smart. Overseas Development Institute, Working and discussion papers, October 2015.

Qatar Financial Center (QFC), (2012) Public private partnerships: a vehicle of excellence for the next wave of infrastructure development in the GCC http://online.wsj.com/adimg/assetmanagement-related_ppp.pdf (Last Accessed 12 October 2016).

Schaeffer, R, Szklo AS, Pereira de Lucena AF, Borba BS, Nogueira LP, Fleming FP, Troccoli A, Harrison M, Boulahya MS (2012) Energy sector vulnerability to climate change. Energy 38 (1): 1–12.

Sieber, J. (2013) Impacts of, and adaptation options to, extreme weather events and climate change concerning thermal power plants. Climatic Change 121 (55): 55–66.

Tolani, O (2013) An examination of risk allocation preferences in public-private partnerships in Nigeria, Afe Babalola University: Journal of Sustainable Development Law and Policy 2 (1): 206–221.

World Bank, (2016) Public-Private Partnerships http://www.worldbank.org/en/topic/publicprivatepartnerships. (Last Accessed 12 October 2016).

Zawahri N, (2010) 2nd UNITAR-Yale Conference on Environmental Governance and Democracy, September, Building institutional adaptive capacity to respond to climate change in the Middle East. pp. 17–9. https://www.unitar.org/event/full-catalog/2nd-unitaryale-conference-environmental-governance-and-democracy-strengthening (Last Accessed 12 October 2016).

Zereini F, Hötzl H (2008) Climatic changes and water resources in the Middle East and North Africa. Springer, Heidelberg, pp. 29–45.

Urban Stormwater Infrastructure Planning: The Contribution of a Set of Universities to the Sustainable Management of Stormwater in Baltic Cities

Walter Leal Filho, Ana Maria Jones, Anne Paavolainen, Tuomas Valtonen, Maret Merisaar, Piia Leskinen and Gunta Slihta

Abstract Universities are known to be able to provide an important contribution to climate change adaptation. The aim of this paper is to present an overview of current state, challenges, and shortcomings in urban stormwater management in the Baltic cities of Turku, Tallinn, Hamburg and Riga, outlining the contribution of the partner universities to this process. Based on practical examples implemented in cities in and outside the Baltic Sea Region (BSR), this paper makes the case for the need to integrate green infrastructure-related strategies as ecosystem services that can yield multiple benefits and present opportunities for the benefit and well-being of the community. This paper suggests that green strategies are key methods to boost livability and the development of an integrative urban stormwater strategy by the corresponding authorities in charge of implementation efforts.

Keywords Water resource efficiency · Stormwater management · Green infrastructure · Urban rainwater recycling · Sustainable development · Green growth · Baltic cities

---

W. Leal Filho (✉) · A.M. Jones
Faculty of Life Sciences, Hamburg University of Applied Sciences,
Ulmenliet 20, 21033 Hamburg, Germany
e-mail: walter.leal@haw-hamburg.de

A. Paavolainen · T. Valtonen
Technology Research Center, Joukahaisenkatu 3 Stair B 5. Floor ICT-City,
20520 Turku, Finland

M. Merisaar
Tallinn University of Technology, Ehitajate Tee 5 U03-236,
19086 Tallinn, Estonia

P. Leskinen
Turku University of Applied Sciences, Lemminkäisenkatu 30,
20520 Turku, Finland

G. Slihta
Institute of Physical Energetics, Aizkraukles 21, Riga 1006, Latvia

W. Leal Filho (ed.), *Climate Change Research at Universities*,
DOI 10.1007/978-3-319-58214-6_34

1 Introduction: Climate Change and Urban Sustainable Development

Climate change is known to pose challenges to cities across the world. Current demographic trends in global urbanization indicates that cities are preferred areas for living. Half of the world's population already lives in urban areas and it is projected that by 2030 the urban world's population would reach 60% (OECD 2010). Effective management of essential water resources is an issue of transnational interest for urban areas striving to provide livability and sustainable growth. Flooding caused during extreme wet weather events is a major issue as excessive rainwater clogs combined sewer systems and pipes where stormwater and wastewater is discharged, creating overflows and further contamination of water streams. The trend toward the use of excessive impervious surfaces for the sake of development can also exacerbate the generation of excessive water runoff and as such, further the pollution of receiving water bodies (Hoyer et al. 2011). Additional negative effects associated with sealed surfaces involved soil sustainability and soil degradation. Studies have shown that soil sealing has a direct impact in the landscape and in the ability of a given site to perform natural soil functions of water absorption, retention, storing, filtering and transforming nutrients and substances (McBratney et al. 2014). Excessive development has an impact on the urban climate; thus, the Institute for Advanced Sustainability Studies (2013) found that sealed surfaces could be 20 °C warmer than unsealed surfaces shaded by vegetation. Further, the trend towards an increase in global temperatures as a result of climate change (IPCC 2013) continues to put cities at a higher level of vulnerability and risk.

Green infrastructure (GI) is a multi-functional approach to landscape ecology management that is being implemented in many cities globally through various systems and technologies intended to perform in a decentralized manner and with the main purpose of enhancing and restoring natural ecosystems and water cycles. The United States Environmental Protection Agency, EPA, defines green infrastructure from a city and neighborhood perspectives stating that:

> At the scale of a city or county, green infrastructure refers to the patchwork of natural areas that provide habitat, flood protection, cleaner air, and cleaner water. At the scale of a neighborhood or site, green infrastructure refers to stormwater management systems that mimic nature by soaking up and storing water (EPA 2016).

The components of green infrastructure are strongly linked to ecosystem services, biodiversity, and well- being. Ecosystems services in urban areas include areas like parks and gardens that yield benefits in the form of air quality, water regulation, local climate regulation, cultural heritage, recreation, and education. The benefits that are derived from the interconnection with these systems, is better known as ecosystem services (EEA 2010). The basis for the provision of ecosystem services lies in biodiversity—the foundation of ecosystem services to which human well-being is said to be intimately linked. The absence of sufficient interconnection between natural ecosystems and humans in urban areas is a challenge in today's urban development as the availability of landscape areas competes with accelerated need for urbanization. It has been noted that losses in biodiversity and associated changes in ecosystem services have caused people to experience a decline in well-being, and to some social groups to be pushed into poverty. Studies by the EEA (2010) found that, exposure to nature has the potential to reduce blood pressure, improve cognitive abilities, and increase the feeling of happiness. Further, a study of the domestic garden and its contribution to urban green infrastructure, Cameron et al. (2012) found that gardening as a pastime is seen as a retreat that relates to ownership, identity and the ability to interact with nature; thus in this context, the act of gardening puts in perspective, the deep connection that exist between greenery (and its properties), and human health. On the opposite side, findings of the links between biodiversity, ecosystem services and well-being, also explained that biodiversity loss and deterioration of ecosystem services, contribute (directly or indirectly) to worsening health, higher food insecurity, increasing vulnerability, lower material wealth, worsening social relations, and less freedom of choice. Thus, green infrastructure offers a multi-scale approach. The application of GI can encompass larger rural areas and landscapes from a regional, national, and transnational approach; and it can also be implemented at a neighborhood or at the building level. However, the effective connectivity of the various systems across regions and urban areas is what can give green infrastructure a larger meaning and impact on the land.

Across the board, Baltic Sea Region cities are being confronted with individual challenges from urbanization, geological and climatic aspects, diverse typologies, a lack of applicability of resource efficiency and planning measures in some regions-to deficiencies in the effective management and regulation of stormwater resources across regions. Against this backdrop, this paper presents an up-to-date view and current status of the urban stormwater management situation in the Baltic. The paper takes into consideration the current state of affairs in the selected cities, and brings to light the constrains and bottlenecks hindering development, to then discuss methods towards a common strategy for development to support sustainable growth and promote best practices in BSR cities.

2 The Need for Urban Stormwater Management

Previous studies have found two main activities contributing to pollution and deterioration of surface waters, namely, urban runoff and combined sewer overflows (CSOs).

Practitioners have suggested various approaches as alternatives to deal with urban stormwater management, including Sustainable Urban Drainages Systems (SUDS), Water Sensitive Urban Design (WSUD), Low Impact Design (LID), and Best Management Practices (BMPs). These concepts have a common approach to deal with urban stormwater at its source. They suggest decentralisation of excessive rainwater as a solution to tackle various urban challenges that range from stormwater design and management aspects, to volume reduction, social and economic aspects. The need for urban stormwater management is then based on multiple urban challenges and the hydrological capacities of the city's infrastructure in addition to aspects concerning livability.

3 Current Trends and Challenges in Urban Stormwater Implementation in Baltic Cities

The goal of this section is to identify challenges and constrains surrounding the planning, management, and implementation of urban stormwater in the various cities participating in the Balticflows project. Cities in the Baltic region are faced with social, economic, and environmental challenges including vulnerability to flooding which creates the need for a collaborative effort to protect the public and to minimise damage to urban infrastructure. Thus, through this discussion, Baltic cities could benefit from the knowledge created between regions, to make concrete efforts toward solving issues concerning rainwater management. Here, four Baltic cities including Turku, Tallinn, Hamburg, and Riga will present an overview of their current state of affairs, priorities, and challenges hindering development:

3.1 Turku

The city of Turku is located in the South-Western coast of Finland, where both the climate and the soil properties set special requirements to the stormwater management systems. The winter mean temperatures in Turku are below zero between December and March (Pirinen et al. 2012). The soil freezes during the winter months, but due to the vicinity of the sea, precipitation also occurs, falling on frozen ground during the winter months. The climate change predictions state that precipitation is likely to increase during winter months and that especially heavy rain events will become more frequent (BACC 2008; Jylhä et al. 2009). The fact that the soil type in Turku is either clay or bed rock, further complicates the planning of SUDs. Thus, due to soil conditions, the systems of soil infiltration are generally not functional in Turku.

The South West coast of Finland has been classified as one of the flood risk areas in Finland. The floods occur mostly during low pressure storms, as a combined result of the high stormwater flows and sea level rise. The scarcity of water resources is not a problem in Turku and there is no urge for stormwater reuse as people generally use potable water even for irrigation and car washing. Rather, the stormwater is seen as a threat, potentially causing moist problems in the buildings and people want to conduct it outside of the property as soon as possible. The problems with stormwater are often linked with the quality issues. Turku is surrounded by the Archipelago Sea, which is a shallow and closed part of the Baltic Sea. Thus, the impacts of even relatively small amounts of nutrient loading or hazardous substances can be seen in the water quality of the coastal areas.

The General Plan of Stormwater systems for the Turku region was drafted in 2013. In the Turku city area a number of sites that require implementation of improved stormwater management systems were identified and solutions were proposed. In some areas the municipal plan includes requirements for private properties to have a retention tank for the rain water, proportional to the impermeable surface in the property. However, the realization of the sites proposed in the General plan of Stormwater systems has taken a slow start, as the costs are high and the municipal authorities do not have enough evidence of the benefits of stormwater investments. Further, there is a lack of experiences and research information on the functionality of different types of SUDs in cold climate sites with impermeable soil type.

So far, only few SUDS, such as retention basins below shopping malls and small retention dams in city streams have been constructed in the city of Turku. New testing and pilot sites representing different types of SUDs would be needed to determine the performance, cost, and capacities of the methods and to develop strategies to be adapted to local conditions in the Turku region. In order to create genuinely innovative test sites with direct impacts on water management practices, these would need to be realized in close co-operation between municipal authorities, research institutions, regional authorities and citizens. This kind of co-operation would enable creation of inclusive solutions that can further motivate the public to be active participants in an urban stormwater management strategy and water reuse.

3.2 Tallinn

The city of Tallinn faces land use-related challenges regarding proper planning, allocation and management of urban landscape and open areas. Thus, highlighting shortcomings in the planning and protection of green networks is essential for stormwater control. Attention needs to be given to land governance, public policy, city planning, building codes as well as the need for alternative materials for the protection of the soil and natural preservation of ecosystem services.

Estonia is the most northern of the three Baltic republics (Estonia, Latvia and Lithuania). Its territory is 45,215 km^2 and population about 1.3 million. According to the Tallinn Development Plan 2014–2020 (Tallinna Arengukava 2013) one third of the population, ca 0.44 million people, live in the capital, Tallinn (178 km^2).

The main strategy document giving guidance on how to avoid pollution of the runoff stormwater and how to prolong the retention time of the water in case of intensive precipitation, is the “Tallinn Stormwater Strategy until 2030”

(Tallinna sademevee strateegia 2013). This document contains the analysis of the mosaic geological structure of the ground as well as the recommendations for the reconstruction of the combined and separate sewage pipelines. The strategy also mentions infiltration of the excess water into soil as a potential solution for stormwater control. At the same time it states, that in certain areas, with very thin soil layer on limestone klint and depth of groundwater layer only 2-m, infiltration is not recommended. Today the relevance of this possibility is not understood yet by decision-makers and planning authorities. The renovation of streets has closed a lot of open ditches into underground pipelines and the areas of impermeable surfaces are extending very rapidly. One of the reasons is the need for more parking areas as the number of cars is growing. According to Parts (2012) the number of cars in Tallinn has grown for 2.5 times during the last ten years.

In the new housing areas modern planning principles are taken into account. But in the apartment house districts that were built in the sixties, various problems remain today, mainly due to the outdated building codes and lack of room being allocated for parking space. Thus, in present development, existing green lawns are being transformed into parking areas. One of the compromises from the point of view of the need to infiltrate stormwater, would be usage of permeable tiles, but there are several reasons why this option may not be applied:

(i) Low awareness on the capacity of soil bacteria to break down oil products from the stormwater;
(ii) The maintenance of apartment houses and their yards is totally on the shoulders of local inhabitants and of flat owners
(iii) There is not enough building regulation enforcement in place neither financing mechanisms for innovative solutions and actions plans that take into account local situation.

The Development Plan for Tallinn 2014–2020 (p. 66) states that "the activities foreseen in the stormwater strategy have to be implemented" but this wording leaves a lot of room for interpretation.

On 7 January 2016 the plans of measures for mitigating flood risks (Üleujutusohuga seotud riskide maandamiskavad 2016) were adopted. Assessment of alternative measures (Green Infrastructure) are mentioned there as advisory and location specific possibilities.

On 8 August 2016 the Ministry of Environment finalised the draft "Strategy for Adaptation to Risks from Climate Change" (Kliimamuutustega kohanemise strateegia 2016) that will be presented to the government in this autumn. Flooding in urban areas is mentioned in two different chapters: the one on better planning and the one on water and sewage infrastructure. The planned measures include decrease of the impermeable areas by developing public transport (starting from 2017) and increase of green areas in city planning taking into account the need to infiltrate stormwater (starting from 2021). Responsible actors are local governments and the cost of reconstructions of parking lots is estimated to be 100,000 euros from 2017 to 2100. Until the year 2020, eight million euros is planned to be invested in the whole action plan from which 805,000 euros is meant to be used towards better land use planning.

More binding and detailed legal requirements on innovative measures for stormwater management would be very supportive for the leaders of the apartment owners' unions, who can be educated to understand the significance of green areas in providing ecosystem services.

3.3 Hamburg

The current challenges when dealing with the management of stormwater/rainwater resources in Hamburg, are strongly linked to the city's set priorities for sustainable urban development and "spatial vision" (Freie und Hansestadt Hamburg 2007) that collapse with urbanization growth and migration; limitations on the capacity of existing sewer networks; and the likelihood of flooding.

As the second largest city in Germany, Hamburg has a population of almost 1.8 million people (BUE 2016). The number of new residents moving to live and stay in Hamburg is growing and it is estimated that its population can increase to up to 1.85 million people by the year 2030 (BUE 2014). Migration issues pose challenges for meeting Hamburg's high standards for liveability as in the year 2015, at least 22,000 refugees were found to required living accommodation (BUE 2016). The adjustments needed to integrate newcomers will very likely translate into the need to expand citywide infrastructure and as a result, the capacity of the public sewer.

Hamburg has a dual system of water discharge for inner and outer city areas categorised as combined and separate systems (Bertram et al. 2015). For the inner city areas, the sewer system is mainly combined, while separate networks are often found in the outer areas where the amount of space and green areas is also larger (Bertram et al. 2015). More urbanization means less flexibility and fewer options for separate (decentralised) systems of stormwater discharge this putting the city and its infrastructure at a higher risk of environmental and economic damage. In Hamburg, combined sewer overflows (CSOs) in the inner city areas have been contained; however, main challenges remain for surrounding areas where separate systems are implemented and where more needs to be done to ensure that good water quality standards are met. Further integration of localised innovative solutions that mitigate flooding of streets and parks is a priority and on this issue, the adaptation measures like that of the Rainwater Infrastructure Adaptation project (RISA) is an example. RISA is a major attempt by the local Ministry of Environment and Energy (BUE) to design a legal framework for implementation of an integrated stormwater/rainwater management plan that first identifies flood-prompt areas to then carry out implementation of a flood mitigation plan (Waldhoff et al. 2012). These areas that include roads, buildings, parks and other urban infrastructure are being mapped out through the use of geographic information system-based tools following an implementation plan for the most appropriate drainage method. The project has exposed the need for further structure in water management based on cooperation among actors-this as an essential tool due in part to the different priorities, tasks, and budgets that exists among decision makers.

The handling of stormwater in Hamburg is regulated by the Ministry of Environment and Energy, the responsible public authority in charge of setting up policies and strategies for the protection of surface waters. The BUE works actively in cooperation with the state-owned water utility- Hamburg Wasser (HW) and with universities and other stakeholders to find solutions that better integrate the issues of stormwater into urban development. Challenges remain on the planning an distribution of responsibilities as well as shared costs issues needed to be addressed for the legal implementation of a stormwater action plan for Hamburg.

3.4 Riga

Long-term climate trends pose environmental challenge for the city of Riga and increases the vulnerability to flooding. The daily rainfall observation data obtained for the years 1924–2006 indicate that the medium air temperature of the year and months have become higher in Riga. During the period of 83 years, the rainfall sum has increased in Riga both in cold and warm seasons. There are 684 mm rainfall on the average, incl. about 33% in the cold season (November-March) and 67% in the warm season (April-October). Riga is rich in rainfall in July (83 mm/month) and August (77 mm/month). February is poor in rainfall with average sum 35 mm/month. Rainfall probability once in 5 years is 677 mm/year. Rainfall probability in 20 years is 820 mm/year. Maximum rainfall amount in 24 h is 14 mm with probability 99%, 34 mm with probability 50% and 49 mm with probability 10% (Ziemelnieks and Tilgalis 2008).

The main challenge in urban stormwater management in Riga is the existing old sewage co-system that has insufficient capacity during heavy rainfall. Overflowing of streets and pumping stations occurs if rain is pouring in the entire territory of Riga, at least for 1–3 h, periodically becoming stronger with impulsive strength maintaining for several minutes at 1–2 mm min^{-1}, which is more characteristic of Latvia's summer months June, July and August. By reviewing data of a 27-year period, the average intensity in 3 h at moderate rain intensity was observed to be 0.07 mm min^{-1}, but in 9 h on average 0.02 mm min^{-1} (Ziemelnieks et al. 2008).

Since the estimated costs for separation of rainwater collectors and household sewage collectors in all districts of Riga are about 100 million EUR, the implementation of this solution is not economically feasible. The management of stormwater infrastructure in Riga requires individual, sustainable and cost-effective solution for each part of the stormwater system. The solution for optimal stormwater management system in Riga will include reduction of maximum discharge by adjusting, accumulating and infiltrating in the ground, drainage to the green zone (Ziemelnieks et al. 2010).

Riga municipality is taking important steps to meet existing challenges in urban stormwater management. The municipality has adopted Sustainable Development Strategy of Riga until 2030 (hereafter—the Strategy). The Strategy is a planning document for long-term territory development of Riga municipality. One of four long-term development objectives included in the Strategy is convenient, safe, and pleasant urban environment that includes modernisation of the existing water drainage and treatment infrastructure and building of new engineering communications (City of Riga 2014).

Development Programme of Riga for 2014–2020 is a medium-term planning document that provides 19 actions for reaching the objectives defined in the Strategy, including the priority to ensure high-quality water management services. Riga city plans to carry out 15 activities until 2020 to improve urban stormwater system in the city by renovating the existing rainwater sewage network, building new collectors and separating the rainwater and household sewage network. Main funding sources for the implementation of these activities are the European Union Structural and Investment Funds and the municipal budget (Riga City Council No. 2491, 2015).

The most important challenge from regulatory perspective is the lack of definition who bears the responsibility for funding the rainwater management system. The Parliament of Latvia is working to elaborate the final reading of the Law on Water Management Services (hereafter—the Law). The Law will define the liability and rights of municipalities who ensure the collection, treatment and conducting of waste water and storm water. The Law will enable municipalities to charge natural persons and legal persons for water management services. Thus, the municipalities will be able to increase the investment in public water management infrastructure, and the users of rainwater sewage network will pay for using the service. The introduction of rainwater tariff system will become an important part of sustainable urban planning as long as the municipalities find an optimal solution for the implementation of the rainwater tariff system in a sustainable way.

Crucial part of the implementation of sustainable rainwater management is the awareness and active participation of the society. Public acceptance plays important role in the implementation of national legislative initiatives. The municipality should work in close cooperation with non-governmental organisations, academia, entrepreneurship associations and other relevant stakeholders to raise awareness of stormwater management in society.

4 Towards the Implementation of and UWM Strategy in Baltic Cities: The Role of the Joint Action Plan

The Joint Action Plan developed as part of the BalticFlows 7th framework Regions of knowledge project, is a living document and as such there is a commitment across the participating regions of the project and partners to review the plan on a regular basis, alongside the Smart Specialisation Strategies, to ensure that the actions are taken forward. Europe is in the midst of transformation.

Globalisation is placing increasing pressure upon competitiveness in all sectors of European industry. At the same time, we must pay close attention to the well-being of our environment, and the livability and safety of our cities and regions, as short-term financial gains may bear a heavy long-term cost upon that most valuable to us. We cannot wage war against a deteriorating environment, we cannot force economies to prosper, and there is no single measure to raise European competitiveness beyond global competition. Not only are new solutions urgently needed, but also a new way of comprehending our environment and the fundamental mechanisms driving change. In the Baltic Sea Region, we are committed to facing the challenge of transformation, to seek and discover our latent regional strengths, and build future prosperity upon these.

In the Baltic Sea Region, by including citizens in both environmental activities and development of regional strategies, we aim at strengthening citizen empowerment and the social innovation in our regions. In the Baltic Sea Region, we will strive to include citizens in environmental monitoring and preservation, urban and stormwater management planning, as well as in the formulation of regional smart specialisation strategies. The Joint Action Plan is meant to set guidelines towards a sustainable Citizen Union with safe, liveable cities and active citizen involvement via three basic pillars:

- Innovation pillar. Gathering and processing new ideas from citizens on how to aesthetically increase the amount of permeable surfaces in urban habitats, potentially leading to new social innovations.
- Sustainability pillar. Providing information to citizens on the environmental impact and aesthetic significance of alternatives when making choices during building design, yard planning, stormwater system planning, surface material selection, etc.
- Engagement pillar. Identifying the key driving motivation and relevant incentives for each citizen, aiming at active citizen involvement in improving urban environments.

An Engagement and Inspiration Plan for Europe without support from each city and region—from the local government, academia and industry but before all from the citizens—remains only a plan. Together we will need to make the plan to become reality.

5 Conclusions

As this chapter has shown, there are significant efforts from the side of the cities of Turku, Tallinn, Hamburg and Riga in respect of pursuing the means via which they can implement integrative urban stormwater strategies, and hence adapting to climate chanfe. By doing so, they can reduce the many adverse impacts on floods on the environment including hydrological disruption, groundwater depletion, downstream flooding, water quality degradation, and stream ecosystem damage. The continue use of green infrastructures by urban stormwater managers is also valuable to promote and even restore hydrological functions of the landscape, and reduce surface runoff. Existing stormwater governance, however, is often centralized and structured to support the conventional systems. They should be adjusted and involves multiple stakeholders, not only government agencies, but also non-governmental organizations and the private sector.

References

Bertram, N.-P., Waldhoff, A., & Bischoff, G. (2015). Rain InfraStructure Adaption (RISA) in the City of Hamburg and suitable funding for Stormwater Management (p. 8). Hamburg. Retrieved from http://www.risa-hamburg.de/Downloadbereich.html.

Cameron, R. W. F., Blanusa, T., Taylor, J. E., Salisbury, A., Halstead, A. J., Henricot, B., & Thompson, K. (2012). The domestic garden – its contribution to urban green infrastructure. Urban Forestry & Urban Greening, 9.

European Environmental Agency. (2010). 10 Messages for 2010 Urban Ecosystems. Copenhagen: European Environment Agency. Retrieved from http://www.eea.europa.eu/publications/10-messages-for-2010-urban-ecosystems.

Freie und Hansestadt Hamburg (2007). *The Spatial Vision (draft abridged version)*. Freie und Hansestadt Hamburg, Hamburg.

Behörde für Stadtentwicklung und Umwelt (BSU) (2014). Green, inclusive, growing city by the water Perspectives on Urban Development in Hamburg. Retrieved July 12, 2016, from http://www.hamburg.de/contentblob/4357518/data/broschuere-perspektiven-englisch).pdf.

Behörde für Umwelt und Energie (BUE)(2016). Hamburg - European Green Capital: 5 Years On. Retrieved June 12, 2016, from http://www.hamburg.de/bue/6408388/environment-hamburg-european-green-capital-five-years-om/.

Hoyer, J., Dickhaut, W., Kronawitter, L., & Björn, W. (2011). Water Sensitive Urban Design. Hamburg: jovis Verlag GmbH and HafenCity Uniersität Hamburg.

Institute for Advanced Sustainability Studies e.V. (2013). Soil Sealing. Retrieved from http://globalsoilweek.org/wp-content/uploads/2014/11/GSW_factsheet_Sealing_en.pdf.

IPCC. (2013). Climate Change 2013: The Physical Science Basis. Retrieved from http://www.climatechange2013.org/images/report/WG1AR5_ALL_FINAL.pdf.

Jylhä, K., Ruosteenoja, K., Räisänen, J., Venäläinen, A., Ruokolainen, L., Saku, S. ja Seitola, T., Arvioita Suomen muuttuvasta ilmastosta sopeutumistutkimuksia varten, ACCLIM-project report 2009.

Kliimamuutustega kohanemise strateegia (2016) (Estonian National Strategy for adaptation to climate change, http://www.klab.ee/kohanemine/arengukava/ksh/.

The BACC Author Team, 2008: Assessment of Climate Change for the Baltic Sea Basin. Springer Verlag, 474 s.

McBratney, A., Field, D. J., & Koch, A. (2014). The dimensions of soil security. Geoderma, 213, 203–213. doi:10.1016/j.geoderma.2013.08.013.

OECD. (2010). OECD Factbook 2010: Economic, Environmental and Social Statistics. October (p. 292). doi:10.1787/factbook-2010-en.

Pirinen, P., Simola, H., Aalto, J., Kaukoranta, J-P., Karlsson, P., Ruuhela, R. (2012) Tilastoja uomen ilmastosta 1981–2010. (Climatological statistics of Finland 1981–2010) Ilmatieteen laitos, Helsinki. Ilmatieteen laitoksen raportteja 2012:1. 83 s.

Riga City Council (2015) Resolution No.2491 on 21 April 2015, 119 pp. (in Latvian) Retrieved June 6, 2015, from http://www.rdpad.lv/wp-content/uploads/2015/05/RICIBAS_PLANS_2014_2020.pdf.

City of Riga (2016) Sustainable Development Strategy of Riga until 2030 and Development Programme of Riga for 2014–2020. (2014). Summary. Retrieved July 26, 2016, from http://www.rdpad.lv/wp-content/uploads/2014/11/ENG_STRATEGIJA.pdf.

U.S. Environmental Protection Agency (2016) What is EPA Doing to Suppor t Green Infrastructure? Green Infrastructure| US EPA. Retrieved June 29, 2016, from https://www.epa.gov/green-infrastructure/what-epa-doing-support-green-infrastructure.

Waldhoff, A., Ziegler, J., Bischoff, G., & Rabe, S. (2012). Multifuctional Spaces for Flood Management - an Approach for the City of Hamburg, Germany. *SCIENCE*, (International), 5.

Ziemeļnieks R., Tilgalis E. (2008) Calculations of lasting rainfall in Riga. Journal Ecology & Safety. International Scientific Publications, Vol. 2, Part 1. Published by Info Invest, ISSN 1313–2563, Bulgaria, pp. 24-30. Retrieved June 5, 2015, from http://www.scientificpublications.net/download/ecology-and-safety-2008.swf.

Ziemelnieks R., Tilgalis E., Juhna V. (2008) The Effect of Rainfall on Activity of the Sewage Co-system in Riga. Research for Rural Development 2008 Annual 14th International Scientific Conference Proceedings. Jelgava, Latvia, 194–199. Retrieved June 5, 2015, from http://www2.llu.lv/research_conf/Proceedings/14th.pdf.

Ziemelnieks R., Tilgalis E., Juhna V. (2010) Household and Rainwater Sewage System Separation Possibility in Riga. Research for Rural Development 2010 Annual 16th International Scientific Conference Proceedings. Jelgava, Latvia, 115–120. Retrieved June 6, 2015, from http://www.llu.lv/getfile.php?id=27927.

Tallinna arengukava aastateks 2014 kuni 2020 (Tallinn Development Plan 2014–2020), In: State Bulletin, (RT) IV, 235.06.2013, 41 (82 pp, in Estonian).

Tallinna sademevee strateegia aastani 2030. (Tallinn Stormwater Strategy until 2030), In: State Bulletin (RT) IV, 09.03.2013, 41 (83 pp, in Estonian) Geological map on p 21.

Parts, E.. (2012) Economic growth and developments in Estonia in European and global context. Presentation held in Tartu University on 9 February 2012.

Üleujutusriskidega seotud alade maandamiskavad (2016). (Activity plans for mitigating the risks on flooded areas) http://www.envir.ee/et/uleujutused retrieved on 11 August 2016.

Escaping the Economist's Straightjacket: Overcoming the Free-Rider Mentality Which Prevents Climate Change from Being Effectively Addressed

Gherardo Girardi and Gian Lorenzo Preite

Abstract Economists' ability to propose radical solutions to the problem of climate change is severely limited by one of the assumption they normally make about human nature, namely that it is fundamentally selfish and that this defining characteristic cannot be changed. An important consequence of this characteristic is known as free-riding, which occurs when people do not bother to take eco-friendly actions as these are costly to them, preferring to wait for others to take such actions and enjoy the resultant benefits. The problem is that, if all behave in this way, social paralysis may easily occur and climate change is likely to remain unaddressed, or to be addressed only superficially. Using a simplified version of George (2001)'s model of first and second order preferences, this paper considers ways of motivating people at a deeper (second order) level to take actions (at the first order level) that are eco-friendly. These include reducing market pressure/advertizing, promoting contact with nature and relying more on spirituality and mindfulness.

Keywords Free-rider · First and second order preferences · Climate change

1 Introduction

It is likely that addressing climate change will require finding solutions that appeal both to people's self-interest and their sense of the common good. In economics, we find that it is self-interest that is generally assumed to motivate people. This self-interest is of a particular kind in that it does not value the importance of human relationships per se, but rather focuses on material goods and services and sees relationships—people—as a means to obtain these goods and services. This

G. Girardi (✉)
London Guildhall School of Business and Law, London Metropolitan University, 84 Moorgate, London EC2M 6SQ, UK
e-mail: g.girardi@londonmet.ac.uk

G.L. Preite
Good Works, Impact Hub, 34b York Way, King's Cross, London N1 9AB, UK

W. Leal Filho (ed.), *Climate Change Research at Universities*,
DOI 10.1007/978-3-319-58214-6_35

superficial type of self-interest is no doubt common among people, and leads to the problem of free-riding, whereby some, perhaps many people would rather wait for others to take an action that will benefit society so as to enjoy the benefits of that action without incurring any of the cost themselves. An example could be someone who chooses not to recycle as this is "costly" in some way (effort, time, etc.) and just waits for others to do the recycling themselves.

The discipline of economics studies how to address free-riding taking as given that people are self-interested. Take pollution—this can be seen as a manifestation of free-riding where, for example, (a) a car manufacturer does not wish to take on the extra cost of creating a cleaner engine, and (b) a car buyer does not wish to compensate the manufacturer for creating a cleaner engine, leaving it to others to take on the burden of cutting pollution. The standard remedy in front of pollution is to introduce a tax or a cap on emissions which will bring into alignment private interests and social well-being. Another, similar way of attaining this alignment is, as explored by Nordhaus (2015) in a paper on international climate change agreements, to create clubs of countries that undertake harmonized emission reductions whilst penalizing non-participating countries by means of tariffs on imports into the participants' clubs. Whether it is through taxes or penalties, either way the solution involves hurting polluting members of society.

However, it is unlikely that major social change can be achieved by relying purely on people self-interest and ways to align that with the social good. Indeed, for many ecologists and religious leaders, it is just as necessary for the public to undergo a change of attitude away from seeking simply its private interest and towards seeking the common good of society—indeed, of all species and of the planet. It is for this reason that Pope John Paul II and all successors have been calling on all people of goodwill to undergo an "ecological conversion" (see Pope Francis' encyclical letter Laudato Sí, 2015).

This paper considers strategies to encourage people to stop being free riders and so take eco-friendly actions. The paper starts by recognizing that many, if not most or even all people have deep preferences about how they would like to be, and, given the choice, would prefer to care about the environment than not. This implies that, deep down, they already care about the environment,[1] consistently with the well-known *biophilia* hypothesis described below. The question then becomes how to encourage these deeper motivations in people to emerge. To conceptually underpin the main argument, the paper begins by describing a theoretical framework by George (2001) that considers superficial (first order) and deep (second order) preferences. It then proposes strategies to enable people's deeper motivations to emerge. A general discussion and a conclusion follow.

[1]These deep motivations are consistent with a deeper kind of self-interest, which, unlike its superficial counterpart mentioned at the start of the introduction, acknowledges that people, to be happy, need to live in harmonious relationships with the ecosystem and with one another.

2 First and Second Order Preferences

David George, in a series of articles, considered the fact that individuals are able to express a judgement with regards to what they like. For example, many who smoke understand the significant health and monetary costs of smoking and can say "I wish I did not like smoking". Indeed, the philosopher Frankfurt (1971, in George 2001; see also George 1993) believes that the ability of making such a judgement is what makes human beings unique. Animals, like humans, can express preferences, but not preferences or judgements about preferences, and so cannot feel the tension that humans feel within themselves when their behaviour is different from how they would like it to be.

Using George's terminology, the consumer has a first order (or superficial, one could say) preference to smoke, but a second order (or deeper) preference not to smoke.[2] The consumer might behave in this way because he or she is addicted to nicotine. This will make the consumer dissatisfied in George's model.

Using a different example that is more closely related to climate change, the consumer may opt to not recycle an empty carton even though he wished deep down that he did. He might do so because he might otherwise have had to carry the carton with himself for a couple of minutes before finding a recycling bin, when in fact, had he stopped and given the voice of his conscience due consideration, he might have chosen to carry the carton with him for a while and then recycled it.

If an individual wanted his or her second order preferences to emerge (and so not smoke rather than smoke, recycle rather than not recycle, etc.), society at large, including teachers, churches, NGO's and policy-makers, arguably has a responsibility to make it possible for him to undergo this transition. One way of doing this would be raise the perceived benefit of recycling at his first order preference level so as to align his first and second order preferences.[3] How this can be done is the main subject of this paper and is explored in Sect. 3 below.

Raising the perceived private benefit can allow individuals and groups of individuals to escape the so-called prisoner's dilemma.[4,5] In the context of climate

---

[2]More precisely, the consumer may have a first order preference to smoke over not to smoke, but a second order preference such that preferring not to smoke over smoking is preferred to preferring to smoke over not smoking. In the main text we drop this precision in favour of a more fluid discussion.

[3]There is also the issue of willpower, but aligning perceived benefits (and costs) should help greatly in this respect. Having said that, willpower may still be weak after alignment, as a result of habit or addiction.

[4]The name prisoner's dilemma comes from the idea that a prisoner, or someone accused of a crime who may go to prison, may have an incentive to confess his crime and that of his partner no matter whether the partner confesses or not. However, if they both confess, they both go to prison.

[5]The material on the prisoner's dilemma has been included here following comments by participants at the Universities and Climate Change: the Role of Higher Education Institutions in Addressing the Mitigation and Adaptation Challenges (Manchester, September 2016), who were for the most part non-economists and expressed an interest in the material.

		Country A	
		Cut emissions	Do not cut emissions
Country B	Cut emissions	10,10	11,3
	Do not cut emissions	3,11	4,4

Fig. 1 The prisoners' dilemma in the context of climate change negotiations

		Country A	
		Cut emissions	Do not cut emissions
Country B	Cut emissions	*12,12*	11,*5*
	Do not cut emissions	*5*,11	4,4

Fig. 2 The prisoner's dilemma undone: climate change negotiations with higher perceived benefits of cutting emissions

change, the prisoner's dilemma is probably best understood with reference to climate change negotiations. Here, a country may not wish to cut its emissions for fear of losing economic competitiveness, for example as a result of imposing a tax on its firms. Not cutting emissions may be the "best" (or "dominant" in game theory parlance) strategy no matter what the other country does. However, in this way neither country will take action to cut its emissions (a behaviour which resembles that of a free-rider), when in fact they would have been better off if both had cut their emissions, without affecting their (relative) competitiveness. This is illustrated in the example below, where the numbers indicate perceived benefits, and where the first number in each pair is country A's payoff and the second number is country B's payoff (Fig. 1).[6]

If, as a result of the individual's deeper preferences surfacing, the perceived benefit of cutting emissions rises, the prisoner's dilemma may disappear, as illustrated in Fig. 2 (numbers in italics indicate that the perceived benefit of cutting emissions has risen).

In the example of Fig. 2, the effect of increased competitiveness on one country if another country cuts its emissions is still visible, but it is dominated by the improvement to society as a result of lower emissions, so that each country's best—or dominant—strategy is to cut emissions no matter what other countries decides to do. It would seem that some counties regard themselves as being in a prisoner's

[6]These payoffs are entirely for illustration purposes and have no particular units attached (an economist may wish to regard them as expressed in "utils", or units of happiness).

dilemma and so take little action to cut emissions, whilst other countries are less inclined to see themselves in such a dilemma and are therefore much more pro-active in cutting emissions. The example works just as well if, instead of raising the perceived benefit of cutting emissions, the perceived benefit of not cutting emissions is lowered; this is shown in the Appendix.

3 Ways to Enable Second Order Preferences to Emerge

How can second order preferences be encouraged to emerge? Looking at George's model, two paths appear to stand out, namely through the weakening of external forces that cause a mis-alignment of first and second order preferences, and through the strengthening of the internal forces by which the re-alignment is stimulated. Three strategies are described below that fall into one of these two paths: (a) reduced market pressure and advertising, which belongs to the category of weakening external forces; and (b) greater contact with nature and (c) greater reliance on spirituality, which belong to the category of strengthening internal forces.

3.1 Reduced Market Pressure/Advertising

George accuses the market of distorting people's preferences: "market forces fail to create in agents those preferences that they would prefer to have" and "the market too often changes them [i.e. preferences] for the worse and too infrequently changes them for the better (as judged by the agents whose preferences are being altered)" (George 2001, p. 463; see also George (1993)). A good example might be food, which in the UK (and in some other countries) is loaded with sugar, salt and fat in quantities deemed excessive by Public Health England, an agency of the Department of Health (The Guardian 2014). In the context of climate change, an example could be advertising of cars that are not eco-friendly, flights that do not allow for the planting of trees in sufficient quantities and so are not carbon neutral, etc. In general, the market induces the public not to be satisfied with what they currently have so as to sell more to them, often at the expense of the environment. Here there is a role for the state to step in and substantially reduce the market's profit-driven stimulation of people's preferences, so as to reduce the noisy signals which prevent (at least some) people from experiencing the emergence of their second order, eco-friendly preferences. This would then amount to experiencing an increased perceived benefit of behaving in an eco-friendly manner.

There is certainly no lack of support in the literature for the view that markets distort preferences. As Dauvergne (2010) and Ericson et al. (2014) point out, the market, largely through advertising, generates or exacerbates "perceived obsolescence" of goods and services, making it difficult for the public to accurately assess what is genuinely valuable to them. In a similar vein, Galbraith (2007) discusses the manipulation of consumers' desires by corporations. As a consequence, the public often fails to reflect on what is most important in the context of their lives (Shapiro et al. 2006). A number of studies find that individuals with "self-enhancing, materialistic values and goals directed toward achievement, money, power, status, and image also have more negative attitudes towards the environment and are less likely to be involved in environmentally friendly behaviors" (Ericson et al. (2014), p. 76, who draw from Brown and Kasser (2005) and Crompton and Kasser (2009)).

3.2 Greater Contact with Nature

According to the *biophilia* hypothesis (Kellert and Wilson 1995, in Tam 2013), humans are biologically disposed to be attracted to nature and to depend on it. Urban life comes at the cost of not being able to enjoy an intimate relationship with nature and all the benefits that come with it (see Bartlett 2008 and Vining et al. 2008, both in Tam 2013). Following the concise survey of the literature by Zelenski and Nisbet (2014, p. 4), these effects can be identified: outdoor walks in nearby nature give rise to pleasant moods (Nisbet and Zelenski 2011); natural images promote prosocial aspirations and caring attitudes (Weinstein et al. 2009); nature may help recover self-regulation (Kaplan and Berman 2010), restore attention (Berman et al. 2008), and reduce crime and aggression (Kuo and Sullivan 2001a, b); nature improves health (Frumkin 2001; Van den Berg 2005), and may even reduce the mortality risk associated with income inequality (Mitchell and Popham 2008).[7]

The biophilia hypothesis can explain the emergence of second order preferences, particularly when the external negative influences of market and other pressures are not so strong. Indeed, the harmonious integration of first and second order preferences constitutes deep healing and may be interpreted as the development of what ecologists call "the ecological self" (see Bragg 1996). A manifestation of the need for healing is that nature, which used to be seen as a source of awe and wonder, is now seen largely as a source of scientific knowledge (see Bartlett 2008); there is a strong need, therefore, to reconcile these two paradigms (see Berman 1981), which may be roughly described as 'intuitive' and 'rational' and so understood as complementary.

Researchers have experimented with various measures of a sense of connection to nature and found these to predict ecological behaviour. Following Tam (2013), these include emotional affinity to nature (Kals et al. 1999), inclusion of nature in the self (Schultz 2001), commitment to nature (Davis et al. 2009), and explicitly multi-dimensional concepts such as nature relatedness (Nisbet et al. 2009). Perhaps not surprisingly, nature relatedness is correlated with happiness and well-being (Zelenski and Nisbet 2014), suggesting that (small) sacrifices in the form of

[7]Mayer et al. (2009) show that even short exposure to nature increases connectedness to it.

activities like walking instead of driving one's car need not be a burden (see Girardi 2016; Frantz and Mayer 2009), and suggest instead there may be a "happy path to sustainability" (Nisbet and Zelenski 2011). This raises the possibility of a "double reward" whereby more nature relatedness is associated with both more sustainable behaviour and greater happiness. This has important implications for those engaged in the promotion of ecological behaviour, including NGO's, educators and policy makers: in the words of Zelenski and Nisbet (2014, p. 18), "some people might be more persuaded to protect the natural environment by understanding how connecting with nature can contribute to their personal well-being".

3.3 Greater Reliance on Spirituality and Mindfulness

One of the most fundamental ways to raise the perceived benefit of taking eco-friendly action is for people who practice a religion or faith to discover, or better understand, what their religion says about protecting the environment. Most faiths, and certainly those faiths with millennial traditions, promote the protection of the environment. In Judaism, Islam and Christianity, this is a fundamental duty of humans as stewards of God's creation. In our times, there is an urgent need for this point to be brought to the fore by religious leaders and put into practice by believers. A recent example is the book, known as encyclical, by Pope Francis entitled Laudato Sí, which was released in May 2015 and is likely to have helped motivate the successful round of climate change negotiations in Paris. In this book, the Pope speaks with admiration of Saint Francis of Assisi, a medieval saint who lived a simple and joyful life in harmony with nature. He felt a strong bond of affection with his fellow creatures, using terms such as 'brother wolf', 'sister water' (as well as 'mother Earth'). Pope Francis says in the encyclical: "If we approach nature and the environment without this openness to awe and wonder, if we no longer speak the language of fraternity and beauty in our relationship with the world, our attitude will be that of masters, consumers, ruthless exploiters, unable to set limits on their immediate needs" (Pope Francis, p. 35).[8,9,10]

The academic literature in the area of spirituality and ecology discusses the practice of mindfulness, which has its roots in Buddhism (Ericson et al. 2014), itself a millennial tradition. Ericson et al. (2014, p. 74) explain that mindfulness "involves paying attention to one's whole experience in the moment. […] Regardless of whether these experiences are pleasant or unpleasant, scary or seductive,

[8]As Zhang et al. (2014) point out, being able to perceive natural beauty is associated with extraversion and gratitude (Diessner et al. 2013), two key determinants of subjective well-being (DeNeve and Cooper 1998; Zhang and Howell 2011).

[9]According to philosopher Murdoch (1967), beauty can lead to going beyond self-interest and being generous and kind.

[10]There are loud echoes here of the native American Indians' traditional attitude towards the land, which, in so far as it was "mother", could not be sold; see Szeghi (2012).

one simply pays attention to one's experiences, from moment to moment. A cousin technique of mindfulness meditation is loving-kindness meditation. This form of meditation [...] aims to evoke positive emotions more directly, and is used to increase feelings of warmth and caring for self and others (Fredrickson 2009)". There is some evidence that mindfulness is associated with environmental behaviour. Brown and Kasser (2005), for example, found a positive association between mindfulness and ecological behaviour (as well as well-being). Amel et al. (2009) found that acting with awareness, which is closely related to mindfulness, is correlated with environmental behaviour. There is evidence that spirituality and mindfulness mediate the relation between connectedness to nature and well-being described in Sect. 3.2 (Howell and Passmore 2013; Kamitsis and Francis 2013; both in Zhang et al. (2014)).[11]

More generally, mindfulness can help the mind defend and detach itself from the aggressive messages that the market launches at it, and so "serve as an antidote to consumerism" (Rosenberg 2004, in Ericson et al. 2014). The market causes stress as it induces in many people a range of desires that are collectively too expensive to satisfy, and stress makes it harder to devote attention to issue of social concern such as climate change (Ericson et al. 2014; see Shah et al. 2012). Mindfulness is designed to reduce this stress and can help people avoid the hedonic treadmill (Fredrickson et al. 2008), which is the tendency displayed by people to quickly return to a given level of happiness when major events take place in their lives, be they pleasant or not.

4 The Economics Syllabus

If economics courses are to support the emergence of second order—or ecological—preferences, there needs to be a fundamental review of the economics syllabus. A number of authors express disappointment with the current state of the syllabus. Green (2012), for example, finds that only 3.2% of the printed area of Ec101 textbooks is devoted to externalities, public goods and other issues related to the environment.[12] This is a worrying state of affairs considering that as many as 40% of North American students take an introductory course in economics, but the vast majority will not take other economics course at a more advanced level and so will not be able to familiarize themselves with the economics of the environment (Green 2013). Green (2013) reports that students are dissatisfied with the present

[11]Zhang et al. (2014) find that connectedness with nature leads to well-being only when individuals are also emotionally attuned to natural beauty.

[12]See also Plumridge (2010).

economics syllabus, whilst (Green 2015) finds that many lecturers are often not aware of their own universities' sustainability commitments[13] and don't think it a priority to revise the syllabus to take into account environmental issues.

Previous studies which have looked critically at how economics textbooks deal with environmental issues have found that textbooks tend to take a normative stand in that they favour consumerism and material growth, whilst downplaying the importance of the environment (see Green 2012; Nelson 2009, both in Green 2013). Green (2013) also finds evidence that these textbooks normalize selfish behaviour at the expense of the environment, and even appear to suggest that the economy can exist in isolation from the environment.

There is no doubt as to the importance of education in promoting the spread of sustainable behaviour across all areas of life, including business and economics. The UN Global Compact initiative (2007) states: "Any meaningful and lasting change in the conduct of corporations toward sustainability must involve the institutions that most directly act as drivers of business behaviour, especially academia. Academic institutions help shape the attitudes and behaviour of business leaders through business education, research, management development programs, training, and other pervasive, but less tangible, activities, such as the spread and advocacy of new values and ideas. Through these means, academic institutions have the potential to generate a wave of positive change, thereby helping to ensure a world where both enterprises and societies can flourish". There is a widespread consensus among promoters of sustainability education that sustainability should be taught in an interdisciplinary fashion that cuts across and at the same time integrates different study programmes (see Green 2015; Figueiró and Raufflet 2015). Such an approach, it is hoped, will foster the creative process by which humanity is seeking solutions to the unprecedented problem of anthropogenic climate change, which in turn will requires identifying new economic objectives and models (Hueting 2010; Speth 2008, 2009).

In economics education, a specific barrier preventing the reform of the economics syllabus is the fact that sustainability is not generally regarded as a profit-making activity in business (see Boxer 2008; Wheeler et al. 2005), together with the fact that mainstream economics assumes that firms' objective is to maximize profits.[14] However, the effect of sustainability practices on profitability does not generally turn out to be negative in academic studies; indeed recent evidence is indicative of a positive relationship (Khan et al. (2015) and Oxford University/Arabesque Partners (2015); see also Cheng et al. (2014)). In this case,

---

[13]Although a considerable number of universities have made great progress in meeting their sustainability targets, paradoxically the economics curriculum is incorporating sustainability at a worryingly slow pace (see Green 2013). This is even more perplexing considering that, as Green (2015) points out, much environmental change is the result of economic factors (see also Dietz et al. 2007).

[14]In addition, as Velazquez et al. (2005, p. 386) explain, "professors are learning and teaching about sustainability at the same time. In fact, few educators are being taught to teach sustainability".

the argument that countries cannot cut emissions as doing so would imply a loss in economic competitiveness loses bite; indeed it could no longer be represented by the prisoner's dilemma.[15]

A further barrier to the reform of the economics syllabus lies in the assumptions that the neoclassical school of thought, which dominates the syllabus, makes about human nature, in particular that human beings are fundamentally selfish and rational. Both assumptions are problematic. Consider rationality: if humanity is rational, why are we doing so little to address climate change? Why have we not yet changed our lifestyles to any considerable degree so as to reduce the massive pressure that we place on the environment? The discipline of behavioural economics, which lies at the intersection of economics and psychology, deals with all sorts of examples of human irrationality which generally speaking economics textbook do not cover, or cover only superficially. The assumption of selfishness is problematic in a different sense: its unquestioned adoption may encourage students to identify with and promote selfish economic ideologies, and to become selfish themselves (see Frank et al. (1993) on the possibility that studying economics may stifle co-operation). Yet common sense suggests that sometimes people behave altruistically, and that many tend to reciprocate; indeed there is a well-established literature on this topic (see, for example, the review by Fehr and Schmidt 2006), which tutors can teach students so as not to indoctrinate them into thinking that humanity is fundamentally selfish, which may encourage them to free ride and not behave in an ecologically responsible way.

5 Discussion

Bateson (1972)'s framework, as explained in Sterling (2011) and Sidiropoulus (2014), can be usefully applied here to understand the proposal being put forward in this paper. Mainstream economics assumes that people are fundamentally selfish and indicates strategies that can implemented given this particular view of human nature, such as pollution taxes, markets in pollution permits, etc. The branch of economics known as environmental economics falls into this category, which corresponds to level 1 in Bateson's framework, or "doings things better". Level 2, which is described as "doing better things", could be understood to coincide with ecological economics, which represents an improvement over environmental economics in that it seeks to focus attention on the fact that current consumption patterns are globally unsustainable.

Level 3 is a step higher still in so far as, in Sidiropoulos (2014, p. 475)'s words, it involves "'seeing things differently' and leads to paradigmatic change".

[15]Bassi and Duffy (2016) find that current policies to tackle climate change have not damaged the competitiveness of businesses, and that climate change policies may increase the competitiveness of the UK in the long term by encouraging greater innovation and efficiency.

The fundamental idea in this paper is that addressing climate change will require appealing not only to people's self-interest, as mainstream economics does, but also to people's sense of the common good. It will require digging below the surface and allowing people's deeper preferences to surface above their first order preferences, using the terminology of David George. The emergence of deeper preferences involves seeing the natural world in a new light, not unlike the light that inspired Saint Francis, who entertained a deep sense of affection towards his fellow creatures. This change of perspective can be very profound and represents a paradigmatic shift away from the more materialistic outlook that is typical of mainstream economic thinking. It is in actual fact an experience of transformation and renewal, which people across the world—and in particular those societies that impose heavy burdens on the environment—are in urgent need of undergoing if humanity is to successfully address the problem of climate change.

6 Conclusion

This free-rider problem stems from a selfish attitude which many, possibly most individuals display. In the context of the environment, the free-rider problem involves not taking eco-friendly actions and leaving it to others to incur the cost of eco-friendly activities, for example by not recycling. Whilst offering incentives that appeal to individuals' self-interest is certainly necessary to effectively address the free-rider problem, these are unlikely to be sufficient, and should be accompanied by strategies that appeal to people's sense of the common good. George (2001)'s model of first and second order preferences acknowledges that people can harbour contradictory preferences, at the same time wishing to be eco-friendly (act in the interest of the common good) and not (free ride). This paper argues that generally speaking people's deeper (or second order) preference is to be eco-friendly and that this position is consistent with the biophilia hypothesis, according to which people need to live in contact and in harmony with nature to be able to lead genuinely fulfilling lives.

The paper goes on to suggest various ways by which second order preferences can be made to emerge, or, in economics parlance, by which the perceived benefit of acting ecologically at the first level of preferences can be increased. These include dumbing down the market signal to "buy, buy and buy again", connecting with nature, and relying more on spirituality and mindfulness. All three strategies are likely to enable people to better connect with, and to practice more, their pursuit of the common good. At the same time, these strategies are consistent with a deeper, more enlightened notion of individual self-interest, as, for example, when an individual comes to understand the deeper, personal benefits of living in greater contact and harmony with nature.

The discussion in this paper of the remedies to address the free-rider problem in the context of climate change is intended to be the start of a larger conversation which is likely to include the identification of additional remedies. This exercise

		Country A	
		Cut emissions	Do not cut emissions
Country B	Cut emissions	10,10	*9,3*
	Do not cut emissions	*3,9*	*2,2*

Fig. 3 The prisoner's dilemma undone: climate change negotiations with lower perceived benefits of not cutting emissions (numbers in italics indicate that the perceived benefit of not cutting emissions has fallen)

would greatly benefit from the input of experts from different disciplines; among these, psychology and spirituality would seem to be particularly relevant, since the free-rider problem is essentially the expression of a fundamentally selfish attitude.

Acknowledgements We thank participants for valuable questions and comments at the conference Universities and Climate Change: the Role of Higher Education Institutions in Addressing the Mitigation and Adaptation Challenges, held at Manchester Metropolitan University on 1 and 2 September 2016.

Appendix

See Fig. 3.

References

Amel, E.L. et al (Manning, C.M. and Scott, B.A.), 2009. Mindfulness and sustainable behavior: Pondering attention and awareness as means for increasing green behavior. *Ecopsychology, 1* (1), pp. 14–25.

Bateson, G., (1972). Steps to an Ecology of Mind. San Francisco: Chandler.

Bartlett, P.F. (2008). Reason and reenchantment in cultural change: sustainability in higher education. *Current Anthropology*, 49, pp. 1077–1098.

Bassi, S. and Duffy, C., 2016. *UK climate change policy: how does it affect competitiveness?* Centre for Climate Change Economics and Policy, http://www.cccep.ac.uk/publication/uk-climate-change-policy-how-does-it-affect-competitiveness/.

Berman, M.G., Jonides, J. and Kaplan, S., 2008. The cognitive benefits of interacting with nature. *Psychological science*, *19*(12), pp. 1207–1212.

Berman, M., 1981. *The reenchantment of the world.* New York, NY: Cornell University Press.

Boxer, L., 2008. Preparing leaders to deal with sustainability. *Journal of Management Education*, 1, pp 1–19.

Bragg, E.A., 1996. Towards ecological self: Deep ecology meets constructionist self-theory. *Journal of environmental psychology*, *16*(2), pp. 93–108.

Brown, K.W. and Kasser, T., 2005. Are psychological and ecological well-being compatible? The role of values, mindfulness, and lifestyle. *Social Indicators Research*, *74*(2), pp. 349–368.

Cheng, B., Ioannou, I., & Serafeim, G., 2014. Corporate social responsibility and access to finance. *Strategic Management Journal*, *35*(1), pp 1–23.
Crompton, T. and Kasser, T., 2009. *Meeting environmental challenges: The role of human identity*. Godalming, UK: WWF-UK.
Davis, J.L., Green, J.D. and Reed, A., 2009. Interdependence with the environment: Commitment, interconnectedness, and environmental behavior. *Journal of Environmental Psychology*, *29*(2), pp. 173–180.
DeNeve, K.M., & Cooper, H., 1998. The happy personality: A meta-analysis of 137 personality traits and subjective well-being. *Psychological Bulletin*, *124*, 197.
Dauvergne, P., 2010. The problem of consumption. *Global environmental politics*, *10*(2), pp. 1–10.
Diessner, R., Iyer, R., Smith, M.M. and Haidt, J., 2013. Who engages with moral beauty?. *Journal of Moral Education*, *42*(2), pp. 139–163.
Dietz, T., Rosa, E.A. and York, R., 2007. Driving the human ecological footprint. *Frontiers in Ecology and the Environment*, *5*(1), pp. 13–18.
Ericson, T., Kjønstad, B.G., and Barstad, A., 2014. Mindfulness and sustainability. *Ecological Economics*, *104*, pp. 73–79.
Fehr, E. and Schmidt, K., 2006. The Economics of Fairness, Reciprocity and Altruism, in Kolm, S. and Ythier, J. (ed's) *The Handbook of the Economics of Giving, Altruism and Reciprocity*, Volume 1, pp. 615–691, Elsevier.
Figueiró, P.S. and Raufflet, E., 2015. Sustainability in higher education: a systematic review with focus on management education. *Journal of Cleaner Production*, *106*, pp. 22–33.
Francis, Pope, 2015. *Encyclical Letter Laudato Si' of the Holy Father Francis: On Care for Our Common Home*. Vatican Web Site. http://w2.vatican.va/content/francesco/en/encyclicals/documents/papa-francesco_20150524_enciclica-laudato-si.html.
Frank et al (R.H., Gilovich, T. and Regan, D.T.), 1993. Does studying economics inhibit cooperation?. *The Journal of Economic Perspectives*, *7*(2), pp. 159–171.
Frankfurt, H.G., 1971. Freedom of the will and the concept of a person. *Journal of Philosophy*, 68, pp. 5–20.
Frantz, C.M. and Mayer, F.S., 2009. The emergency of climate change: Why are we failing to take action?. *Analyses of social issues and public policy*, *9*(1), pp. 205–222.
Fredrickson, B.L., Cohn, M.A., Coffey, K.A., Pek, J. and Finkel, S.M., 2008. Open hearts build lives: positive emotions, induced through loving-kindness meditation, build consequential personal resources. *Journal of personality and social psychology*, *95*(5), p. 1045.
Fredrickson, B.L., 2009. *Positivity*. Crown Publishers, New York.
Frumkin, H., 2001. Beyond toxicity: human health and the natural environment. *American journal of preventive medicine*, *20*(3), pp. 234–240.
Galbraith, J.K., 2007. *The new industrial state*, Princeton University Press, Princeton.
George, D., 1993. Does the market create preferred preferences?. *Review of Social Economy*, *51* (3), pp. 323–346.
George, D., 2001. Unpreferred preferences: unavoidable or a Failure of the Market?. *Eastern Economic Journal*, *27*(4), pp. 463–479.
Girardi, G., 2016. Changes in Attitude Towards Climate Change and Transformative Learning Theory. In *Implementing Climate Change Adaptation in Cities and Communities* (pp. 321–352). Springer International Publishing.
Green, T.L., 2012. Introductory economics textbooks: what do they teach about sustainability?. *International Journal of Pluralism and Economics Education*, *3*(2), pp. 189–223.
Green, T.L., 2013. Teaching (un) sustainability? University sustainability commitments and student experiences of introductory economics. *Ecological Economics*, *94*, pp. 135–142.
Green, T.L., 2015. Lecturers' perspectives on how introductory economic courses address sustainability. *International Journal of Sustainability in Higher Education*, *16*(1), pp. 44–56.
Guardian, The, 2014. *Too much sugar, salt and fat: healthy eating still eluding many Britons*, 14 May.

Howell, A.J. and Passmore, H.A., 2013. The nature of happiness: Nature affiliation and mental well-being, in Keyes, C.L.M. (ed), *Mental well-being* (pp. 231–257). Springer Netherlands.

Hueting, R., 2010. Why environmental sustainability can most probably not be attained with growing production. *Journal of Cleaner Production*, *18*(6), pp. 525–530.

Kamitsis, I. and Francis, A.J., 2013. Spirituality mediates the relationship between engagement with nature and psychological wellbeing. *Journal of Environmental Psychology*, *36*, pp. 136–143.

Kaplan, S. and Berman, M.G., 2010. Directed attention as a common resource for executive functioning and self-regulation. *Perspectives on Psychological Science*, *5*(1), pp. 43–57.

Kellert, S.R. and Wilson, E.O., 1995. *The biophilia hypothesis*. Island Press.

Khan, M. et al (Serafeim, G., & Yoon, A. (2015)). Corporate sustainability: First evidence on materiality. *The Accounting Review*. http://hbswk.hbs.edu/item/corporate-sustainability-first-evidence-on-materiality.

Kals, E., Schumacher, D. and Montada, L., 1999. Emotional affinity toward nature as a motivational basis to protect nature. *Environment and behavior*, *31*(2), pp. 178–202.

Kuo, F.E. and Sullivan, W.C., 2001a. Aggression and violence in the inner city effects of environment via mental fatigue. *Environment and behavior*, *33*(4), pp. 543–571.

Kuo, F.E. and Sullivan, W.C., 2001b. Environment and crime in the inner city: does vegetation reduce crime?. *Environment and behavior*, *33*(3), pp. 343–367.

Mayer, F.S., Frantz, C.M., Bruehlman-Senecal, E. and Dolliver, K., 2009. Why is nature beneficial? The role of connectedness to nature. *Environment and Behavior*, 41, pp. 607–643.

Mitchell, R. and Popham, F., 2008. Effect of exposure to natural environment on health inequalities: an observational population study. *The Lancet*, *372*(9650), pp. 1655–1660.

Murdoch, I., 1967. *The Sovereignty of Good Over Other Concepts*. New York: Cambridge University Press.

Nelson, J.A., 2009. The principles course, in Reardon, J. (Ed.), *The Handbook of Pluralist. Economics Education*. Routledge, pp. 57–68.

Nisbet, E.K., Zelenski, J.M. and Murphy, S.A., 2009. The nature relatedness scale: Linking individuals' connection with nature to environmental concern and behavior. *Environment and Behavior*, 41, pp. 715–740.

Nisbet, E.K. and Zelenski, J.M., 2011. Underestimating nearby nature affective forecasting errors obscure the happy path to sustainability. *Psychological science*, *22*(9), pp. 1101–1106.

Nordhaus, W., 2015. Climate clubs: overcoming free-riding in international climate policy. *The American Economic Review*, 105(4), pp. 1339–1370.

Oxford University/Arabesque Partners (2015), *From the stockholder to the stakeholder* http://www.arabesque.com/index.php?tt_down=51e2de00a30f88872897824d3e211b11.

Plumridge, A., 2010. Costing the earth: The economics of sustainability in the curriculum in Jones, P., Selby, D. and Sterling, S. (eds). *Sustainability Education: Perspectives and Practice Across Higher Education*, Earthscan: London and New York, pp. 273–293.

Rosenberg, E.L., 2004. Mindfulness and consumerism, in Kasser, T. and Kanner, A.D. (eds). *Psychology and Consumer Culture: the Struggle for a Good Life in a Materialistic World.* American Psychological Association, Washington, D.C., pp. 107–125.

Schultz, P.W., 2001. The structure of environmental concern: Concern for self, other people, and the biosphere. *Journal of environmental psychology*, *21*(4), pp. 327–339.

Shah, A.K., Mullainathan, S. and Shafir, E., 2012. Some consequences of having too little. *Science*, *338*(6107), pp. 682–685.

Shapiro, S.L., Carlson, L.E., Astin, J.A. and Freedman, B., 2006. Mechanisms of mindfulness. *Journal of clinical psychology*, *62*(3), pp. 373–386.

Sidiropoulos, E., 2014. Education for sustainability in business education programs: a question of value. *Journal of cleaner production*, *85*, pp. 472–487.

Speth, J.G., 2008. *The bridge at the edge of the world: Capitalism, the environment, and crossing from crisis to sustainability*. Yale University Press.

Speth, J.G., 2009. The Coming Transformation Values to Sustain Human and Natural Communities, in Kellert, S.R. and Speth, J.G. (eds). *The Coming Transformation Values to*

Sustain Human and Natural Communities. Yale School of Forestry & Environmental Studies, New Haven, Connecticut.

Sterling, S., 2011. Transformative learning and sustainability: sketching the conceptual ground. *Learning and Teaching in Higher Education*, *5*(11), pp. 17–33.

Szeghi, S., 2012. Lessons in Equality, Social Justice, and Ecological Balance from Indigenous Cultures, https://gcgij.com/archives/.

Tam, K.P., 2013. Concepts and measures related to connection to nature: Similarities and differences. *Journal of Environmental Psychology*, *34*, pp. 64–78.

United Nations Global Compact, 2007. Principles for responsible management education. https://www.unglobalcompact.org/docs/news_events/8.1/PRME.pdf.

Van den Berg, A.E., 2005. *Health impacts of healing environments; a review of evidence for benefits of nature, daylight, fresh air, and quiet in healthcare settings*. Groningen, Netherlands: Foundation 200 years University Hospital Groningen.

Velazquez, L., Munguia, N. and Sanchez, M., 2005. Deterring sustainability in higher education institutions: An appraisal of the factors which influence sustainability in higher education institutions. *International Journal of Sustainability in Higher Education*, *6*(4), pp. 383–391.

Vining, J., Merrick, M.S. and Price, E.A., 2008. The distinction between humans and nature: Human perceptions of connectedness to nature and elements of the natural and unnatural. *Human Ecology Review*, *15*(1), p. 1.

Weinstein, N., Przybylski, A.K. and Ryan, R.M., 2009. Can nature make us more caring? Effects of immersion in nature on intrinsic aspirations and generosity. *Personality and Social Psychology Bulletin*, *35*(10), pp. 1315–1329.

Wheeler, D., Zohar, A. and Hart, S., 2005. Educating senior executives in a novel strategic paradigm: early experiences of the Sustainable Enterprise Academy. *Business Strategy and the Environment*, *14*(3), pp. 172–185.

Zelenski, J.M. and Nisbet, E.K., 2014. Happiness and feeling connected the distinct role of nature relatedness. *Environment and Behavior*, *46*(1), pp. 3–23.

Zhang, J.W. and Howell, R.T., 2011. Do time perspectives predict unique variance in life satisfaction beyond personality traits?. *Personality and individual differences*, *50*(8), pp. 1261–1266.

Zhang, J.W., Howell, R.T. and Iyer, R., 2014. Engagement with natural beauty moderates the positive relation between connectedness with nature and psychological well-being. *Journal of Environmental Psychology*, *38*, pp. 55–63.

The manufacturer's authorised representative in the EU is Springer Nature Customer Service Centre GmbH, Europaplatz 3, 69115 Heidelberg, Germany. If you have any concerns regarding our products, please contact ProductSafety@springernature.com

Printed and bound by CPI Group (UK) Ltd, Croydon, CR0 4YY
30/09/2026
02225229-0019